Selected papers from the 2019 IEEE International Workshop on Metrology for AeroSpace

Selected papers from the 2019 IEEE International Workshop on Metrology for AeroSpace

Editors

Pasquale Daponte
Eulalia Balestrieri

MDPI • Basel • Beijing • Wuhan • Barcelona • Belgrade • Manchester • Tokyo • Cluj • Tianjin

Editors
Pasquale Daponte
University of Sannio
Italy

Eulalia Balestrieri
University of Sannio
Italy

Editorial Office
MDPI
St. Alban-Anlage 66
4052 Basel, Switzerland

This is a reprint of articles from the Special Issue published online in the open access journal *Sensors* (ISSN 1424-8220) (available at: https://www.mdpi.com/journal/sensors/special_issues/MetroAeroSpace).

For citation purposes, cite each article independently as indicated on the article page online and as indicated below:

LastName, A.A.; LastName, B.B.; LastName, C.C. Article Title. *Journal Name* **Year**, *Volume Number*, Page Range.

ISBN 978-3-0365-0058-4 (Hbk)
ISBN 978-3-0365-0059-1 (PDF)

© 2021 by the authors. Articles in this book are Open Access and distributed under the Creative Commons Attribution (CC BY) license, which allows users to download, copy and build upon published articles, as long as the author and publisher are properly credited, which ensures maximum dissemination and a wider impact of our publications.
The book as a whole is distributed by MDPI under the terms and conditions of the Creative Commons license CC BY-NC-ND.

Contents

About the Editors

Pasquale Daponte, Full Professor at University of Sannio, was born in Minori (SA), Italy, on March 7, 1957. He obtained his bachelor's degree and master's degree "cum laude" in Electrical Engineering in 1981 from University of Naples, Italy. He is a Full Professor of Electronic Measurements at University of Sannio, Benevento. He is Immediate Past Chair of the Italian Association on Electrical and Electronic Measurements and Past President of IMEKO. He is a member of the Working Group of the IEEE Instrumentation and Measurement Technical Committee N°10 Subcommittee of the Waveform Measurements and Analysis Committee and the IMEKO Technical Committee TC-4 "Measurements of Electrical Quantities", and he serves on the Editorial Board of Measurement, Acta IMEKO and Sensors. He is Associate Editor of IET Science, Measurement and Technology. He is a member of the Board of Armed Forces Communications and Electronics Association (AFCEA) Naples Charter. He has organized some national and international meetings in the field of Electronic Measurements and European co-operation, and he was General Chairman of the IEEE Instrumentation and Measurement Technical Conference for 2006 and Technical Programme Co-Chair for I2MTC 2015. He was a co-founder of the IEEE Symposium on Measurement for Medical Applications MeMeA; now, he is the Chair of the MeMeA Steering Committee, memea2018.ieee-ims.org. He is the co-founder of IEEE Workshop on Metrology for AeroSpace, www.metroaerospace.org; IMEKO Workshop on Metrology for Archaeology and Cultural Heritage, www.metroarcheo.com; IMEKO Workshop on Metrology for Geotechnics, www.metrogeotechnics.org; IMEKO Workshop on Metrology for the Sea, www.metrosea.org; IEEE Workshop on Metrology for Industry 4.0 and IoT, www.metroind40iot.org; and IEEE Workshop on Metrology for Agriculture and Forestry, www.metroagrifor.org. He is involved in some European projects. He has published more than 320 scientific papers in journals and at national and international conferences on the following subjects: measurements and drones, ADC and DAC modeling and testing, digital signal processing, distributed measurement systems. He has received the following awards and honors: in 1987, from the Italian Society of Ophthalmology, the award for the researches on the digital signal processing of the ultrasounds in echo-ophthalmology; in 2009, the IEEE Fellowship; in 2009, the Laurea Honoris Causa in Electrical Engineering from Technical University "Gheorghe Asachi" of Iasi (Romania); the "The Ludwik Finkelstein Medal 2014" from the Institute of Measurement and Control of United Kingdom; in May 2018, the "Career Excellence Award" from the IEEE Instrumentation and Measurement Society "For a lifelong career and outstanding leadership in research and education on instrumentation and measurement, and a passionate and continuous service, international in scope, to the profession."; in September 2018, the IMEKO Distinguished Service Award; and in 2020, the AESS Outstanding Organizational Leadership Award "For contributions to metrology for aerospace applications".

Eulalia Balestrieri joined the Department of Engineering, University of Sannio, Benevento, Italy in 2018 as an Assistant Professor in electric and electronic measurement, where she has been involved in the research activities carried out at the Laboratory of Signal Processing and Measurement Information. She received the M.S. degree in software engineering and the Ph.D. degree in information engineering from the University of Sannio in 2003 and 2007, respectively. Her research interests include digital signal processing for measurement in telecommunications, data converter characterization and medical measurements. Dr. Balestrieri is a member of the IMS TC-10 and a member of the working group developing the new jitter standard.

Preface to "Selected papers from the 2019 IEEE International Workshop on Metrology for AeroSpace"

This Special Issue book offers extended versions of selected papers from the 2019 IEEE International Workshop on Metrology for AeroSpace. It is devoted to recent developments of instrumentation and measurement techniques applied to the aerospace field, including new technology for metrology-assisted production in aerospace industry, aircraft component measurement, sensors and associated signal conditioning for aerospace as well as calibration methods for electronic test and measurement for aerospace. The aerospace environment is very diversified and is the subject of all those activities that involve flying in the atmosphere of the Earth and surrounding space of interest for a multitude of military, industrial and commercial applications. Metrology and measurements have an essential role for all the aerospace applications, being indispensable for developing and/or using aerospace electronic instrumentation, automatic test equipment, sensors and wireless sensor networks, attitude and heading reference systems, monitoring systems, navigation and precise positioning, data fusion techniques and air traffic management as well as flight testing instrumentation and techniques. The contributions of research in the field of aerospace metrology are constantly increasing, and it is important to update the state of the art achieved in order to understand how to maintain and improve the results reached and to propose new solutions.

Manned or unmanned aircraft are complex technical objects whose airworthiness capabilities must be assured to prevent them from critical damages as well as from potential material losses on the ground. Despite rapid advances in material engineering and mechanic strength analysis methods in modern software, the importance of experiments remains crucial for critical safety applications. This theme is addressed in the Special Issue book by discussing the preparation and execution of on-ground static and engine load tests for the composite unmanned aerial vehicle (UAV). In recent years, a highly creative approach for the construction of UAVs has arisn. This has led to the increased importance and complexity of UAV testing, a topic addressed in the book by presenting a highly customizable platform allowing an in-depth analysis of the UAV software and algorithms to be carried out. It is thus possible to provide researchers with numerous degrees of freedom and high flexibility in proposing and testing new solutions, in a simpler way, in a short time and without significant financial outlays. Together with testing, maintenance of the UAVs is also of considerable importance. By means of improved and additional monitoring and diagnosis ability, UAV maintenance may become more effective. An example that can be found in the book aims to develop a useful diagnostic concept of the rotor-bearing system for UAV jet engine maintenance. In particular, the new concept of a rotor suspension system with magnetic bearings for a mini turbojet engine rotor is described, allowing the efficient reduction of the transverse vibration amplitude and the negative performance features of a classical bearing system and enabling continuous diagnostics of rotor engines. Research is also oriented to significantly enhance the level of autonomy of UAVs, in terms of their guidance, navigation and control functions.

Magnetometers play a key role in UAV navigation performance. An interesting contribution in the book presents an innovative method for the calibration of low-cost magnetometers for small/micro UAVs aimed at improving the accuracy in the magnetic heading estimate in presence of onboard magnetic fields during nominal flight conditions. The method works by collecting calibration data in flight and adopting a cooperative approach. The undisputed advantages of UAVs, including their ability to be used in high-risk situations as well as in inaccessible areas, at low altitude and at flight profiles close to the objects where manned systems cannot be flown, justify their diffusion in various applications, including remote sensing applications. The possibility of using a small autonomous helicopter to perform tasks using a remote sensing system to monitor gas transmission and distribution networks is discussed in the book, showing the most effective way to properly set up autopilot and to process its validation during flight tests.

A special category of UAVs is the one called miniature UAVs or mini-UAVs, widely used in civilian enterprises related to the activities of both state institutions and private enterprises. The implementation of a totally new mini-UAV design is presented, focusing on the efforts taken to develop an original but easy-to-use system and to integrate subsystem elements and test them so to prove their functionality and reliability.

The global navigation satellite system (GNSS) continues to attract research attention, for example, to satisfy the strict safety and accuracy requirements aimed at assuring the position solution integrity, availability, accuracy and reliability, by means of the GNSS hybridization with other technologies. Regarding this topic, an interesting starting point for research is the comparison of different approaches to the design of hybrid positioning solutions for the railway sector, under the perspective of the uncertainty evaluation of the attained results and performance, that can be found in the book. GNSS can be also successfully used to improve the performance of airplane takeoff and landing operations that can be compromised by the airfield characteristic and conditions.

In particular, a global navigation satellite system/inertial navigation system (GNSS/INS) sensor and another in which airplane ground speed was measured with the use of an optical noncontact sensor have shown to be capable of measuring the data to determine the required parameters of the ground distance of takeoff and landing. Another important GNSS application concerns its use for accurate time and frequency dissemination, synchronization and remote clock comparison. Recent research results have shown the advantages of three updates of the Atomium precise point positioning (PPP) software developed by the Royal Observatory of Belgium. In particular, the update carried out by combining the Galileo and global positioning system (GPS) signals for PPP time transfer has led to a further improvement of the frequency stability inside the computation batch.

The improvement of knowledge of the planets represents another important research topic. Starting from the fact that airborne dust and regional or global dust storms cause changes in the atmospheric thermal structure of Mars, a consortium of Italian research institutes and Spain's space agency has developed MicroMED (Micro Martian Environmental Dust systematic analyzer). This instrument is capable of carrying out measurements that have not been performed before on Mars, namely the abundance and size distribution of atmospheric dust close to the surface of the red planet, where dust lifting takes place. The book contains a summarized description of the thermo-mechanical design of the instrument, the manufacturing of the flight model and its successful qualification in expected thermal and mechanical environments. Moreover, information about the computational fluid dynamics (CFD) simulation campaign made to solve the MicroMED design criticalities is also provided together with the experimental validation of the results of the analysis.

Great interest is also devoted to future qualification procedures and practical usage of small, cheap and easily moveable radars for the detection of low-observable air targets in difficult ground areas. For this purpose, a set of procedures, included in the book, tested both theoretically and experimentally are presented to calibrate the radar sensor using standard scatterers, to measure the multipath effect and compensate for it and to create "true radar cross section" maps of the area of interest for both points and distributed clutter by using a commercial X-band radar. Due to their characteristic of being a reliable and performant solution for target localization and identification, especially in environments with weak or denied GNSS performance, secondary radar systems continue to capture research interest. The book presents a detailed analysis and implementation of secondary radar beacons designed for a local ad hoc localization and landing system (LAOLa) to support the navigation of autonomous ground and aerial vehicles.

Research methods related to Polish radars developed for the needs of the Polish Armed Forces can be also found, including methods for testing radars with the function of tracking small high-speed ballistic objects along with examples of results of observations of combat ammunition.

Light detection and ranging (LiDAR)-based systems are going to play an increasingly more important role because of their higher capability to discriminate smaller objects with respect to other active systems, such as radar systems. A detailed analysis of LiDAR systems, pointing out the performance that can be achieved through the use of LiDAR data when combined with data from other sources, is presented in the book. Moreover, two case studies, based on the use of two- and three-dimensional LiDAR systems, are also provided and analyzed in an extensive way.

Focusing instead on the research aimed at sensor topic, interesting contributions are provided in the book, starting with the presentation of a novel horizon sensor prototype that leverages the availability of low-cost, highly miniaturized infrared cameras having increasingly higher image resolution to enhance its accuracy. The preliminary design activities accomplished to define analytical sensors (or virtual ones) for angle of attack and angle of sideslip estimation exploiting simulated flight data are also shown. This research work is aimed at carrying out a reliable and ready-to-be-flown air data system able to measure data from the external environment, within the MIDAS project funded in the frame of the Clean Sky 2 program.

Another research objective related to sensors is the measurement of forces and moments that act on a landing gear wheel, by means of a wheel force sensor designed and built based on strain gauge technology and then embedded in the left landing gear wheel of a test aircraft. The designed sensor is capable of simultaneously measuring three perpendicular forces and three moments and wirelessly sends data to a handheld device.

Piezoelectric materials, although widely used as sensors in a variety of aerospace applications, as proven by many research contributions, can also be used as actuators or energy harvesters. An example can be found in the electromechanical impedance-based structural health monitoring (SHM) system presented in the book, in which, during flight, when normal loadings are applied, the patches work as energy harvesters by maintaining or restoring the battery charge of the stand-by SHM electronic board. However, if relevant or abnormal loadings occur, the patches instead act as sensors for the power-up SHM board by acquiring in real-time relevant data on the component usage. Finally, during maintenance, the patches, operated by external high-voltage electronic equipment, can also work as actuators, to stress the structure with predefined load profiles, as well as sensors, to monitor the structural response.

Recently, the research on the framework of the aerospace applications has been devoted to CubeSats because they are small and compact, characterized by low launch costs, and suitable to create satellite constellations that can improve the terrestrial network infrastructure, providing better global coverage. An example of research contribution in this field is the experimental results of a test campaign performed on the radio-frequency (RF) receiver prototype, designed and fabricated for CubeSat applications. The prototype testing has been carried out through a board-level test approach for the verification of the functional requirements and a component-level one for specific characterization measures. The presented design approach allows testing the components individually after they are mounted on the board, which could be used to perform radiation tests and to discover the behavior of the COTS RF components for debugging purposes. Performing the component-level radiation tests through a unique board-level radiation test is extremely attractive to reduce design costs and avoid the purchase of expensive evaluation boards.

Nondestructive testing (NDT) for inspection, sizing, localization and repair of damages on aircraft composite structures continues to gain importance and to be explored by several researchers. In recent years, increasing use of ultrasonic phased array has emerged in comparison to conventional single transducer ultrasonic inspections for NDT, due to their flexibility, speed of operation and good imaging performance. The challenges and achievements in the development and use of a compact ultrasonic phased array (PA) module, with signal processing and imaging technology, for autonomous nondestructive evaluation of composite aerospace structures can be found in the book as a good example of the research trend in this field.

Aerospace research considers the safety of not only aerospace structures but also those on board, including pilots. High-frequency electromagnetic fields can have a negative effect on both the human body and electronic devices. The devices and systems utilized in radio communications constitute the most numerous sources of electromagnetic fields. The book presents an interesting research investigation focused on the values of the electric component of electromagnetic field intensification determined with the ESM 140 dosimeter during the flights of four aircraft from the airport in Depultycze Krolewskie in Poland. The point of reference for the obtained results was the normative limits of the electromagnetic field that can affect a pilot in the course of a flight.

An onboard communication system includes any onboard architecture that permits performing information transmission on board a spacecraft and can be implemented with different technologies, physical architectures and redundancy schema. This communication system is currently demanded to withstand substantial data rates, mainly due to the usage of image-based sensors. Aerospace missions can require a large bandwidth in the order of one or even several Gbps to perform the routing of payloads related to data traffic throughout the rest of the spacecraft.

Therefore, to improve onboard communication, communication protocols such as SpaceFibre and SpaceWire protocols have been developed. An innovative, proper and complete instrument is presented in the book for the scientific community and the space industry, offering tools and advanced features for the test and development of new SpaceFibre devices and at the same time supporting the previous SpaceWire standard and cross-communications.

The papers collected in this book face challenges related to several different themes touched by the most recent research, offer insights into new solutions to key measurement problems for aerospace applications and shed light on unique opportunities. The research activities presented in the book may be of interest not only for people working in the specific field of metrology for aerospace but also for researchers interested in advanced measurement topics.

Pasquale Daponte, Eulalia Balestrieri
Editors

Article

Numerical and Experimental UAV Structure Investigation by Pre-Flight Load Test

Artur Kurnyta *, Wojciech Zielinski, Piotr Reymer, Krzysztof Dragan and Michal Dziendzikowski

Air Force Institute of Technology, Airworthiness Division, 01-494 Warsaw, Poland; wojciech.zielinski@itwl.pl (W.Z.); piotr.reymer@itwl.pl (P.R.); krzysztof.dragan@itwl.pl (K.D.); michal.dziendzikowski@itwl.pl (M.D.)

* Correspondence: artur.kurnyta@itwl.pl; Tel.: +48-261-851-628

Received: 16 April 2020; Accepted: 21 May 2020; Published: 26 May 2020

Abstract: This paper presents the preparation and execution of on-ground static and engine load tests for the composite unmanned aerial vehicle (UAV). The test was conducted for pre-flight structural strength verification of the remotely piloted aerial target named HORNET, after introducing some structural modifications. The ground tests were performed before the flight test campaign, to ensure the strength and operational safety of the modified structure. The panel method and Computer Aided Design (CAD) modelling were adopted for numerical evaluation of aerodynamic and inertial forces' distribution to simulate loading scenarios for launch, flight and parachute deploying conditions during the static test. Then, the multi-stage airframe static test was prepared and executed with the use of a designed modular test rig, artificial masses, as well as a wireless strain measurement system to perform structure verification. The UAV was investigated with 150% of the typical load spectrum. Furthermore, an engine test was also conducted on a ground test stand to verify strain and vibration levels in correspondence to engine speed, as well as the reliability of data link and the lack of its interferences with wireless control and telemetry. In the article, data achieved from the numerical and experimental parts of the test are discussed, as well as post-test remarks are given.

Keywords: UAV; load test; structural integrity; numerical simulation; panel method

---

1. Introduction

The aircraft, manned or unmanned, is a complex technical object, which needs to maintain integrity in the wide-load spectrum. For that reason, before putting any flying object in the air, its airworthiness capabilities must be proven, both to prevent critical damage of the aircraft as well as potential material losses on the ground. Despite rapid advances in material engineering and mechanic strength analysis methods in modern software, the importance of experiments remains crucial for critical safety applications. This is true for common materials as well as for new materials and technologies, e.g., glass and carbon composites. Structural analysis and testing of a composite wing for an ultralight unmanned aerial vehicle (UAV) was considered by Sullivan et al. [1]. Structural integrity of some composite parts of a high-altitude long-endurance (HALE) class UAV was also evaluated by Frulla et al. [2]. The necessity of performing aircraft fatigue and static tests for structure strength design and evaluation was investigated by Jian Wu et al. [3]. Load distribution measurements are also conducted in other branches of industry, e.g., Lei Gao et al. [4] for large diameter pile. The overall reliability and maintenance analysis of UAV is delivered in Reference [5].

The first experimental step to verify structural strength is static loading tests, performed on the ground. This paper presents the preparation and execution of that type of pre-flight verification experiment on an example of a UAV used in the Polish Air Force [6] as remotely piloted aerial system (RPAS). Although that UAV has already been used for many years, proposed structural modifications and their influence on the fuselage integrity need to be verified before flight. The main modifications were: different winglet construction and their attachment to the wing, as well as front fuselage part prolongation to improve balance of the UAV. After positive results of the static test, the UAV will be prepared for in-flight campaign with a sufficient real time load monitoring system (example in Reference [7] for manned fighter-bomber). The operation of monitoring system elements will also be tested during static investigation.

To ensure high correspondence between loading for the static test and the operational spectrum, numerical modelling was utilized. Nowadays, the wide choice of numerical methods for modelling of aircraft aerodynamics allows to increase the model's fidelity at the expense of computational time, starting with very basic methods, such as classic potential theory, up to Direct Numerical Solution (DNS) of the turbulent flow. The choice of method results is always a trade-off between accuracy of results and computational cost. The current problem of establishing pressure distribution on surfaces of the aircraft in various flight conditions requires (in some regions, fine-) grid discretization. On the other hand, exact values of aerodynamic coefficients, such as drag, are not required.

2. Materials and Methods for Test Preparation

2.1. Remotely Piloted Aerial System (RPAS) Platform Description

The RPAS named HORNET (Figure 1, Table 1) is part of the Aerial Target System, developed and deployed for operation in the Polish Army in 2005. Flying air targets are intended for training and performing artillery and artillery-rocket shooting by anti-aircraft defense forces.

Figure 1. HORNET remotely piloted aerial system (RPAS) in-flight.

Table 1. HORNET RPAS basic parameters (min = minimum, max = maximum).

Length/wing span	1.7 m/3.2 m
Starting weight/payload	38 kg/5 kg
Min/cruise/max speed	85/150/230 km/h
Operating range	40 km
Max climb speed	16 m/s (at 140 km/h)
Parachute drop velocity	5.5 m/s

The Aerial Target System consists of:

- Controlled maneuvering reusable air targets HORNET, equipped with a towed sleeve with an acoustic miss distance indicator and a towing line development system.
- Flight control and control systems allowing for remote control of "visual visibility" and command control, with the use of an on-board automatic control system.
- Starting launcher driven by rubber bands, with energy about 8 kJ.
- Transport system and a set of operating equipment.

From the start of the project, 47 pieces of the aircraft were produced. At present, due to airframe modifications to enhance performance and changes in the manufacturing process to increase product repeatability and precision, HORNET's structure needs to be verified during ground and flight tests.

The flying wing aerodynamic configuration ensures high operational resistance in difficult conditions, as well as good performance at higher speeds and low mass of the structure. The maximum take-off weight is approximately 38 kg, including 5 kg of payload. To ensure the required performance, a 116 cm^3, 11 house power, two-stroke two-cylinder spark-ignition piston engine was chosen. The take-off is carried out from the catapult (launcher) which ensures acceleration of the airframe with maximum take-off mass to the speed of about 74 km/h, with the maximum acceleration of about 150 m/s^2.

The airframe of the aerial target is made using composite technology with sandwich structures formed in negative molds using vacuum techniques. The main material of the reinforcement is glass fiber, carbon cloth is used only for elements requiring stiffness and lightness (e.g., hatch cover, wing spar belts). Fuselage is circular and accommodates the engine, and fuel tanks are placed in the front as well as the battery for on-board systems. The middle part of the fuselage is a compartment for task equipment with servo-driven covers and a rolled-up shooting sleeve with the hit sensor. In the rear part of the fuselage, there is a parachute compartment with a servo-controlled cover. For transportation purposes, the plane is divided into the fuselage, wings and vertical stabilizers. The 140 cm long wings are connected to the fuselage using a circular connector made of duralumin. That duralumin connector in carbon mounting is an inner part of a front spar, until the second of 3 ribs. The back spar closes the wing profile with a working skin (Figure 2). There are also two fittings, placed on the leading and rear section of the socket ribs to transmit moments of forces.

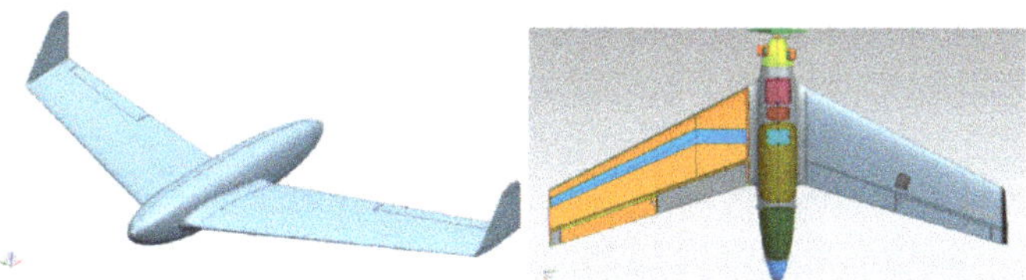

Figure 2. Geometry of "HORNET" unmanned aerial vehicle (UAV): skin (**left**) and internal structure (**right**).

2.2. UAV Numerical Model Preparation

The first part of the work was dedicated to preparation of the model, for reliable determination of aerodynamic forces on the 'HORNET' UAV. The data is required for evaluation of loads acting on the structure, so that a static experiment can be designed. Obtaining a good model requires performing a sequence of steps regarding geometry transformation into computational mesh. The process of model creation includes choice of physics, software, geometry and mesh generation, and finally, verification of the chosen method. The choice of potential flow method, also known as the panel method, was dictated by current needs. Often used for aircraft conceptual design process [8], the panel method's advantages include high accuracy–computational cost ratio, ability to model required wing geometry and good prediction of pressure coefficient on wing surface.

On the other hand, there are some limitations of potential flow for modelling of aerodynamics. They are a result of assumptions, which panel codes make concerning flow physics [9], which are:

- Flow is steady,
- Viscosity is zero,
- Flow is irrotational,
- Flow is incompressible.

A number of panel method implementations are available for free, e.g., XFLR5, Panukl, etc., which are based on two-dimensional (2D) potential flow code XFoil [10–12].

"HORNET" is an UAV in a single wing configuration. The fuselage has conventional, streamlined geometry and is smoothly blended with the wing. The wing is tapered, swept, twisted and ends with a winglet. The NX CAD geometry is presented in Figure 2.

The bases for operational loading spectrum determination to conduct the static test were aerodynamic and inertial forces' distributions from the CAD geometrical model (Figure 2). Additional information was achieved from collection of an on-board data recorder to specify the setting of the control planes for individual calculation variants.

Panukl allows to split the whole UAV geometry into smaller, easy to model parts, which can be merged afterwards. Fuselage is defined using point sets defining consecutive cross-sections. These can be directly imported from NX. The fuselage discretization is directly influenced by number of cross-sections, and points in a cross-section. Hence, the areas of higher curvature require finer spacing. The part of the geometry, which was simplified in this step, is the fuselage–wing blend. The main wing of "HORNET" was designed using NACA (National Advisory Committee for Aeronautics) profiles: root chord as NACA1410 and tip chord as NACA0008 with linear transition. The exact geometry was imported directly from NX via coordinates of points of key features:

- Wing sections at aileron start and end,
- Non-linear twist, resulting from constant leading-edge z-coordinate and linearly increasing trailing edge z-coordinate,
- Fuselage cross-sections.

The last part, namely wingtip, is modeled similarly to the main wing: NACA0006 profile is constant, and winglet is mounted at 90°.

Control surfaces are modeled by modification of airfoil, by rotating the trailing edge region around the hinge axis. This is a typical way of modelling control surfaces and was shown to give reliable results [13].

Based on the above data, a family of models was created that responded to different configurations of aircraft aileron. Due to program limitations in the aspect of fuselage aerodynamic characteristics, it was modeled as a wing element, which can be seen in Figure 3. Then, the grid was developed and calculations of pressure and lift force distributions were calculated in correspondence to Figure 4.

Figure 3. Isometric view of HORNET and fuselage with cross-sections defining its geometry.

Inertia loads were determined on the basis of mass distribution along the wingspan. The basis for these activities was a detailed HORNET model prepared in the NX software, illustrated in Figure 2. Using geometry from the CAD model as well as material data of individual elements, the distribution of masses was determined by dividing the wing in accordance with the geometry of the computational grid for aerodynamics. All concentrated masses and equipment built in the wing were also taken into account. Due to the distribution of the torsional moment of the wing, not only the distribution of masses along the wingspan but also the location of the centers of gravity of individual masses in the X-direction was taken into consideration in the numerical model.

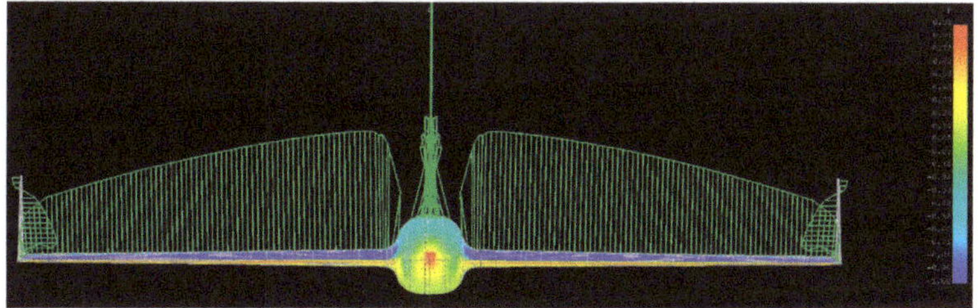

Figure 4. Lifting force distribution of the HORNET UAV.

Having all the necessary models, static test conditions were determined. Based on archival data from flights of the pre-modified version of the HORNET, the configuration of aileron during flights, including overload, was established. As a result of these analyses, it was determined that the most common values of the aileron is −5° for steady flights, while for stable flight fragments with Nz (vertical acceleration) above 5, the swing oscillates slightly around −10°. The influence of the aileron deflection on the aerodynamic loading distribution of the wing is presented in Figure 5. It can be noticed that for a higher aileron angle, the lifting force decreased for the outer part of the wing and the pressure point moved forward along the chord of the wing.

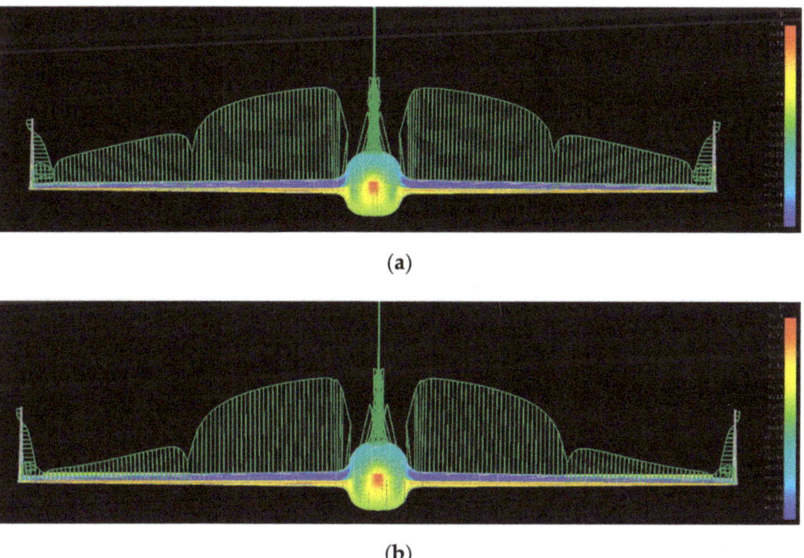

Figure 5. Aileron deflection influence on the aerodynamic loading distribution of the wing for (**a**) −5°, and (**b**) −10°.

Another aspect taken into account was the effect of geometric dislocation and angle of attack on the load distribution. To obtain the maximum load states, in particular for bending moment, the maximum angle of attack from the aerodynamic model of 10° was taken for calculations. This allowed to achieve maximum loads on the outer parts of the wing and hence, the bending moment. Adoption of smaller angles would be associated with an almost complete loss of lift in the area of ailerons due to their upward deflection and negative geometrical dislocation of the wing, with a value of 3°. On the other hand, none of the above factors affected inertial loads.

After determining conditions to be met by the test, aerodynamic and mass loads were extracted from the models and the distribution of forces and moments derived from the assumed loads was specified. Afterwards, they were digitized to cross-sections in which the location of clamps was anticipated. The position of the clamps was selected on the distribution of the wing's strength elements based on the CAD model (Figure 6). The use of a larger number of clamps was considered unnecessary due to the impossibility of mapping small loads on the external parts of the wing and limiting the transmission of concentrated forces by the wing itself.

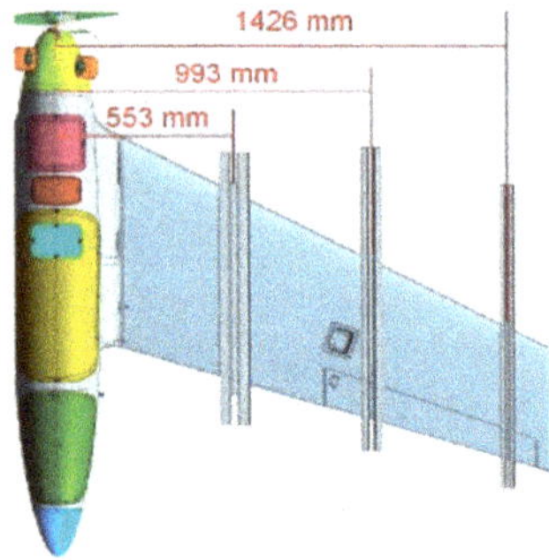

Figure 6. Position of the wing clamps.

Numerical analysis of aerodynamic and inertial forces as well as geometry from the CAD model allowed for calculating their continuous distribution for the wing. A set of substitute masses has been selected for each UAV acceleration condition, which reflect the distribution of continuous forces with sufficient accuracy. The distributions were optimized for the largest possible matching of the bending moment distribution and not exceeding the loads at the base of the wing due to the load capacity of the fasteners. The comparison of continuous and applied loading is presented for shear forces, bending and twisting moments respectively, in Figures 7–9. Load distribution for shear forces was calculated for the front spar, and the bending moment is transmitted by both front and back spar as well as by the working skin. The wing caisson experienced the twisting moment.

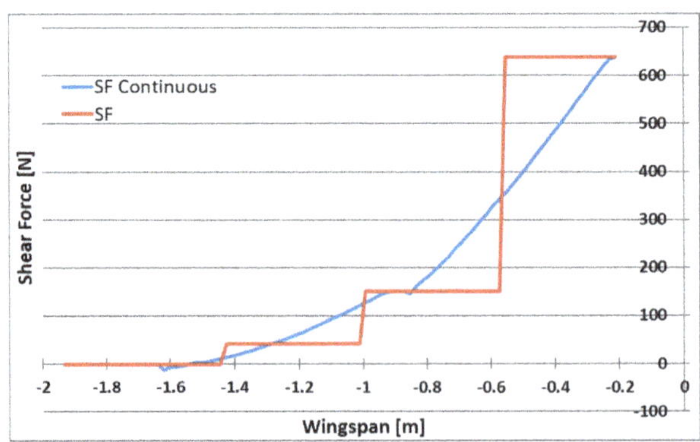

Figure 7. Shear force distribution for continuous and applied loading.

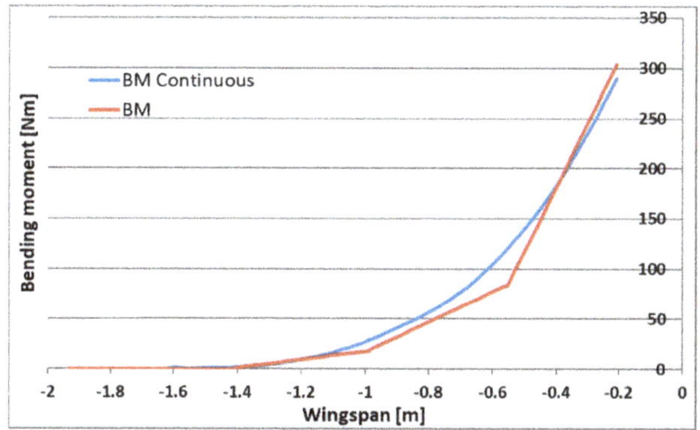

Figure 8. Bending moment distribution for continuous and applied loading.

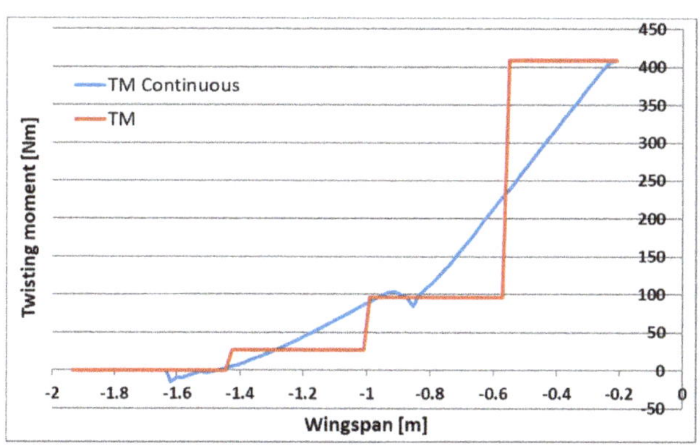

Figure 9. Twisting moment distribution for continuous and applied loading.

2.3. Preliminary Selection of Critical Areas of the Structure

The structure must be able to carry loads during landing at the permissible wind speed (hence the underdog system is not susceptible to damage). The most frequently occurring and significant structural damage of the HORNET aircraft is due to landing impact loads, including the parachute deployment phase. Due to unpredictable landing position and lack of landing gear, various areas are subject to the adverse effects of the landing impact, depending on the aircraft configuration during landing. The following areas are the most critical with respect to landing impact:

1. The propeller mounting frame (Figure 10a,b).
2. Parachute mounting points' elements (Figure 10c,d).
3. Stabilizers (winglets) and wind-to-stabilizer mounting points (Figure 10e).
4. Wing structural elements—wing spar, especially wing-to-fuselage mounting points (Figure 10e,f).

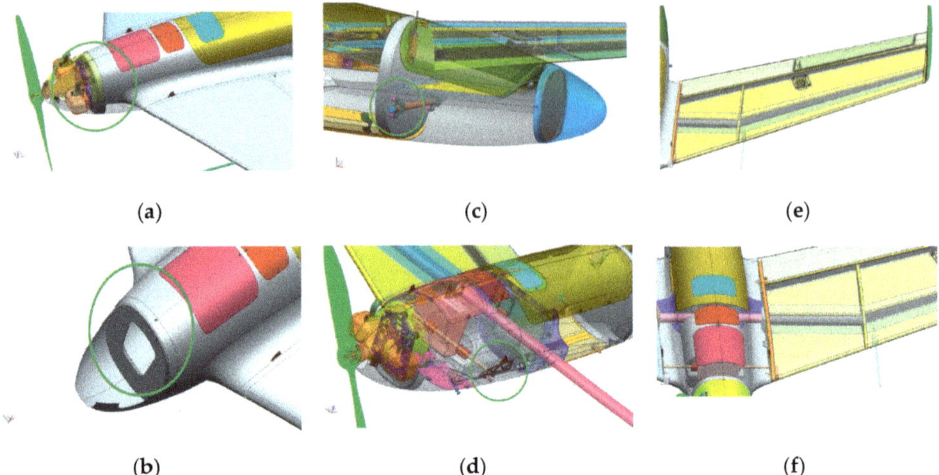

Figure 10. Critical areas of the Hornet UAV structure (a) propeller mounting area (b) propeller mounting frame (c) rear parachute mounting points (d) front parachute mounting points (e) wing with stabilizer (winglet) (f) wing structural elements.

2.4. Static Load Test Preparation

In order to ensure safety during flight tests on modified RPAS, pre-flight ground static tests were prepared and performed. These tests can successfully reveal structural design issues which can arise during flight tests and lead to potential failure and damage. The first critical part of the ground test is to apply static loads to the modified HORNET structure in order to investigate its capability to withstand loads expected to occur during flight. The load spectrum envelope was evaluated on the basis of historical flights of the standard version of HORNET as well as numerical research. The modular test rig was designed to enable investigation in various flight configurations (Figure 11). In addition, the static loads test of the HORNET structure was also used for the following tasks:

- Validation of the finite element model of the HORNET aircraft,
- Calibration and verification of components of the load monitoring system and acquisition of data useful for sensor network design during flight tests [14,15].

Figure 11. Test bed for static loading ground test configured for aerodynamical loads representation.

Airframe loading magnitude and the spread of strain and stresses during the static test correspond to those gained in the real flight. Apart from aerodynamic forces, another source of stress are vibrations due to a working engine. An additional quasi-dynamic test was also conducted to verify airframe response to vibration as a separate activity, which is also described further in this paper.

For the static on-ground testing, the investigation object was equipped with a sensor network based on foil strain gauges mounted to the platform skin for local measurements of the structure strain field. The suitable bridge configuration was chosen for the type of loads acting at a particular location and structural element [16].

In Figure 12a, the location of the strain gauges on the HORNET wing is presented with top and rear views. Three measuring sections are marked, in which bending of the wing will be measured along the main spar. Bending is the structure's reaction to lifting force during flight and maneuvers with vertical acceleration. The other three measuring sections are located directly on the main spar from the inner side of the wing, before the element will be fully manufactured. Those channels will allow for measuring front-to-rear bending of the wing, which occurs during take-off as a reaction to drag of the aircraft elements. Also, the change in torque of the wing is measured when the loading is set to several points along the chord of the airfoil section.

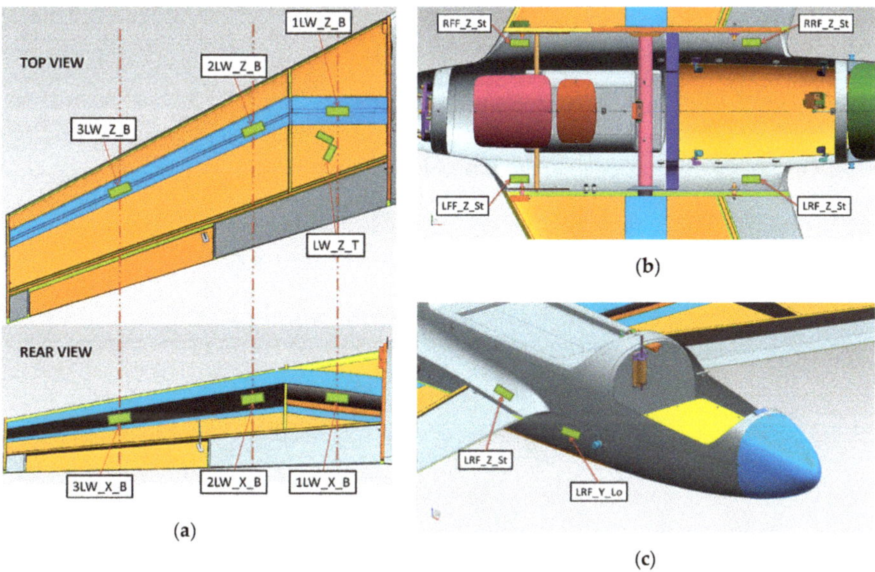

Figure 12. Strain gauges location (**a**) on the wing, (**b**) on the top fuselage and (**c**) on the side fuselage.

In Figure 12b, the location of the strain gauges on the top of the fuselage are presented. These front and rear measuring points will give the multi-type strain information during take-off and parachute deployment simulation. Additionally, the parachute fixing point is located near the front wing–fuselage connection points, which can produce significant strain on the fuselage during maneuvers. Sensors bonded behind rear wing-fuselage connection points should produce significant strain on the fuselage during maneuvers and take-off from the launcher. In Figure 12c, the location of the strain gauges on the side of the fuselage is presented. The measuring points will give the axial-longitudinal strain information during take-off and parachute deployment simulations.

In total, 20 analog input channels for strain gauges' measurements were used, connected to wireless measuring modules by LORD Sensing MicroStrain to avoid routing long cables and ensure safety for test staff [17]. The sensors network was configured to investigate the magnitude and correctness of the loads at selected critical hot-spots during take-off, landing and parachute release. Additionally, the magnitude and correctness of the loads' spacing along the airplane wingspan was also measured, using the optical three-dimensional (3D) scanning for deflection measurement of structural elements.

Several loading scenarios were planned for that test regarding load distribution and weight value to acquire discrete multipoint strain/load characteristics. In particular, the following type of loads will be represented on the HORNET structure.

2.5. Static Load Test Execution

The static loads test was divided into three stages:

- Stage 1—launch loads of the structure during take-off,
- Stage 2—flight loads of the structure during flight,
- Stage 3—parachute loads.

The static loads were exerted to the airframe mounted in a modular test rig. Therefore, it allowed for installing the object in different positions according to the test stage (Figure 13). Since large loads are imposed on the structure during the launch by the four connection pins, these pins were used to mount the test specimen to the rig. No other support points on the fuselage were taken into consideration (besides a wooden block under the middle section of the fuselage, with the role of preventing the fuselage from critically deforming under higher loads). The total loads exerted on the aircraft structure are a superposition of aerodynamic loads and inertial loads due to maneuvers. The HORNET UAV structure is designed to be operated within <0; 6> vertical g-level. During the ground test, investigation was conducted to 150% of the maximum flight load, which corresponds to vertical load of order 9. The aerodynamic forces corresponding to the chosen flight conditions (in particular, corresponding with high vertical overload maneuvers) were computed using fluid dynamics. The main loading components taken into consideration to achieve real load patterns were: wing bending moment and shear force (which were controlled by adjusting weights attached to the clamps) as well as wing torque (which was controlled by location of the weight mounting point along the clamp). The following figures show the test rig in different configurations, corresponding to launch loads (Figure 13a,b) flight loads (Figure 13c,d) and parachute release loads (Figure 13e,f).

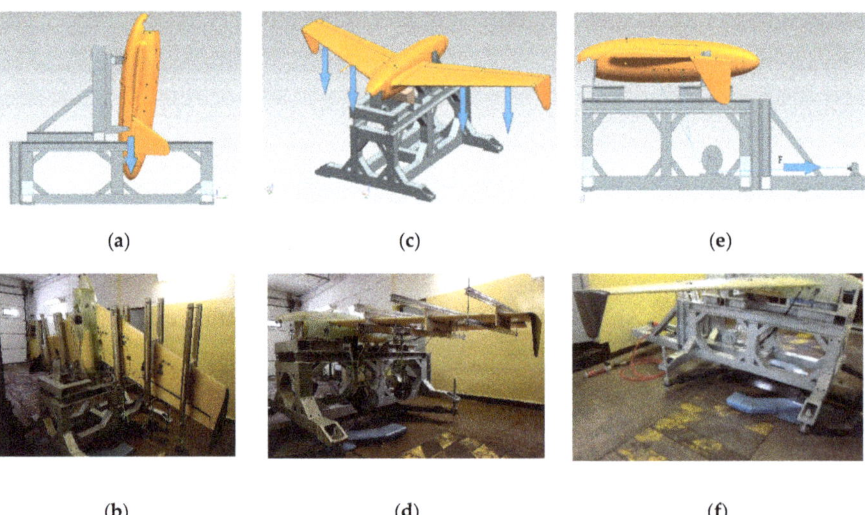

Figure 13. Test rig and RPAS configuration for all stages of static test (**a**) take-off (model) (**b**) take-off (execution) (**c**) flight (model) (**d**) flight (execution) (**e**) parachute release (model) (**f**) parachute release (execution).

Together with strain measurements, the test object was covered with an array of markers (special stickers with contrast pattern) which allowed for optical displacement measurements of the structure under loads. The deformations were measured using an ATOS III digital scanner and are presented as displacements of the markers (Figure 14).

(a) (b)

Figure 14. Structural deformation measurement with optical three-dimensional (3D) scanning system (**a**) camera (**b**) structure with reference point.

2.6. Engine Test Execution

After static test execution, an additional test was conducted with an UAV equipped with engine, avionics and electronics for control and telemetry. The engine test allowed for simultaneous determination of loading and vibrations from the engine as well as electromagnetic compatibility and lack of interferences between wireless communication of RPAS control and measuring equipment for in-flight load monitoring. This test will also allow to establish signal-to-noise ratio in the worst case scenario, when the aircraft is standing on the ground (Figure 15). Measurements were conducted for various engine speeds in the range of 1800–6400 revolutions per minute (RPM). The influence of vibration on equipment and its fixing was qualitatively evaluated. In the further stage of investigation during the flight test, this will lead to separate vibrations being derived from the engine and those from another source, like flatter. To do so, strain gauges bonded for the static test and additional MEMS (microelectromechanical system) accelerometers with a high sampling rate were used. For the sake of comparative character of research and type of used sensors, they were identical with anticipated flight configurations. In total, there were two 3-axis MEMS accelerometers as wireless modules, and 12 strain gauge channels grouped in 3 analog wireless modules (Figure 16).

(a) (b)

Figure 15. Hornet on the test stand for the engine test (**a**) bottom view, and (**b**) top view.

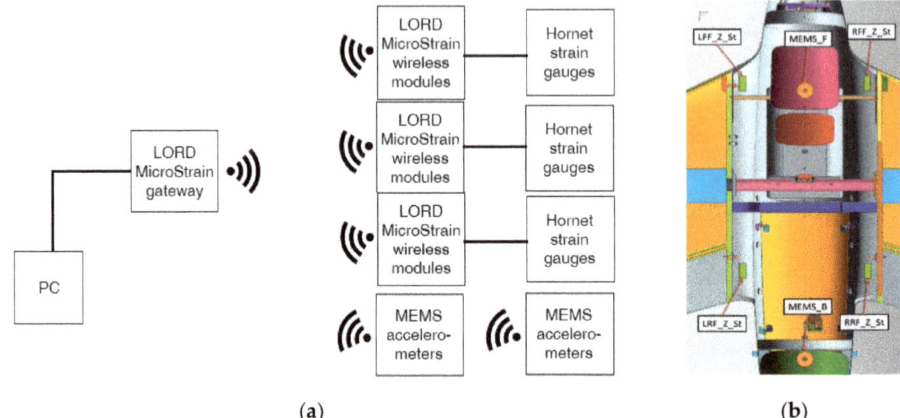

Figure 16. Engine test, (**a**) block diagram of wireless strain and acceleration measurement system, and (**b**) location of MEMS accelerometers on the fuselage.

An additional benefit from that test was a quality check of compatibility and a lack of electromagnetic interferences from RPAS control, telemetry and navigation with wireless measurement system. Both are wireless and use the frequency of 2.4 GHz. If both systems co-operate fine on their maximum emitted power with a presence of vibration on the ground, there should not be any interferences in the air.

In order to conduct the strain measurements for the aircraft structure both during on-ground and in-flight tests, wireless measuring modules by LORD MicroStrain V-Link and V-Link-200 were used. They are equipped with analog inputs and excitation voltage output sets for measuring bridges in which strain sensors are included. MEMS-type accelerometers are suitable for both ground and flight tests, because of their small size and low power requirement with integrated signal conditioning and temperature compensation. LORD MicroStrain G Link 200 contains 3-axis accelerometers of a range of up to 40 G and a 20-bit analog-digital converter acquisition system with a set of options (digital high- and low-pass filters) were utilized.

The presented measuring modules, independently from type, can be wirelessly connected to the data aggregator/base station. That device ensures wireless data acquisition from measurement modules and can be communicated through Ethernet or USB with PC via specialized software, Autonomous operation of the device and recording of the data on the internal memory of 16 GB is also possible.

3. Test Results and Discussion

3.1. Static Load Test

During the test, weights corresponding to a particular load condition were hung on the clamps installed on the wings symmetrically on each side. For the flight load stage of the test, the aircraft was placed backwards in the horizontal position on the rig (Figure 13c,d). That configuration of the object allowed for use of only inertial forces acting on the structure, in order to achieve the lower vertical load conditions with no additional masses or clamps. The detailed load conditions used in this stage of the test are listed in Table 2. Results of strain measurements carried out in this stage for modified Hornet structure are presented in Figures 17 and 18.

Table 2. Load conditions used in Stage 2—flight loads.

Load Condition	Inner Clamps (kg)	Middle Clamps (kg)	Outer Clamps (kg)
	Total Weight	Total Weight	Total Weight
nz = 1	7.17	0.00	0.00
nz = 2	14.92	2.87	0.00
nz = 3	22.67	4.43	1.80
nz = 4	31.17	6.43	2.15
nz = 5	39.42	8.43	2.65
nz = 6	47.67	10.18	3.40
nz = 7	55.92	12.18	4.15
nz = 8	64.17	13.93	4.90
nz = 9	72.42	15.93	5.65

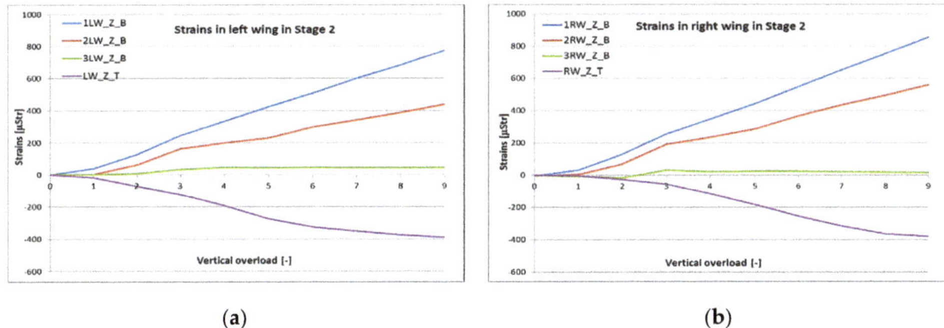

(**a**) (**b**)

Figure 17. Strain measured in the (**a**) left and (**b**) right wing (bending in the Z-axis and torque) during Stage 2.

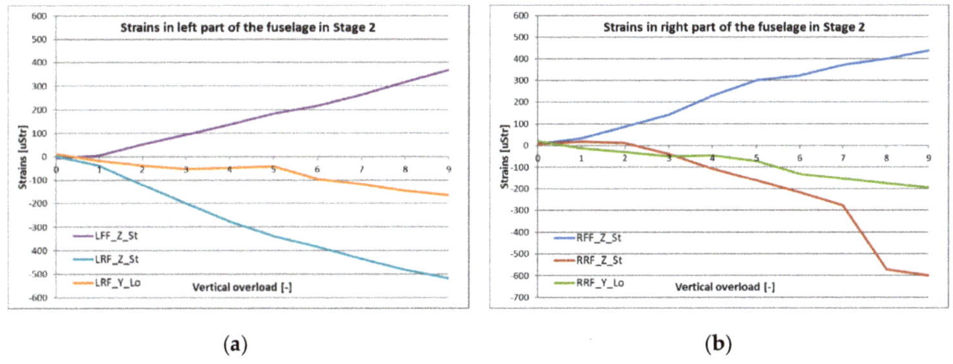

(**a**) (**b**)

Figure 18. Strain measured in the (**a**) left and (**b**) right side of the fuselage during Stage 2.

It can be noticed that changes in strain during this stage are considerable due to applied load. The linear increase of strain was measured with added mass in all cross-sections of the wing. Peak–peak value of strain from bending for inner and middle locations was around 800 and 500 microStrains (uStr), respectively (Figure 17). The decrease in peak–peak strain level corresponds with the modelling, where the strain levels are lower for the middle cross-section in correspondence with the inner one.

The median difference during all loading conditions was around 0.43 for this stage of the test. For the outer section, strain change was much lower and remained almost constant after reaching nz = 3.

For the front and rear fuselage measurement points, similar positive (for tension) and negative (for compression) points can be seen. Strain value increase is linear with the loading condition, reaching around 400 uStr for front, and more than −500 uStr for the rear wing-fuselage fixing point (Figure 18). The exception with the rear right point could be explained by temporary keying of the structure in the test rig or a structure delamination (not confirmed by visual inspection).

After installation of test specimens in the test rig and after applying desired loads to the structures, displacement measurements were carried out with the ATOS III scanner. A series of digital scans were performed allowing to determine the actual location of each marker. Using the baseline as well as location of markers placed on the rigid test rig (for solid reference), structural deflection was defined under a certain loading condition. The comparison for states nz = 1 and nz = 5 are illustrated in Figure 19. Especially in the horizontal view, the scale of wing deflection can be observed for those two loading conditions, reaching more than 50 mm for the outer end of the wing. There is almost no deflection of fuselage during that stage of the test.

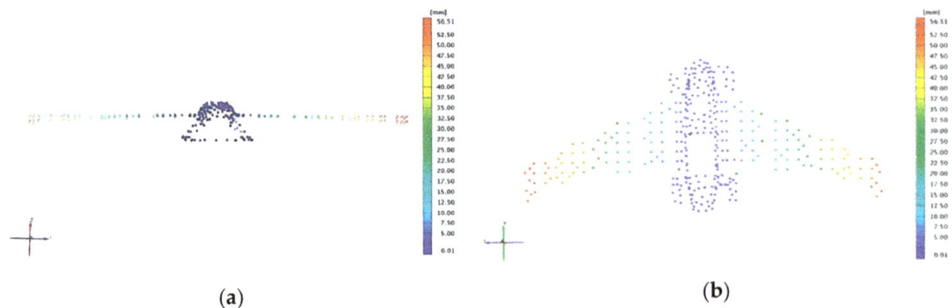

(a) (b)

Figure 19. Comparison of structure displacements for nz = 1 and nz = 5 in the (**a**) horizontal and (**b**) vertical plane.

Stage 3 was designed to verify the ability of the structure to withstand loads from deployment of the onboard parachute. Parachute mounting points in the front of the fuselage were used to attach lines to impose the loading. At this stage, the displacements were not measured with the optical system due to safety reasons. Figure 20 illustates strain data which were achieved for the modified HORNET structure until nz = 10. Despite the fact that the test was carried out to reach nz = 15 (Table 3), the hydraulic actuator was not able to stably maintain such high pressure for a prolonged time and strain measurements were not exact. However, the structure was able to withstand the test.

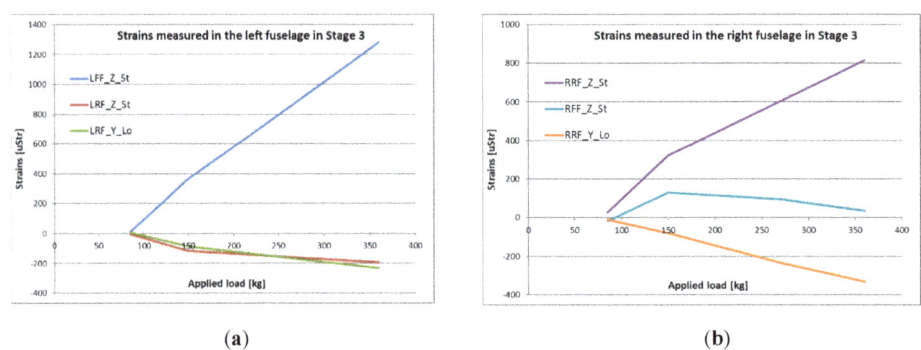

(a) (b)

Figure 20. Strain measured in the (**a**) left and (**b**) right side of the fuselage during Stage 3.

Table 3. Load conditions used in Stage 3—parachute release.

Load Condition	nz = 2.5	nz = 5.0	nz = 7.5	nz = 10.0	nz = 12.5	nz = 15.0
F (kg)	85	150	270	360	450	540

During the "parachute" stage of the test, the hydraulic actuator was pulling the parachute cords through a set of shackles and a force sensor. Several loading conditions were performed, applying the force in the range from 85 to 540 kg. The highest measured strain values were obtained for the front fixing points of the parachute lines with a magnitude of around 1000 uStr for nz = 10. Some differences in measurements between the left and right side of the fuselage can be noticed, but the character of those changes can be explained with a little asymmetry of the fuselage and test rig.

The main goal for the test was to evaluate structural strength for pre-flight validation. The structure withstood the applied loads with no critical damages to the structure. However, during visual inspection after the test, some damages were found, e.g., local delamination and cracks in the fuselage root rib area or cracks in the fuselage access. The damage was not critical, since the residual strength of the structure was high enough to carry the imposed loads (the failure was not critical). Those areas will be corrected for the in-flight campaign specimen to ensure structural integrity of the modified HORNET.

3.2. Engine Test

The engine test was performed on a test rig, with the mounting point similar to those from the UAV launcher. The test was conducted for various engine speeds in range of 2000–6400 RPM. For each engine speed, approximately 10 s of data was taken for further analysis, in which speed remained constant. Data from transient states have been removed to allow for comparison of time and frequency domain results for chosen engine speeds.

Strains measured on the wing and fuselage as well as corresponding engine speed are illustrated in Figure 21. For measurements from wing bending in the YZ plane (Figure 21a,b), some value alteration can be noticed. Strain alteration for the middle cross-sections 2LW_Z_B and 2RW_Z_B was about 150 uStr, with respect to RPM change. The highest peak–peak value can be seen for 3900 RPM for all channels. Regarding strain data from the inner cross-section, peak–peak range was as such for all RPM, but the median value increased with respect to RPM. Again, the highest range can be noticed for 3900 RPM.

Regarding measurements from wing bending in the horizontal plane (Figure 21c), the strain alteration with respect to RPM was almost negligible for all of the measurement channels. Peak–peak values were alike for all engine speeds. The slight change in mean level can be observed for the wing inner cross-section in the lower engine speed range. But, the overall change was approximately 10 uStr.

Alteration of strain gauges from the right parachute attachment point was relatively small, independent of engine speeds below 100 uStr peak–peak (Figure 21d). Strain gauge from the left front fixture showed a change of about 200 uStr.

The results from all measurement channels revealed some changes in both strain magnitude and its mean value, depending on RPM. The alteration was much higher for lower engine speeds, below 4000 RPM, even 2–3 times higher, in comparison to higher speeds, e.g., 6200 RPM. It can be explained by the high impact of the ground for the test object during that type of test as well as more "smooth" engine operation with higher RPM.

Acceleration data illustrated in Figure 22 give information of vibration level in a function of engine speed from the front (Figure 22a) and rear (Figure 22b) 3-axis accelerometer. For RPM 2000–4000 peak–peak range for the rear of the fuselage, the acceleration is ±7.5 g for the z-axis, ±5 g for the y-axis and ±4 g for the x-axis. Above 4000, the envelope for each axis decreased by approximately 50%.

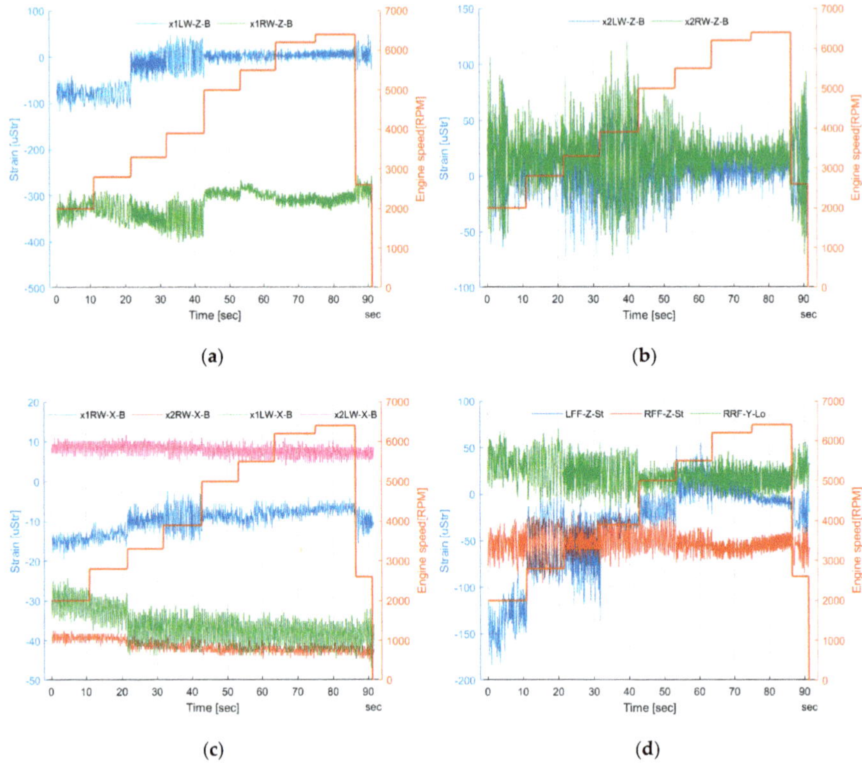

(a) (b)

(c) (d)

Figure 21. Strain measured on the wing and fuselage during the engine test (**a**) wing bending in the vertical plane—inner section (**b**) wing bending in the vertical plane—middle section (**c**) wing bending in the horizontal plane—inner and middle section, (**d**) fuselage parachute attachment points.

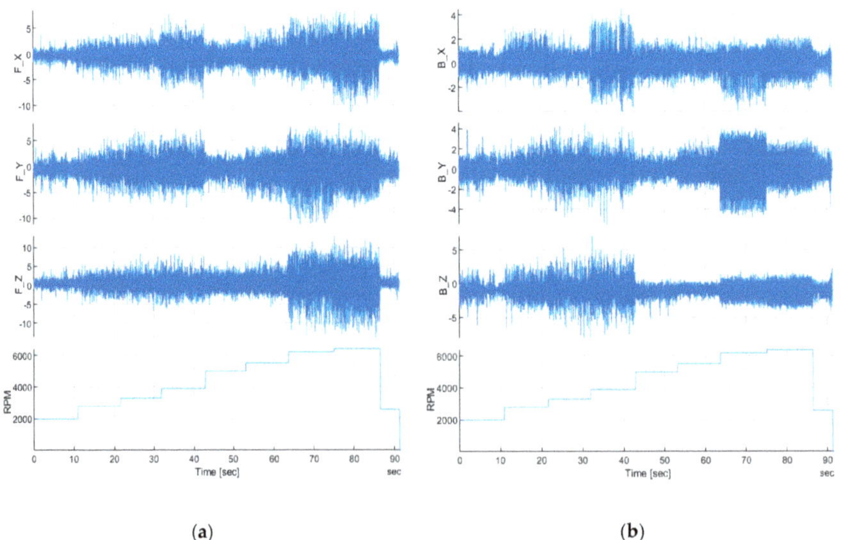

(a) (b)

Figure 22. 3-axis acceleration data versus engine speed for the (**a**) front, and (**b**) rear sensor.

Regarding the MEMS accelerometer from the front of the fuselage (Figure 23a), an engine operation effect can be seen, especially for higher RPM. For 2000 and 2800 RPM, peak–peak value for all axes are quite similar, in contrast to the rear sensor, where the z, y and x axes sensed significantly different acceleration. The lowest envelope can be noticed for 5000 RPM and is fitted in the range of [+6; −5] for the z-axis and [+4; −5] for the y and x axes. Above 6000 RPM, a clear increase in the vibration envelope can be noticed for all sensing directions and was the highest for the test. Respectively, for the x-axis, in a range of [+8; −12] g, the y-axis [+8; −8] g and [+13; −15] g for the z-axis (Figure 23b).

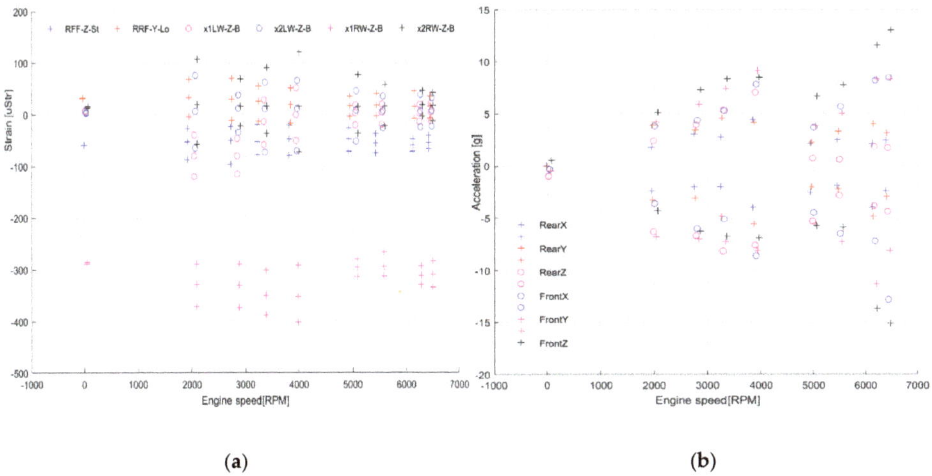

(a) (b)

Figure 23. Max/median/min values during engine test for (**a**) strain sensors, and (**b**) accelerometers.

Although the front sensor was mounted much closer to the UAV center of gravity than the rear one, the acceleration level was much higher and almost constantly increasing with respect to engine RPM. Acceleration measurements from the rear sensor confirmed (together with strain data) that higher structural vibration occurs on the ground for engine speeds lower than 4000 RPM. Also, the negative influence of engine operation can be decreased by mounting accelerometers closer to the rear end of the fuselage, so behind the UAV center of gravity. If not, a special dumping platform should be considered to reduce engine vibration on acceleration measurement.

In Figure 24, amplitude spectrums based on Fast Fourier Transform (FFT) for both front and rear 3-axis acceleration sensors are illustrated for selected engine speeds. The spectrum decomposition and frequency analysis confirmed remarks from time domain signals. For the rear sensor, there were significant vibrations for 2800 and 3900 RPM (Figure 24b,d) within the range 1–25 Hz, mostly from the z-axis. Additionally, some resonance effects can be observed for the speed of 3900 RPM, as a delta-type envelope around the dominant, rotational frequency for all 3 axes. Above that speed (5000 RPM and 6200 RPM), the spectrum become more flat, and low-frequency components were strongly decreased, similarly for all axes. Sharp peaks on the charts correspond with dominant, rotational frequency and its subparts (Figure 24f,h).

Regarding the front accelerometer, some frequency components in a range below the dominant one for engine speed increase in the amplitude spectrum, in line with rotational frequency. For lower engine speeds (2800 and 3900 RPM), the envelope for all 3 axes was similar in that range. For the main dominant frequency, the y-axis component had the highest level. With the further increase of engine speed, the z-axis envelope became significantly more elevated in regions outside the rotational frequency than the x-axis and y-axis envelope. The character of frequency domain data affirms that the front acceleration sensor is strongly affected by engine operation. The level of vibration on frequencies other than rotational speed can have several sources: moving parts in the two-stroke, two-cylinder, spark-ignition piston engine itself, local resonance of the structure (frames, internal shelfs) due to external excitation, as well as multibody effects from a test stand–UAV connection. But, on the contrary to the rear sensor spectrum, those effects occur and are stronger with the increase of engine speeds.

Summing up, the engine test revealed that there is no significant change in strain of wing bending in the XY plane corresponding to engine speed. Bending in the YZ plane showed some alterations with respect to engine speed, which is higher for the inner cross-section. On the other hand, peak–peak range changed significantly for acceleration and the envelope was higher for lower RPM for rear fuselage. In each case, the z-axis showed the biggest alteration. For the front fuselage, the engine effect can be seen, as in the lower RPM region, the envelope was alike for all axes. The envelope for all sensing directions increased significantly with respect to engine speed. Furthermore, the operation reliability of the measurement system was confirmed in the presence of external disturbances from engine vibration.

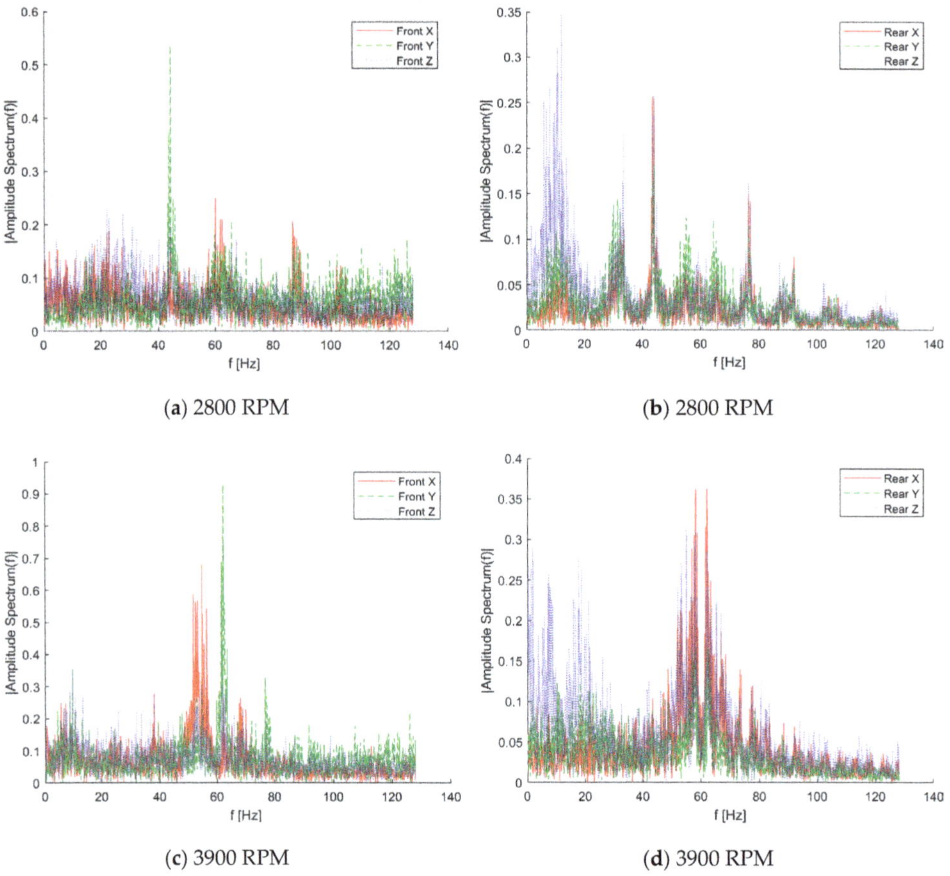

(**a**) 2800 RPM (**b**) 2800 RPM

(**c**) 3900 RPM (**d**) 3900 RPM

Figure 24. *Cont.*

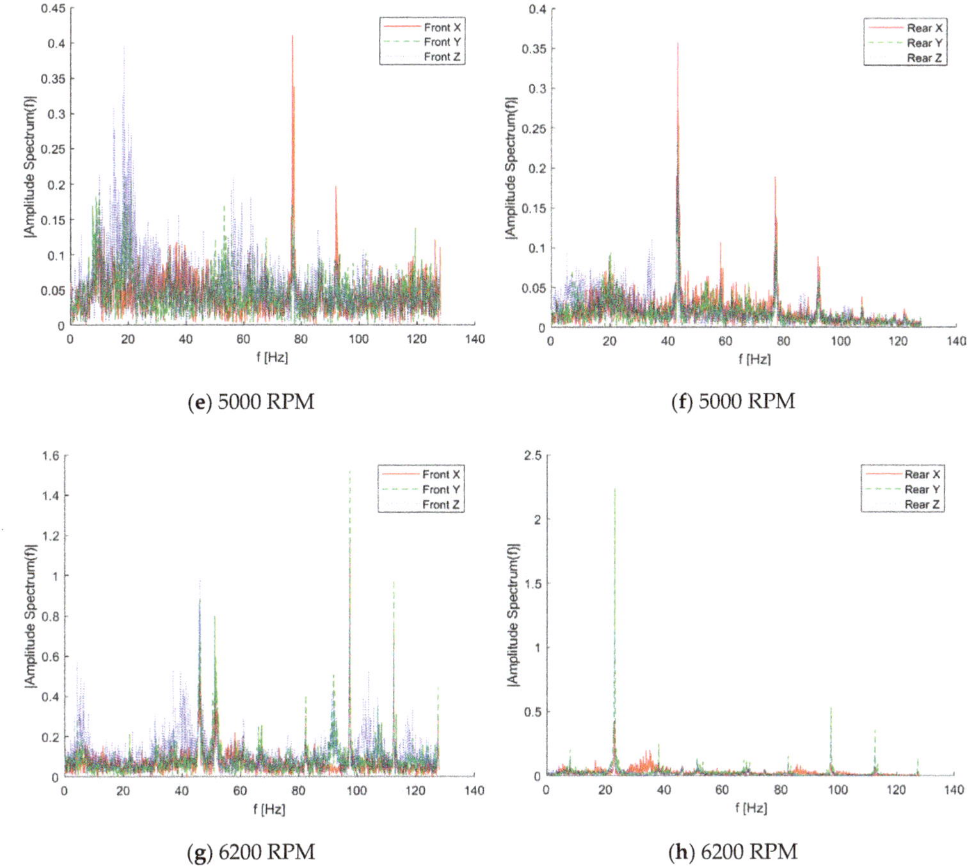

(e) 5000 RPM (f) 5000 RPM

(g) 6200 RPM (h) 6200 RPM

Figure 24. 3-axes acceleration for selected engine speed for sensor (**a**) front—2800 RPM (**b**) rear—2800 RPM (**c**) front—3900 RPM (**d**) rear – 3900 RPM (**e**) front—5000 RPM (**f**) rear—5000 RPM (**g**) front—6200 RPM (**h**) rear—6200 RPM.

The supplementary information from the engine test was the evaluation of the joint operation of two wireless data links: one for UAV control and telemetry and the second for a strain and acceleration monitoring system for future activities during the flight test. Real-time load monitoring system implementation for the aircraft is currently one of the main tasks to ensure the optimal maintenance and utilization strategy as a part of health and usage monitoring for technical objects [7,18] for the individual aircraft tracking (IAT) concept. Regarding wireless data links, one of the concerns was the fact that both are using the same frequency of 2.4 GHz. During the engine test, simultaneously with strain and acceleration data, the Received Signal Strength Indicator (RSSI) parameter was acquired, which is the measure of the power present in a received radio signal. The RSSI value is measured in dBm, in their response packet. For the chosen measurement system, this signed 1-byte value can range from −95 dBm to +5 dBm. The Node RSSI is the signal strength that the node received the command from the base station. The Base Station RSSI is the signal strength that the base station received the response back from the node. Both Node and Base RSSI values were within the range of −14 to −30 dBm and were unaffected by the operation of the engine. The test also proved no negative effect on RPAS on-board primary equipment due to wireless measurement system utilization.

The short- (few meters) and long-range (100 m) test of control equipment was performed and no interferences within both wireless data links were experienced.

4. Conclusions

This paper presented an on-ground static load test as well as engine test preparation and execution for UAV pre-flight structural strength verification and wireless real time monitoring system reliability. The loading spectrum for the static test was evaluated using historical data of a non-modified version of the HORNET UAV and numerical modelling with the use of the panel method. The novel approach during that activity was that the fuselage was modeled as a part of the wing. As a consequence, better estimation of load distribution was achieved in the wing–fuselage connection area. Furthermore, the influence on aerodynamic load distribution along the wingspan due to different aileron angles was taken into consideration. This is also quite important, when investigating UAV in tailless configuration, where lift is generated on wings.

The ground test was performed with the use of a specially designed modular test bed, which could be quickly adapted to carry out test scenarios for different flight stages. The structure load spectrum was prepared to be higher than the expected envelope during flights up to the ultimate load for that airframe. Strain measurements from the wing bending confirmed load distribution, in which strain magnitude was significantly higher for the inner part of the wing during the static test, but low values of strain for the wing outer section were caused by the load application method. Under actual operational aerodynamic loads, these values should be enhanced, which must be taken into account while designing winglet fixing points. Additionally, utilization of a 3D optical scanner together with strain measurement gave the information of continuous deflection of the fuselage and wings during loading. The 3D scanner can also be useful to monitor pre- and post-test object geometry, which could disclose potential plastic deformation of the airframe. Based on the visual inspection, no critical damage was noticed after the test was completed. The airframe withstood the applied loads in all 3 test stages.

Based on the results of the performed tests, the following conclusions can be drawn:

- Structural damages found after completion of the tests were not critical, however would probably propagate during operation. These locations (Critical Points) should be monitored during regular operation.
- Strain gauges intended to measure strains resulting from front-to-rear wing bending showed relatively low values. This might be caused by wing design, in which loads are mainly transferred by the leading and trailing edge of the wing.
- The engine is the main source of vibration for the structure and can significantly influence sensors' reading, especially for those located in front of the fuselage. Sufficient damping platforms should be applied to reduce that effect, in particular, for acceleration sensors.
- Simultaneous operation of two wireless data links, for control and telemetry and for the strain and acceleration monitoring system, was confirmed to be reliable and can be safely used during flight tests in the near future.

Based on the positive results of presented tests, the next stage of the research will cover the flight test campaign. That will be the final test for HORNET UAV structure as well as for the integrated load monitoring system, which can become a permanent part of the on-board equipment of that aircraft.

Author Contributions: Methodology A.K. and W.Z.; software W.Z., M.D. and P.R.; validation, W.Z. and P.R.; formal analysis, A.K., W.Z. and P.R.; investigation, A.K., W.Z. and P.R.; resources, K.D. and M.D.; writing—original draft preparation, A.K., W.Z. and M.D.; writing—review and editing, A.K. and M.D.; supervision, K.D. and M.D. All authors have read and agreed to the published version of the manuscript.

Funding: The research tests described in this paper were conducted under the European Defense Agency and financed by the Italian and Polish governments as a project EDA Ad Hoc Category B, Projects and Programmes—n° B—1404-ESM2-GP, "SHM application to Remotely Piloted Aircraft Systems—SAMAS".

Acknowledgments: The authors would like to acknowledge all team members who contributed to the project.

Conflicts of Interest: The authors declare no conflicts of interest.

References

1. Sullivan, R.W.; Hwang, Y.; Rais-Rohani, M.; Lacy, T. Structural Analysis and Testing of an Ultralight Unmanned-AerialVehicle Carbon-Composite Wing. *J. Aircr.* **2009**, *46*, 814–820. [CrossRef]
2. Frulla, G.; Cestino, E. Design, manufacturing and testing of a HALE UAV structural demonstrator. *Compos. Struct.* **2008**, *83*, 143–153. [CrossRef]
3. Wu, J.; Yuan, S.; Zhou, G.; Ji, S.; Wang, Z.; Wang, Y. Design and Evaluation of a Wireless Sensor Network Based Aircraft Strength Testing System. *Sensors* **2009**, *9*, 4195. [CrossRef] [PubMed]
4. Gao, L.; Yang, K.; Chen, X.; Yu, X. Study on the Deformation Measurement of the Cast-In-Place Large-Diameter Pile Using Fiber Bragg Grating Sensors. *Sensors* **2017**, *17*, 505. [CrossRef] [PubMed]
5. Petritoli, E.; Leccese, F.; Ciani, L. Reliability and Maintenance Analysis of Unmanned Aerial Vehicles. *Sensors* **2018**, *18*, 3171. [CrossRef] [PubMed]
6. Kurnyta, A.; Zieliński, W.; Reymer, P.; Dziendzikowski, M.; Dragan, K. UAV Pre-flight Structural Strength Verification during On-ground Static Load Test. In Proceedings of the 2019 IEEE 5th International Workshop on Metrology for AeroSpace (MetroAeroSpace), Torino, Italy, 19–21 June 2019; pp. 272–277. [CrossRef]
7. Kurnyta, A.; Zieliński, W.; Reymer, P.; Dziendzikowski, M. Operational load monitoring system implementation for Su-22UM3K aging aircraft. In Proceedings of the Structural Health Monitoring, Brisbane, Australia, 5–8 December 2017.
8. Mieloszyk, J.; Goetzendorf-Grabowski, T.; Mieszalski, D. Rapid geometry definition for multidisciplinary design and analysis of an aircraft. *Aviation* **2016**, *20*, 60–64. [CrossRef]
9. Available online: https://en.wikipedia.org/wiki/Aerodynamic_potential-flow_code (accessed on 20 November 2019).
10. Drela, M. XFOIL: An analysis and design system for low Reynolds number airfoils. In *Low Reynolds Number Aerodynamics*; Springer: Berlin/Heidelberg, Germany, 1989; pp. 1–12.
11. PANUKL Potential Solver, Software Package, Warsaw University of Technology. Available online: http://www.meil.pw.edu.pl/add/ADD/Teaching/Software/PANUKL (accessed on 20 November 2019).
12. Deperrois, A. XFLR5—Analysis of Foils and Wings Operating at Low Reynolds Numbers—Guidelines for XFLR5 v6.03. Technical Report. 2011. Available online: https://engineering.purdue.edu/~{}aerodyn/AAE333/FALL10/HOMEWORKS/HW13/XFLR5_v6.01_Beta_Win32%282%29/Release/Guidelines.pdf (accessed on 20 November 2019).
13. Girardi, R.M.; Cavalieri, A.V.; Araujo, T.B. Experimental Determination of the Aerodynamic Characteristics and Flap Hinge Moment of the Wing Airfoil used at ITA's Unmanned Aerial Vehicle (UAV). In Proceedings of the 19th International Congress of Mechanical Engineering, Brasília, Brazil, 5–9 November 2007; pp. 5–9.
14. Molent, L.; Aktepe, B. Review of fatigue monitoring of agile military aircraft. *Fatigue Fract. Eng. Mater. Struct.* **2000**, *23*, 767–785. [CrossRef]
15. Airoldi, A.; Marelli, L.; Bettini, P.; Sala, G.; Apicella, A. Strain field reconstruction on composite spars based on the identification of equivalent load conditions. In Proceedings of the SPIE—The International Society for Optical Engineering, Bellingham, WA, USA, 9–13 April 2017.
16. Hoffmann, K. *An Introduction to Measurement Using Strain Gages*; Hottinger Baldwin Messtechnik GmbH: Darmstadt, Germany, 1987.
17. LORD MicroStrain V-Link-LXRS Datasheet. Available online: http://files.microstrain.com/V-Link_2.4GHz_Datasheet_Rev_10.05jLXRS.pdf (accessed on 20 November 2019).
18. Frövel, M.; Carrión, G.; Pintado, J.M.; Cabezas, J.; Cabrerizo, F. Health and usage monitoring of Spanish National Institute for Aerospace Technology unmanned air vehicles. *Struct. Health Monit.* **2017**, *16*, 486–493. [CrossRef]

© 2020 by the authors. Licensee MDPI, Basel, Switzerland. This article is an open access article distributed under the terms and conditions of the Creative Commons Attribution (CC BY) license (http://creativecommons.org/licenses/by/4.0/).

Article

The Design and Implementation of a Custom Platform for the Experimental Tuning of a Quadcopter Controller †

Michał Waliszkiewicz [1], Konrad Wojtowicz [1,*], Zdzisław Rochala [1] and Eulalia Balestrieri [2]

1 Faculty of Mechatronics and Aerospace, Military University of Technology, 00-908 Warsaw, Poland; michal.wa94@gmail.com (M.W.); zdzislaw.rochala@wat.edu.pl (Z.R.)

2 Department of Engineering, University of Sannio, 82100 Benevento, Italy; balestrieri@unisannio.it

* Correspondence: konrad.wojtowicz@wat.edu.pl; Tel.: +48-261-839-851

† This paper is an extended version of the paper entitled "Experimental method of controller tuning for quadcopters" presented at 2019 IEEE 5th International Workshop on Metrology for AeroSpace (MetroAeroSpace) Conference, Torino, Italy, 19–21 June 2019.

Received: 23 January 2020; Accepted: 26 March 2020; Published: 30 March 2020

Abstract: This paper describes the development process of the quadcopter-based unmanned flying platform, designed for testing and experimentation purposes. The project features custom-made hardware, which includes the prototype quadcopter frame and the flight controller, and software solutions, such as control loop setup. The article specifies the controller tuning used for the initialization of the flight stabilization system and presents the final results of the quadcopter performance evaluation.

Keywords: quadcopter; flight controller; regulator tuning; control loop; UAV

---

1. Introduction

Unmanned aerial vehicles (UAVs) have become a vital part of many industries [1,2]. The usage of the so-called drones has become widespread, which made the various technologies that drive them commercial, prominent, and easily accessible. These factors, in turn, allow for a highly creative approach towards construction of UAVs and their application [3,4].

Hand in hand with the rapid development and increasing number of new UAV implementations and applications, UAV testing has gained increased importance, while at the same time becoming more complex. Different testing solutions have been proposed in the scientific literature, focusing on the tests of the UAV as a whole system or its specific components and systems, including multiple drones. Testing is carried out in all the stages required for UAV development and use. The results provided by the test performed in the intermediate UAV development stages have essential repercussions on the subsequent steps and on the costs to be faced. Particularly crucial is the test carried out in the initial stage of development because, in this stage, the feasibility of the project is decided, and the UAV development costs, times, and difficulties to be faced in the subsequent stages depend on that decision.

UAV testing can involve even new challenges. A good example is the drone testing in the field, wherein the environment is not entirely controllable. The environmental conditions are not repeatable, numerous, and very different from each other. Another problem concerns the UAV regulations that must be met, even though they are continuously evolving and sometimes different depending on the place considered.

Moreover, it should be highlighted that UAV performance strictly depends on the onboard algorithms that can concern its stabilization, control, and navigation. Therefore, it is substantial that these algorithms are adequately tested entirely and faithfully with the target field and application environment. However, this need has not yet found a suitable solution, representing a further challenge for drone testing.

The paper outlines the development of a quadcopter-based unmanned flying platform built using custom UAV hardware and software aimed at providing an efficient solution for custom algorithms in-flight testing in a real environment at the initial and further stages of development.

The paper expands upon several topics already presented in the conference paper [5], including the implementation of the attitude estimation algorithm; hardware design; and execution, the introduction of software functionality, and additional experimental data.

The paper begins with a state-of-the-art analysis followed by a presentation of the platform conception and flight stabilization fundamentals. Section 2 describes the following subsystems of the platform: the airframe's construction and equipment, and the flight controller's electronic hardware and software. It goes on with the sample algorithm design, which is developed to be tested with the presented platform. The general information on control loop designing, controller implementation, and attitude estimation is followed by a section highlighting the versatility of the platform as a test tool. Then the details of the sample tuning method are presented, including inner loop and outer loop tuning in the cascade controller. The section ends with a description of the system's initialization procedure and of the test program. Section 3 presents the result of the tests conducted on the test platform with the sample control algorithms employed. Finally, the results presented in the paper are discussed and the future research direction mentioned.

1.1. Systems for Quadcopters Testing

The growing market of small drones involves the development of test solutions for supporting the designing and maintenance of the aircraft.

Among the different existing test-bench solutions, it is possible to distinguish between two main groups. In the first group, benches are providing comprehensive stationary solutions for testing the whole UAV, unmanned aerial system (UAS), which contains UAVs, the ground control station, and accessories, or even the set of UASs. In the second one, there are those designed for testing specific UAV components.

The stationary test-bench is typically designed to test the parameters of the drones mounted in the test-bench [6–8]. The drone being tested has to be complete. Therefore, the frame, electronic hardware, and software have to be integrated before the test.

An example of this kind of test bench named DronesBench (University of Sannio, Benevento, Italy) has been proposed in [7,8]. This system allows testing of a UAS as a whole platform but in a controlled environment. In particular, DronesBench measures the attitude of the UAS in terms of pitch, yaw, and roll angles; accelerations; power consumption; and thrust force exerted by UAS [7,8]. A new version of the DronesBench system allowing remote control of the testing procedure over the Internet has also been proposed. With this drone bench, the researchers can control the test simultaneously from various laboratories [6].

Test bench solutions belonging to the second class are mainly focused on components such as the UAV propulsion system, the inertial navigation system, the control system algorithms, and the software, and on a specific related parameter or phenomenon, such as the ground effect or optical payload [5–8].

An example belonging to the first solution category is the test-bench presented in [9] and designed for the ground effect testing of a quadrotor with eight propellers arranged in four coaxial pairs. A series of experiments have been performed in that research. First, propellers have been tested on a test stand, and then the vehicle has been tested in flight to compute the in-ground and out-of-ground thrusts. In this way, an explicit relationship between the thrust produced by a single propeller while operating in-ground-effect and the thrust out-of-ground effect has been determined as a function of the normalized propeller height above ground [9].

Other critical parameters to be measured in the propulsion system are the thrust forces to the motor speeds, the motor speed response time, and the power efficiency. There is a simple test bench commonly used for the static thrust force measurements. The primary sensor on the bench is an electronic weight balance [10]. The system consists of a rigid rod with a 1° of freedom pivot in the middle. A propulsion system and an electronic balance are fixed on the opposite ends of the rode. Additionally, a tachometer is attached to the drive shaft of the propeller. The motor rotates a propeller and generates the thrust, which is measured by the electronic balance. In this way, the test bench measures the thrust of the propeller depending on the engine speed, which can be applied in the process of the control system design and optimization.

Another example of a test bench is presented in [10] in which the shared driveshaft connects a motor to a load motor. The load motor emulates the operation of the propeller. The more complicated solution also presented in [10] is a balancing test bench with a wind tunnel. This measurement system can determine a relationship between the power consumption of the propulsion system and the thrust in various environmental conditions and different aircraft airspeeds.

A comprehensive UAV testing also requires analyzing it in a real environment. For this reason, the UAV test site-based solutions have been proposed.

An example of a test site simulating the real environmental conditions for a drone along with a test procedure has been presented in [11]. The test site has been designed to test a drone for taking samples of water from a river. The flight can be disrupted by the gusts of wind emulated in the site. The test site consists of a pool which emulates the river and fans for generating wind gusts. There are also test sites simulating the environmental conditions for testing multiple drones and focused on commanding all of them at the same time during a long period [12,13]. The project is based on the simulation software and on the real sandbox for testing ground and unmanned aerial vehicles. That sort of test site has caught the interest of researchers for years. A similar test site had been developed earlier at Stanford University [14]. It is focused mostly on the development and testing of the collision avoidance algorithms for drones. At a similar time, Illinois University researchers presented a successful implementation of the early-stage test site for remote controlling unmanned robots via WIFI [15].

UAV control and navigation algorithm design and testing is a significant challenge due to the more and more stringent performance requirements and the existence of new and very different flight scenarios. For these reasons, researchers' interest has also been focused on the design and implementation of the UAV testing platform aimed explicitly at UAV control and navigation algorithm testing.

An example of a platform to simulate and test the control system of a tail-sitter UAV has been presented in [16] as an efficient tool for developing the control system. Through the component breakdown approach, a mathematical model of dynamics covering the full angle-of-attack range has been derived along with an environmental model to provide essential information about the environmental influence on the vehicle, such as wind field estimation and wind disturbance control. A commonly used open-source flight controller has been embedded in the proposed framework. The proposed platform has been validated with the typical complete flight scenarios of a tail-sitter, including hovering, forward transition, cruise, and back transition [16]. However, the choice of using the off-the-shelf Pixhawk (controller designed in the team led by Lorenz Meier in Zurich, Switzerland) as a flight controller unit allows only the Pixhawk controller parameters to be tuned. Implementing the user's algorithm with the original firmware is not possible without replacing all the firmware. Although developing user custom software for Pixhawk is possible, it is a complex and not easy task.

An experimental platform using an evolutionary algorithm for the automatic tuning of the gains of a cascaded proportional–integral–derivative (PID) quadcopter controller has been proposed in [17]. By means of the presented platform, the UAV automatic tuning is allowed, precluding the necessity of a physical operator, domain knowledge, or a model. Tuning to different payloads is also possible [17]. The platform provides an automatic, repeatable, and nondestructive hardware testbed for UAV control, and it has been validated in a sample experiment that automatically tunes a PID controller for quadcopter hover. However, in the research work proposed, the only option is parameter tuning for the embedded PID algorithm. The authors focused less on the issue related to the platform airframe and electronics.

The problem of UAV controller tuning has also been faced in [18]. In particular, the performances of the most used algorithms to tune controllers, and the two new iterative algorithms proposed by the authors, have been evaluated. The performances of the algorithms have been compared using Matlab (MathWorks, Natick, USA) simulations, a quadcopter located in a virtual environment and field tests using an experimental platform consisting of a custom-built quadrotor with an on-board control system, including a MikroKopter (HiSystems GmbH, Moormerland, Germany) flight controller and speed controllers. The proposed iterative algorithms have shown a significant improvement concerning the PID controller tuning performance. However, the tuning based on the results of the in-flight tests has not been considered in the analysis [18].

To sum up, there are many examples of test-benches for drone components and payload testing. Further, there are comprehensive test-sites available for testing complete drones or systems of multiple drones. The researchers have paid particular attention to UAV control and navigation algorithms. As a consequence, some testing platforms have been proposed to solve the problems related to obtaining efficiently and completely satisfactory UAV control. However, UAV dynamics are often complex and sometimes unknown, especially when operating in real environments.

However, there is no platform for testing custom algorithms in-flight, in a real environment and at the initial stage of development. Employing a highly customizable platform and in-depth analysis of the UAV software and algorithms can be carried out, providing researchers with numerous degrees of freedom and high flexibility in proposing and testing new solutions, in a simpler way, in a short time, and without significant financial outlays. The possibility of carrying out in-flight tests in the real environment can provide a better knowledge of UAV dynamics and the factors influencing the performance. Finally, the possibility of carrying out tests already from the initial stage of the UAV development can sharply reduce its costs and required time and significantly optimize the results obtained in the later stages.

1.2. Concept

The concept behind the proposed platform is to create a highly customizable, inexpensive, unmanned aerial platform for testing and experimentation. The presented platform delivers all the components to test algorithms and software at the initial stage of development.

The main novelty is an idea of a test platform for hardware-in-the-loop tests of the on-board system algorithms. The platform consists of a frame, propulsion system, electronic hardware, and software that provides diagnostic tools ready for custom algorithm implementation. Stabilization, control, or navigation algorithms can be tested on the platform with the diagnostic tools. With the tools, researchers can monitor the parameters during flight and acquire data for postflight analysis. The tests can be conducted either on the test-stand with limited possible movements of the platform and reduced risk at the initial step of tuning or on the free-flying platform in specified environmental conditions.

The algorithm test starts commonly with the software-in-the-loop simulation. The researcher can move to a hardware-in-the-loop test with simulated sensors when the target on-board computer is ready, and further, to the test on the test-bench when the frame and electronics are prepared. Finally, he can carry out a comprehensive test on a test-site when all the drone systems are operational. Although the systems covering some of the mentioned stages of tests have been developed and described in the literature [6–18], a platform allowing free-flight tests with the custom user algorithms implemented has not been developed yet. The presented platform offers in-flight tests of the algorithms at the initial stage of development.

Moreover, with this platform, a researcher can also test the software solution in further stages by replacing more and more pieces of software with the new code. In this way, one of the algorithms of stabilization, control, and navigation only can be customized or all of them simultaneously. The frame of the platform is 3D printable and fully reproducible with commonly available 3D printers. In the event of a failure, most parts of the airframe can be 3D printed on-site. Additionally, the frame structure has been designed to be easily transportable. The principle of operation of the folding arms and their construction were designed and tested during the project.

The software of the controller is totally open-source and well documented so that the user can quickly implement his custom control algorithms. The quadcopter is controlled by a custom flight controller operated by the author's software to achieve optimal flexibility and potential for experimental purposes, backed by the principles presented in [19]. The software provides the embedded in-flight data acquisition tools for in-flight and after-flight diagnostics and analysis. An example process of control algorithms development has been presented to show the consecutive steps of designing, implementing, and testing the algorithms with the platform using embedded diagnostic tools.

1.3. Flight Stabilization

The flight controller is an essential part of every quadcopter as its purpose is to maintain the aircraft's orientation about the ground. This process is often referred to as attitude stabilization and is critical for the aircraft to perform flight successfully [20–22]. Every system of this kind must be able to perform the following tasks, necessary to perform flight stabilization:

- Calculate aircraft attitude;
- Receive commands from the pilot (usually via radio receiver);
- Run flight stabilization loop at high frequency;
- Control propulsion.

Reliable acquisition of information about the quadcopter attitude is essential for the flight stabilization to be performed. The attitude can be defined as the relative angular position of two reference frames:

- Quadcopter frame of reference;
- A ground frame of reference.

Both frames have their centers located in the quadcopter center of gravity (i.e., in the sensor location). The ground frame vertical axis is collinear to the Earth gravity vector, while the longitude axis position either is irrelevant or pointing towards the local magnetic north. It depends on the usage of a magnetometer or other means of heading acquisition. The quadcopter frame of reference is tied to its structure, where the longitude axis usually points towards the front of the aircraft.

The attitude information can be described with a variety of methods such as quaternions or direct cosine matrix [23]. For data presentation, Euler angles—roll pitch and yaw—are most commonly used due to their simplicity. In the context of this project, Euler angles are used to describe the relative angular position of the quadcopter frame of reference (X, Y, Z) concerning the ground frame (Xg, Yg, Zg). The corresponding angular velocities around each axis are represented as $\dot{\phi}$, $\dot{\theta}$, $\dot{\psi}$, (Figure 1).

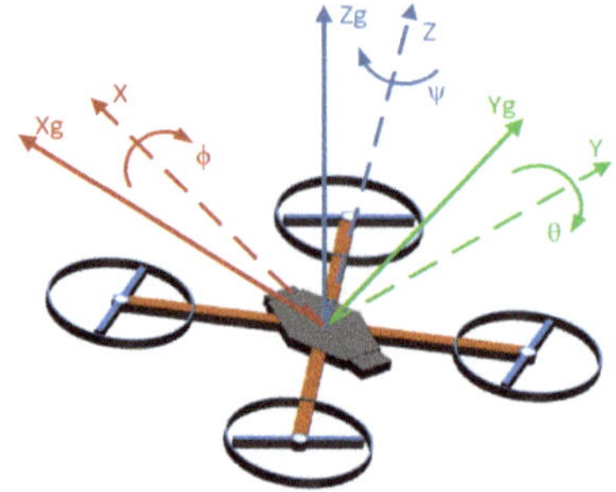

Figure 1. The representation of quadcopter attitude.

The primary purpose of the flight controller is to control the quadcopter motor set to achieve a desirable attitude. This process is formed by comparing the measured attitude information with the set of desired flight parameters, in this case—generated by the pilot (ϕ_d, θ_d, ψ_d, $\dot{\phi}_d$, $\dot{\theta}_d$, $\dot{\psi}_d$). In the current context, throttle information is not included in the comparison as the implemented solution does not include autonomous altitude control.

In order to acquire the above set of information, the AHRS (Attitude Heading and Reference System) is most often used. It is comprised of a sensor package—IMU (inertial measurement unit) and a processing unit, responsible for data filtering and calculations, with the latter usually being the flight controller's main microchip. The IMU contains a set of three-axis gyroscopes and accelerometers, sometimes extended with input from a magnetometer. Data from IMU can be used to acquire a complete minimum set of necessary attitude information required for flight stabilization.

Gyroscopes are used to track rapid changes to the quadcopter attitude. Despite them being prone to drift, their resilience to linear accelerations makes them indispensable in-flight applications. In order to counteract the gyro vulnerabilities, accelerometers are used with a twofold role. Firstly, the accelerometers are responsible for maintaining the connection to the ground frame of reference through the measurement of the gravity vector. Secondly, the gathered information is used by the filtering algorithm to diminish the effects of drift from the result of attitude estimation.

The most optimal way to implement this process takes the form of a "best of both worlds" approach, where data from each sensor is processed in a specific way to provide the most accurate and useful information possible. The process is a fundamental principle of a complementary filter, with its basic block diagram presented in Figure 2.

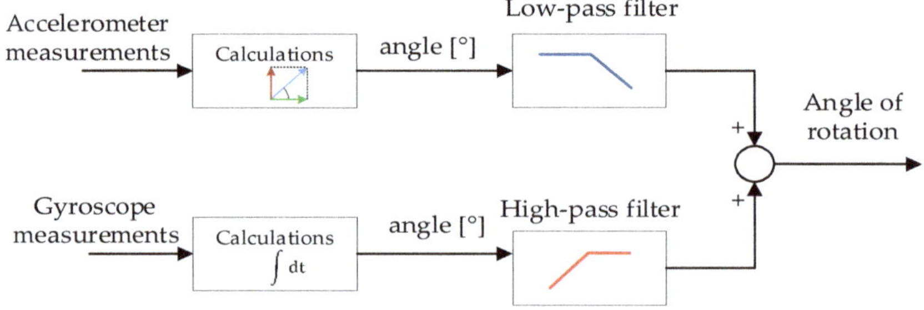

Figure 2. Complementary filter block schematic.

Attitude derived from the accelerometer is processed using a low-pass filter to minimize the presence of vibration-induced noise in the result. In contrast, the angle derived by integrating the gyro is processed using a high-pass filter, due to the presence of drift. Such a solution can be written down and implemented with the following formula:

$$\text{angle}_{est} = (1-\alpha) * \left(\text{angle}_{est\_prev} + \int_{t-1}^{t} gyro_{dps} * \text{dt}\right) + \alpha * \text{angle}_{acc}, \quad 0 < \alpha < 1 \tag{1}$$

where: angle_{est}—is the estimated orientation angle for a given rotation axis, angle_{est_prev}—estimated orientation angle from the previous iteration, *gyro*—are the gyroscope readings for given axis (scaled degrees per second), angle_{acc}—the orientation angle calculated based acceleration vector measurement, α—is the weight coefficient, used to tune the filter.

Like almost any kind of data processing, such a filter must be run continuously in a loop to work correctly. During each iteration, the weight coefficient, here represented by α, manages the amount of contribution of each sensor in the result. Finding the optimal value of α is key to achieving the most promising results out of this fusion technique. As a rule of thumb, the optimal α value is that which would allow making the lowest possible accelerometer contribution, while being sufficient to compensate for gyro drift. If this contribution were too low, the angle calculated by the gyro would converge too slowly onto the accelerometer-derived angle, and drift would likely cause the estimation error to increase over time. On the other hand, with α value being too high, due to excessive accelerometer contribution, the effects such as vibrations or undesirable linear accelerations would cause issues too significant to consider the filter a reliable source of information.

This technique, while useful in its simplicity, has its limitations uniquely when combining corresponding data from more than two sources. Still, it can be viewed as a primary archetype for much more advanced solutions such as Mahony's and Madgwick's filters [24], the latter of which was ultimately integrated into this project flight controller software as means of attitude determination.

2. Materials and Methods

2.1. Platform Construction

The quadcopter frame is made out of 3D-printed components made using FDM (Fused Deposition Modeling) method. Each element was designed beforehand using a Computer-Aided Design (CAD) software (Figure 3a) [20]. This approach allowed us to achieve the type of frame that fully meets the project requirements, most prominent of which are related to high capacity and customizability. The frame geometry was made to be fully compatible with the preselected off-the-shelf electronic equipment with a limited degree of miniaturization, mostly intended for hobby usage, hence the emphasis on high capacity.

(a) (b)

Figure 3. **(a)**—The 3D model of a quadcopter frame design [5], **(b)**—fully assembled quadcopter.

The frame structure was designed with attention to mass distribution, with the final components layout already present during the early stages of development. Another valuable aspect behind the usage of 3D printing was access to spare parts. In case the frame would suffer any significant damage during its use, newly printed parts could replace the broken components in a relatively short amount of time and at a low cost.

The completed quadcopter (Figure 3b), utilizes 10-inch propellers mounted on four ABC-Power A2212 1400 KV brushless motors, each of which is rated for 5.884 N of thrust. The frame opposite corner motor distance measures at approximately 400 mm.

The frame portability was increased substantially by the inclusion of a foldable arms mechanism that uses a simple spring-based locking solution that allows the arms to be folded manually by rotating them inwards (Figure 4). This feature enables the quadcopter to be conveniently packed for transport or storage without the need for removing the propellers.

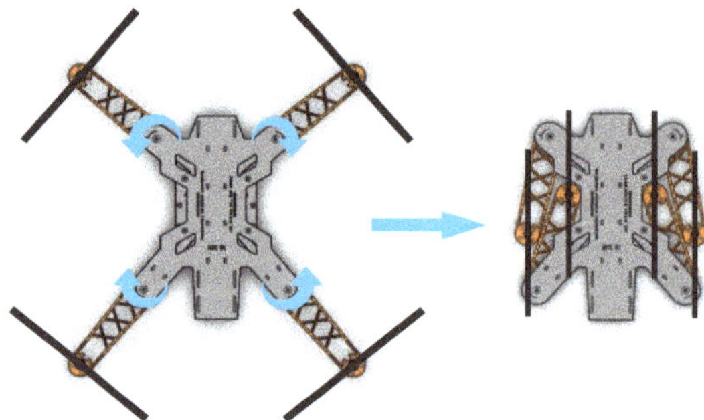

Figure 4. Operating principle of the frame foldable arms [5].

The prototype of the frame was printed using PLA (polylactide). Still, the material was later replaced with PETG (polyethylene terephthalate glycol-modified) using a 40% rectangular infill to improve further the structural integrity and the quality of the printed parts, due to the latter showing better adhesive properties.

The minimal set of electronics includes the flight controller, radio receiver, and power distribution board, with some space left available for future upgrades such as additional sensors or a camera. The takeoff mass was determined to be about 1.3 kg, which is comparable to medium-sized UAVs available on the market.

2.2. Flight Controller—Hardware

The developed prototype flight controller was intended to be fully open for both hardware and software modifications. With that in mind, components selected for its construction were chosen with an emphasis on accessibility and configurability. It was deduced that for the system to be equipped adequately for experimental purposes, some means of on-board flight data transmission and recording would have to be present.

After examining the solutions available on the market, the ESP32 series microcontroller [25] was picked as the central processing unit for the flight controller. The chip features an integrated wireless communication module, which allowed the implementation of a fully functional, close-range telemetric data transmission system, without the need for 3rd party equipment. Furthermore, the chip uses a 32-bit dual-core processor, 4 MB of FLASH, and 520 KB SRAM memory blocks and numerous data-bus interfaces: 4 SPI, 2 I2S, 2 I2C, 3 UART, and a CAN bus controller [26]. In terms of software execution, the module offers multi-threading capabilities through the use of the FreeRTOS operating system, and the support of core Arduino libraries.

The initial flight controller prototype was provided with a basic MEMS sensor package. The choice came in the form of MPU6050 by InvenSense, which already proved itself in numerous UAV applications. The sensor combines an integrated 3-axis accelerometer and gyroscope modules that provide the data necessary for accurate attitude estimation.

In order to uphold the notion of hardware accessibility, the above components were integrated using off-the-shelf modules designed for prototyping. The ESP32 comes with Lolin32 prototyping made by Wemos, while the MPU6050 was supplied on a small circuit board manufactured by DFRobot. The flight controller was put together using a custom PCB (Figure 5).

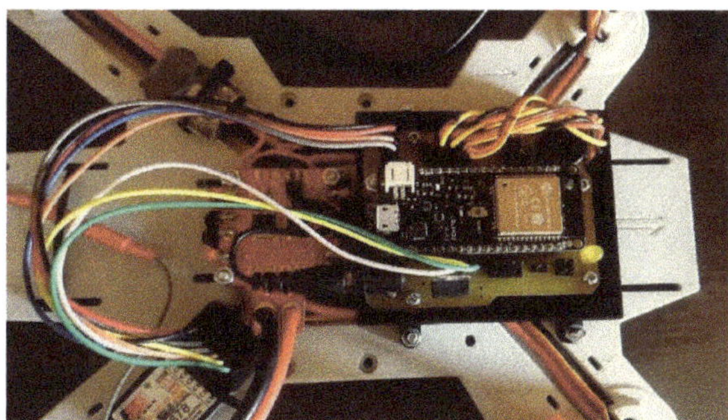

Figure 5. The prototype flight controller mounted on board [5].

The board provides all the necessary connections and mounting points for electronic components and a buzzer, led indicator, and a connector for additional power supply. A simple 3D-printable case was designed as well to provide the board with proper protection from external damage.

2.3. Flight Controller—Software

The software was written in C++, with most of the hardware functionality being accessed through the use of Arduino core libraries, provided by Espressif. The code was developed using PlatformIO IDE plugin for Visual Studio Code.

The program's main loop is executed at a frequency of 200 Hz. It consists of typical flight controller functionality, as described in [27], which includes sensor reading, attitude calculations, control loop pipeline, and motors control. For debugging and ad-hoc configuration, the software contains various subroutines accessible through both USB and wireless interfaces, which allow for the modification of the device operational parameters without the need for re-uploading new firmware. However, if necessary, the module is capable of wireless firmware uploads, which is not common in embedded devices.

The wireless telemetry capabilities are provided via an integrated Wi-Fi module that can function both as an access point and a client for the host station and is compatible with virtually any modern computer and smartphone device. The telemetric data can be streamed on-demand, with the use of the UDP (User Datagram Protocol) protocol—the solution chosen mostly due to its high performance and ease of use. UDP does not provide handshaking before or after the data exchange, which has a positive impact on the overall bandwidth. On the other hand, TCP (Transmission Control Protocol) is a slower but more reliable counterpart because it verifies the retrieval of each packet by the recipient device.

2.4. Control Loop Design

The flight controller was initially developed with a basic sensor package in the form of a single inertial sensors module. Therefore, in the initial version of the control algorithm, the control loop is limited to attitude stabilization based solely on the pilot's input in reference to the estimated attitude obtained from AHRS.

A handful of designs was considered for the example feedback loop setup. The simplest was direct angular stabilization in reference to the ground. The most advanced was linear–quadratic regulator (LQR). The final choice came in the form of cascaded PID (Proportional Integral Derivative) feedback loop, which was considered the viable option to implement, given the project already present complexity. Such solutions are proven to be very effective in multirotor applications [27–31].

The control loop design, as mentioned above, provides a dual set of PID regulators for roll and pitch axis, respectively, consisting of the "inner" and the "outer" loop control loops. The vertical yaw axis is stabilized individually by one controller belonging to the inner loop (Figure 6). In this case, the rotation speed around the yaw axis is directly proportional to the position of the control stick on the radio controller.

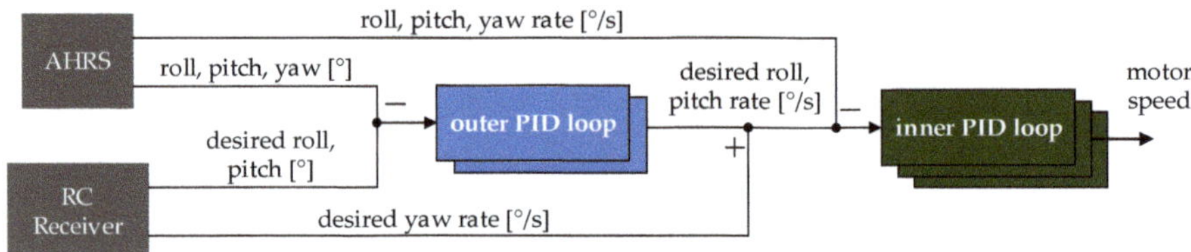

Figure 6. Simplified control loop diagram [5].

The outer PID loop function is to calculate the optimal target rotation speed in the given axis that needs to be applied to decrease the input error. Reducing error is performed by subtracting the desired angular position in the given axis, provided by the pilot, from the values estimated by the AHRS. The calculated angular velocity is then relayed to the inner loop regulator as one of its inputs.

The inner loop is responsible for controlling the motors, based on the information received from the previous PID set. Similarly, the calculated angular velocities are compared with the readings obtained from the gyroscope module. The computed output signal is then used to generate the pulse-width modulation (PWM) wave with a given duty cycle, which drives the brushless DC motors at the desired speed.

In other words, the inner loop stabilizes the quadcopter angular velocity with the use of gyro readings since the outer loop sets a target to the ground in an angular position desired by the pilot. Such setup is potentially effective for fast rejection of disturbances and reduction of their impact on the rest of the control loop [32].

The inner loop output is comprised of three signals, each of which corresponds to the given rotation axis. These signals need to be appropriately combined (mixed) to generate the appropriate output for each motor.

The quadcopter momentary attitude variations can be viewed as the sum of rotations around its roll, pitch, and yaw axis. Since the system must be able to monitor and act upon all three rotation speeds simultaneously, the output signal for each motor needs to be a combination of all three output values generated by the inner loop regulators: $R_{rate\ out}$ (roll axis), $P_{rate\ out}$ (pitch axis) and $Y_{rate\ out}$ (yaw axis). The math behind the process is based on the quadcopter flight behavior characteristics (Figure 7) and the position of a given motor (Figure 8).

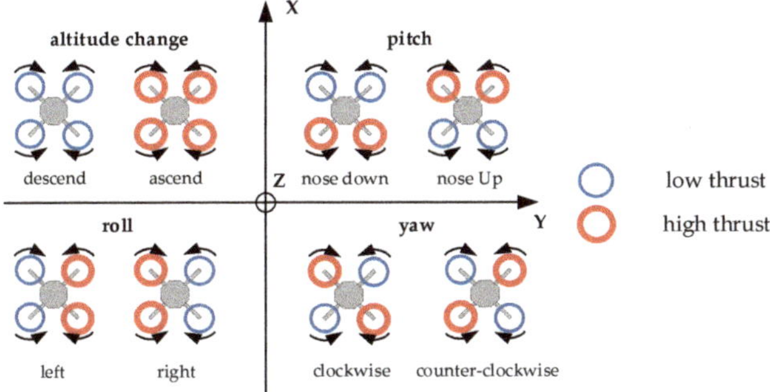

Figure 7. Basic principles of quadcopter maneuvering.

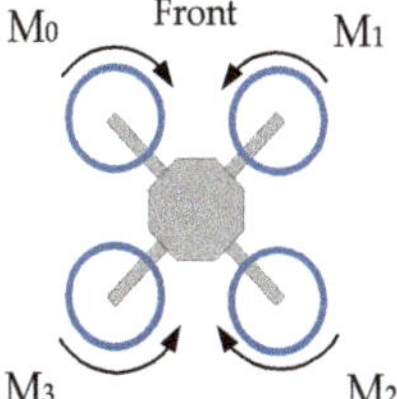

Figure 8. Motor markings.

For example, for the aircraft to rotate around the longitudinal axis (X—roll), the motors on the one side of this axis must increase the generated thrust, while at the same time, motors on the other side reduce power. Therefore, the roll output for the motors on the one side must be negated. Similarly, to start rotating around its vertical (Z—yaw) axis, the motors spinning in one direction must generate more thrust while the motors rotating in the opposite direction should lower their speed to generate the asymmetry in the overall momentum applied to the frame. For the project, the motors were numbered in the order starting clockwise from the front left motor—M_0, to the rear left motor—M_3.

By using the above markings, the proposed signal mixing formula can be presented by the following set of equations:

$$\begin{aligned} M_0 &= T_g + R_{rate\ out} + P_{rate\ out} + Y_{rate\ out} \\ M_1 &= T_g - R_{rate\ out} + P_{rate\ out} - Y_{rate\ out} \\ M_2 &= T_g - R_{rate\ out} - P_{rate\ out} + Y_{rate\ out} \\ M_3 &= T_g + R_{rate\ out} - P_{rate\ out} - Y_{rate\ out} \end{aligned} \tag{2}$$

where: $M_0 \dots M_3$—motor signals, T_g—pilot's throttle signal, $R_{rate\ out}$, $P_{rate\ out}$, $Y_{rate\ out}$—inner loop output for roll, pitch, and yaw controllers.

The mixing equation presented above, albeit very basic, is what essentially makes it possible for the quadcopter to autonomously decrease its attitude error with the use of its four motors, which in turn makes the aircraft controllable.

Overall, the control loop operation can be configured by a total of 15 parameters listed in Table 1.

Table 1. Full set of control loop adjustable parameters.

Axis	Inner Loop			Outer Loop		
roll	$P_{rate\ roll}$	$I_{rate\ roll}$	$D_{rate\ roll}$	$P_{angle\ roll}$	$I_{angle\ roll}$	$D_{angle\ roll}$
pitch	$P_{rate\ pitch}$	$I_{rate\ pitch}$	$D_{rate\ pitch}$	$P_{angle\ pitch}$	$I_{angle\ pitch}$	$D_{angle\ pitch}$
yaw	$P_{rate\ yaw}$	$I_{rate\ yaw}$	$D_{rate\ yaw}$	-	-	-

The control loop contains five individual regulators—three belonging to the inner loop and two residing within the outer control loop, with the yaw axis being stabilized using a singular rate controller (Figure 6). With each one having a specific P, I and D gain values assigned to it. Finding the optimal values for these gains was a crucial task that required establishing a tuning method that would provide freedom of experimentation while maintaining proper safety measures.

The proposed method takes from the heuristic Ziegler–Nichols approach, by which the regulator gains are acquired with the use of measuring the oscillatory system response. The measured period of the system response is compared to the standardized tabular values upon which the gains are calculated.

2.5. Controller Implementation

It was decided that proportional gain would be separated from integral and derivative blocks to achieve distinctiveness between the contributions of each gain. The following equation describes the controller in its discrete form:

$$u(n) = K_p e(n) + \frac{\Delta T}{K_i}\sum_{j=0}^{n} e(j) + \frac{K_d}{\Delta T}[e(n) - e(n-1)] \quad (3)$$

where: K_p—proportional gain, K_i—integral gain, K_d—derivative gain, n –sample number, $u(n)$—output signal, $e(n)$—error, ΔT—sampling period.

Furthermore, the regulator operates in the digital domain. Therefore, the controller was designed with its derivative gain input being connected directly to the sensor measurement data—w(t) instead of the classical approach where the regulator is provided with a single error input—e(t), as presented in Figure 9. It is a form of protection against rapid increases of derivative value due to signal spikes in the controller input.

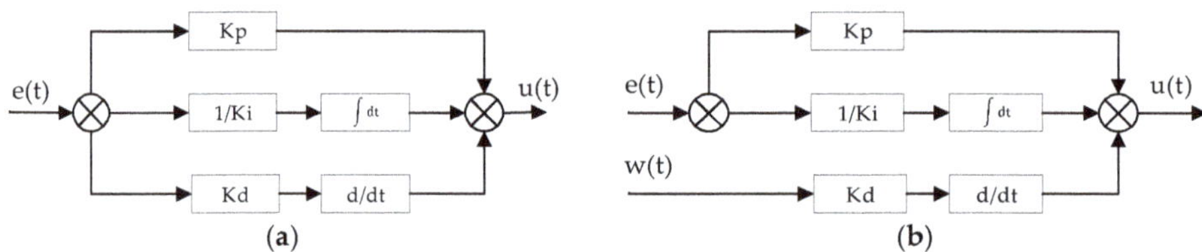

Figure 9. Comparison of basic PID regulator types. (**a**)—the classic approach with single error input, (**b**)—regulator with derivative gain input connected directly to sensor measurements (used in the project), where e(t)—input error, w(t)—sensor measurement, u(t)—regulator output.

Interestingly enough, such undesirable conditions were accidentally encountered during software development, long before the flight trials could begin. The cause was discovered within the on-board radio receiver, used as means of remotely steering the aircraft. The device used in the project was programmed to update its output registers with a frequency of 50 Hz, which is a quarter of the frequency of control loop execution (200 Hz). It caused the derivative gain to generate a spike in its output every time the receiver would modify its output signals, which was the reason why the above solution was implemented [33,34].

The set of controllers was implemented in the form of a custom-written class library containing numerous functions that allow for adjusting the control parameters of each regulator individually at any time during the control loop operation. The example code can be found in Appendix A. Both the inner and the outer control loop regulators have the same structure, with the difference being the input and gain settings provided separately, depending on the controller function.

Anti-windup functionality was implemented as well, with its logic being run outside the main class. This functionality is controlled by a process that tracks the throttle value set by the pilot. Through a set of logic, tests determine the current anti-windup setting by comparing the throttle value to a specified set of boundaries. The throttle range was divided into three sections. The lower part intended for landing, the transition zone, and the flight range, which ends at a 100% setting (Figure 10). The transition zone works as a buffer. By operating similarly to a hysteresis loop, it prevents the safeguard from activating unexpectedly during flight. The boundary values for the transition range were estimated based on the thrust range depicted in the datasheet of the motors and then adjusted after the first test flight.

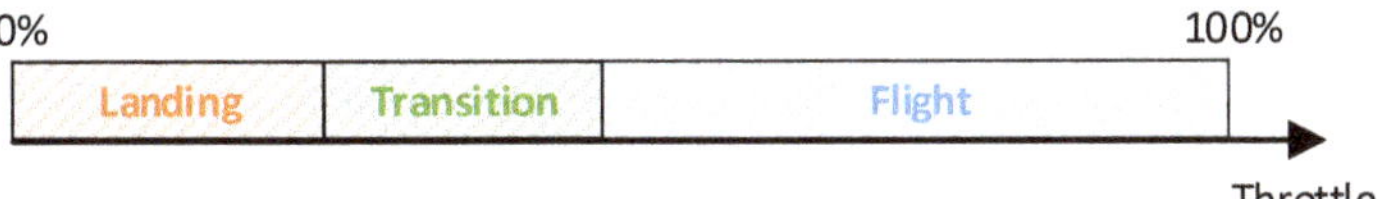

Figure 10. Anti-windup throttle range.

The anti-windup is switched on as default and remains active until the throttle value goes outside the lower range. Shifting the throttle value in or out of the transition zone triggers the anti-windup module to change a mode of operation. Transition into the flight range indicates takeoff in which case safeguard is turned off, making the controllers fully operational. The lower range is used mostly for landing. In that instance, the anti-windup is turned on, the cumulated errors are nullified, and integral gains are left disabled for all controllers.

2.6. Attitude Estimation

As for the method by which the quadcopter attitude is calculated, while the MPU6050 does provide an onboard proprietary attitude calculation algorithm due to its limited customizability and various software-related issues, an alternative solution was used instead. The implemented algorithm is based on S. Madgwick's filter, more comprehensively described in [35]. The algorithm outputs information about the quadcopter attitude about the ground in the form of Euler angles, using the measurement data provided by the inertial sensor (Figure 11).

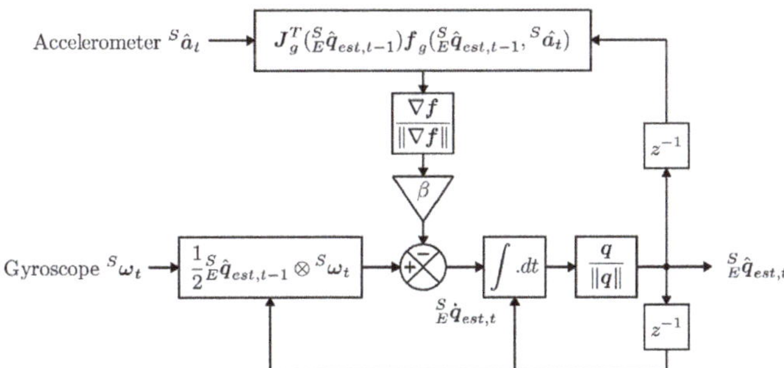

Figure 11. Block diagram of Madgwick's filter in basic IMU (inertial measurement unit) application [35].

This particular filter was chosen due to its availability in the form of a C-compatible library and a widely proven performance record. Moreover, the filter is well suited for aircraft applications as it allows for the addition of magnetometer in the future development to aid in the quadcopter heading acquisition.

The filter was supplied as a library with two operating modes available—with or without a magnetometer contribution. The algorithm can be configured during initialization by providing it with two operational parameters: sampling frequency and beta coefficient—β, with the latter one requiring further explanation.

Madgwick's filter utilizes quaternions to describe the sensor (S) orientation relative to earth (E)—${}^{S}_{E}q_{\text{est, t}}$. This method uses the so-called gradient descent algorithm to remove the impact of the gyroscope error on the computation resulting. Gyroscope measurements form the basis of attitude tracking, while subsidiary sensors such as an accelerometer and also magnetometer are used to estimate the gyroscope error. The basis for this process can be written as the following formulas:

$$ {}^{S}_{E}q_{\text{est, t}} = {}^{S}_{E}\hat{q}_{\text{est, t-1}} + {}^{S}_{E}\dot{q}_{\text{est, t}}\Delta t \tag{4} $$

$$ {}^{S}_{E}\dot{q}_{\text{est, t}} = {}^{S}_{E}\dot{q}_{\omega,\text{ t}} - \beta {}^{S}_{E}\dot{\hat{q}}_{\epsilon,\text{ t}} \tag{5} $$

Quaternion ${}^{S}_{E}\dot{\hat{q}}_{\epsilon,\text{ t}}$ represents the estimated gyroscope error calculated by the gradient descent algorithm. The filter uses data gathered from the accelerometer with the optional addition of a magnetometer to calculate the estimated gyroscope error in the form of a quaternion derivative. This derivative is then used to compensate for the gyroscope error by removing it from the gyroscope measurement quaternion ${}^{S}_{E}\dot{q}_{\omega,\text{ t}}$ as it is in equation 2.5. The presence of β enables us to manually fine-tune this process to achieve the best possible performance out of the available hardware. The author proposes that the β coefficient may be calculated based on mean zero gyroscope measurement error—$\widetilde{\omega}_{\beta}$ using the following formula:

$$ \beta = \sqrt{\frac{3}{4}}\widetilde{\omega}_{\beta} \tag{6} $$

Although the method of gyroscope drift nullification in Madgwick's filter differs significantly in its inner workings compared to a complementary filter by experimenting with different β values, one might notice that β has similar properties to that of an α coefficient of a complementary filter. With β set too low, the drift correction becomes suppressed to the point of inefficiency, and the filter becomes unusable due to continuously increasing error caused by gyro drift. When the β value is set too high, the filter overcompensates for the drift and output unreliable results. Based on numerous tests with different sensors, it was deduced that typical β values range between 0.02 and 0.10. Cases that required higher β values were resolved by improving the sensor offsets calibration technique. During the following tests, the filter was operating with β = 0.05.

2.7. Platform Versatility

The proposed platform has been designed to be versatile in terms of software and hardware configurability alike. The usage of open-source solutions and custom quadcopter frame design makes the platform suitable for custom modifications, allowing the inclusion of new components and control solutions to the aircraft's current feature set.

The frame itself, being designed in CAD software, can be easily modified to feature any form of sensors, by suitably changing its source files. Moreover, it is possible to alter a single subpart of the frame, for instance, the lower deck, to feature custom mounting points for a new type of sensor/device while maintaining compatibility with the default components. It is also possible to modify the arms to feature the mounting points for new, more efficient motors. With the current accessibility of 3D printing technology, the freedom of frame geometry customization is unmatched in comparison to any mass production solution.

The flight controller has been designed to have a multi-use potential being achieved through the software design and putting the emphasis on the high openness and customizability requirements. The code is fully modifiable as the software functionality can be changed in all aspects, beginning from the primary control loop settings to rewriting the entire subroutines.

Code portability is assured through the use of Arduino libraries as its hardware abstraction layer. This portability makes the control pipeline executable on other common microcontroller-based hardware platforms. The software was developed with a "block design" approach. In essence, subroutines are grouped based on their functionality, which makes it convenient to add new features without significantly impacting the overall code structure. In its current form, the software features a simple, yet efficient control loop design, which makes it perfect as a learning platform, while being open for improving upon the existing hardware and software solutions.

Lastly, due to the way the platform has been conceived initially, the solution fosters creativity above all. The platform seems to be suitable for experimenting with a nonconventional approach to traditional ideas when the use of commercial equipment is either not viable or too restrictive.

2.8. Tuning Method

Since the Ziegler–Nichols (Z-N) method can only be applied to a moving system, a decision was made to construct some form of a test-stand for limiting the degrees of freedom of the quadcopter.

The test-stand was made as a rigid metal frame mounted on top of a square wooden platform. The frame is equipped with two hooks used to mount the quadcopter between the two poles using wire. The fixing enables the quadcopter to rotate around its longitudinal axis without obstruction while keeping its center of mass in a fixed position. The adjustment of the control loop parameters during the procedure was made via the already mentioned subroutines contained within the flight controller software, which were accessed wirelessly using a dedicated terminal application.

The tuning process was introduced with the focus on the following set of assumptions:

- The acquired gains are considered as initial values with the expectation that further adjustment has to be conducted in flight,
- The Ziegler–Nichols tuning method is used to tune the inner control only,
- The outer loop is tuned manually with the proportional gains being adjusted first,
- The gains acquired for the roll axis is initially applied to the quadcopter pitch axis, further adjustment to follow if necessary.

The control loop gains affect the drone's airworthiness and its behavior concerning the pilot's commands. Therefore, the final parameters' values can be chosen after the first set of flights is completed, and it depends on the pilot's preferences.

To achieve the best results, we decided that the influence of each gain of the inner loop controllers would be adequately studied to gather necessary information about its impact on the controller's overall performance. The tests were performed using the already mentioned test-bed, with the cascade of roll axis controllers being the object of the experiments. The outer loop proportional gain setting was left intact ad matched those resulting from the Z–N method. The pitch and yaw controllers were left disabled for the duration of the test and throttle was limited to 40% for safety. The proportional and derivative gains were tested by applying a 35°-step signal in place of the pilot's roll axis input to the controller using a software subroutine activated with one of radio receiver auxiliary channels.

In contrast, the integral gain was tested during hover flight. After each test, a given gain factor was incrementally increased, and the procedure was repeated. Data were recorded using the ESP32 as a wireless telemetry module, and the PC application was used for logging the data. The results are presented in Section 3.1.

2.9. Inner Loop Tuning

The inner loop tuning procedure was performed using the already mentioned test-bed. The regulators responsible for lateral and vertical stabilization were left disabled for the duration of the process. The measurement recording began after applying 40% of the entire throttle range to the motors and giving the quadcopter a gentle push to shake it out of balance using the stick on the radio receiver (Figure 12).

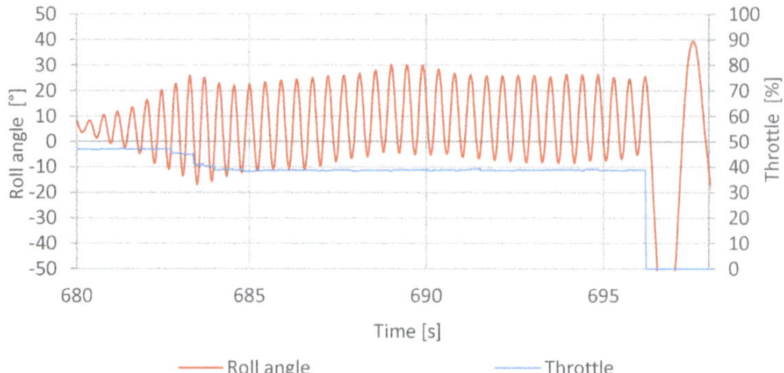

Figure 12. Data gathered during the Z–N (Ziegler–Nichols) tuning procedure—$K_u = 1.1$ [5].

The data were acquired using the aforementioned integrated telemetry system and a simple PC application running on the receiving end.

The proportional was slowly increased using the text commands sent to the flight controller via the text terminal application until the ultimate was defined as l $K_{u\ roll} = 1.1$ with the registered oscillatory period of $T_{u\ roll} \approx 400\ ms$. These values were applied to the tabular data [36] to calculate the controller gains using the following equations:

$$\begin{gathered} K_{p\ rate\ roll} = K_{p\ rate\ pitch} = 0.6 * K_{u\ roll} = 0.66 \\ K_{i\ rate\ roll} = K_{i\ rate\ pitch} = 0.5 * T_{u\ roll} * K_{p\ rate} = 0.14 \\ K_{d\ rate\ roll} = K_{d\ rate\ pitch} = \\ = 0.125 * T_{u\ roll} * K_{p\ rate} = 0.034 \end{gathered} \quad (7)$$

2.10. Outer Loop Tuning

While the inner loop tuning needed to be performed during its operation, due to its direct connection to the propulsion system, the external loop gain could be computed numerically. The calculation is an estimation of the maximum desired rotation speed resulting from the pilot's input and applying it to a simple multiplication. The regulator proportional gain output is based on this estimation.

The initial values for the outer loop proportional gains were set at $P_{angle\ roll} = P_{angle\ pitch} = 4.0$.

Based on the already described outer loop functionality, this produces a relation where 45° difference between measured and desired angles results in the target angular speed of 180 °/s. This setting was further verified in flight and determined to be sufficient. Furthermore, the remaining I and D gains were deemed unnecessary, with the steady error nullification and derivative response already present in the inner loop regulators.

At that point, the quadcopter possessed the ability to perform basic hover flight, with the vertical axis stabilization remaining unadjusted. The inner loop yaw axis proportional gain was tuned experimentally, between flights. The most optimal gain was quickly determined to be $K_{p\ rate\ yaw} = 3.0$. Similar to the outer loop, the remaining gains were left disabled. The gains can be left disabled since the quadcopter propellers demonstrate natural dampening effects, which diminish any disturbances affecting the quadcopter rotation around its vertical axis.

2.11. Sensor Initialization

Sensors were initialized with the use of a particular subroutine embedded within the software. Each axis was initialized with offsets calculated by averaging a sum of 100 measurements taken over 10 s, during which the quadcopter remained immobile in a horizontal position. The offset values were then saved to the microcontroller non-volatile memory so that they could be loaded during boot-up. The measurements were taken in stable conditions in a static room temperature of 29 °C. The calculated offsets are presented in Table 2.

Table 2. Offset values of MPU6050 used in the flight controller.

	X-axis	Y-axis	Z-axis
Gyroscope [°/s]	34.58	33.88	37.26
Accelerometer [m/s²]	−0.37	10.61	−25.82

After the initialization, a series of tests were conducted to verify the validity of calculated offsets and examine the behavior of sensor readings over more extended periods. The first test has been taken for 25 min to observe any deviations from the initial values. It consisted of taking attitude measurements and IMU sensor readings, including the temperature of the sensor. The environment conditions matched the conditions during the initialization. The results are presented below in Section 3.1.

The sensor package was reconfigured to increase its resilience to vibrations introduced by the motor operation as well. The integrated digital low-pass filter was put in use to counter the influence of high-frequency disturbances on obtained sensor measurements. This feature turned out to be very convenient as it makes software filtering redundant, thus reducing the necessary microcontroller workload. Cutoff frequencies of 44 Hz and 42 Hz were chosen for the accelerometer and gyroscope, respectively, based on the sensor documentation (Figure 13).

DLPF_CFG	Accelerometer (F_s = 1kHz)		Gyroscope		
	Bandwidth (Hz)	Delay (ms)	Bandwidth (Hz)	Delay (ms)	Fs (kHz)
0	260	0	256	0.98	8
1	184	2.0	188	1.9	1
2	94	3.0	98	2.8	1
3	44	4.9	42	4.8	1
4	21	8.5	20	8.3	1
5	10	13.8	10	13.4	1
6	5	19.0	5	18.6	1
7	RESERVED		RESERVED		8

Figure 13. Inertial sensor's digital low pass filter register map [37].

The setting was later tested during quadcopter operation. The test was done by mounting the quadcopter in a fixed position and taking measurements of the flight controller estimated attitude angles while applying 15% of thrust to the engines for a specific time. Despite the flight controller housing containing a set of spring-based dampers upon which the PCB was mounted to the frame, the effect of vibrations on attitude estimation exceeded the initial expectations (Figure 14a).

Further tests showed that the effects of vibrations decreased proportionally as the cutoff frequency was lowered. Ultimately the 44/42 Hz frequency set was chosen as the optimal setting, as it demonstrated a compromise between the filter efficiency and introduced delays in sensor response (Figure 14b). The improvement in accuracy is visible, with the deviation range being reduced to ±0.5° about the real value of 0°. The presented register settings provided the best possible results during the tests and were implemented into the AHRS to improve its accuracy.

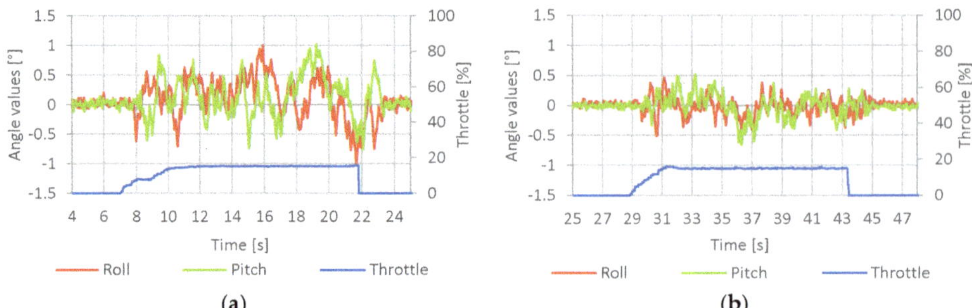

Figure 14. Effects of motor vibrations on the estimated attitude: (**a**) no filtering, (**b**) low-pass filtering with 44/42 Hz cutoff frequencies [5].

2.12. System Evaluation

A series of tests have been performed to check the quality of the designed system. All the presented tests were taken on the test platform using embedded software tools for tracing and acquisition of the parameters during operation. First, the sensor offsets were examined for their effectiveness during the system's operation. The purpose of the test was to determine a drift of the sensors and measurement systems in time. In this test, the platform was fixed in a stand, and the on-board system has been started for 20 min. Since the platform was fixed, no input was exerted. The output parameters measured were measurement system temperature, angular speed indicated by gyroscopes, acceleration indicated by accelerometers, pitch, roll, and yaw angles calculated by the inertial measurement unit. Inertial sensors, i.e., a gyroscope and an accelerometer reading, were examined to verify the calculated offsets over the prolonged activity (Section 3.1).

The controller tuning methods have been tested consecutively for tuning in the inner and outer control loops. Separately, the system has been examined against various gains values in each axis since these parameters are considered crucial for successful control (Section 3.2). The purpose of this test was to determine the step response of the controller to the input angular movement exerted consecutively in each axis with various gain values. In this test, the platform was fixed to allow angular movement around one axis in one test only. The input was an angular movement of 35 degrees exerted as a step consecutively around axis x and y. The input was set to 35 degrees because this is an angular boundary position of the drone during its operation. The output was a response from the controller.

Next, the dynamic test of the controller response in each axis with various values of the K_i the coefficient was performed. The purpose of this test was to determine a response of the controller to the dynamically changing demanded angular position command with various values of the K_i coefficient. The input was a demanded value of the angle in a specified axis. The input was changed randomly by the operator via a radio controller. The output was an angle position in this axis of the drone.

On the pre-tuned controller, a hover stability test and a responsiveness test have been carried out to check the performance of the system. The purpose of this test was to determine the response of the controller to the natural disturbance during hovering. The shape of the response chart is connected to platform stability. Hover stability was examined in an open area with no wind. After takeoff, the quadcopter was placed at an altitude of around 3 m, after which the telemetry recording was started. Due to the flight controller lacking the ability to maintain its position autonomously, minor periodic course corrections were manually applied. Despite this drawback, it was possible to record a 20-s long autonomous flight during which the flight controller attempted to stabilize the quadcopter horizontally. During that time, the quadcopter remained within a radius of two meters from its takeoff location. The input was a demanded value of an angle in a specified axis. The demanded value was set zero degrees as for the hover of the drone. The output was an angle position in this axis of the drone.

The system response time was verified in similar environmental conditions by applying a series of rapid maneuvers to register any significant delays between the pilot's commands and their execution by the flight control system. The input was a demanded value of an angle in a specified axis changing rapidly. The input was changed randomly by the operator via a radio controller. The output was a delay of the angle position change in this axis of the drone (Section 3.3).

The wireless telemetry system was mostly tested for its maximum operational range capabilities. The purpose of this test was to determine the maximum efficient range of the test platform telemetry. The experiment was conducted in an open area, with the quadcopter and laptop maintaining a direct line of sight between each other for its entire duration, with an active connection. The terminal application was used to register and log the number of received telemetry frames in a given second. The procedure involved carrying the quadcopter and going away from the start point at a steady speed until the distance of around 300 m was reached. The time from the beginning of the test was also measured. The input was a distance between the flying platform and the ground control station. The initial value of the input was set to zero meters, and it was being increased to 180 m because, at this distance, a number of successfully transferred data packets diminished significantly. The output was a number of data packets transferred in time. We performed the test to process the gathered data afterward in the way to synchronize the incremental changes in the distance with the recorded log using a unified time scale (Section 3.4).

3. Results

3.1. Sensor Offsets Evaluation

The first test is a static test showing the temperature of the measurement system, drift of the sensors, and the cumulating error of the attitude calculation in the same period of operation. The purpose of the test was to determine a relation between the temperature of the system and drift of the sensors and the attitude calculation error resulted from the drift in time. The IMU temperature throughout the test increased by approximately 10 °C (Figure 15), which had a visible impact on the sensor readings taken during that time (Figures 16–18), which resulted in estimated roll and pitch angle values drifting from their initial values proportionally to the increasing sensor temperature (Figures 19 and 20).

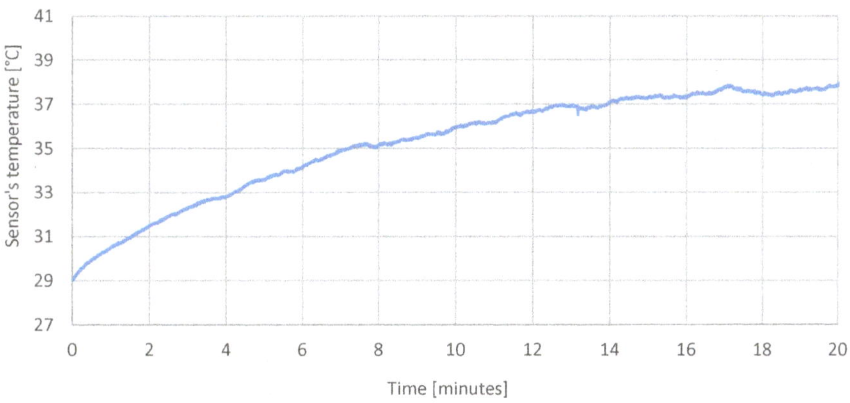

Figure 15. Sensor temperature during the test.

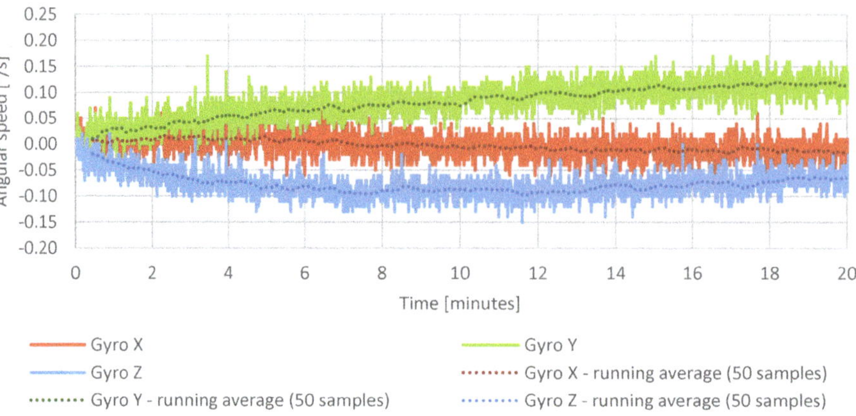

Figure 16. Stationary measurements of gyroscope drift.

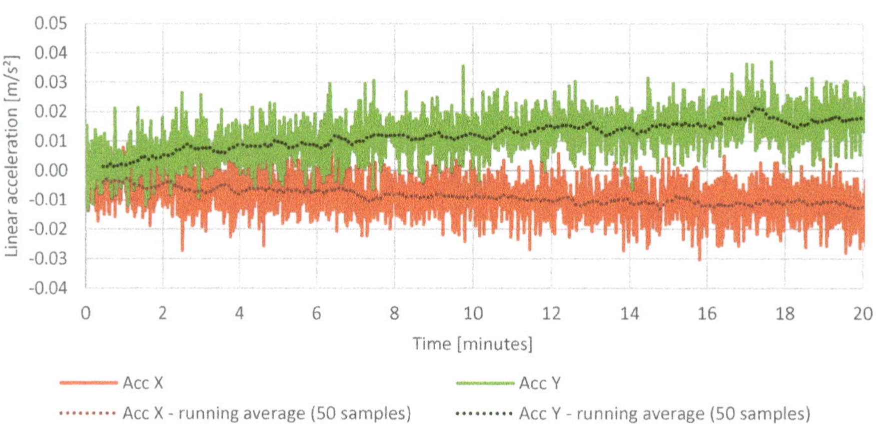

Figure 17. Stationary measurements of linear accelerations.

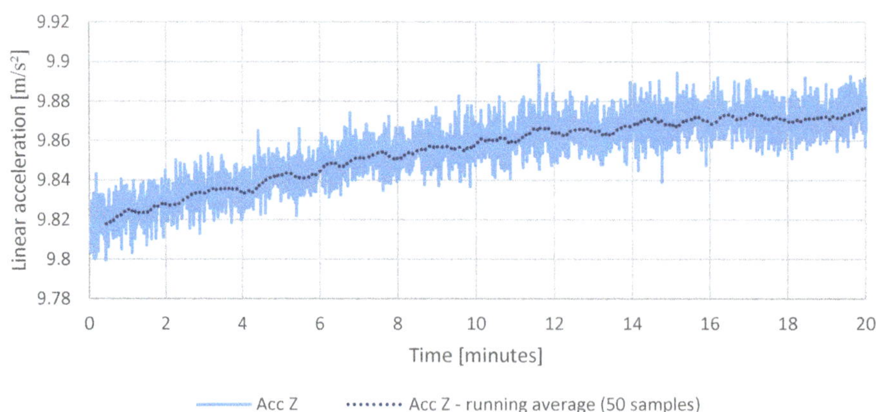

Figure 18. Stationary measurements of linear acceleration in the sensor vertical axis.

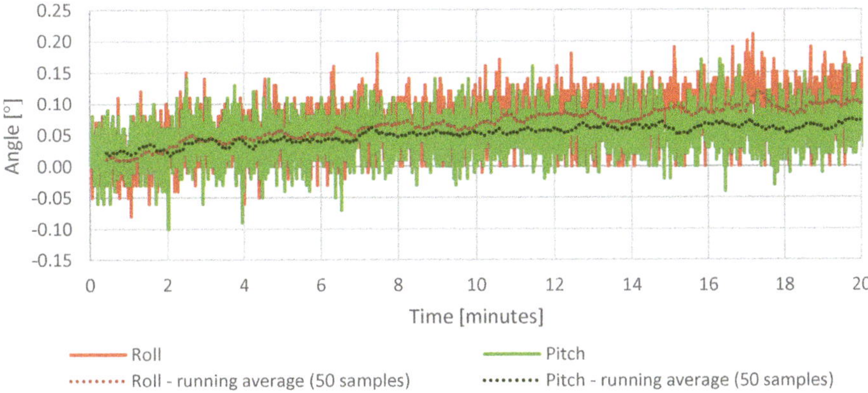

Figure 19. Stationary measurements of estimated roll and pitch angles.

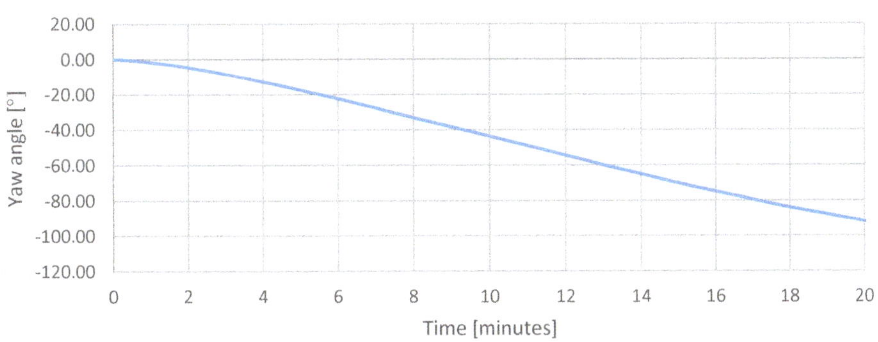

Figure 20. Stationary measurements of estimated yaw angle.

Figure 15 shows the increase in the temperature of the measurement system in 20 min test.

Figure 16 shows the increase of the gyroscopes drifts in 20 min test following the increase of the temperature of the measurement system.

Figures 17 and 18 show the changes in reading of the accelerometers in axis x, y, z during 20 min test following the changes in temperature of the measurement system.

Figures 19 and 20 show a cumulating error in calculated attitude. The pitch and roll calculated values increase in time, and the yaw calculated value decrease while the platform stays fixed.

The gathered readings show that the effectiveness of calculated offsets is correlated with the sensor temperature at the time of calibration. On the other hand, the estimated angle variations in roll and pitch are relatively small—up to +0.10 degrees deviation in roll and +0.05 in pitch, respectively (Figure 19).

The most affected sensor axis is the vertical axis responsible for tracking the yaw angle. Since the magnetometer was not yet implemented at the time of this test, whenever the sensor was operating horizontally to the ground, the yaw angle estimation relied solely on gyroscope measurements, which resulted in a high contribution of drift in the estimated rotation.

In practice, with aerial navigation not being one of the goals of the project, this issue could be neglected, especially in case of manual piloting, since it could be easily compensated using the trim function of the radio receiver. However, it does not diminish the need for proper temperature compensation in the future.

3.2. Controller Gains Examination

The second test was the controller step response test to the angular position step change with various values of the controller gain parameter. The purpose of this test was to determine the step response of the controller to the input angular movement exerted consecutively in each axis with various gain values.

The results of controller validation tests with selected gain settings are presented below. The results were used to evaluate both the validity of controller gains obtained using the Ziegler Nichols method and the response of implemented controllers to the introduced gain adjustments.

The proportional gain was determined to be the most crucial for the operation of the inner loop controller. After decreasing the $K_{p\ rate\ roll}$ value, the quadcopter reactions became slower and less responsive (Figure 21). Since the output is directly proportional to the input error multiplied by the $K_{p\ rate\ roll}$ gain, the motors did not generate enough power required to position the quadcopter at the desired angle. On the other hand, after increasing the $K_{p\ rate\ roll}$ beyond the value of $K_{p\ rate\ roll} = 0.9$, steady oscillations were observed with its amplitude increasing proportionally, which is a typical example of an overshoot. On that basis, it was determined that $K_{p\ rate\ roll} \approx 0.66$, calculated using the Z–N method, is close to optimal.

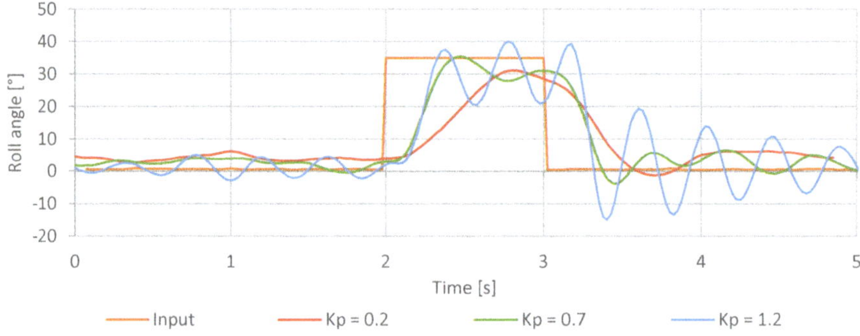

Figure 21. Step response of the roll axis controller for different Kp values.

The derivative gain evaluation was performed similarly. After evaluating the data (Figure 22), it was noted that derivative gain improves the overall system precision. It does that by reducing the overshoot and increasing the overall stability. This effect can be observed during flight—with higher $K_{d\ rate\ roll}$ values, the aircraft became "stiffer" and more resilient to air disturbances. During the stationary tests, $K_{d\ rate\ roll} = 0.055$ turned out to be the optimal setting, which is a slight deviation from the results of the Z–N tuning.

Further experimentation showed that the range of useful derivative gain settings is much more restrictive when compared to proportional gain. Values reaching beyond $K_{d\ rate\ roll} = 0.075$, caused derivate gain to display its dampening properties, by slowing down the quadcopter reaction to the pilot's input. Increasing the $K_{d\ rate\ roll}$ even further—beyond the value of $K_{d\ rate\ roll} = 0.110$ introduced rapid oscillations in the rotation of the aircraft, with their frequency being much higher compared to the overshoot recorded during proportional gain tests. In this case, an over-amplification of the derivative gain causes the motor response with much more energy up to a point where the aircraft becomes uncontrollable.

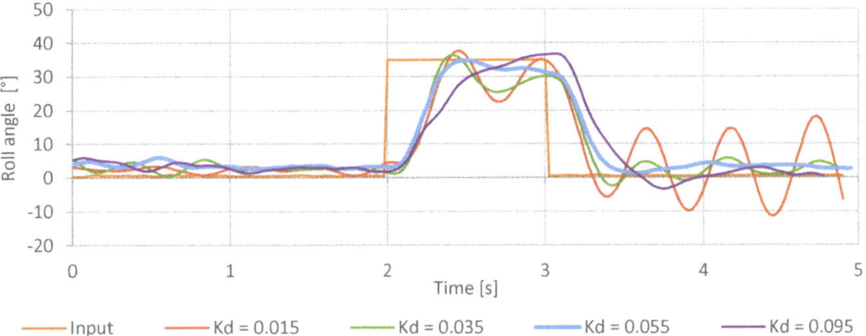

Figure 22. Step response of the roll axis controller for different Kd values.

The value of the K_i was tuned by testing a controller response to the random input roll angle changes demanded by the operator. The purpose of this test was to determine a response of the controller to the dynamically changing demanded angular position command with various values of the K_i coefficient. An analysis of the data gathered during the hover stability test showed that $K_{i\ rate\ roll}$ has a noticeable impact on the quadcopter performance during hover. By disabling this gain, the quadcopter became more prone to drifting away from its takeoff location and required constant corrections from the pilot to stabilize its position. We expected this behavior as the purpose of the integral component is to nullify the steady-state error, which was mostly responsible for the drift. After increasing the integral gain value to $K_{i\ rate\ roll} = 0.2$ the quadcopter stability improved as the desired angular position was maintained more effectively (Figure 23a,b). The integral gain also showed more resilience to overshoot as steady oscillations appeared beyond the value of $K_{i\ rate\ roll} = 1.5$. This result, however, does not diminish the importance of anti-windup or other similar and potentially applicable safeguards as this test was conducted with the quadcopter being in-flight, during which the system was fully capable of maintaining its cumulated error within correctable boundaries.

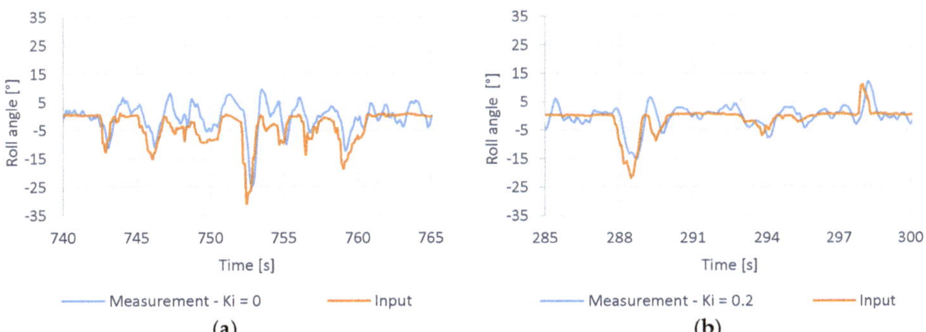

Figure 23. Hover stability with different Ki values, (**a**) measurement—Ki = 0, (**b**)—measurement—Ki = 0.2.

As planned, the final adjustment to the PID gains was conducted between flights based on the observation of the quadcopter response to the pilot's input and hover stability. The inner loop parameters were slightly re-adjusted to suit the pilot's preferences better. The full set of final controller gains is presented in Table 3.

Table 3. Final control loop settings.

Axis	Inner Loop			Outer Loop		
	$K_{p\ rate}$	$K_{i\ rate}$	$K_{d\ rate}$	$K_{p\ angle}$	$K_{i\ angle}$	$K_{d\ angle}$
roll	0.7	0.15	0.035	4.0	0.0	0.0
pitch	0.75	0.20	0.040	4.0	0.0	0.0
yaw	3.0	0.0	0.0	-	-	-

Following the tuning procedure, a series of tests were conducted to verify its efficacy. The trials were performed in flight, with a fully enabled controller set.

3.3. Control Loop Performance Test

The control loop performance test was a test of the stability of the drone. The attitude of the drone in relation to the demanded null value of angular angles pitch and roll was recorded. The purpose of this test was to determine the response of the controller to the natural disturbance during hover.

Figure 24 contains the hovering stability test results, during which the quadcopter was given null target roll and pitch angles to examine its ability to maintain horizontal orientation during hover flight.

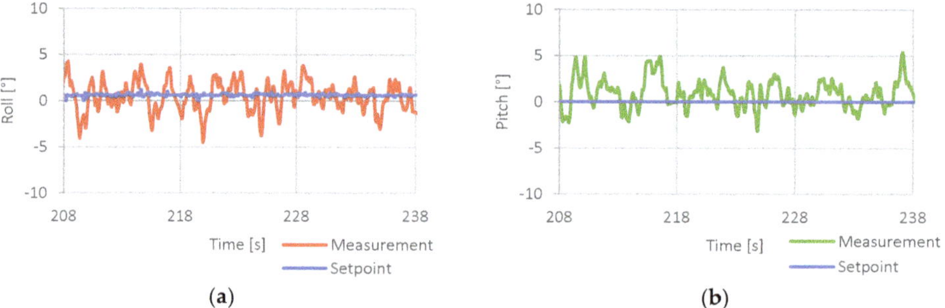

Figure 24. Hover stability test results, (**a**)—roll axis response, (**b**)—pitch axis response [5].

The results proved that in its current state and applied settings, the developed system could achieve horizontal angular accuracy of ± 4° during hover flight. Due to the quadcopter high inertia, such oscillation had little effect on its position across the accounted timespan.

Next, the response time was verified in similar environmental conditions by applying a series of rapid maneuvers to register any significant delays between the pilot's commands and their execution by the flight control system. The purpose of this test was to measure a delay of the drone position changing after the angular position change command receiving.

The results of control loop response time evaluation test are presented in Figure 25.

The system response delay was measured to be around 200 ms. This result was considered acceptable since it can hardly be observed with the naked eye. The quadcopter behavior was appropriate, considering its takeoff mass and the use of large propellers.

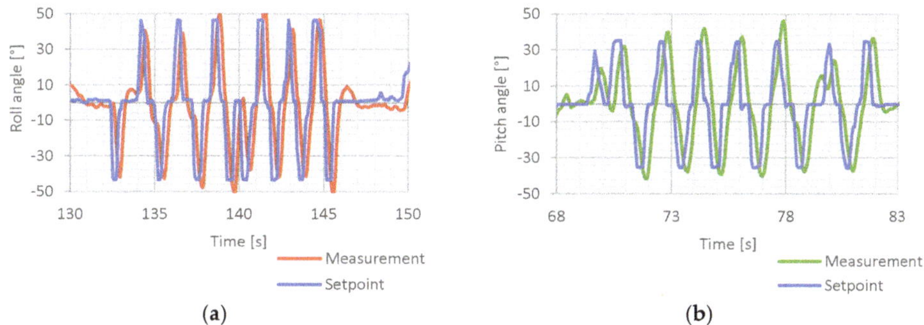

Figure 25. System response time test—$K_u = 1.1$, (**a**)—roll axis, (**b**)—pitch axis [5].

3.4. Wi-Fi Telemetry Range Test

Wi-Fi telemetry range test was a test of successfully transferred data packets along with the increasing distance between the ground control station and the drone. The purpose of this test was to determine the maximum efficient range of the test platform telemetry.

Figure 26 shows the number of data packets from the quadcopter received by the ground station to the distance between network devices.

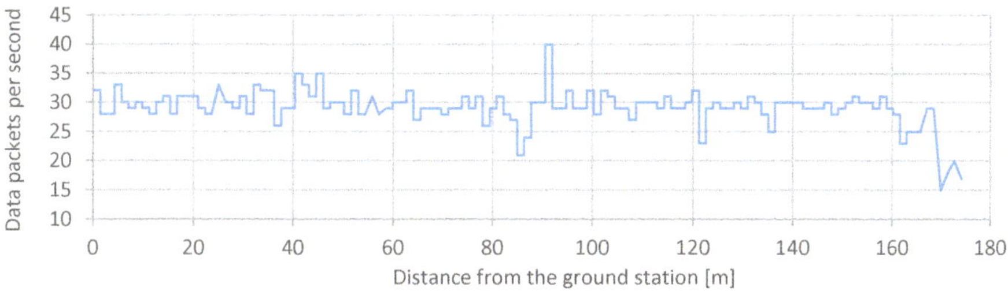

Figure 26. Wi-Fi telemetry system maximum operating range test [5].

The maximum operation range of the ESP32 chip used in the flight controller construction was rated as approximately 150 m. This outcome is satisfactory, based on the fact that this functionality utilizes a miniature transmitter integrated within the chip structure. The device is well fit for this project use-case, being that of a close-range data transmission system.

4. Discussion

The custom quadcopter frame, with its foldable arms design, turned out to be a beneficial asset suitable for experimenting with custom avionics.

The proposed solution provides a tester with a mechanical part, which is a 3D printed frame with off-the-shelf motors and regulators. The frame is easily transportable and serves its purpose of protecting the electronic equipment mounted inside its central hub. In case of permanent damage, the parts can be quickly reprinted and replaced. This design could be modified in the future, preferably to reduce its size and mass while keeping the present functionality mostly intact.

The electrical part is ESP32 based flight controller with sensors and a software part with a code delivered for customization. The main chip of the flight controller, ESP32, turned out to be fast and efficient enough to complete the given tasks. The microcontroller, along with openly available libraries proved to be fully featured for flight controller applications. Its easily accessible wireless transmission capabilities make it a handy device for experimentation and data gathering, with some of its features remaining to be explored, the most prominent one being its multi-threading capabilities.

All the elements come together with a manual for the implementation of the users' algorithms into the software code. In this way, the platform can be a reasonable solution for testing the control system algorithms at the initial stage of the UAV development, when the physical components are still not ready, and there is the need for the control and navigation algorithm testing against the requirements. Furthermore, the proposed platform can also find applications during the whole UAV lifecycle to face the need for the control algorithms upgrade. The idea of a customizable platform with the software open for implementing new algorithms allows testing various solutions in a short time. Moreover, the platform, thanks to its fast-folding frame, turns out to be highly portable, allowing the tests to be performed in a remote area.

The designed example control loop performed with satisfactory results considering its simplistic sensor input. The project suffered from initial drawbacks related to the underestimation of the influence of motor vibrations affecting the quadcopter's estimated orientation, which resulted in the platform's inability to maintain stable flight and occasional loss of control. Fortunately, the problem was reduced to tolerable levels with the help of an inertial sensor's integrated low pass filter and through providing the sensor's mounting point with additional isolation. The appropriate sensor configuration was a starting point for an example controller tuning process. The example algorithm of the cascade PID controller was developed and tested on the platform in consecutive steps described in the paper. Eventually, the process of tuning and testing resulted in a controller algorithm allowed for achieving the performance similar to that of off-the-shelf platforms of this type and size.

Since the software was designed with openness in mind, further hardware could be integrated without restrictions. The future addition of a magnetometer, barometer, and a GPS receiver could further increase the quadcopter capabilities with the potential introduction of essential flight autonomy.

The tuning method based on the Ziegler–Nichols approach proved to be sufficient in this example case, with minor corrections applied to the controller gain settings during flight tests to suit better the pilot's flight style (Figure 27).

Figure 27. The quadcopter during flight.

With its current sensor package, MPU6050, combined with the Madgwick's AHRS filter, the quadcopter was fully capable of performing the flight using developed algorithms. The accuracy of this system, which directly relates to flight stability, could be increased with better vibration dampening and sensor temperature compensation. The latter should be feasibly done by mapping the changes in offset values together with recorded temperature readings and generating an approximated function through this series of data points. This function could then be integrated into the software to be used for calculating the current offset based on the recorded sensor temperature. By repeating this procedure for every sensor axis of both gyroscopes and accelerometers, it should be possible to mitigate the effects of temperature variations on the sensor readings.

The project has been successfully completed, and the initial goals have been achieved. A result is a functional medium-sized, customizable and easily accessible aerial platform, designed for experimental purposes. The task included the construction of the quadcopter itself, flight controller design and execution, control loop design, software development, and a practical implementation of the proposed example tuning procedure, used for the control loop setup.

The control loop tuning method, which was applied as an example, proved adequate to the point of providing controller gains, which enabled the quadcopter to take flight. The acquired settings were verified and tested during the series of trials, which confirmed their correctness. The platform with integrated diagnostic software turned out an appropriate tool for stabilization and control algorithm tuning.

The project also introduced a successful implementation of a short-range wireless telemetric data transmission module, integrated into the flight controller's main processing chip, which further enhanced the platform's capabilities and provided an additional source of research data during the project's evaluation phase.

Author Contributions: Conceptualization: M.W., Z.R., and E.B.; methodology, software, visualization, resources: M.W.; validation, investigation: K.W.; writing—original draft preparation, review, and editing: M.W., K.W., and E.B.; supervision, formal analysis: Z.R., E.B. All authors have read and agreed to the published version of the manuscript.

Funding: This research received no external funding.

Conflicts of Interest: The authors declare no conflict of interest.

Appendix A

PID regulator implementation—the following method calculates and returns the regulator output value

```
float PID::Compute(float mSetpoint, float mInput)
{
float mError = mSetpoint - mInput; // input error value
unsigned long tn = millis(); // current time in [ms]

float dt = (float)(tn - tp); // step interval
float P = (float)(kp * mError); // P gain
float I = (float)(Ip + ki * mError * dt / 1000.0); // I gain
float D = (float)(-1 * (kd * (mInput - pInput) * 1000.0 / dt)); // D gain (measurement derivative)

if (anti_windup) // anti-windup (clearing the integral value)
{
I = 0;
}
float U = P + I + D; // sum of all components

pInput = mInput; // history
pError = mError;
Ip = I;
tp = tn;

return U; // output value
}
```

References

1. Kuantama, E.; Tarca, R.; Dzitac, S.; Dzitac, I.; Vesselenyi, T.; Tarca, I. The Design and Experimental Development of Air Scanning Using a Sniffer Quadcopter. *Sensors* **2019**, *19*, 3849. [CrossRef] [PubMed]
2. Kanistras, K.; Martins, G.; Rutherford, M.J.; Valavanis, K.P. Survey of Unmanned Aerial Vehicles (UAVs) for Traffic Monitoring. In *Handbook of Unmanned Aerial Vehicles*; Valavanis, K.P., Vachtsevanos, G.J., Eds.; Springer: Berlin/Heidelberg, Germany, 2015; pp. 2643–2666.
3. Olejnik, A.; Rogólski, R.; Kiszkowiak, Ł.; Szcześniak, M. Specific Problems of Selecting and Integrating Equipment Components in the Course of Developing a Technology Demonstrator for the mini-UAV. In Proceedings of the 2019 IEEE International Workshop on Metrology for AeroSpace, Torino, Italy, 19–21 June 2019; pp. 284–289.
4. Olejnik, A.; Kiszkowiak, Ł.; Rogólski, R.; Chmaj, G.; Radomski, M.; Majcher, M.; Omen, Ł. Precise Remote Sensing Using Unmanned Helicopter. In Proceedings of the 2019 IEEE International Workshop on Metrology for AeroSpace, Torino, Italy, 19–21 June 2019; pp. 544–548.
5. Waliszkiewicz, M.; Wojtowicz, K.; Rochala, Z. Experimental method of controller tuning for quadcopters. In Proceedings of the 2019 IEEE 5th International Workshop on Metrology for AeroSpace, Torino, Italy, 19–21 June 2019; pp. 165–170.
6. Brzozowski, B.; Daponte, P.; De Vito, L.; Lamonaca, F.; Picariello, F.; Pompetti, M.; Tudosa, L.; Wojtowicz, K. A remote-controlled platform for UAS testing. *IEEE Aerosp. Electron. Syst. Mag.* **2018**, *33*, 48–56. [CrossRef]
7. Daponte, P.; Lamonaca, F.; Picariello, F.; Riccio, M.; Pompetti, L.; Pompetti, M. A measurement system for testing light remotely piloted aircraft. In Proceedings of the IEEE International Workshop on Metrology for AeroSpace (MetroAeroSpace), Padua, Italy, 21–23 June 2017; pp. 21–23.
8. Daponte, P.; De Vito, L.; Lamonaca, F.; Picariello, F.; Riccio, M.; Rapuano, S.; Pompetti, L.; Pompetti, M. DronesBench: An innovative bench to test drones. *IEEE Instrum. Meas. Mag.* **2017**, *20*, 8–15. [CrossRef]
9. Sharf, I.; Nahon, M.; Harmat, A.; Khan, W.; Michini, M.; Speal, N.; Trentini, M.; Tsadok, T.; Wang, T. Ground effect experiments and model validation with Draganflyer X8 rotorcraft. In Proceedings of the 2014 International Conference on Unmanned Aircraft Systems (ICUAS), Orlando, FL, USA, 27–30 May 2014; pp. 1158–1166.
10. Faundes, D.N.; Wunsch, V.; Hohnstein, S.; Glass, B.; Vetter, M. Research paper on the topic of different UAV drive train qualification and parameter sets. In Proceedings of the 32nd Digital Avionics Systems Conference (DASC), East Syracuse, NY, USA, 5–10 October 2013.
11. Smith, B.; Stark, B.; Zhao, T.; Chen, Y.Q. An outdoor scientific data UAS ground trothing test site. In Proceedings of the International Conference on Unmanned Aircraft Systems (ICUAS), Denver, CO, USA, 9–12 June 2015; pp. 436–443.
12. Valenti, M.; Bethke, B.; Dale, D.; Frank, A.; McGrew, J.; Ahrens, S.; How, J.P.; Vian, J. The MIT indoor multi-vehicle flight testbed. In Proceedings of the Robotics and Automation 2007 IEEE International Conference, Roma, Italy, 10–14 April 2007; pp. 2758–2759.
13. How, J.P.; Bethke, B.; Frank, A.; Dale, D.; Vian, J. Real-time indoor autonomous vehicle test environment. *Control Syst. IEEE* **2008**, *28*, 51–64.
14. Hoffmann, G.; Rajnarayan, D.; Waslander, S.; Dostal, D.; Jang, J.S.; Tomlin, C. The Stanford testbed of autonomous rotorcraft for multi agent control (STARMAC). In Proceedings of the Digital Avionics Systems Conference 2004, Salt Lake City, UT, USA, 28 October 2004.
15. Vladimerou, V.; Stubs, A.; Rubel, J.; Fulford, A.; Strick, J.; Dullerud, G. A hovercraft testbed for decentralized and cooperative control. In Proceedings of the 2004 American Control Conference, Boston, MA, USA, 30 June–2 July 2004.
16. Sun, J.; Li, B.; Wen, C.Y.; Chen, C.K. Design and implementation of a real-time hardware-in-the-loop testing platform for a dual-rotor tail-sitter unmanned aerial vehicle. *Mechatronics* **2018**, *56*, 1–15. [CrossRef]
17. Howard, D.; Merz, T. A platform for the direct hardware evolution of quadcopter controllers. In Proceedings of the IEEE/RSJ International Conference on Intelligent Robots and Systems (IROS), Hamburg, Germany, 28 September–2 October 2015; pp. 4614–4619.
18. Wright, H.J.; Strydom, R.; Srinivasan, M.V. A generalized algorithm for tuning UAS flight controllers. In Proceedings of the 2018 International Conference on Unmanned Aircraft Systems (ICUAS), Dallas, TX, USA, 12–15 June 2018; pp. 1369–1378.

19. Schacht-Rodriguez, R.; Ortiz-Torres, G.; Garcia-Beltran, C.D.; Astorga-Zaragoza, C.M.; Ponsart, J.C.; Perez-Estrada, A.J. Design and development of a UAV Experimental Platform. *IEEE Lat. Am. Trans.* **2018**, *16*, 1320–1327. [CrossRef]
20. Chiu, C.; Lo, C. Vision-Only Automatic Flight Control for Small UAVs. *IEEE Trans. Veh. Technol.* **2011**, *60*, 2425–2437. [CrossRef]
21. Tožička, J.; Komenda, A. Diverse Planning for UAV Control and Remote Sensing. *Sensors* **2016**, *16*, 2199. [CrossRef] [PubMed]
22. Dolega, B.; Kopecki, G.; Kordos, D.; Rogalski, T. Review of Chosen Control Algorithms Used for Small UAV Control. *Solid State Phenom.* **2017**, *260*, 175–183. [CrossRef]
23. Fernández-Madrigal, J.A.; Blanco, J.L. *Simultaneous Localization and Mapping for Mobile Robots: Introduction and Methods*; IGI Global: Hershey, PA, USA, 2012.
24. IMU Data Fusing: Complementary, Kalman, and Mahony Filter. Available online: http://www.olliw.eu/2013/imu-data-fusing/#chapter22 (accessed on 11 December 2019).
25. ESP32 Series Datasheet. Available online: https://www.espressif.com/sites/default/files/documentation/esp32_datasheet_en.pdf (accessed on 9 January 2020).
26. Cao, N.; Lynch, A.F. Inner–Outer Loop Control for Quadrotor UAVs with Input and State Constraints. *IEEE Trans. Control Syst. Technol.* **2016**, *24*, 1797–1804. [CrossRef]
27. Tavares, J.; Mamede, H.S.; Amaral, P.; Pinto, P. Software-defined controllers: Where are we? In Proceedings of the 2017 12th Iberian Conference on Information Systems and Technologies (CISTI), Lisbon, Portugal, 21–24 June 2017; pp. 1–6.
28. Castillo-Zamora, J.; Camarillo-Gómez, K.A.; Pérez-Soto, G.I.; Rodríguez-Reséndiz, J. Comparison of PD, PID and Sliding-Mode Position Controllers for V–Tail Quadcopter Stability. *IEEE Access* **2018**, *6*, 38086–38096. [CrossRef]
29. Le Nhu Ngoc Thanh, H.; Hong, S.K. Quadcopter Robust Adaptive Second Order Sliding Mode Control Based on PID Sliding Surface. *IEEE Access* **2018**, *6*, 66850–66860. [CrossRef]
30. Mou, Y.; Zhang, Q.; Liu, S.; Liang, K. The flight control of micro quad-rotor UAV based on PID. In Proceedings of the 2016 31st Youth Academic Annual Conference of Chinese Association of Automation (YAC), Wuhan, China, 11–13 November 2016; pp. 353–356.
31. Nguyen Duc, M.; Trong, T.N.; Xuan, Y.S. The quadrotor MAV system using PID control. In Proceedings of the 2015 IEEE International Conference on Mechatronics and Automation (ICMA), Beijing, China, 2–5 August 2015; pp. 506–510.
32. Szafranski, G.; Czyba, R. *Different Approaches of PID Control UAV Type Quadrotor*; Silesian University of Technology: Gliwice, Poland, 2011.
33. Ruoyu, Z. A Drift Eliminated Attitude & Position Estimation Algorithm in 3D. Master's Theses, University of Vermont, Burlington, VT, USA, January 2016.
34. Kotarski, D.; Benić, Z.; Krznar, M. *Control Design for Unmanned Aerial Vehicles with Four Rotors*; Croatian Interdisciplinary Society: Zagreb, Croatia, 2016.
35. Madgwick, S.O.H. *An Efficient Orientation Filter for Inertial and Inertial/Magnetic Sensor Arrays*; Technical Report; University of Bristol: Bristol, UK, 2010.
36. Ziegler-Nichols Tuning Rules for PID. Available online: http://www.mstarlabs.com/control/znrule.html (accessed on 15 October 2019).
37. MPU-6000 and MPU-6050 Register Map and Descriptions Revision 4.2. Available online: https://invensense.tdk.com/wp-content/uploads/2015/02/MPU-6000-Register-Map1.pdf (accessed on 15 October 2019).

© 2020 by the authors. Licensee MDPI, Basel, Switzerland. This article is an open access article distributed under the terms and conditions of the Creative Commons Attribution (CC BY) license (http://creativecommons.org/licenses/by/4.0/).

Article

Measurement System of a Magnetic Suspension System for a Jet Engine Rotor †

Paulina Kurnyta-Mazurek [1,*], Artur Kurnyta [2] and Maciej Henzel [1]

1 Faculty of Mechatronics and Aerospace, Military University of Technology, 00-908 Warsaw, Poland; maciej.henzel@wat.edu.pl

2 Airworthiness Division, Air Force Institute of Technology, 01-494 Warsaw, Poland; artur.kurnyta@itwl.pl

* Correspondence: paulina.mazurek@wat.edu.pl; Tel.: +48-261-839-143

† This paper is an extension version of the conference paper: Kurnyta-Mazurek, P.; Kurnyta, A.; Henzel, M. Concept of wireless measurement system of UAV jet engine rotor. In Proceedings of the 2019 IEEE 5th International Workshop on Metrology for AeroSpace (MetroAeroSpace), Torino, Italy, 19–21 June 2019.

Received: 24 November 2019; Accepted: 27 January 2020; Published: 6 February 2020

Abstract: This paper presents laboratory results on the measurement system of a magnetic suspension bearing system for a jet engine rotor of an unmanned aerial vehicle (UAV). Magnetic suspension technology enables continuous diagnostics of a rotary machine and eliminates of the negative properties of classical bearings. This rotor-bearing system consists of two radial magnetic bearings and one axial (thrust) magnetic bearing. The concept of the bearing system with a magnetically suspended rotor for UAV is presented in this paper. Rotor geometric and inertial characteristics were assumed according to the parameters of a TS-21 jet engine. Preliminary studies of the measurement system of rotor engines were made on a laboratory stand with homopolar active magnetic bearings. The measurement system consisted of strain gauges, accelerometers, and contactless proximity sensors. During the research, strains were registered with the use of a wireless data acquisition (DAQ) system. Measurements were performed for different operational parameters of rotational rotor speed, control system parameters, and with the presence of disturbance signals from the control system. In this paper, obtained operational characteristics are presented and discussed.

Keywords: active magnetic suspension; jet engine; wireless measurement system; UAV

---

1. Introduction

In recent years, magnetic suspension systems have been used in practice to solve certain problems of classical bearing systems that occur during the operation and maintenance of machinery and other technical objects. This technology was successfully implemented for the rail industry in Japan and China. In the track and vehicle, permanent magnets or electromagnets were implemented, generating magnetic levitation forces and providing non-contact movement of the vehicle on the track [1].

Magnetic bearing technology has been adopted in other branches of industry as well. It was developed and commercialized due to its advantages, such as lack of lubrication and no mechanical contact between operational elements [2,3]. Immense efforts have been undertaken to expand the magnetic technology applicability in high-speed machinery due to its invaluable advantages in terms of friction loss reduction, for example, in turbomachinery, compressors, and generators, [4–9]. In comparison with classical mechanical bearings, magnetic ones possess many advantages, [6,10] such as low amplitudes of mechanical vibration, high durability, lack of tribological wears, and ability of the extended operational term at high speed [11–13]. These features of magnetic technology give the immense potential to become a key element in smart and intelligent machines, such as unmanned aerial vehicle (UAV) jet engines [13].

Using active and passive magnetic bearing technologies in high-speed machinery can result in overcoming the physical limitations of classical bearings. This technology decreases stiffness coefficients and increases the damping coefficients of radial bearings, which reduce the value of critical rotor speed. Active magnetic bearings allow precise control of the rotor position and movement and enable "on-line" diagnosing, monitoring, and identification of high-speed machines. Thus, passive magnetic bearings can increase the damping coefficient of the bearing system for high rotor speed. In contrast to active bearings, passive ones do not consume electric energy. This technology uses magnets or superconductors to suspend the rotor of high-speed machinery. The concept of the structurally simple and operationally robust suspension of high-speed rotors in electrodynamic passive magnetic bearings (EDPMB) was presented in [14].

Developing a useful diagnostic concept of the rotor-bearing system is very important in the maintenance process of high-speed rotary machinery, including UAV jet engines. The accurate selection of the high-speed rotor-bearing system is continued with the design process. It is usually based on analysis of the load capacity, durability, bearing operating conditions, and experimental fatigue models of cooperating pairs (used to determine the bearing durability). However, these systems (without former change symptoms) are often damaged [15]. Therefore, an important issue is the development of modern and advanced methods of high diagnosis susceptibility for bearing system analyses. Rotating machinery monitoring methods are well established as well as detection algorithms for failure detection. Generally, rotary machinery monitoring systems are usually based on vibration signal analysis. These signals are essential and vital fault sources and contain enough information about operating states of rotating machinery [16]. The application of magnetic suspension technology in the bearing system of high-speed machine rotors enables one to measure vibrations with the use of eddy current sensors, which are a necessary control system element. Therefore, the possibility of using magnetic bearings creates an additional monitoring and diagnosis ability in comparison to classical rotary machines.

The structure of the diagnosis system correlates closely with the analyzed system, and only its elements have a universal character. Therefore, it is necessary to carry out measurements to establish a database (background) of typical operational parameters of the undamaged machine. Deviation from the initial parameters in the monitoring process constitutes the primary database for evaluating the technical conditions of rotating machinery. Dedicated condition indicators (CIs) are usually developed based on vibration signals, rotational speed values, local temperature gradients, and strain measurements. Changes in those CIs indicate damage progression. In magnetic bearings, the control signals (e.g., control current, air gap value) can also be used to monitor their operational conditions only after proper post-processing.

Preliminary studies of the monitoring system of the UAV jet engine rotor are presented in this paper. In the first section, the production of the magnetically suspended rotor of a small jet engine of a UAV is introduced. It consists of active and passive magnetic bearings as well. The second section presents a measurement system dedicated to the rotor-bearing system. Subsequently, a laboratory stand is presented. Finally, some conclusions and remarks are provided.

2. Design of the Magnetically Suspended Rotor of a UAV

Magnetic suspension technology allows the effective reduction of vibration amplitude and the elimination of friction force and negative operational properties of the classical bearings. It is estimated that the efficiency of magnetic suspension systems is 10% higher than for classical bearing systems, for example, by eliminating critical speeds [17]. Moreover, this solution increases the durability and the reliability of the bearing system and increases the operating range of the shaft's rotational speeds.

Magnetic bearings are classified into two categories, namely passive and active bearings. Passive magnetic suspensions use permanent magnets or superconductors. They do not require a closed-loop control system. On the other hand, active magnetic suspensions use electromagnets to generate magnetic forces and require a feedback control signal (the open-loop control system is unstable) [18].

The fundamental element of an active magnetic bearing is an electromechanical actuator that generates a controlled force between cooperation pair–rotor and stator. There are four types of electromechanical actuators: heteropolar and homopolar electromechanical actuators, homopolar electromechanical actuators with permanent magnets, and homopolar electromechanical actuators with electromagnets [19].

Figure 1 shows the design of the magnetic suspension system of a UAV rotor with active and passive magnetic bearings. Passive solutions effectively damp vibration amplitudes for high rotor speeds, without consuming electric energy. Radial active magnetic bearings generate radial forces in two perpendicular radial axes. The first active magnetic bearing #1 is controlled in the x_1 and y_1 axes' coordinates. The second active radial magnetic bearing #2 is controlled in the x_2 and y_2 axes' coordinates. The axial position in the z-axis (i.e., along the shaft axis) is controlled by axial forces generated by a thrust magnetic bearing. In total, there are five axes, x_1, y_1, x_2, y_2, and z, that are controlled by active magnetic bearing systems [10,18,20]. Radial active magnetic bearings suspend the rotor for low rotor speed, whereas passive ones suspend the rotor in the high-speed range. In this way, less energy is consumed than in the case of using only an active solution.

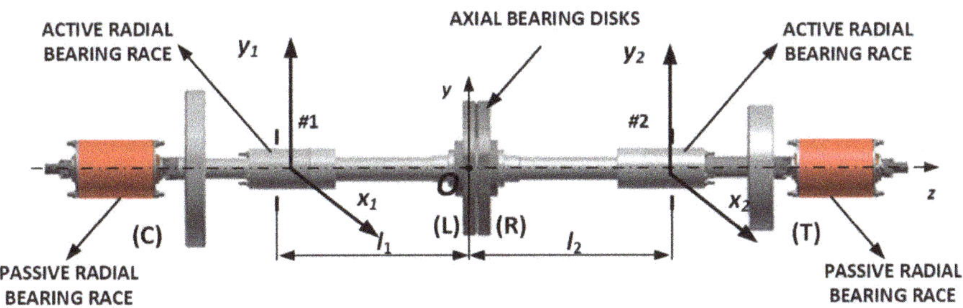

Figure 1. Magnetic suspension system of an unmanned aerial vehicle (UAV) rotor.

The designed magnetic suspension system for the UAV engine was developed for the rotor parameters corresponding to the TS-21 engine. This construction works as a starter on Mig-23 and Mig-27 aircraft, as well as Su-7, Su-20, and Su-22 aircraft. The TS-21 engine is driven by an electric starter with about 3 kW of power. The engine characterizes 1 kN thrust force, 60 ÷ 80 kW power, and 50,500 rpm. The rotor shaft has a 1.22 kg mass and is 15 mm in diameter.

In Figure 1, the elements marked as (C) and (T) are mass equivalents of the TS-21 compressor and turbine. The symbol O denotes the center of mass, which is located 178 mm (l_1) from the "left" active radial bearing #1 and 191 mm (l_2) from the "right" active radial bearing #2.

Figure 2 presents the laboratory stand with the magnetic suspension system. The radial active magnetic bearings are set in the universal grips, and the axial magnetic bearing is dismounted. Air gaps in the radial and axial magnetic bearings are equal to 0.4 mm and 0.5 mm, respectively. Thus, the operating point current for each bearing is 5, and the maximum current for each bearing is 10 A, whereas axial one has 80 coils. Viscoelastic parameters of active magnetic bearings depend on control law parameters as well as current stiffness coefficient and displacement stiffness coefficient, being 3.74×10^5 N/m and 1168 Ns/m, respectively. The axial bearing was designed to carry axial loads of 1.5 kN, because of the engine thrust force, whereas radial bearing was designed to maintain a 100 N load. Mass equivalents of compressor and turbine are marked as (C) and (T), as shown in Figure 1. Eddy current sensors are located in the electromagnet covers. Signals from these sensors provide information about rotor position for the closed-loop control system and could be useful for the diagnosis system as well.

Due to the high level of complexity of the laboratory stand shown in Figure 2, preliminary qualitative studies were carried out on the system with only one active magnetic bearing and one ball bearing, presented in Figure 3. It consisted of an electromechanical actuator, rotor, control and data acquisition system, and an electric drive. The rotor was supported on one side by a classical ball bearing and on the other, by the magnetic bearing. The drive torque was transmitted to the rotor suspended by the magnetic bearing. For that bearing system, a measurement system was designed as a proof-of-concept. It consisted of eddy current sensors, which were a part of the control system and strain sensors.

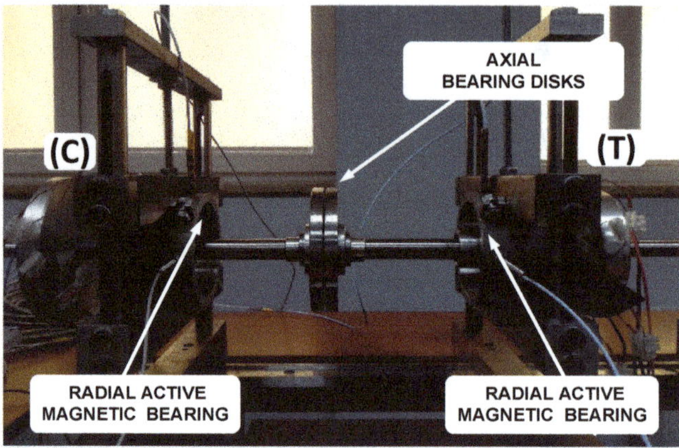

Figure 2. Laboratory stand with the magnetic suspension system of a UAV rotor.

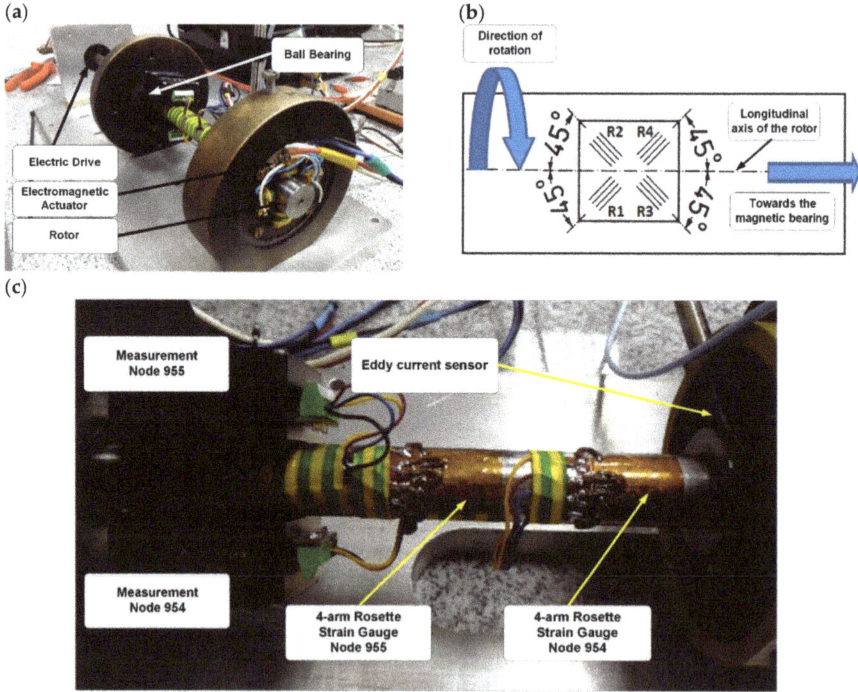

Figure 3. Laboratory stand of the active magnetic bearing with sensors: (**a**) laboratory model, (**b**) strain gauge configuration, (**c**) strain gauges with measurement nodes.

In Figure 3b, the four-arm rosette strain gauge is illustrated with the orientation of its measuring grids (R1 ÷ R4) in comparison to the longitudinal axis of the rotor. The location of individual measurement system elements is shown in Figure 3c. In the next part of the paper, a laboratory stand with measurement system is described. Rotor deformation and rotor displacement are measured by strain gauges and eddy current sensors, respectively.

3. Measurement System

Bearing monitoring problems in turbomachinery, in the jet engines, currently represent significant issues. Monitoring systems should enable to detect defects in the early stages, before the breakdown of the whole system or the occurrence of further damage. Unfortunately, due to bearing operational conditions, monitoring of these parameters is not a simple task. Bearings are often located in places with difficult access. Another critical issue is the construction of moving engine parts, which make the use of even a simple cable connection impossible. It becomes necessary to use a wireless connection between sensors and recording modules. However, the use of wireless sensors can be made difficult due to the electromagnetic noise in a broad frequency spectrum that may occur during engine operation [13].

The laboratory stand provided a wireless data acquisition system to measure rotating shaft strain. Sensors for monitoring the strain under forced rotation were bonded on the shaft in two cross-sections, as shown in Figure 3. Measurement nodes with a transmitter for wireless telemetry were mounted near the ball bearing. Strain measurements were performed in the system build-up of the Wheatstone bridge, amplifier, and signal transmission circuit.

The measurements of the shaft strains were performed for two cross-section areas (Nodes 954 and 955). Strain gauges operated first as a full-bridge and then as a quarter-bridge configuration, connected to the measurement system, as shown in Figure 4. The strain signal was measured and conditioned by nodes and wirelessly transferred to the gateway access point connected to the PC computer with dedicated software. A small, low-power analog sensor node SG-Link® and the gateway WSDA®-200-USB from LORD MicroStrain® Sensing Systems (Williston, VT, USA) were used for wireless measurements. Measurements were performed in two independent channels, namely Nodes 954 and 955. The grid rosettes for measuring the compress and stretch forces of the shaft in the longitudinal direction were in the range of ±45 degrees. Then, the strain gauge R1 was used to measure values of the local strains [21]. In this bridge configuration, the other Wheatstone's bridge arms were supplemented by precision resistors with a resistance of 350 Ω ± 0.1%. The data acquisition system worked with 32 μs accuracy synchronization in a wide wireless range. Conducted measurements allowed to characterize the magnetically suspended rotor by comparing the local strain level in the middle and front parts of the shaft at various rotational speeds. Additionally, the natural feature of that type of bearing system is lack of torsion forces on the rotor. In that case, vibration components can be obtained directly on the rotor by using the above full-bridge strain gauge configuration.

Figure 4. Diagram of the measurement system [13]: (**a**) strain gauge in a full-bridge configuration as a sensor, (**b**) wireless nodes, (**c**) gateway, (**d**) software.

The laboratory stand of the magnetic suspension system was also equipped with contactless position sensors to measure the rotor displacement in the air gap. These sensors were a part of the control system, as shown in Figure 5. The laboratory stand consisted of a control unit, proportional–integral–derivative PID controller, amplifier, electromechanical actuator with the supported rotor, and contactless eddy current sensors.

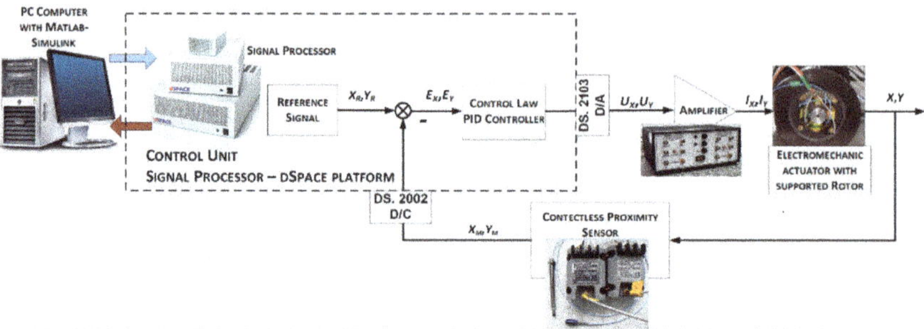

Figure 5. Active magnetic bearing control system.

The control unit was based on a PC and the dSpace platform (dSpace GmbH, PaderbornyGermany) with input/output modules with 16-b A/D and D/A converters. The PID controller was designed in MATLAB Simulink 2010a software (Mathworks, Natick, MA, USA) and then implemented in the dSPACE platform. The control signals from the controller were transferred to two-channel bipolar amplifiers. The amplifiers supplied magnetic bearing electromagnets. Control current signals caused magnetic flux variations, which changed the rotor position. These displacements were measured by contactless eddy current sensors from the Bently Nevada company (Minden, NV, USA). Measured signals were transferred to the control unit via the input module. In this way, the feedback loop was obtained. Contactless proximity sensors can measure the rotor position in the range of 2 mm with an accuracy of 1 μm. Sample characteristics of the control system are presented in Figure 6.

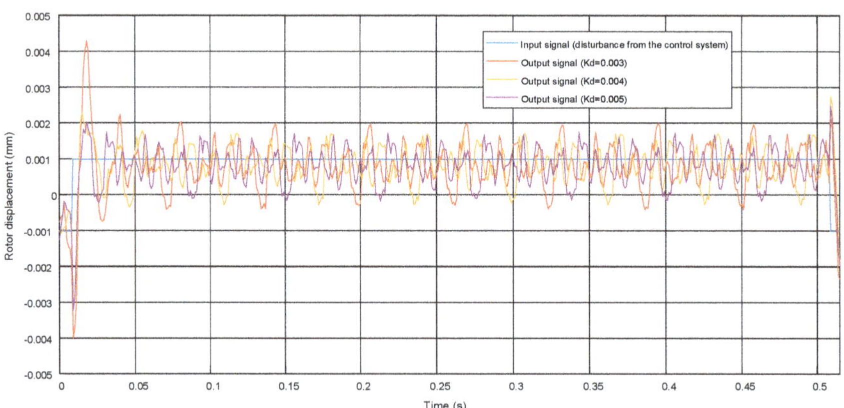

Figure 6. Time characteristics of the control systems with proportional-derivative PD controller for different K_d values and constant K_p equal to 1.75.

The time characteristics of the control systems with the proportional-derivative PD controller are presented in Figure 6. The proportional parameter of the PID controller was marked as K_p and was constant during the presented measurement, whereas the derivative parameter of the PID controller was marked as K_d. The reference signal had the form of a rectangular signal acting on the Ox axis with amplitude and frequency equal to 0.01 V and 1 Hz, respectively. Later in the paper, this reference signal is called disturbance from the control system acting on the Ox or Oy axes. Other signals depicted in the figure present outputs from the control systems with the PD controller in the form of rotor displacements. In this figure, the vibration amplitude of the rotor can be observed. As the derivative parameter of controller K_d increases, the vibration amplitude decreases. The system behaves like an oscillator with one degree of freedom and its transfer function reads as follows:

$$ms^2 + k_i K_d s + K_p k_i - k_x = 0, \tag{1}$$

where m denotes rotor weight, current stiffness coefficient is described by k_i, and displacement stiffness coefficient is expressed by k_x. This equation has the following solutions $p_{1,2}$:

$$p_{1,2} = -\omega_0\zeta \pm i\omega_0\sqrt{1-\zeta^2}, \tag{2}$$

where ω_0 denotes the eigen frequency of the system, $\omega_0 = 3.2/(t_r\zeta)$, t_r is the settling time, and ζ is the dimensionless damping factor.

The location of the closed-loop system poles depends on settling time t_r and the dimensionless damping factor ζ. When the poles p_1 and p_2 are known, PD controller parameter values can be calculated from the following relations:

$$K_d = \frac{(-p_1 - p_2)m}{k_i}, \tag{3}$$

$$K_p = \frac{p_1 p_2 m + k_x}{k_i}. \tag{4}$$

The above characteristic Equation (1) can be expressed as a differential equation of oscillatory motion:

$$m\frac{d^2y}{dt^2} + c\frac{dy}{dt} + ky = 0. \tag{5}$$

Comparing values of the equation parameters, the stiffness coefficient is described by the following:

$$k = K_p k_i - k_x. \tag{6}$$

Finally, the damping coefficient reads as follows:

$$c = k_i K_d \tag{7}$$

From the above considerations about using a PD controller, it is possible to have an effect on stiffness and damping coefficients of the entire active magnetic bearing system.

In conclusion, a measurement system for active magnetic bearings consists of strain gauges and eddy current sensors to measure strains and rotor displacements, respectively. The flowchart of the overall measurement process is presented in Figure 7. At the laboratory stage of the study, comparison results were obtained from strain gauges placed on the rotor surface. The analysis of research results allows the development of the monitoring and diagnosis system for a bearing system for a UAV engine rotor, equipped with passive and active magnetic bearings.

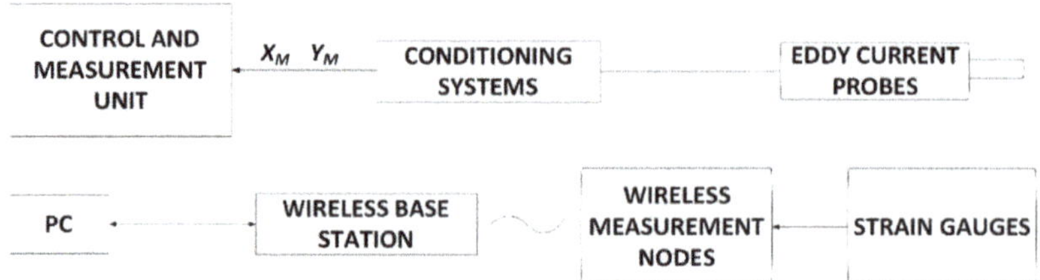

Figure 7. Flowchart with the overall measurement process.

4. Results

The rotary system with an active magnetic bearing, as shown in Figure 3, was investigated to obtain its dynamic behavior with the developed measuring system. This system enabled the continuous measurement of the magnetic bearing operational parameters, such as rotating shaft strains, rotor displacements, and vibrations. During studies, strain measurements with full bridge and quarter bridge were registered by a wireless data acquisition (DAQ) system. The sensors were located in two cross-sections of the magnetically suspended shaft, that is, one in the middle (Node 955) and the second near the magnetic bearing (Node 954). Two different bridge configurations for strain gauge were utilized to measure the local uniaxial strain with a quarter bridge and quasi-vibrations from the rotational motion with a full bridge. In a classical bearing system, a full-bridge configuration allows the measurement of torsion strain, but in the magnetic bearing system, this component is absent.

Strain measurements were conducted for both cross-sections simultaneously for constant parameters of the control system and disturbance signals acting on the Ox and Oy axes. These disturbances were generated by the control system. Strain time-domain signals were acquired using a 512 Hz sampling frequency, for approximately 20 seconds of stable rotational speed. For each rotational speed, three measuring series were made with no disturbance, disturbances added in the Ox axis, and disturbances added in the Oy axis. Fast Fourier transform (FFT) analysis was performed for time-domain signals and the frequency characteristics are presented below.

In Figures 8 and 9, fast Fourier transform (FFT) characteristics of the rotating shaft strain measured for a constant motor rotary speed equal to 25 rev/s for a quarter-bridge configuration at both shaft cross-sections, Nodes 954 and 955, are presented, respectively. These characteristics were registered without and with disturbance from the control system. Measurement with no disturbance is marked by a blue line, and measurements with disturbance added to the control signal in the Ox and Oy axes are indicated by red and orange lines, respectively. In Table 1, the list of significant values from the characteristics shown in Figures 8 and 9 is presented. Cells with peaks from the sampling frequency are marked in blue.

In Figures 10 and 11, fast Fourier transform (FFT) characteristics of the rotating shaft strain measured at the shaft front cross-section (Node 954) for a constant motor rotary speed equal to 30 rev/s for both full- and quarter-bridge configurations are presented, respectively. These characteristics were registered without and with disturbance from the control system. Significant values from obtained characteristics are presented in Table 2.

In Figures 12 and 13 and Table 3, the same set of characteristics as above are presented, but from the middle shaft cross-section.

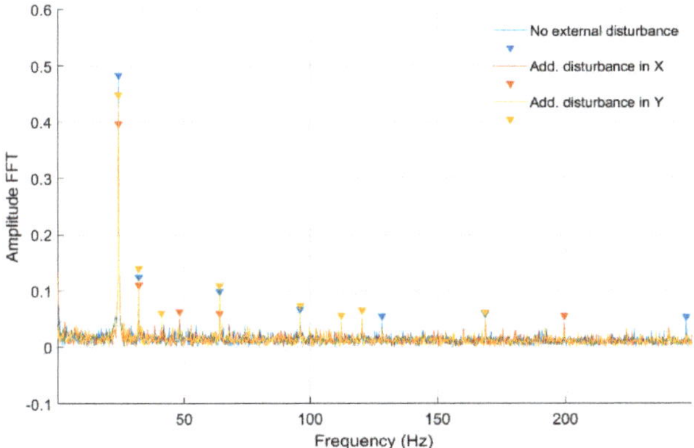

Figure 8. Fast Fourier transform (FFT) of strain-level measurement for constant motor speed equal to 25 rev/s with disturbance from the control system for a quarter-bridge configuration (Node 954).

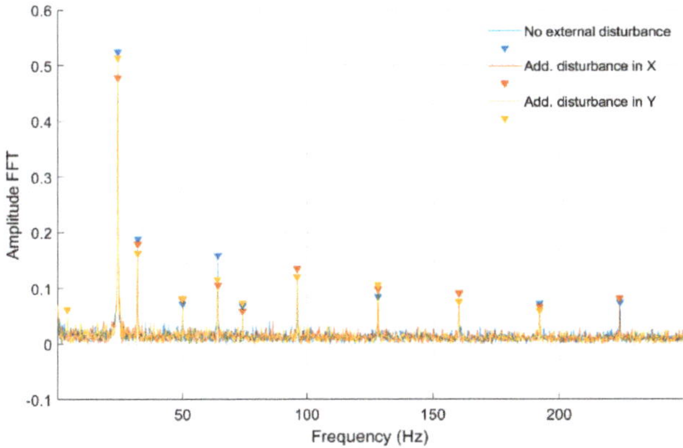

Figure 9. FFT of strain-level measurement for constant motor speed equal to 25 rev/s with disturbance from the control system for a quarter-bridge configuration (Node 955).

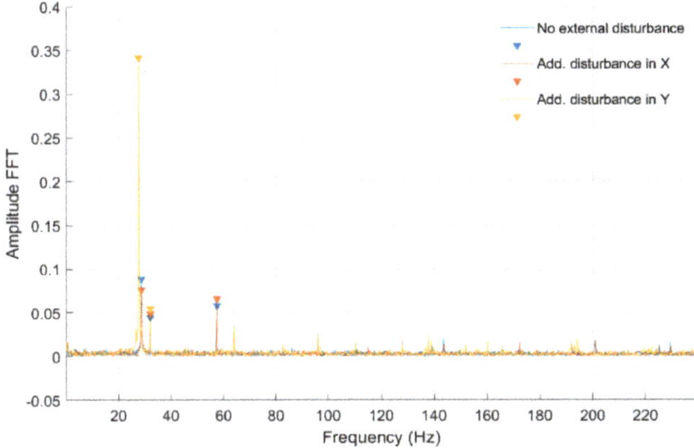

Figure 10. FFT of strain-level measurement for constant motor speed equal to 30 rev/s with disturbance from the control system for a full-bridge configuration (Node 954).

Table 1. List of parameters read from the characteristics shown in Figures 8 and 9.

No external force for Node 954						
FFT Amplitude	0.474	0.116	0.090		0.058	
Frequency (Hz)	24.14	32.05	63.90		95.95	
Additional force in Ox axis for Node 954						
FFT Amplitude	0.385	0.098	0.051		0.048	
Frequency (Hz)	24.14	32.05	48.08		63.90	
Additional force in Oy axis for Node 954						
FFT Amplitude	0.437	0.128	0.048		0.097	
Frequency (Hz)	24.14	32.05	41.00		63.90	
No external force for Node 955						
FFT Amplitude	0.514	0.178	0.062	0.149	0.059	0.110
Frequency (Hz)	24.14	32.05	49.95	64.10	74.09	95.95
Additional force in Ox axis for Node 955						
FFT Amplitude	0.466	0.167	0.068	0.93	0.047	0.123
Frequency (Hz)	24.14	32.05	49.95	64.10	74.09	95.95
Additional force in Oy axis for Node 955						
FFT Amplitude	0.050	0.502	0.151	0.070	0.104	0.062
Frequency (Hz)	4.16	24.14	32.05	49.95	63.90	74.09

Table 2. List of parameters read from the characteristics shown in Figures 10 and 11.

No external force for Node 954, full bridge				No external force for Node 954, quarter bridge								
FFT Amplitude	0.087	0.043	0.056	FFT Amplitude	0.082	0.056	0.276	0.099	0.149	0.060	0.056	0.049
Frequency (Hz)	28.60	32.00	57.40	Frequency (Hz)	0.62	1.87	28.93	32.05	57.65	64.10	86.58	95.95
Additional force in Ox axis for Node 954, full bridge				Additional force in Ox axis for Node 954, quarter bridge								
FFT Amplitude	0.074	0.047	0.064	FFT Amplitude	0.303	0.122	0.157	0.060	0.067	0.066		
Frequency (Hz)	28.60	32.00	57.40	Frequency (Hz)	28.72	32.05	57.65	64.10	95.95	115.30		
Additional force in Oy axis for Node 954, full bridge				Additional force in Oy axis for Node 954, quarter bridge								
FFT Amplitude	0.334	0.047		FFT Amplitude	0.047	0.756	0.127	0.043	0.082	0.062	0.128	
Frequency (Hz)	27.60	32.00		Frequency (Hz)	22.69	27.47	32.05	49.95	64.10	101.35	109.48	

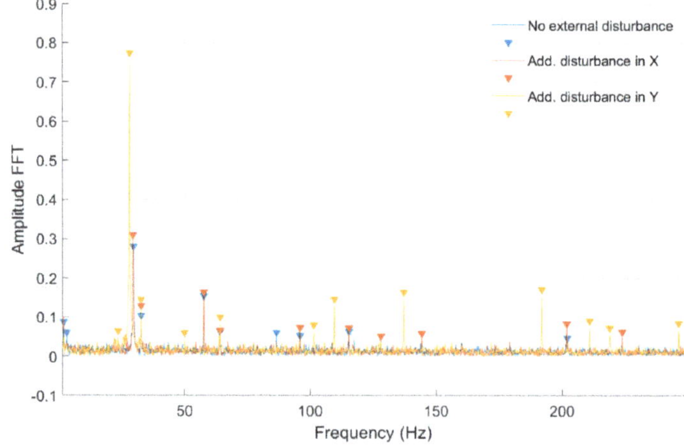

Figure 11. FFT of strain-level measurement for constant motor speed equal to 30 rev/s with disturbance from the control system for a quarter-bridge configuration (Node 954).

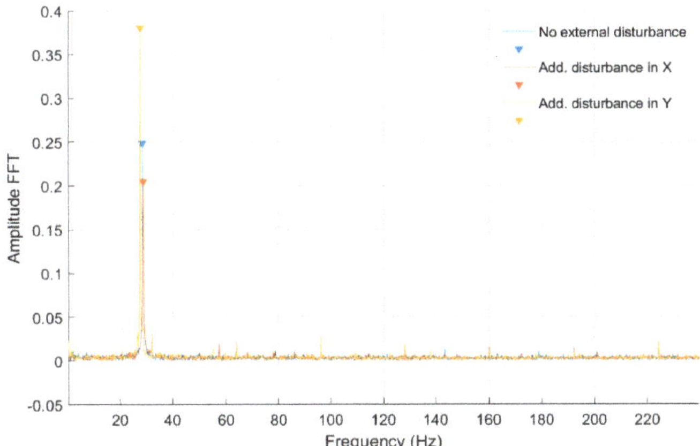

Figure 12. FFT of strain level measurement for constant motor speed equal to 30 rev/s with disturbance from the control system for a full-bridge configuration (Node 955).

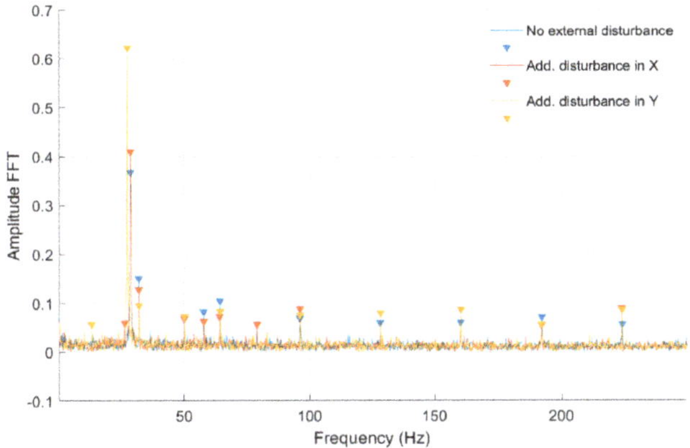

Figure 13. FFT of strain-level measurement for constant motor speed equal to 30 rev/s with disturbance from the control system for a quarter-bridge configuration (Node 955).

Table 3. List of parameters read from the characteristics shown in Figures 12 and 13.

No external force for Node 955, full-bridge		No external force for Node 955, quarter bridge								
FFT Amplitude	0.244	FFT Amplitude	0.361	0.143	0.076	0.098	0.050	0.063		
Frequency (Hz)	28.60	Frequency (Hz)	28.72	32.05	57.65	64.10	78.88	95.95		
Additional force in *Ox* axis for Node 955, full-bridge		Additional force in *Ox* axis for Node 955, quarter bridge								
FFT Amplitude	0.199	FFT Amplitude	0.052	0.402	0.120	0.062	0.055	0.065	0.049	0.080
Frequency (Hz)	28.80	Frequency (Hz)	26.43	28.72	32.05	49.95	57.65	63.90	78.88	95.95
Additional force in *Oy* axis for Node 955, full-bridge		Additional force in *Oy* axis for Node 955, quarter bridge								
FFT Amplitude	0.373	FFT Amplitude	0.043	0.608	0.081	0.059	0.069	0.061		
Frequency (Hz)	27.60	Frequency (Hz)	13.11	27.47	32.05	49.95	64.10	95.95		

In Figures 14–17 and Table 4, fast Fourier transform (FFT) characteristics of the rotating shaft strain measured for a constant motor rotary speed equal to 7 rev/s for quarter- and full-bridge configurations at both front and middle shaft monitoring points are presented.

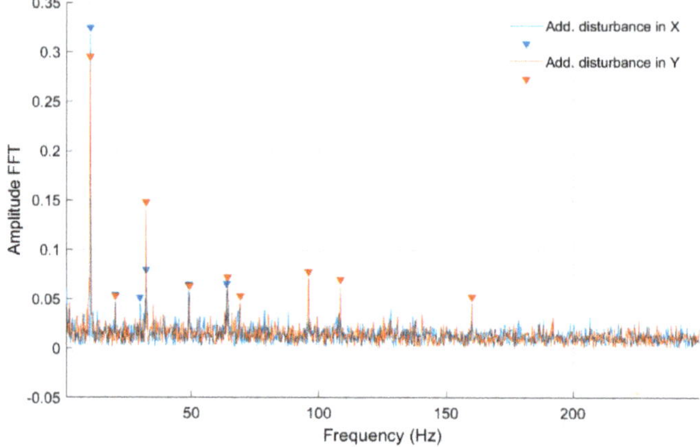

Figure 14. FFT of strain-level measurement for constant motor speed equal to 7 rev/s with disturbance from the control system for a quarter-bridge configuration (Node 954).

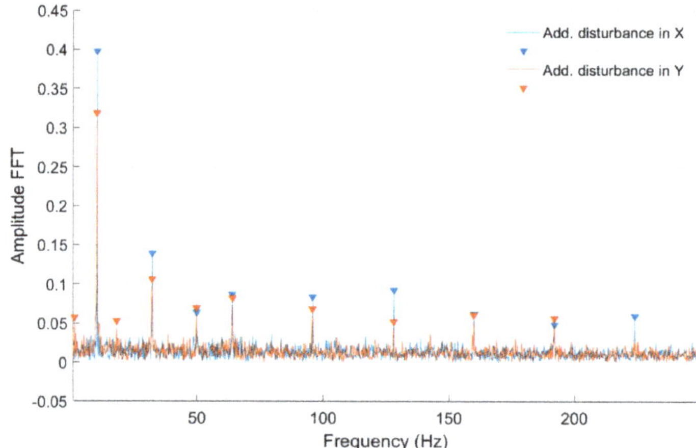

Figure 15. FFT of strain-level measurement for constant motor speed equal to 7 rev/s with disturbance from the control system for a quarter-bridge configuration (Node 955).

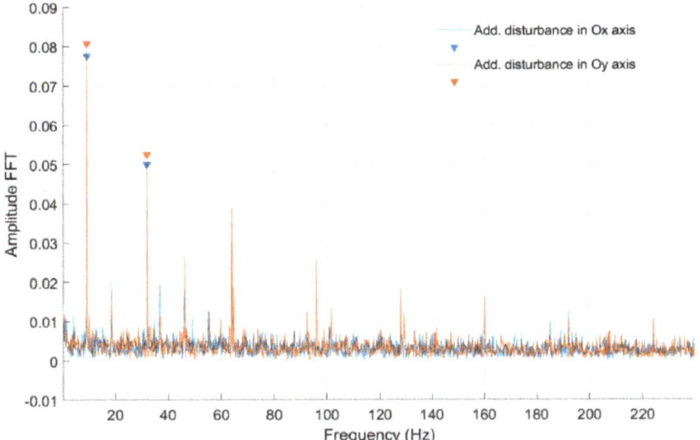

Figure 16. FFT of strain-level measurement for constant motor speed equal to 7 rev/s with disturbance from the control system for a full-bridge configuration (Node 954).

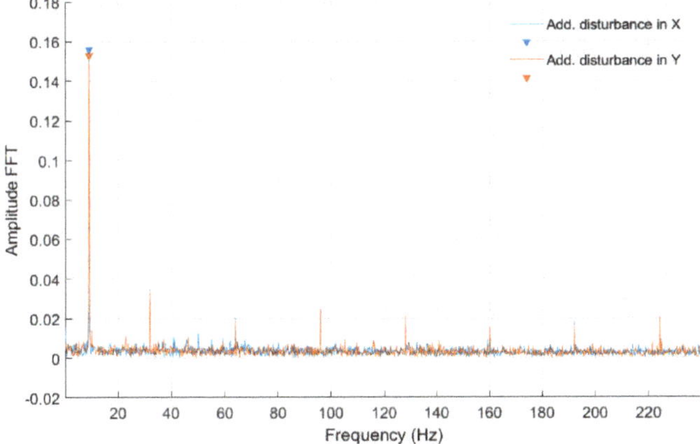

Figure 17. FFT of strain-level measurement for constant motor speed equal to 7 rev/s with disturbance from the control system for a full-bridge configuration (Node 955).

Table 4. List of parameters read from the characteristics shown in Figures 14–17.

Additional force in *Ox* axis for Node 954, full bridge				Additional force in *Ox* axis for Node 954, quarter bridge							
FFT Amplitude	0.076	0.047	0.020	FFT Amplitude	0.318	0.048	0.045	0.073	0.059	0.059	
Frequency (Hz)	9.20	32.00	18.40	Frequency (Hz)	9.78	19.56	29.56	32.05	48.91	63.69	
Additional force in *Oy* axis for Node 954, full bridge				Additional force in *Oy* axis for Node 954, quarter bridge							
FFT Amplitude	0.079	0.047	0.016	FFT Amplitude	0.288	0.046	0.141	0.057	0.065	0.046	0.071
Frequency (Hz)	9.20	32.00	18.40	Frequency (Hz)	9.78	19.77	32.05	49.33	64.10	69.10	95.95
Additional force in *Ox* axis for Node 955, full bridge				Additional force in *Ox* axis for Node 955, quarter bridge							
FFT Amplitude	0.153			FFT Amplitude	0.391	0.133	0.057	0.080	0.077		
Frequency (Hz)	9.20			Frequency (Hz)	9.78	32.05	49.95	63.90	95.95		
Additional force in *Oy* axis for Node 955, full bridge				Additional force in *Oy* axis for Node 955, quarter bridge							
FFT Amplitude	0.149			FFT Amplitude	0.049	0.310	0.044	0.098	0.061	0.074	0.060
Frequency (Hz)	9.20			Frequency (Hz)	0.83	9.78	17.69	32.05	49.95	64.10	95.95

5. Discussion

Presented test results were carried out for various rotational speeds of the magnetically suspended rotor and constant values of controller parameters. Proportional and differential gains of the controller were set to 1.75 and 0.003, respectively, and the amplitude of disturbance was set to 0.01 V with a frequency of 1 Hz. Differences in results for the quarter- and full-bridge strain measurements are attributed to various types of strain state to which the bridge configuration is sensitive, as one or four grids are actively used to obtain data.

In Table 1 were presented the FFT amplitudes and corresponding frequencies of strain-level measurements for an engine speed equal to 25 rev/s without and with external disturbance measured for Nodes 954 and 955 for a quarter-bridge configuration. Cells with the frequency of 24.14 Hz correspond to engine speed, whereas cells marked by blue peaks are the artefacts from the sampling frequency. For the measurement with no disturbance for the front shaft cross-section, the frequency response is clear with no additional peaks. For measurements with external disturbance in the Ox and Oy axes for Node 954, peaks at the frequencies of 48 Hz and 41 Hz occurred, respectively. In the case of Node 955 measurements without and with external disturbance, peaks at 49.95 and 74.09 Hz occurred. It can be clearly seen that an added disturbance signal gives an additional frequency response for both cross-sections. Both peak amplitudes are higher at the same frequencies for Node 955 (middle cross-section of the shaft) in comparison with Node 954 (front cross-section) in both axes. However, the ratio of FFT amplitudes for measurements without and with disturbance is lower for the front cross-section and is 0.81 for Ox and 0.92 for Oy axes, respectively. For the middle cross-section, 0.9 for the Ox axis and 0.97 ratios were obtained. That suggests the higher impact of an added disturbance in both axes for the cross-section closer to the magnetic bearing.

In Tables 2 and 3 were presented the FFT amplitudes and corresponding frequencies of strain-level measurements for an engine speed equal to 30 rev/s without and with external disturbance for Node 954 for quarter- and full-bridge configurations. Cells with the frequency of approximately 28 Hz correspond to engine speed. For the measurement with full-bridge configuration, the frequency characteristics are much clearer with a smaller number of peaks than in the case of the quarter-bridge configuration. The FFT amplitude ratio for measurements without and with disturbance in the Ox axis is similar for the first harmonic in both shaft cross-sections and is equal to 1.10. However, in the second harmonic, the ratio is 0.72 and 1.05 for the middle and front measuring points. Regarding the added disturbance in the Oy axis, the amplitude ratio differs significantly with values of 1.68 for the middle and 2.74 for the front shaft cross-sections. The analysis for full-bridge measurements reveals a similar effect with added disturbance in the Ox and the Oy axes. Here, the influence of added disturbance in the Oy axis caused the further increase of the FFT amplitude ratio to values of 1.53 for the middle and 3.84 for the front measuring points. It can also be seen that the first harmonic frequency with added disturbance in the Oy axis is slightly decreased in comparison to those with no disturbances and added in the Ox axis. The same phenomena can be observed for quarter-bridge measurements, although those were performed separately. That decrease in the frequency characteristic was absent for lower rotational speeds. The specific source of that effect requires further investigation. It is probable that additional deflection modes occurred for the shaft at that speed, when a disturbance in the Oy axis was present. An higher FFT amplitude with further frequency harmonics was measured for the quarter-bridge configuration compared with the full-bridge configuration for that rotational speed. Again, the shaft cross-section closer to the magnetic bearing was influenced more than its middle part in the presence of disturbances.

In Table 4 were presented the FFT amplitudes and corresponding frequencies of strain-level measurements for a low engine speed equal to 7 rev/s with external disturbance for Nodes 955 and 954 for quarter- and full-bridge configurations. For that rotational speed, no noticeable effects were detected for disturbances in the Ox and Oy axes in both bridge configurations. Several frequency peaks were seen for the strain gauge connected to Node 954 in the front of the shaft, which suggests a higher level of strain and vibrations on frequencies above and even below the rotational speed.

Additionally, higher harmonics were visible for the full-bridge configuration at that point. In classical bearing diagnostics, that is an indicator of degradation. With the magnetic bearing, it can be a sign of a rotational speed that is too low to ensure a suitable level of self-centering of the shaft to the working point.

In summary, the performed measurements revealed that the strain gauge monitoring point should be located near the magnetic bearing, as the presence of induced disturbances influenced more that part of the rotating shaft. Because of the lack of physical contact between rotor and stator, the classical approach for bearing diagnostics with the use of accelerometers would not be as efficient. The full-bridge configuration gave a clearer indication of induced disturbances with a lower level of noise in the frequency spectrum. Disturbance in the Oy axis impacted the frequency characteristics more than the equal disturbance in the Ox axis. This is caused by the fact that the Oy axis stabilizes vertically and provides the lifting force for the rotating shaft. Therefore, even in a static condition, forces generated by the lower and upper coils are not equal, as in the Ox axis.

6. Conclusions

This paper presented preliminary studies of a measurement system dedicated to UAV engine rotors. Moreover, the new concept of a rotor suspension system with magnetic bearings for a mini turbojet engine rotor was described. In the proposed support system, both active and passive magnetic bearings were introduced. Magnetic suspension technology allows the efficient reduction of the transverse vibration amplitude and the negative performance features of a classical bearing system [7]. Additionally, this technology enables continuous diagnostics of rotor engines.

In this paper, a wireless measuring system for the rotary machine was introduced. It consisted of strain sensors and measuring modules with a transmitter for wireless telemetry. Measurements of strain gauge were performed in the system build-up of the Wheatstone bridge, amplifier, and signal transmission circuit. During studies, strain measurements with strain gauges located in two cross-sections of the magnetically suspended shaft were registered. Measurements were made for the different rotational speeds of the rotor, different control system parameters, and various disturbances acting on the Ox and Oy axes. Registered characteristics presented strain amplitude in two cross-sections and fast Fourier transform of the strain measurements for constant motor speed, without and with disturbance. The measurement revealed a higher amplitude level with an increase of rotational speed. Additionally, disturbances in the Oy axis were more noticeable compared with the data with no disturbance and those added in the Ox axis, especially in the front cross-section monitoring point.

Developed laboratory stands with the magnetically suspended rotor and the positive effect of preliminary studies have opened new perspectives for development work associated with the monitoring system of UAV rotor engines. Results obtained during studies are the basis for developing an identification and monitoring system of a rotor supported by magnetic bearings. The next stage of studies includes an attempt to compare measurements from strain gauges and eddy current sensors to indicate phenomena related to incorrect rotor operation using sensors integrated with the object.

Author Contributions: Methodology, A.K. and P.K.-M.; software, A.K. and P.K.-M.; validation, A.K. and P.K.-M.; formal analysis, M.H.; investigation, A.K. and P.K.-M.; resources, M.H.; writing—original draft preparation, A.K. and P.K.-M.; writing—review and editing, M.H.; supervision, M.H. All authors have read and agreed to the published version of the manuscript.

Funding: This research was co-funded by the National Center of Research and Development under the grant entitled "The use of surface engineering new technologies and magnetic bearings in the construction of a miniature turbine jet engine". Research presented in the paper was carried out in the Aircraft Propulsion Research Laboratory. This Laboratory was modernized under the project entitled "Modernization and construction of a new scientific and research infrastructure at the Military University of Technology and at the Warsaw University of Technology for joint numerical experimental research of aviation turbine engines in 2009–2015".

Conflicts of Interest: The authors declare no conflict of interest.

References

1. Liu, Z.; Long, Z.; Li, X. *Maglev Trains Key Underlying Technologies*; Springer: Berlin/Heidelberg, Germany, 2015; pp. 29–39.
2. Zhang, L. Hopf Bifurcation and Vibration Control for a Thrust Magnetic Bearing with Variable Load Mass. *Sensors* **2018**, *18*, 2212. [CrossRef] [PubMed]
3. Kurnyta-Mazurek, P.; Kurnyta, A.; Henzel, M. Concept of wireless measurement system of UAV jet engine rotor. In Proceedings of the 2019 IEEE 5th International Workshop on Metrology for AeroSpace (MetroAeroSpace), Torino, Italy, 19–21 June 2019; pp. 539–543.
4. Hung, J.Y.; Nathaniel, G.A.; Xia, F. Nonlinear control of a magnetic bearing system. *Mechatronics* **2003**, *13*, 621–637. [CrossRef]
5. Sawicki, J.T.; Maslen, E.H.; Bischof, K.R. Modeling and Performance Evaluation of Machining Spindle with Active Magnetic Bearings. *J. Mech. Sci. Thchnol.* **2007**, *21*, 847–850. [CrossRef]
6. Schweitzer, G.; Traxler, A.; Bleuler, H. *Magnetlager: Grundlagen, Eigenshaften und Anwendungen berührungsfreier elektromagnetischer Lager*; Springer: Berlin/Heidelberg, Germany, 1992; pp. 27–50, 107–142.
7. Zhang, C.; Tseng, K.J. Design and control of a novel flywheel energy storage system assisted by hybrid mechanical-magnetic bearings. *Mechatronics* **2013**, *23*, 297–309. [CrossRef]
8. Brusa, E. Semi-active and active magnetic stabilization of supercritical rotor dynamics by contra-rotating damping. *Mechatronics* **2014**, *24*, 500–510. [CrossRef]
9. Halminen, O.; Kärkkäinen, A.; Sopanen, J.; Mikkola, A. Active magnetic bearing-supported rotor with misaligned cageless backup bearings: A dropdown event simulation model. *Mech. Syst. Signal Process.* **2015**, *50–51*, 692–705. [CrossRef]
10. Schweitzer, G.; Maslen, E.H. *Magnetic Bearings: Theory, Designs and Application to Rotating Machinery*; Springer: Berlin/Heidelberg, Germany, 2009; pp. 2–26.
11. Polajzer, B. Modeling and Control of Horizontal-Shaft Magnetic BearingSystem. In Proceedings of the IEEE International Symposium on Industrial Electronics (ISIE '99), Bled, Slovenia, 12–16 July 1999; pp. 1051–1055.
12. Motee, N.; Queiroz, M.S.D. Control of Active Magnetic Bearing. In Proceedings of the 41st IEEE Conference on Decision and Control, Las Vegas, NV, USA, 10–13 December 2002; pp. 860–865.
13. Kurnyta-Mazurek, P.; Kurnyta, A.; Pręgowska, A.; Kaźmierczak, K.; Frąś, L. Application concept of the active magnetic suspension technology in the aircraft engine. *Aviat. Adv. Maint.* **2018**, *41*, 161–193. [CrossRef]
14. Szolc, T.; Falkowski, K. The design of a combined, self-stabilizing electrodynamic passive magnetic bearing supporting high-speed rotors. In Proceedings of the 13th International Conference, Dynamics of Rotating Machinery (SIRM 2019), Copenhagen, Denmark, 13–15 Feberary 2019; pp. 272–281.
15. Żokowski, M.; Majewski, P.; Spychała, J. Detection damage in bearing system of jet engine using vibroacustic method. *Acta Mech. Autom.* **2017**, *11*, 237–242.
16. Yan, X.; Sun, Z.; Zhao, J.; Shi, Z.; Zhang, C.A. Fault Diagnosis of Active Magnetic Bearing-Rotor System via Vibration Images. *Sensors* **2019**, *19*, 244. [CrossRef] [PubMed]
17. Żokowski, M.; Falkowski, K.; Kurnyta-Mazurek, P.; Henzel, M. Control of bearingless electric machines dedicated for aviation. *Aircr. Eng. Aerosp. Technol.* **2020**, *92*, 27–36. [CrossRef]
18. Chiba, A.; Fukao, T.; Ichikawa, O.; Oshima, M.; Takemoto, M.; Dorrel, D.G. *Magnetic Bearings and Bearingless Drives*; Elsevier: London, UK, 2005; pp. 1–15.
19. Genta, G. *Dynamics of Rotating Systems*; Ling, F.F., Hart, D.W.H., Eds.; Springer Science & Business Media: New York, NY, USA, 2005.
20. Falkowski, K.; Gosiewski, Z. *Multifunctional Magnetic Bearings*; Institute of Aviation Scientific Library: Warsaw, Poland, 2003.
21. Hoffmann, K. *An Introduction to Measurement Using Strain Gages*; Hottinger Baldwin Messtechnik GmbH: Darmstadt, Germany, 1987.

© 2020 by the authors. Licensee MDPI, Basel, Switzerland. This article is an open access article distributed under the terms and conditions of the Creative Commons Attribution (CC BY) license (http://creativecommons.org/licenses/by/4.0/).

Article

Magnetometer Calibration for Small Unmanned Aerial Vehicles Using Cooperative Flight Data†

Roberto Opromolla

Department of Industrial Engineering, University of Naples Federico II, Piazzale Tecchio 80, 80125 Naples, Italy; roberto.opromolla@unina.it; Tel.: +39-0817683365

† This paper is an extended version of the paper by Opromolla, R.; Esposito, G.; and Fasano, G. In-flight estimation of magnetic biases on board of small UAVs exploiting cooperation. In Proceedings of the 2019 IEEE 5th International Workshop on Metrology for AeroSpace (MetroAeroSpace), Torino, Italy, 19–21 June 2019, pp. 655–660.

Received: 18 November 2019; Accepted: 16 January 2020; Published: 18 January 2020

Abstract: This paper presents a new method to improve the accuracy in the heading angle estimate provided by low-cost magnetometers on board of small Unmanned Aerial Vehicles (UAVs). This task can be achieved by estimating the systematic error produced by the magnetic fields generated by onboard electric equipment. To this aim, calibration data must be collected in flight when, for instance, the level of thrust provided by the electric engines (and, consequently, the associated magnetic disturbance) is the same as the one occurring during nominal flight operations. The UAV whose magnetometers need to be calibrated (chief) must be able to detect and track a cooperative vehicle (deputy) using a visual camera, while flying under nominal GNSS coverage to enable relative positioning. The magnetic biases' determination problem can be formulated as a system of non-linear equations by exploiting the acquired visual and GNSS data. The calibration can be carried out either off-line, using the data collected in flight (as done in this paper), or directly on board, i.e., in real time. Clearly, in the latter case, the two UAVs should rely on a communication link to exchange navigation data. Performance assessment is carried out by conducting multiple experimental flight tests.

Keywords: unmanned aerial vehicles; magnetometers; magnetic bias; magnetic heading; multi-UAV cooperation; vision-based relative sensing; GNSS-based relative sensing; Levenberg-Marquardt

---

1. Introduction

The use of unmanned aerial vehicles (UAVs) has been increasing exponentially in recent years, and thus they are having a strong economic impact on the global market [1]. Indeed, they can be used for a large variety of military and civilian applications [2], such as border patrol [3], search and rescue [4], infrastructure monitoring [5], package delivery [6] and precision agriculture [7]. To fully enable the abovementioned mission scenarios, it is widely agreed that strong research efforts must be dedicated to significantly enhance the level of autonomy of UAVs regarding their guidance, navigation and control functions [8], on the one hand, and on the other hand, to the definition of a framework where they can be safely integrated in the civil airspace (see for example the NASA Unmanned Traffic Management program [9] and Corus project by SESAR [10]).

In this respect, this work lies in the framework of research activities, carried out at the Department of Industrial Engineering of the University of Naples "Federico II", aiming at investigating the possibility to exploit cooperative strategies to improve navigation performance of mini/micro UAVs (MAVs) [11,12]. If these vehicles fly under nominal Global Navigation Satellite System (GNSS) coverage, autonomous and safe navigation is enabled by integrating measurements from low-cost GNSS receivers and commercial-grade micro-electro-mechanical systems (MEMS)-based inertial (i.e., gyroscopes and accelerometers) and magnetic sensors, within Extended Kalman Filters [13,14] or, more recently,

Particle Filters [15], and using either loosely- or tightly-coupled architectures [16]. In this respect, since onboard gyros do not have enough sensitivity to measure the Earth rate vector, magnetometers play a key role, despite being typically characterized by low bandwidth and high measurement noise. Indeed, their capability to measure the Earth magnetic field, and, consequently to identify the North direction, can be used to bound the heading error by including a magnetic heading estimate among the measurements used by the onboard navigation system of the MAV to get its navigation state. Unfortunately, the presence of magnetic sources, either external or placed on board the MAV, can significantly affect the measurements of a three-axis magnetometer. Together with intrinsic sensor error sources (e.g., biases, scale factor deviations, misalignment and non-orthogonality of the sensor axes) for which a proper ground calibration is always required, these magnetic sources generate disturbances which lead to additional systematic errors in the measurements of the components of the Earth magnetic field in the Body Reference Frame (BRF) of the MAV. Consequently, the measured magnetic heading will be referred to an apparent magnetic North direction, thus leading to an angular heading error which may be up to 10°. For these reasons, such standalone navigation solutions can provide limited accuracy levels, i.e., 5–10 m and 1°–5° for MAVs position and attitude, respectively [17]. Although, these accuracy levels enable real-time stabilization and control, they do not allow meeting the requirements imposed by several applications (like 3D mapping [18]), unless ad-hoc solutions are adopted. For instance, multi-antenna-based GNSS techniques allow reaching sub-degree level in the attitude accuracy [17], but they pose installation constraints which may not be met considering the allocation resources of many micro/small UAVs [19]. So, in the most general case, considering performance of low-cost IMUs on board MAV, the uncertainties in the roll and pitch estimates provided by the onboard autopilot are typically bounded at the degree-level [20] (due to the presence of gravity), while the error in the heading angle estimate can be up to several degrees [21].

In order to address this problem, this paper investigates the possibility to improve the accuracy in the estimate of the magnetic heading provided by magnetometers. The calibration of magnetic sensors is typically based on the assumption that the intrinsic error sources can be characterized using an ellipsoidal error model [22,23]. Consequently, the parameters of the model ellipsoid can be computed by applying a non-linear optimization process, which requires in input a set of measurements taken with different pointing conditions of the sensor's axes in order to properly sample the ellipsoid surface. However, these calibration procedures cannot cope with systematic errors induced by onboard magnetic sources which act during nominal flight operations of the MAV [24,25]. A closed-form analytical solution for the calibration problem of magnetometers on board UAV, conceived to potentially account also for the contribution of the electric motors, can be found in [26]. However, this method has not been tested in flight, and no information is provided by the authors regarding the achievable magnetic heading accuracy.

For this reason, an original calibration approach to estimate magnetic biases caused by onboard disturbances is proposed in this paper. This method exploits cooperation between two MAVs, and it requires calibration data to be collected in flight executing a predefined coordinated maneuver. In the following, the vehicle whose magnetometers need to be calibrated is called "chief", while the cooperative vehicle is called "deputy". The chief must be able to detect and track the deputy using a monocular camera (and, consequently, proper vision-based algorithms [27,28]) to estimate the unit vector corresponding to the chief/deputy relative position (i.e., the so-called line-of-sight, LOS) in the camera reference frame. Also, during the calibration flight, both the vehicles must be kept under nominal GNSS coverage to allow estimating their relative position in local (North-East-Down, NED) coordinates. Using the chief/deputy relative position information in camera and NED coordinates, the calibration problem can be mathematically formulated as a system of non-linear equations as described in detail in Section 2. This problem is addressed using a customized implementation of the Levenberg-Marquardt (LM) method [29] (i.e., a state-of-the-art least-squares solver) based on the formulation proposed by Gavin [30]. It is worth outlining that if both the MAVs are equipped with a camera, this approach can be used to calibrate the magnetic sensors on board both the two vehicles, before being able to carry out their cooperative or independent mission. Clearly, although this method is conceived to exploit cooperation between two MAVs, the proposed concept is perfectly applicable even if the deputy is another type of autonomous vehicle (e.g., ground or marine) or even if it is a fixed ground station, provided that it can acquire GNSS information while being visually tracked. Finally, it is also important to underline that the calibration approach can be applied either off-line (as done in this paper), or directly on board, i.e., processing in real time the acquired flight data. Clearly, the applicability of this latter option requires that the two UAV are able to exchange information about their navigation status using a communication link. An experimental campaign of flight tests is carried out to assess the applicability and performance of the proposed cooperative calibration method. The flight tests are carried out using both fixed and flying deputies. Preliminary results of this work were initially presented in a conference paper [31].

The rest of the paper is organized as follows: Section 2 formulates the magnetic bias calibration problem, and it describes in detail the proposed cooperative strategy and the associated LM-based algorithmic solution. Section 3 presents the setup used for data collection and the adopted experimental strategy, and it discusses the related results. Finally, Section 4 provides a conclusion and indications about future works.

2. Magnetometers Calibration Method

2.1. Magnetic Heading Definition

The calibration problem addressed in this work relies on the assumption that magnetic sensor's measurements are not affected by disturbances from the environment where the MAV is operating. Clearly, this condition is not met if the MAV must fly within areas characterized by large external magnetic sources (e.g., close to power plants, power-lines or other similar facilities). Under this assumption, the magnetic field measured by a three-axes magnetometer ($\underline{H}$) can be expressed using Equation (1):

$$\underline{H} = \begin{bmatrix} H_x \\ H_y \\ H_z \end{bmatrix} = \underline{H_E} + \underline{\Delta H} \tag{1}$$

where H_x, H_y and H_z are its components in BRF, $\underline{H_E}$ is the Earth's magnetic field, and $\underline{\Delta H}$ is the magnetic bias. $\underline{\Delta H}$ is caused by residual intrinsic error sources and (mostly) by onboard magnetic disturbances, thus being constant in BRF.

If the attitude of an MAV (namely, the orientation of its BRF with respect to NED) is parametrized by a 321 sequence of Euler Angles, i.e., heading (ψ), pitch (θ) and roll (φ), the Body-Stabilized Reference Frame (BSRF) can be defined as the reference frame obtained by projecting the BRF axes in the North-East plane thanks to the pitch and roll angle estimates provided by the chief onboard autopilot. Thus, the projection of $\underline{H}$ in the BSRF ($\underline{H_s}$) can be computed using Equation (2):

$$\underline{H_s} = \begin{bmatrix} Hx_s \\ Hy_s \\ Hz_s \end{bmatrix} = \left(\underline{\underline{M_\varphi}}\,\underline{\underline{M_\theta}}\right)^{-1} \begin{bmatrix} H_x \\ H_y \\ H_z \end{bmatrix} \tag{2}$$

where $\underline{\underline{M_\theta}}$ and $\underline{\underline{M_\varphi}}$ (i.e., the elemental rotation matrixes corresponding to θ and φ, respectively) can be written as follows:

$$\begin{aligned} \underline{\underline{M_\varphi}} &= \begin{bmatrix} 1 & 0 & 0 \\ 0 & \cos(\varphi) & \sin(\varphi) \\ 0 & -\sin(\varphi) & \cos(\varphi) \end{bmatrix} \\ \underline{\underline{M_\theta}} &= \begin{bmatrix} \cos(\theta) & 0 & -\sin(\theta) \\ 0 & 1 & 0 \\ \sin(\theta) & 0 & \cos(\theta) \end{bmatrix} \end{aligned} \tag{3}$$

Clearly, the in-plane components of $\underline{H_s}$ (namely, Hx_s and Hy_s) can be used to compute the magnetic heading (ψ_M), i.e., the heading angle estimate obtained using the measurements of a three-axes magnetometer, as shown by Equation (4)

$$\psi_M = \tan^{-1}\left(-\frac{Hy_s}{Hx_s}\right) + \delta_M \tag{4}$$

where δ_M is the local magnetic declination, i.e., the angle (on the local horizontal plane) between the magnetic North (N_M) and the true (geographic) North (N). Unfortunately, due to the presence of $\underline{\Delta H}$, such heading estimate differs from the true heading (ψ) as it is referred to an apparent geographic North direction (N_a) rather than to the true one, as depicted by Figure 1. By substituting Equation (1) in Equation (4), ψ_M can be expressed using Equation (5):

$$\psi_M = \tan^{-1}\left(\frac{-\left(H_{E,ys} + \Delta Hy_S\right)}{\left(H_{E,xs} + \Delta Hx_S\right)}\right) + \delta_M \tag{5}$$

where $H_{E,xs}$ (ΔHx_S) and $H_{E,ys}$ (ΔHy_S) are the in-plane components of $\underline{H_E}$ ($\underline{\Delta H}$) in BSRF. Consequently, the proposed calibration approach aims at increasing the magnetic heading accuracy by computing the in-plane components of the magnetic bias expressed in BSRF ($\underline{\Delta H_s}$). Indeed, this allows obtaining a calibrated magnetic heading ($\psi_{M,c}$) as shown by Equation (6):

$$\psi_{M,c} = \tan^{-1}\left(\frac{-Hy_S + \Delta Hy_S}{Hx_S - \Delta Hx_S}\right) + \delta_M \tag{6}$$

Figure 1 provides a simplified representation of the problem addressed in this paper.

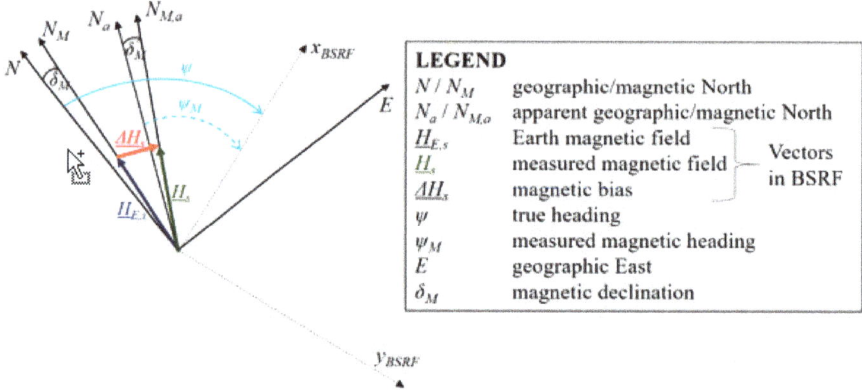

Figure 1. Effect of onboard magnetic sources on the magnetic heading estimation process. They cause the heading angle to be estimated with respect to an apparent geographical North rather than the true North direction. Please note that the figure is not in scale and the sizes of the vectors are chosen for the sake of clarity.

2.2. Cooperative Calibration Strategy

The determination of ΔHx_S and ΔHy_S must rely on data collected in flight, i.e., when all the chief onboard systems operate nominally, thus ensuring that the level of the associated magnetic disturbances is also nominal. To carry out this task, a cooperative calibration strategy is here proposed. Specifically, the two vehicles must execute a coordinated maneuver, during which the visual camera mounted on board the chief is used to detect and track the deputy through a sequence of frames. Two conditions must be strictly met regarding the definition of the flight path for the two vehicles:

- They must fly under nominal GNSS coverage.
- The deputy must stay within the camera Field-of-View (FOV).

Besides these mandatory points, some additional guidelines can be followed regarding the flight path definition to improve accuracy in the result of the calibration process, as better explained in Section 3.

For each frame acquired by the camera on board the chief, the position of the deputy projection on the image plane can be detected. To this aim, if the magnetic calibration is done in real time, an autonomous visual detection and tracking algorithm must be implemented on board. Due to the cooperative nature of the proposed calibration strategy, this image processing task can be carried out exploiting the knowledge of the deputy (e.g., in terms of size, shape and color) as well as its possibility to exchange navigation data with the chief by means of a communication link, as proposed in [27] and [28]. Instead, if the calibration is carried out off-line (i.e., after flight data acquisition), the visual detection task can be realized using a supervised approach (i.e., the position of the deputy projection on the image plane is extracted manually by a human operator). In both the cases, the image position information can be used to compute the unit vector ($\underline{u}_{CRF}$) representing the deputy LOS in the camera reference frame (CRF) using the intrinsic camera calibration parameters [32]. If autonomous visual algorithms are used, the angular accuracy in the LOS estimate can be of the order of the camera instantaneous FOV [28], while a supervised (off-line) approach can even lead to better accuracy due to the possibility to identify the deputy position at sub-pixel level.

Since the two vehicles are both flying under nominal GNSS coverage, the relative position vector and, consequently, the unit vector representing the deputy LOS in NED with respect to the chief ($\underline{u}_{NED}$) can also be determined. In this respect, it is worth outlining that the level of accuracy characterizing the estimates of $\underline{u}_{NED}$ depends on the adopted relative positioning method. Specifically, the chief-deputy relative position vector can be obtained either computing the difference between the two position fix or applying Differential Global Positioning System (DGPS) [33], or Carrier-Phase DGPS (CDGPS) [34] algorithms. Clearly, in the latter case, GNSS raw data, namely pseudo-range and carrier-phase measurements, must be available. With regards to the guidelines for flight path definition mentioned earlier, the angular accuracy in $\underline{u}_{NED}$ can be also improved by keeping the deputy as far as possible from the chief. Of course, the maximum allowable distance is determined by the size of the deputy, the camera resolution and the local background (which can be more or less cluttered) in the image plane.

At this point, if the visual and positioning data are properly synchronized (meaning that a GNSS-based relative position measurement is associated to each visual frame), the relationship between $\underline{u}_{CRF}$ and $\underline{u}_{NED}$ can be written as follows:

$$\underline{u}_{CRF} = \underline{\underline{M}}_{M} \underline{\underline{M}}_{\psi\theta\varphi} \underline{u}_{NED} \tag{7}$$

where $\underline{\underline{M}}_{M}$ is the camera-body extrinsic calibration matrix and $\underline{\underline{M}}_{\psi\theta\varphi}$ is the UAV attitude matrix (i.e., the full rotation matrix corresponding to the 321 sequence of Euler angles defined in Section 2.1). If the camera is mounted according to a strapdown configuration, $\underline{\underline{M}}_{M}$ (representing the orientation of the CRF with respect to the BRF) is constant and it can be computed by carrying out an ad-hoc extrinsic calibration procedure either off-line [35] or on-line [36]. Using the pitch and roll angles available from the chief onboard navigation system, $\underline{\underline{M}}_{\psi\theta\varphi}$ can be expressed as a function of the unknown calibrated magnetic heading as shown by Equation (8):

$$\underline{\underline{M}}_{\psi\theta\varphi}(\psi_{M,c}) = \underline{\underline{M}}_{\varphi} \underline{\underline{M}}_{\theta} \begin{bmatrix} \cos(\psi_{M,c}) & \sin(\psi_{M,c}) & 0 \\ -\sin(\psi_{M,c}) & \cos(\psi_{M,c}) & 0 \\ 0 & 0 & 1 \end{bmatrix} \tag{8}$$

Consequently, if Equations (6) and (8) are substituted within Equation (7) for k time instants at which synchronized visual and GNSS data have been acquired, the magnetic bias calibration process can be formulated as a system of $3k$ non-linear equations in the two unknowns, i.e., ΔHx_S and ΔHy_S. In fact, for each of the k time instants, a triplet of equations is obtained by setting to zero the residual vector function ($\underline{r}$) as shown by Equation (9):

$$\underline{r} = f(\Delta Hx_S, \Delta Hy_S) = \underline{u}_{CRF} - \underline{\underline{M}}_M \underline{\underline{M}}_{\psi\theta\varphi} \underline{u}_{NED} = 0 \tag{9}$$

Hence, the estimation of the unknown magnetic bias components is entrusted to a customized implementation of the LM algorithm, which is an iterative technique to solve non-linear least squares problems. In the following, the symbol $\underline{r}$ is used to indicate directly the $[3k \times 1]$ residual vector obtained writing (9) for all the selected images. A block diagram which summarizes the proposed cooperative calibration approach is depicted in Figure 2.

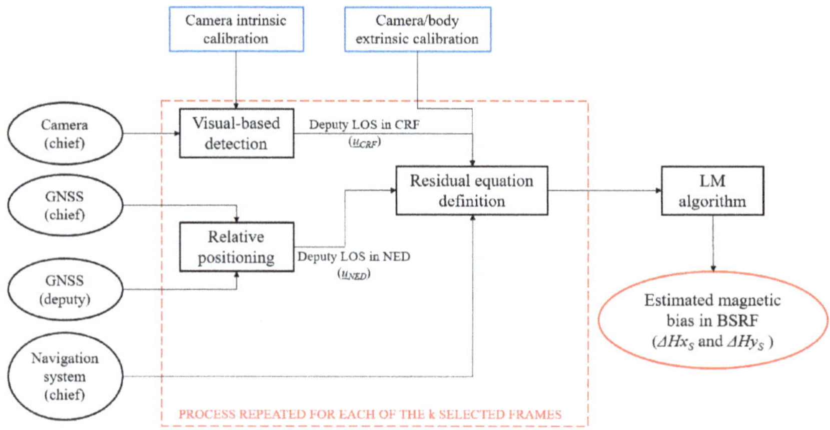

Figure 2. Cooperative strategy for magnetic sensor calibration. The camera and camera/body calibrations (blue rectangular boxes) are carried out offline. The final output is enclosed within an ellipse red box.

2.3. LM-Based Solution

The LM algorithm allows formulating the least-squares minimization of the residual vector function in Equation (9) as a scalar problem by introducing an equivalent scalar cost function, namely the chi-squared error (χ^2). This cost function is the sum of the weighted squares of the errors (also called residuals) between the measured data, i.e., the left-hand side in Equation (7) in this case, and the curve-fit function, i.e., the right-hand side in Equation (7) in this case, and it can be defined as follows:

$$\chi^2(\Delta Hx_S, \Delta Hy_S) = \underline{r}^T \underline{\underline{W}} \underline{r} \tag{10}$$

where $\underline{\underline{W}}$ is a $[3k \times 3k]$ diagonal weight matrix, whose elements are the reciprocal of the weights representing the level of confidence associated to each measured vector ($\underline{u}_{CRF}$). Here, $\underline{\underline{W}}$ is set to be an identity matrix, meaning that the same weight/importance is assigned to each visual estimate of the deputy LOS in CRF. This choice is in line with the fact that the accuracy in the estimate of the position of the deputy projection on the image plane does not vary significantly with factors like relative distance and local background (which can mostly influence the detection probability, if the image processing task is carried out on line, rather than its accuracy [27]).

In this work, the minimization of χ^2 is carried out exploiting the numerical formulation of the LM technique proposed by [30]. According to this formulation, an updated estimate of the unknown in-plane components of $\underline{\Delta H}$ in BSRF is obtained at each LM iteration using Equation (11),

$$\begin{bmatrix} \Delta Hx_S{}^j \\ \Delta Hy_S{}^j \end{bmatrix} = \begin{bmatrix} \Delta Hx_S{}^{j-1} \\ \Delta Hy_S{}^{j-1} \end{bmatrix} + \underline{h}_{LM} \tag{11}$$

where j is the iteration counter, and $\underline{h}_{LM}$, i.e., the LM-based correction vector, is computed using the Marquardt's relationship [29]:

$$\underline{h}_{LM} = \left(\underline{\underline{J}}^T \underline{\underline{W}}\underline{\underline{J}} + \lambda D_{JTWJ}\right)^{-1} \underline{\underline{J}}^T \underline{\underline{W}}\underline{r} \tag{12}$$

where λ is the LM damping parameter, $\underline{\underline{J}}$ is the $[3k \times 2]$ Jacobian matrix defined as in Equation (13), and D_{JTWJ} is a $[2 \times 2]$ matrix containing only the diagonal elements of $\underline{\underline{J^T W J}}$:

$$\underline{\underline{J}} = \begin{bmatrix} \frac{\partial \underline{r}^1}{\partial \Delta Hx_S} & \frac{\partial \underline{r}^1}{\partial \Delta Hy_S} \\ \vdots & \vdots \\ \frac{\partial \underline{r}^k}{\partial \Delta Hx_S} & \frac{\partial \underline{r}^k}{\partial \Delta Hy_S} \end{bmatrix} \tag{13}$$

Considering that the LM algorithm is a combination of two minimization techniques, namely the gradient descent method and the Gauss-Newton method, the value of the damping factor is a key parameter to determine its convergence. In fact, for large values of λ, the correction vector obtained from Equation (12) is closer to the one provided by the gradient descent method, meaning that χ^2 is reduced in the direction opposite to its gradient. Instead, for small values of λ, the correction vector obtained from Equation (12) is closer to the one provided by the classical Gauss-Newton method.

Once the LM-based correction is computed, the updated solution provided by Equation (11) is not passed at the next step of the iterative process unless a specific condition on the associated variation on χ^2 is met. The adopted confirmation criterion is given by the following relation [30]:

$$\frac{\chi^2\left(\Delta Hx_S{}^{k-1}, \Delta Hy_S{}^{k-1}\right) - \chi^2\left(\Delta Hx_S{}^{k}, \Delta Hy_S{}^{k}\right)}{2\underline{h}_{LM}\left(\lambda \underline{h}_{LM} + \underline{\underline{J}}^T \underline{\underline{W}}\underline{r}\right)} > \varepsilon_0 \tag{14}$$

where ε_0 is a positive threshold set by the user. For instance, this condition allows discarding an update which would cause an increase in χ^2. Based on the result of Equation (14), the value of the damping factor is changed at each iteration to accelerate convergence. Specifically, if the condition in Equation (14) is met (not met), λ is decreased (increased) using the scaling factor λ_{DN} (λ_{UN}). Clearly, large scaling factors can provide convergence in fewer iterations, but the process may encounter numerical instabilities.

With regards to the initialization of the iterative procedure, the initial values of ΔHx_S and ΔHy_S are set to zero, though a non-zero initial guess may be used if available, e.g., obtained by an on-ground calibration. The initial value of λ (λ_0) is also freely selected by the user.

Finally, the iterative process is ended if at least one out of the three criteria shown in Equation (15) is met, unless the iteration counter reaches a maximum value (e.g., 100):

$$\max\left(\left|\underline{\underline{J}}^T \underline{\underline{W}}\underline{r}\right|\right) < \varepsilon_1; \quad \max\left(\left|\frac{\underline{h}_{LM}}{\underline{\Delta H}}\right|\right) < \varepsilon_2; \quad \frac{\chi^2}{k-1} < \varepsilon_3 \tag{15}$$

It is also worth highlighting that although all the (dimensionless) user parameters, i.e., ε_0, ε_1, ε_2, ε_3, λ_0, λ_{UN} and λ_{DN}, defined earlier in this section are common to any LM implementation, their selection must be done in view of the specific application.

3. Experimental Flight Test Campaign

An experimental campaign of flight tests has been carried out to validate and assess the performance of the proposed cooperative strategy for the calibration of magnetic sensors installed on board MAVs.

Specifically, the main performance parameter is the accuracy level charactering the estimation of the calibrated magnetic heading obtained applying Equation (6). To evaluate this accuracy, a reference estimate of the heading angle must be derived to be used as a ground truth (as described in detail in the following). This section is divided in two parts corresponding to different experimental tests carried out using, as deputy, a fixed ground station and a flying vehicle, respectively.

3.1. Experiment with a Ground-Fixed Deputy

3.1.1. Experimental Setup

For the first experiment, the chief is a customized version of the Pelican^{TM} quadrotor from Ascending Technologies^{TM} (Krailling, GermanY; http://www.asctec.de/) while the role of the deputy is played by a fixed ground station, i.e., a laptop connected via serial RS-232 and a serial-USB adapter to a GNSS antenna (AV59) and receiver (BD960) both produced by Trimble^{TM} (Sunnyvale, CA, USA; https://www.trimble.com/). The chief UAV and the deputy ground station are depicted in Figure 3.

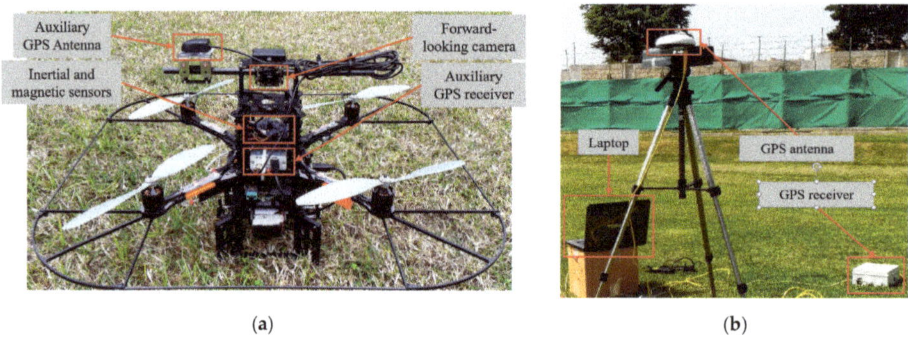

(**a**) (**b**)

Figure 3. (**a**) Pelican quadcopter used as chief UAV. (**b**) Fixed ground station used as deputy.

The chief uses a 752 × 480 miniaturized CMOS forward looking camera, i.e., the mvBlueFOX-MLC-200wc produced by Matrix Vision^{TM} (Oppenweiler, Germany; https://www.matrix-vision.com/), to image the deputy. Since the Pelican autopilot does not give access to the GNSS raw data (which are required to enable DGPS or CDGPS processing), an auxiliary antenna and GNSS receiver (LEA 6T) produced by uBlox^{TM} (Thalwil, Switzerland; https://www.u-blox.com/en) are installed on the chief. Inertial and magnetic sensor measurements are collected at a frequency of 40 Hz, while images and GNSS data are gathered simultaneously at a frequency of 1 Hz. To ensure the synchronization of GNSS, visual and inertial/magnetic data, they must be referred to the same time scale. This task is achieved by using an acquisition software [11] developed to save all the sensor data with an accurate time-tag based on the CPU clock. This time-tag is associated with GNSS measurements, which include the GPS time, with very small latency. Moreover, a precise visual/GNSS synchronization is attained since the image acquisition is triggered by the reception of the first GPS package. The deputy ground station also collects GNSS raw data at 1 Hz.

Despite a single deputy suffices for the application of the proposed calibration approach, an additional ground station is positioned on the test field to collect GNSS raw data. Indeed, if both the ground stations are contained in the field of view of the camera on board the chief during the entire flight test, the DGPS-vision algorithm described in [11] can be adopted to obtain the desired reference solution for the heading angle of the chief. This quantity, indicated as $\psi_{DGPS\text{-}VISION}$ in the following, has an average accuracy of the order of 0.5° (which is far better than the heading accuracy provided by the Pelican onboard autopilot) [11].

In this work, the proposed magnetic calibration strategy is carried out using off-line the data collected in flight. Of course, to enable an autonomous real-time implementation, a Vehicle-2-Vehicle

(V2V) communication link is required to exchange the navigation data. In this respect, it is worth outlining that the amount of information to be exchanged (i.e., the GNSS data) is limited to a few Kbits per second, thus being compatible with the specifications of standard V2V communication systems [37].

3.1.2. Results

The magnetic calibration strategy is applied over data collected during a single flight test, lasting around 8 minutes (although this relatively long time duration is not strictly required for the applicability of the method). The flight path travelled by the chief on the testing site is depicted in Figure 4, where the positions of the two ground stations are also indicated. Among them, the farthest one with respect to the chief (blue dot in Figure 4) is selected as deputy for the cooperative calibration process to limit the uncertainty in the GNSS-based estimation of $\underline{u}_{NED}$. The other ground station (green dot in Figure 4) is indicated as auxiliary deputy as it is used to allow the estimation of the reference heading solution ($\psi_{DGPS\text{-}VISION}$) according to the DGPS-vision algorithm in [11].

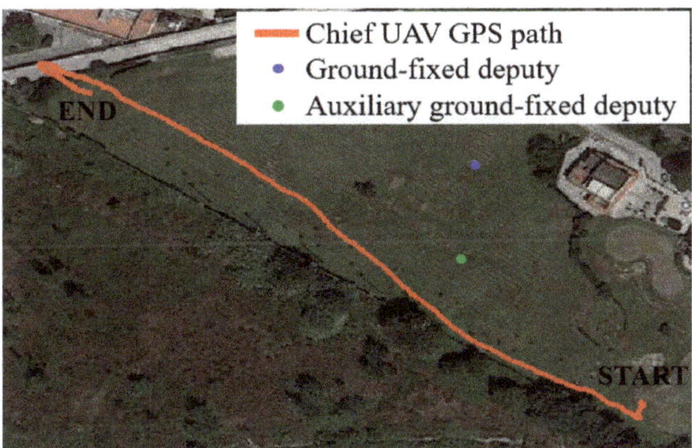

Figure 4. Flight path of the chief (UAV to be calibrated) during the experimental test. The locations of the deputies (ground stations) on the test site are also highlighted.

The flight path is assigned to provide a significant variation of the heading of the chief. This aspect is important to ensure an adequate sampling of the residual vector function for the minimization process. The time variation of the non-calibrated magnetic heading, obtained applying (4), is depicted in Figure 5. For this computation, the local magnetic declination is obtained using an on-line free software [38], which requires in input the location of the test site (latitude, longitude and altitude) and the time of the experiment. In this case, δ_M is 3.07°.

To highlight the effect caused by the presence of onboard magnetic disturbances on the magnetic heading estimation, the heading error (ψ_{ERR}), measured as the difference between $\psi_{DGPS\text{-}VISION}$ and ψ_M, is depicted in Figure 6. Specifically, ψ_{ERR} is expressed as a function of time and of the reference $\psi_{DGPS\text{-}VISION}$ estimate in Figure 6a,b, respectively.

The starting time in Figure 6a is 20 s. In fact, the time interval during the power on of the electric engines and the takeoff can be neglected as it does not correspond to a nominal flight condition (and, consequently, nominal value of the magnetic bias caused by onboard sources). It is worth outlining that the presence of gaps in the time variation of ψ_{ERR} in Figure 6a is due to temporary absence of the reference heading solution ($\psi_{DGPS\text{-}VISION}$), mainly occurring if the ground stations are not both enclosed in the camera FOV. By looking at both Figure 6a,b, it is clear how the time variation of ψ_{ERR} is correlated to the variation of the true heading. This is due to the magnetic bias vector ($\underline{\Delta H}$) being constant in BRF. Specifically, the sign of the magnetic heading estimation error is positive for negative heading, while it becomes negative as the chief is pointed with a positive heading.

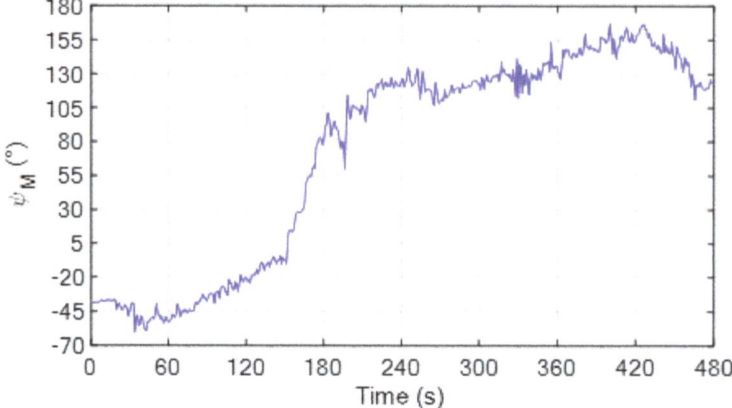

Figure 5. Time variation of the non-calibrated magnetic heading (ψ_M), obtained applying (4), during the flight test with a fixed deputy. The local magnetic declination at the test site and date of the experiment is 3.07°.

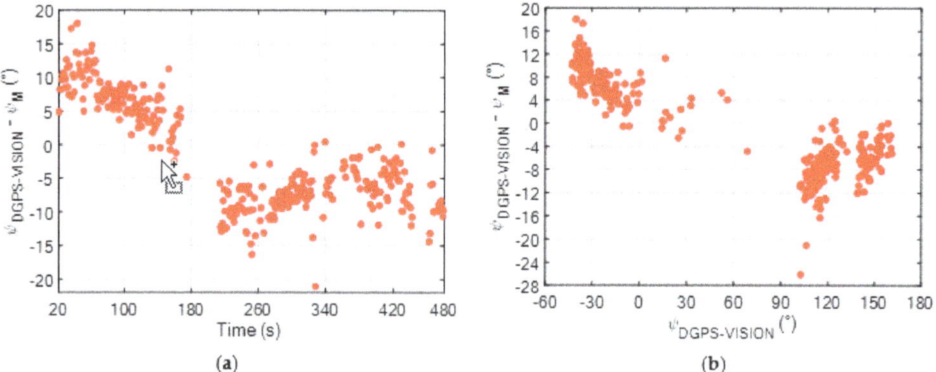

Figure 6. (**a**) Time variation of ψ_{ERR} for the non-calibrated magnetic heading (ψ_M) during the flight test; (**b**) Variation of ψ_{ERR} for the non-calibrated magnetic heading (ψ_M) as a function of the reference heading solution obtained using the DGPS-vision algorithm [11].

Both for the computation of ψ_M as shown in Figure 5, and for the calibration procedure, the raw magnetic sensor measurements must be filtered to cope with the associated high level of noise. Here, this is done by applying a moving average operator to these noisy magnetic data. A smoothing time of 2.5 s is set when applying the moving average operator considering the relatively smooth flight dynamic and the low magnetometer bandwidth. The result of this operation for the components of $\underline{H}$ along the x and y axes of BRF is shown by Figure 7.

At this point, the LM-based calibration algorithm can be run. To this end, a key role is played by the selection of the k time instants at which the visual and GNSS data are used. This is done by considering two main guidelines:

- Time instants at which the yaw rate, evaluated considering the gyro measurement along the z axis of BRF, is larger than a threshold ($\tau_{\omega,z}$) should be excluded.
- Time instants at which the chief-deputy relative distance is lower than a threshold (τ_R) should be excluded.

The first condition is set to minimize the heading variation associated to the selected time instants considering the low bandwidth of the magnetometer. The second condition allows limiting the DGPS uncertainty which characterizes the estimation of $\underline{u}_{NED}$. In this work, $\tau_{\omega,z}$ is set to 0.25°/s, while τ_R is set to 70 m, thus allowing the selection of k = 18 time instants. The heading value at the selected frames varies from −38.5° to 125.3°, while the mean chief-deputy relative distance is 105.12 m. With regards to the selection of the user's parameters for the LM implementation (defined in Section 2.3), the adopted setting is reported in Table 1.

Table 1. Dimensionless parameters characterizing the LM-based magnetic calibration algorithm.

ε_0	ε_1	ε_2	ε_3	λ_0	λ_{UN}	λ_{DN}
10^{-4}	10^{-10}	10^{-4}	10^{-6}	10^{-5}	11	9

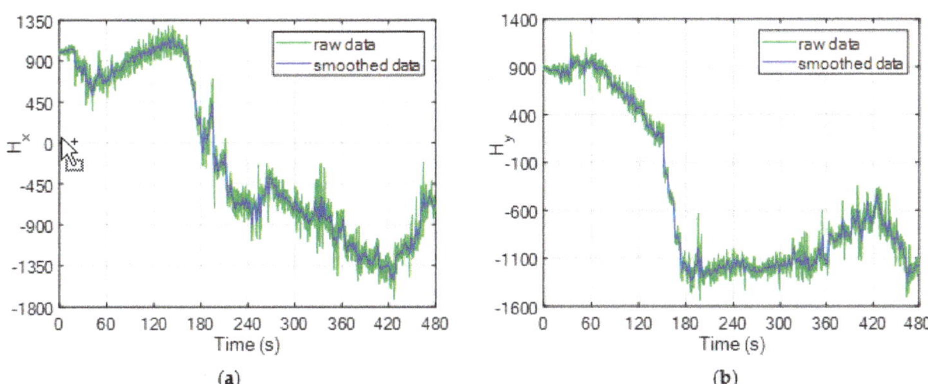

Figure 7. Time variation of the magnetometer measurements during the flight test. The raw data (green line) are smoothed using a moving average operator obtaining a lower-noise dataset (blue line). (**a**) Component of H along the x axis of BRF; (**b**) Component of H along the y axis of BRF. Please note that these magnetometer measurements, provided by the autopilot installed on board the chief by Ascending Technologies^{TM}, are dimensionless quantities. In this respect, consider that the norm of the Earth magnetic field vector at date and location of the flight test is 46,126 nT.

The initialization of the damping factor is chosen to favor the Gauss-Newton minimization process with respect to the gradient descent approach, while the values set for λ_{UN} and λ_{DN} have shown to provide good convergence in the LM optimization process [30]. With regards to the convergence thresholds, ε_2 is the largest (10^{-4}) since it depends on the desired accuracy in the estimated value of ΔHx_S and ΔHy_S. In this respect, a variation of ΔHx_S and ΔHy_S in the order of ε_2 must have a negligible effect on the value of $\underline{r}$. Given this parameters' setting, the convergence is reached simultaneously for both the first (1×10^{-11}) and second (8×10^{-5}) condition in Equation (15) after five iterations, while the value of the third convergence metric, i.e., $\chi^2/(k-1)$, stops at (5×10^{-4}). The resulting values of ΔHx_S and ΔHy_S are −267.53 and −4.17, respectively.

Using these components of the magnetic bias vector in BSRF, the calibrated magnetic heading can be computed using (6), and the resulting estimation error is reported in Figure 8, again as a function of time and $\psi_{DGPS\text{-}VISION}$.

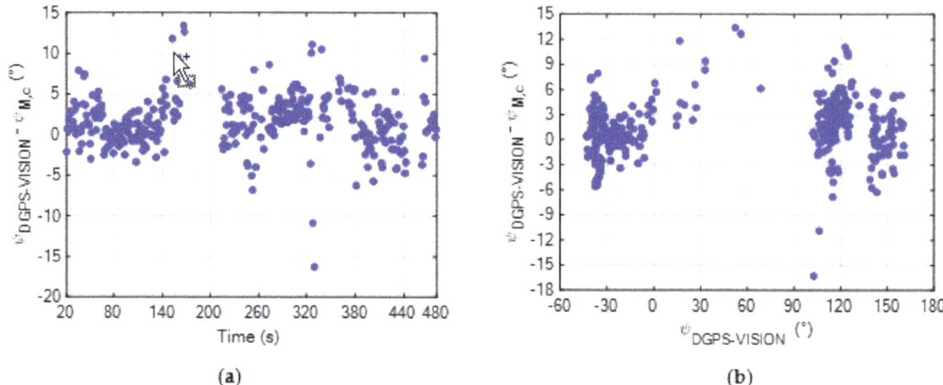

Figure 8. (**a**) Time variation of ψ_{ERR} for the calibrated magnetic heading ($\psi_{M,c}$) during the flight test; (**b**) Variation of ψ_{ERR} for the calibrated magnetic heading ($\psi_{M,c}$) as a function of the reference heading solution obtained using the DGPS-vision algorithm [11].

By looking at Figure 8b, it can be stated that the proposed calibration approach allows removing the bias induced in the magnetic heading estimation by onboard sources, as the resulting error is independent of $\psi_{DGPS\text{-}VISION}$ (unlike in Figure 6b for the non-calibrated magnetic heading). A statistical comparison between the accuracies characterizing ψ_M and $\psi_{M,c}$ is reported in Table 2, considering the absolute value of ψ_{ERR} as a performance metric.

Table 2. Time statistics of the absolute value of the magnetic heading estimation error. Calibrated vs. non-calibrated case. The symbol |·| indicates the absolute value operator.

Magnetic Heading	$\lvert\psi_{ERR}\rvert$ (°)	
	Mean	**Standard Deviation**
ψ_M	7.4	3.6
$\psi_{M,c}$	2.7	2.4

The statistics reported in the table above quantitatively show the achievable improvement in the magnetic heading accuracy. Specifically, the absolute value of the bias reduces by about 5°, while the standards deviation reduces by about 1°.

3.2. Experiment with a Flying Deputy

3.2.1. Experimental Setup

Additional flight tests are conducted using a flying vehicle as deputy for the magnetic calibration process. The advantages of using another UAV as deputy, instead of a fixed ground station, are mainly related to the possibility to realize more quickly the data collection process (while meeting all the identified guidelines regarding the true heading, the range and the yaw rate), especially if the pre-defined coordinated maneuver is executed autonomously. Moreover, as already mentioned, if each UAV is able to visually detect and track the other, the collected data can be used for the calibration of both their corresponding magnetic sensors.

The chief UAV is the same described in Section 3.1.1. Instead, the role of the deputy is played by a X8+ octocopter produced by 3D Robotics™ (Berkeley, CA, USA; https://3dr.com/) depicted in Figure 9. Just like the chief, the deputy is customized by the installation of an auxiliary GNSS receiver and antenna to enable the acquisition of the GNSS raw data. The GNSS receiver is connected via USB to an embedded CPU, i.e., the Odroid XU4^{TM} now produced by Hardkernel (Anyang, South Korea; https://www.hardkernel.com/), for data acquisition and storage.

Figure 9. Customized X8+ by 3D Robotics used as flying deputy.

Besides the deputy, the same ground stations as described in Section 3.1.1 are also installed on the test field and they are always kept in the field of view of the camera to allow the estimation of the reference heading solution. Finally, the same acquisition software and data synchronization strategy described in Section 3.1.1, have been used for these flight experiments. The only difference lies in the fact that inertial data are collected at a higher data rate (i.e., around 76 Hz).

3.2.2. Results

Three flights are carried out at a different test site, so the local value of δ_M at the time of the experiments is 3.21°. The goal of these additional tests is twofold. On one side, the collected data can be used to further validate the reliability of the proposed method by applying the values of ΔHx_S and ΔHy_S computed in Section 3.1.2 to correct the bias affecting the estimated magnetic heading. On the other side, the LM-based calibration algorithm can be applied again to check the consistency of the method.

With regards to the first analysis, the time variation of ψ_{ERR} during the three flight tests for both the calibrated and non-calibrated magnetic heading is reported in Figure 10. Please note that for all the three cases, the time intervals during which the engines are not operating nominally, e.g., before takeoff and after landing, are not considered. Also, during the second and third flight, a set of 360° turns is commanded to the chief, so in the corresponding time interval the ground stations disappear from the FOV and the reference $\psi_{DGPS\text{-}VISION}$ cannot be computed.

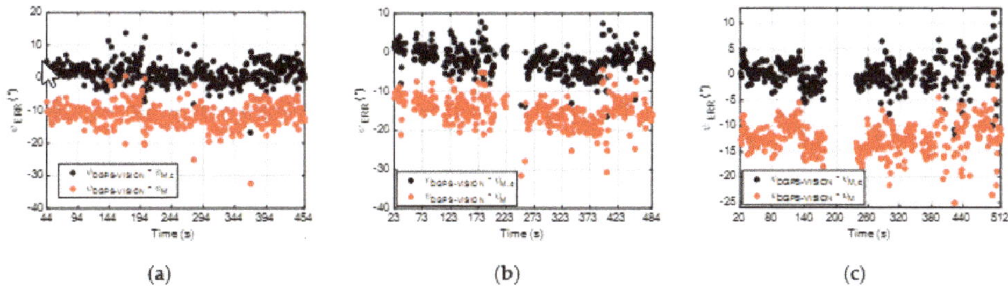

Figure 10. Time variation of ψ_{ERR} for the calibrated ($\psi_{M,c}$) and non-calibrated (ψ_M) magnetic heading during the three flight tests conducted with a flying deputy. $\psi_{M,c}$ is obtained applying (6) with the values of the magnetic bias vector components obtained from the calibration results described in Section 3.1.2, i.e., $\Delta Hx_S = -267.53$, $\Delta Hy_S = -4.17$. (**a**) Flight test 1; (**b**) Flight test 2; (**c**) Flight test 3.

First, it is interesting to note that, since the true heading is positive during the three flight tests, the bias in ψ_M is always negative. This is in line with the dependence of the magnetic heading error on the true heading observed in Figure 6b. Overall, Figure 10 shows that the magnetic bias vector components derived thanks to the calibration process described in Section 3.1.2 can be used to correct the magnetometers measurements relative to these new flight tests, thus significantly reducing the systematic error in $\psi_{M,c}$ with respect to ψ_M. To synthetize these results, a statistical comparison between the accuracies characterizing ψ_M and $\psi_{M,c}$ is reported in Table 3, considering the absolute value of ψ_{ERR} as the performance metric.

Table 3. Time statistics of the absolute value of the magnetic heading estimation error for the three tests with a flying deputy. Calibrated vs. non-calibrated case. The symbol $|\cdot|$ indicates the absolute value operator. The magnetic heading is corrected using the magnetic bias vector components in BSRF as computed in Section 3.1.2 ($\Delta Hx_S = -267.53$, $\Delta Hy_S = -4.17$).

Magnetic Heading (Flight Test)	$\lvert\psi_{ERR}\rvert$ (°)	
	Mean	**Standard Deviation**
ψ_M (1)	11.5	3.4
$\psi_{M,c}$ (1)	2.5	2.2
ψ_M (2)	15.2	3.6
$\psi_{M,c}$ (2)	3.2	2.5
ψ_M (3)	12.5	3.2
$\psi_{M,c}$ (3)	2.1	1.9

By looking at these statistics, first, it is important to highlight that the noise in the non-calibrated magnetic heading (measured by the standard deviation of $|\psi_{ERR}|$) has the same level for all the flight tests analyzed in this work. Instead, the bias (measured by the mean of $|\psi_{ERR}|$) is larger in the latter three flight tests (i.e., it varies from 11.5° to 15.2°) than in the one analyzed in Section 3.1.2 (i.e., 7.4°). This occurs since the constancy of $\underline{\Delta H}$ in BRF induces a bias in the magnetic heading which is a function of the true heading (as highlighted, for instance, in Figure 6b). In the case of the three flights analyzed in this section, the average pointing of the chief (determined by the main direction of the test field with respect to the geographic North) provides a less favorable condition for the vector composition of $\underline{H_E}$ and $\underline{\Delta H}$ (considering that $\underline{\Delta H}$ has its main component along the x axis of BSRF). However, by applying the correction on the magnetic sensor measurements computed by the proposed calibration approach, both the bias and the noise in the estimated magnetic heading are significantly reduced.

Finally, the proposed calibration approach is applied again, but using the data collected during one of the flight tests performed with a flying deputy, in order to show the consistency of the method. In this respect, the third flight is selected as it is characterized by the larger variability in the true heading (which allows better sampling the residual vector function). As $\tau_{\omega,z}$ and τ_R are again set to 0.25°/s, and 70 m, respectively, the visual and GNSS data required for the calibration are relative to $k = 14$ time instants. The heading value at the selected frames varies from 47.4° to 87.2°, while the mean chief-deputy relative distance is 107.71 m.

By keeping the same LM parameters listed in Table 1, the resulting values of ΔHx_S and ΔHy_S are −275.53° and 41.16°, respectively. A statistical comparison between this solution and the one obtained in Section 3.1.2 is carried out in terms of $|\psi_{ERR}|$ computing the calibrated magnetic heading for the third flight test. The results, reported in Table 4, confirm that the same level of accuracy is kept, despite a slight change in the component of $\underline{\Delta H}$ along the y axis of BSRF (which is still much lower than ΔHx_S).

Table 4. Time statistics of the absolute value of the estimation error on the calibrated magnetic heading for the third flight test with a flying deputy. Comparison between two solutions for the magnetic bias vector components.

Magnetic Bias Vector Components in BSRF	$\lvert\psi_{ERR}\rvert$ (°)	
	Mean	**Standard Deviation**
Flight test with a fixed deputy (Section 3.1.2) $\Delta Hx_S = -267.53$, $\Delta Hy_S = -4.17$	2.08	1.94
Flight test 3 with a flying deputy $\Delta Hx_S = -275.53$, $\Delta Hy_S = -41.16$	2.02	1.94

4. Conclusions

An innovative method for the calibration of low-cost magnetometers for small/micro unmanned aerial vehicles was presented. It aims at improving the accuracy in the magnetic heading estimate by correcting the systematic error caused by the presence of onboard magnetic fields during nominal flight conditions (e.g., generated by the engines and, potentially, by other electric devices). This problem was addressed by collecting calibration data in flight and adopting a cooperative approach. In fact, the proposed method requires the unmanned aerial vehicle to be calibrated (chief) to visually detect and track a cooperative deputy, both placed under nominal Global Navigation Satellite System coverage to allow relative positioning. If the deputy is a flying vehicle a coordinate flight maneuver must be executed. Using the collected data, the calibration problem can be formulated as a system of non-linear equations solved applying the Levenberg-Marquardt algorithm. An experimental campaign of flight test was conducted to assess the feasibility of the proposed approach and to quantitatively evaluate the achievable magnetic heading accuracy, using either a fixed ground station or a flying vehicle as deputy. Results showed that onboard magnetic sources, being constant in the body reference frame of the chief, induce a systematic error in the estimated magnetic heading, whose entity and sign are a function of the true heading of the chief. This effect is removed by applying the magnetic bias vector components estimated by the calibration procedure to correct the magnetometers measurement. In fact, the residual error in the calibrated magnetic heading is reduced down to two degrees in all the analyzed tests, and it is independent of the true heading of the chief. Besides the systematic error, the noise in the estimated magnetic heading is also slightly reduced after the calibration.

Future research activities will be dedicated to analytically deriving the sensitivity of the calibration approach with respect to the uncertainty in the calibration data (both visual and relative positioning ones). This can also lead to the definition of a standard procedure for data collection which can be efficiently executed exploiting fully autonomous flight mode of the chief and deputy unmanned aerial vehicles. Finally, additional efforts can be addressed to develop a generalized calibration approach able to account also for magnetic disturbances caused by external sources.

Funding: This research received no external funding.

Conflicts of Interest: The author declares no conflict of interest.

References

1. Wargo, C.; Snipes, C.; Roy, A.; Kerczewski, R. UAS industry growth: Forecasting impact on regional infrastructure, environment, and economy. In Proceedings of the 2016 IEEE/AIAA 35th Digital Avionics Systems Conference, Sacramento, CA, USA, 25–29 September 2016; pp. 1–5.
2. Valavanis, K.P.; Vachtsevanos, G.J. *Handbook of Unmanned Aerial Vehicles*; Springer: Dordrecht, The Netherlands, 2015.
3. Koslowski, R.; Schulzke, M. Drones Along Borders: Border Security UAVs in the United States and the European Union. *Int. Stud. Perspect.* **2018**, *19*, 305–324. [CrossRef]

4. Qi, J.; Song, D.; Shang, H.; Wang, N.; Hua, C.; Wu, C.; Qi, X.; Han, J. Search and rescue rotary-wing uav and its application to the lushan ms 7.0 earthquake. *J. Field Rob.* **2016**, *3*, 290–321. [CrossRef]
5. Greenwood, W.W.; Lynch, J.P.; Zekkos, D. Applications of UAVs in civil infrastructure. *J. Infrastruct. Syst.* **2019**, *25*, 04019002. [CrossRef]
6. Bamburry, D. Drones: Designed for product delivery. *Des. Manage. Rev.* **2015**, *26*, 40–48.
7. Hernandez, A.; Murcia, H.; Copot, C.; De Keyser, R. Towards the development of a smart flying sensor: illustration in the field of precision agriculture. *Sensors* **2015**, *15*, 16688–16709. [CrossRef] [PubMed]
8. Kendoul, F. Survey of advances in guidance navigation and control of unmanned rotorcraft systems. *J. Field Rob.* **2012**, *29*, 315–378. [CrossRef]
9. Kopardekar, P.; Rios, J.; Prevot, T.; Johnson, M.; Jung, J.; Robinson, J.E. Unmanned aircraft system traffic management (UTM) concept of operations. In Proceedings of the AIAA Aviation and Aeronautics Forum, Washington, DC, USA, 13–17 June 2016.
10. *Concept of operations for European UTM systems*. Available online: https://www.eurocontrol.int/project/concept-operations-european-utm-systems (accessed on 30 October 2019).
11. Vetrella, A.R.; Fasano, G.; Accardo, D.; Moccia, A. Differential GNSS and vision-based tracking to improve navigation performance in cooperative multi-UAV systems. *Sensors* **2016**, *16*, 2164. [CrossRef] [PubMed]
12. Vetrella, A.R.; Opromolla, R.; Fasano, G.; Accardo, D.; Grassi, M. Autonomous Flight in GPS-Challenging Environments Exploiting Multi-UAV Cooperation and Vision-aided Navigation. In Proceedings of the AIAA Information Systems-AIAA Infotech@Aerospace 2017 Conference, Grapevine, TX, USA, 9–13 January 2017.
13. Kim, J.H.; Sukkarieh, S.; Wishart, S. Real-time navigation, guidance, and control of a UAV using low-cost sensors. In *Field and Service Robotics, Springer Tracts in Advanced Robotics*; Yuta, S., Asama, H., Prassler, E., Tsubouchi, T., Thrun, S., Eds.; Springer: Berlin/Heidelberg, Germany, 2006.
14. Wendel, J.; Meister, O.; Schlaile, C.; Trommer, G.F. An integrated GPS/MEMS-IMU navigation system for an autonomous helicopter. *Aerosp. Sci. Technol.* **2006**, *10*, 527–533. [CrossRef]
15. Cappello, F.; Sabatini, R.; Ramasamy, S.; Marino, M. Particle filter based multi-sensor data fusion techniques for RPAS navigation and guidance. In Proceedings of the 2015 IEEE Metrology for Aerospace (MetroAeroSpace), Benevento, Italy, 4–5 June 2015; pp. 395–400.
16. Falco, G.; Pini, M.; Marucco, G. Loose and tight GNSS/INS integrations: Comparison of performance assessed in real urban scenarios. *Sensors* **2017**, *17*, 255. [CrossRef] [PubMed]
17. Hirokawa, R.; Ebinuma, T. A low-cost tightly coupled GPS/INS for small UAVs augmented with multiple GPS antennas. *J. Inst. Navig.* **2009**, *56*, 35–44. [CrossRef]
18. Eling, C.; Wieland, M.; Hess, C.; Klingbeil, L.; Kuhlmann, H. Development and Evaluation of UAV based Mapping Systems for Remote Sensing and Surveying Applications. *Int. Arch. Photogramm. Remote Sens. Spatial Inf. Sci.* **2015**, *40*, 233. [CrossRef]
19. Suzuki, T.; Takahashi, Y.; Amano, Y. Precise UAV Position and Attitude Estimation by Multiple GNSS Receivers for 3D Mapping. In Proceedings of the 29th International Technical Meeting of The Satellite Division of the Institute of Navigation (ION GNSS+ 2016), Portland, OR, USA, 12–16 September 2016; pp. 1455–1464.
20. Brown, A.K. Test results of a GPS/inertial navigation system using a low cost MEMS IMU. In Proceedings of the 11th Annual Saint Petersburg International Conference on Integrated Navigation System, Saint Petersburg, Russia, 24–26 May 2004.
21. Nuske, S.T.; Dille, M.; Grocholsky, B.; Singh, S. Representing substantial heading uncertainty for accurate geolocation by small UAVs. In Proceedings of the AIAA Guidance, Navigation, and Control Conference, Toronto, ON, Canada, 2–5 August 2010.
22. Fang, J.; Sun, H.; Cao, J.; Zhang, X.; Tao, Y. A novel calibration method of magnetic compass based on ellipsoid fitting. *IEEE Trans. Instrum. Meas.* **2011**, *60*, 2053–2061. [CrossRef]
23. Vasconcelos, J.F.; Elkaim, G.; Silvestre, C.; Oliveira, P.; Cardeira, B. Geometric approach to strapdown magnetometer calibration in sensor frame. *IEEE Trans. Aerosp. Electron. Syst.* **2011**, *47*, 1293–1306. [CrossRef]
24. Gebre-Egziabher, D.; Elkaim, G.H.; David Powell, J.; Parkinson, B.W. Calibration of strapdown magnetometers in magnetic field domain. *J. Aerosp. Eng.* **2006**, *19*, 87–102. [CrossRef]
25. Chodnicki, M.; Mazur, M.; Nowakowski, M.; Kowaleczko, G. Algorithms for the Detection and Compensation Interferance of Magnetic Field Measurement. In Proceedings of the 2018 5th IEEE International Workshop on Metrology for AeroSpace (MetroAeroSpace), Rome, Italy, 20–22 June 2018; pp. 495–499.

26. Liu, H.; Liu, M.; Ge, Y.; Shuang, F. Magnetometer Calibration Scheme for Quadrotors with On-Board Magnetic Field of Multiple DC Motors. In Proceedings of the 2013 Chinese Intelligent Automation Conference, Berlin, Heidelberg, 23–25 August 2013; pp. 409–419.
27. Opromolla, R.; Fasano, G.; Accardo, D. A Vision-Based Approach to UAV Detection and Tracking in Cooperative Applications. *Sensors* **2018**, *18*, 3391. [CrossRef] [PubMed]
28. Opromolla, R.; Inchingolo, G.; Fasano, G. Airborne Visual Detection and Tracking of Cooperative UAVs Exploiting Deep Learning. *Sensors* **2019**, *19*, 4332. [CrossRef] [PubMed]
29. Marquardt, W. An algorithm for least-squares estimation of nonlinear parameters. *J. Soc. Ind. Appl. Math.* **1963**, *11*, 431–441. [CrossRef]
30. Gavin, H. The Levenberg-Marquardt method for nonlinear least squares curve-fitting problems. Available online: http://people.duke.edu/~{}hpgavin/ce281/lm.pdf (accessed on 30 October 2019).
31. Opromolla, R.; Esposito, G.; Fasano, G. In-flight estimation of magnetic biases on board of small UAVs exploiting cooperation. In Proceedings of the 2019 IEEE 5th International Workshop on Metrology for AeroSpace (MetroAeroSpace), Torino, Italy, 19–21 June 2019; pp. 655–660.
32. *Camera Calibration Toolbox for Matlab*. Available online: http://www.vision.caltech.edu/bouguetj/calib_doc/ (accessed on 30 October 2019).
33. Gross, J.N.; Gu, Y.; Rhudy, M.B. Robust UAV relative navigation with DGPS, INS, and peer-to-peer radio ranging. *IEEE Trans. Autom. Sci. Eng.* **2015**, *12*, 935–944. [CrossRef]
34. Vetrella, A.R.; Causa, F.; Renga, A.; Fasano, G.; Accardo, D.; Grassi, M. Multi-UAV Carrier Phase Differential GPS and Vision-based Sensing for High Accuracy Attitude Estimation. *J. Intell. Rob. Syst.* **2019**, *93*, 245–260. [CrossRef]
35. *The Kalibr Visual-Inertial Calibration Toolbox*. Available online: https://github.com/ethz-asl/kalibr/ (accessed on 30 October 2019).
36. Yang, Z.; Shen, S. Monocular visual–inertial state estimation with online initialization and camera–imu extrinsic calibration. *IEEE Trans. Autom. Sci. Eng.* **2016**, *14*, 39–51. [CrossRef]
37. Eichler, S. Performance Evaluation of the IEEE 802.11p WAVE Communication Standard. In Proceedings of the 2007 IEEE 66th Vehicular Technology Conference, Baltimore, MD, USA, 30 September–3 October 2007; pp. 2199–2203.
38. *Magnetic Declination Calculator*. Available online: http://www.geomag.nrcan.gc.ca/calc/mdcal-en.php (accessed on 30 October 2019).

© 2020 by the author. Licensee MDPI, Basel, Switzerland. This article is an open access article distributed under the terms and conditions of the Creative Commons Attribution (CC BY) license (http://creativecommons.org/licenses/by/4.0/).

Article

The Use of Unmanned Aerial Vehicles in Remote Sensing Systems †

Aleksander Olejnik [1], Łukasz Kiszkowiak [1,*], Robert Rogólski [1], Grzegorz Chmaj [2], Michał Radomski [1], Maciej Majcher [1] and Łukasz Omen [1]

1 Faculty of Mechatronics & Aerospace, Military University of Technology, 00-908 Warsaw, Poland; aleksander.olejnik@wat.edu.pl (A.O.); robert.rogolski@wat.edu.pl (R.R.); michal.radomski@wat.edu.pl (M.R.); maciej.majcher@wat.edu.pl (M.M.); lukasz.omen@wat.edu.pl (Ł.O.)

2 DRI Solutions Sp. z o.o, 00-580 Warsaw, Poland; office@drisolutions.eu

* Correspondence: lukasz.kiszkowiak@wat.edu.pl; Tel.: +48-261-837-374

† This paper is an expanded version of conference paper by Olejnik, A; Kiszkowiak, Ł; Rogólski, R; Chmaj, G; Radomski, M; Majcher, M; Omen, Ł; "Precise Remote Sensing Using Unmanned Helicopter". Published in Proceedings of 5th IEEE Workshop on Metrology for AeroSpace, Torino, Italy, 19–21 June 2019.

Received: 7 February 2020; Accepted: 1 April 2020; Published: 3 April 2020

Abstract: This paper describes the possibility of using a small autonomous helicopter to perform tasks using a remote sensing system. This article further shows the most effective way to properly set up autopilot and to process its validation during flight tests. The most important components of the remote sensing system are described and the possibilities of using this system to monitor gas transmission and distribution networks are presented.

Keywords: remote sensing; UAV; helicopter; image acquisition; precise flight

---

1. Introduction

Remote sensing refers to the process of detecting and monitoring the physical characteristics of an area by measuring reflected and emitted radiation at a certain distance from the targeted area. In general, remote sensing applies to all measurements that are conducted remotely. Remote sensing methods are divided into active and passive, whereby active remote sensing describes when a signal is sent from the sensor and is reflected back from the object to be received and analyzed, whereas passive remote sensing methods are based on the analysis of signals emitted by the observed object [1,2]. Satellites and manned aircrafts that are used to collect data remotely are effective but relatively expensive, and these technologies tend to have low spatial and temporal resolutions. They also tend not to be flexible enough to accommodate the needs of many organizations working with these applications [3,4]. To overcome these limitations, unmanned aerial vehicle (UAV) technologies have been developed for remote sensing applications. The main advantage of UAVs is that they can be used in high-risk situations without endangering human life. UAVs can also be utilized in inaccessible areas, at low altitude, and at flight profiles close to the objects where manned systems cannot be flown [5]. In particular, unmanned helicopters have been investigated for several decades [6,7], with the rapid development of this type of flying object now being observed [8,9]. This paper describes the integration process of adapting an autonomous helicopter to perform data gathering in a remote manner. Such tasks require precise flights, for example, through predefined waypoints, while maintaining a given flight speed and altitude. So-called stable flights are especially important when a platform is flying very close to the ground, i.e., 2–3 meters high. In many cases, some additional above-ground level (AGL) sensors are required to provide accurate distance to ground in autopilot [10,11].

The following sections of this paper describe the main components of a helicopter-based remote sensing vehicle that was developed by the Military University of Technology in Poland. Moreover, this paper presents the most effective way of autopilot configuration and the process of its validation during flight tests. Section 2 contains a description of the proposed system in terms of its fields of application. Section 3 presents the obtained results, and the last sections conclude the paper with a discussion. In comparison to the paper entitled *Precise Remote Sensing Using Unmanned Helicopter* that was presented during the 2019 IEEE 5th International Workshop on Metrology for AeroSpace (MetroAeroSpace) held in Torino in June 2019 [1], this paper contains the latest research data from flight tests and a more detailed description of the proposed uses of the developed system. One of the most valuable parts of this paper includes a detailed description of the work performed to achieve in-flight UAV stability at the level allowing for the detection, recognition, and identification of a person from 500 m.

2. Materials and Methods

2.1. General Description of UAV WABIK

The main element of developed remote sensing system is the unmanned aerial vehicle WABIK (English: LURE), shown in Figure 1, a 35 kg autonomous helicopter with a 12 kg payload capacity. WABIK is driven by a 9 HP two-stroke piston engine. The basic flight time with standard fuel tanks is 60 minutes, which can be doubled after installing additional fuel tanks. The maximum flight speed in autonomous mode is 35 m/s, the maximum vertical take-off speed is 8 m/s and the maximum flight altitude is 2000 m AMSL. WABIK was equipped with the MP2128Heli [12] autopilot enabling autonomous take-off, landing, and flight over predefined waypoints. Autopilot also allowed control of the system in semi-autonomous mode, where the movements of the RC radio stick were treated by the autopilot as simple commands, such as flying forward at a given speed, increasing the flight altitude, or landing, while autopilot dealt with flight stabilization.

Figure 1. Unmanned aerial vehicle (UAV) WABIK.

According to the classification [13] shown in Table 1, alongside the maximum take-off weight (MTOW), the range, the duration of the flight, and the maximum ceiling, WABIK is classified in the mini-UAV category. WABIK was designed so that 5 kg of payload must be mounted at the front of the frame to achieve mechanical balance. Remote sensing from onboard payload is one of WABIK's tasks. This payload could be an optoelectronic system used for observation and surveillance, a LIDAR system used to construct digital 3D representations of ground objects, a methane detection system used to detect leaking pipes or a spray system used for the agriculture industry [14,15].

Table 1. International classification of unmanned aerial vehicles (UAVs) [13].

Category	MTOW [kg]	Range [km]	Maximum Ceiling [m]
Micro	<5	<10	250
Mini	<25/30/150	<10	150/250/300
Short Range	25 ÷ 150	10 ÷ 30	3000
Medium Range	50 ÷ 250	30 ÷ 70	3000
Long Range	>250	>70	>3000

2.2. Flight Control System

The WABIK was equipped with the MP2128Heli [12] flight control system. The sensory equipment of the autopilot included a GPS receiver, an ultrasonic height sensor, a barometric sensor, a digital magnetic compass, and diagnostic sensors to monitor the temperature of the piston engine and the fuel levels in the tanks. The autopilot architecture and autopilot installation are shown in Figures 2 and 3, respectively.

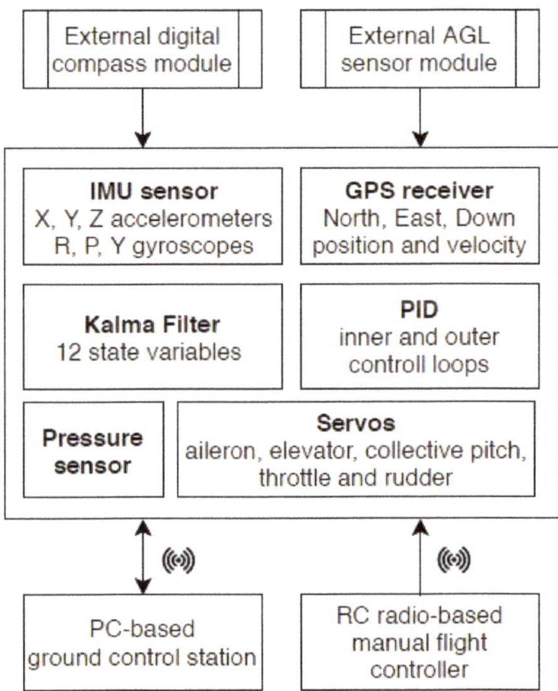

Figure 2. Flight control system architecture.

The MP2128Heli autopilot is a combination of a number of sensors, namely, the Global Navigation Satellite System (GNSS) software and a control system controlling aircraft actuators. The main task of this system is to estimate the following characteristics of the aerial vehicle [16]:

The position: Latitude, longitude, meters above sea level;

The speed: North speed, east speed, vertical speed;

The orientation: Roll, pitch, yaw.

In order to estimate these parameters, the autopilot was integrated with the below-mentioned sensors:

The Inertial Measurement Unit (IMU), a combination of gyro sensors and accelerometers measuring angular velocity and linear acceleration along the X, Y, Z axes and in the X, Y, Z directions, respectively;

The Attitude and Heading Reference System (AHRS), a combination of the IMU system with a magnetic earth field sensor providing information regarding the direction of flight;

The Global Navigation Satellite System (GNSS), which provides location, speed, and altitude information.

The use of GNSS in the autopilot system was carried out in a loosely connected configuration (loosely coupled INS/GPS) [17]. The WABIK flight control was made available to the operator through the Horizon software, as shown in Figure 4.

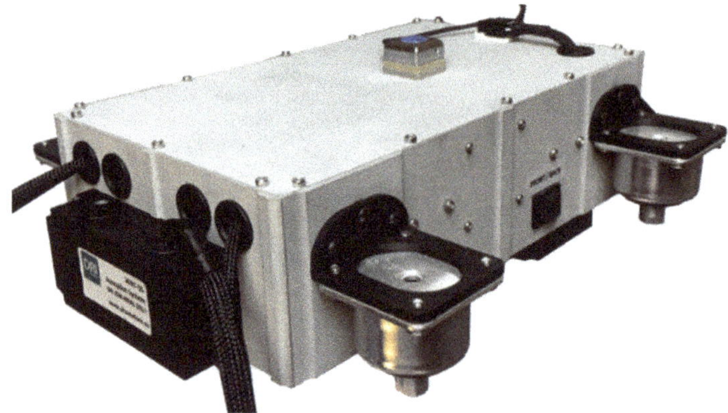

Figure 3. Autopilot installation.

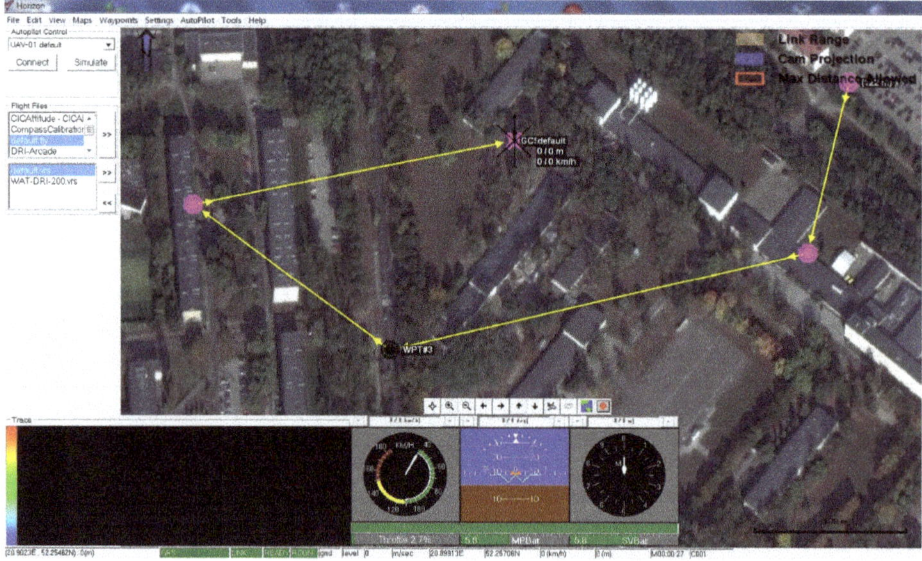

Figure 4. Horizon software interface used in flight control.

The autopilot contained six implemented PID controllers that were responsible for controlling various flight functions depending on the selected flight mode, including:

Controller 1—aileron from roll: Controlled aileron deflection to ensure that the difference between the specified pitch and the pitch estimated by the KF was minimal;

Controller 2—elevator from pitch: Controlled the rudder deflection to ensure that the difference between the specified slope and the slope estimated by the KF was minimal;

Controller 3—hover rudder from heading: Controlled the overall pitch of the tail rotor disc to ensure that the difference between the given direction of flight and the direction estimated by the KF was minimal;

Controller 4—throttle from altitude: Controlled the throttle deflection and the overall stroke of the control disc to ensure that the difference between the set flight altitude and the ceiling estimated by the KF was minimal;

Controller 5—hover pitch from X velocity: Controlled the rudder deflection to ensure that the difference between the set speed in the X direction and the speed estimated by the KF was minimal;

Controller 6—hover roll from Y velocity: Controlled the aileron deflection to ensure that the difference between the set speed in the Y direction and the speed estimated by the KF was minimal.

For example, when a helicopter hovered in a configured control type to maintain a set position, the autopilot set the desired speed in the X and Y directions to reach the set hover point (fixed GPS position set). Based on the difference between the set speed and the actual speed (estimated by the KF) in the X and Y directions, the required pitch and roll angles were used according to the input of the above-mentioned Controllers 1 and 2, respectively. All available control modes implemented in the autopilot and their connections with controllers used in given control modes are presented in Table 2 [16].

Table 2. Available control modes implemented in autopilot [16].

No.	Control Modes	Controller Number
1	CIC Attitude	1, 2, 3
2	CIC Position	1, 2, 3, 5, 6, optional 4
3	CIC Velocity	1, 2, 3, 5, 6, optional 4
4	CIC Altitude	1, 2, 3, 4
5	Full CIC	1, 2, 3, 4, 5, 6

For example, in the CIC attitude control mode, changing the direction stick of the RC radio (corresponding to the change in the angle of attack of the tail blades) gave a new desired direction, and the flight control system was responsible for obtaining and maintaining the desired direction. The above control modes could also be combined with each other. For example, in the CIC altitude position control type, the right stick of the RC transmitter shifted the waypoint (directions X and Y) to a new desired position, with the flight control system ensuring that the WABIK system moved to that point. The left stick of the RC transmitter set a new flight altitude and new direction, while the flight control system was responsible for maintaining a new altitude, direction, and position. The CIC altitude velocity type was similar to the CIC altitude position type, except that the right stick on the RC radio changed the desired flight speed in the same way as the CIC velocity type.

The most advanced and most frequently used flight mode in WABIK is a fully autonomous flight with navigation points. The predefined flight route consisting of a number of waypoints, can be modified during the flight via the ground flight control station. The start of the autonomous flight is carried out by a button integrated into the ground base software in the HORIZON program [16].

2.3. Optoelectronic Equipment

WABIK is capable of carrying an 8″ optoelectronic head equipped with a thermal imaging camera and a daylight camera, as shown in Figure 5. It allows remote detection, recognition, and identification and is equipped with a CCD day camera and a longwave infrared thermal camera with radiometry. Target tracking, geo-pointing, an 18x optical continuous zoom, radiometry and isotherm functionality, robust stabilization, and an integrated INS/GPS allow this system to be used in a wide range of military and industrial applications [18,19].

Figure 5. Optoelectronic head developed especially for use with WABIK.

Examples of WABIK's remote detection capability (500 m away from the target) using a daylight camera with a 15x optical zoom are presented in Figure 6. Full functionality of the developed optoelectronic head was made available to the operator through software installed at the ground flight control station, the interface of which is shown in Figure 7.

This software allows the following features:

Display of video transmitted in real time from an unmanned helicopter;

Control of the optical zoom;

Application of defined color palettes to the image;

Application of subtitles to the image (OSD);

Definition of isotherms or spot temperature measurements of the observed objects.

The video stream from the vision sensors (VIS and MWIR) was coded, compressed, and encrypted using a cryptographic module mounted on an unmanned helicopter. The video stream prepared in this way was sent via the TCP/IP protocol to the transmission module mounted on the helicopter, where it was streamed using Real Time Protocol (RTP). The ground base station was equipped with a receiving module, in which the process of decrypting and decoding the video signal was carried out.

The WABIK system was further equipped with three communication links. The first communication link operated in the UHF band and implemented a helicopter connection with the flight station ground base. The second communication link operated in the S band and was used for operating the on-board observation head. The third communication link was used for data transmission sent from a conventional RC radio to the autopilot.

Figure 6. Real-time video stream from a daylight camera with a 15x optical zoom.

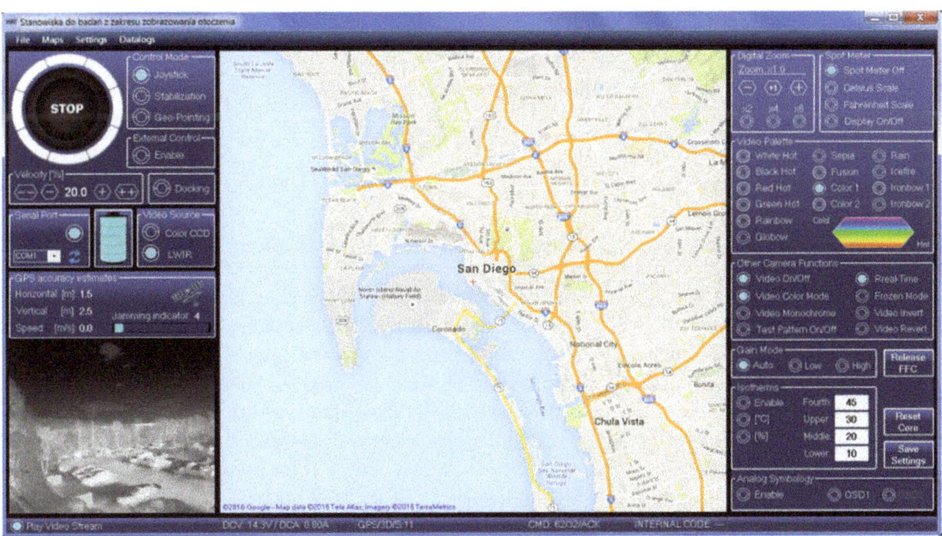

Figure 7. Optoelectronic head control software interface.

3. Results

For UAVs based on helicopter airframes, it is critical to achieve stable hovering according to a given altitude, a stable tail during flight, and stable forward flight before installing any sensors onboard the vehicle. For instance, automated power line inspection tasks require the use of close-range sensors, e.g., thermal cameras, to detect broken insulators. Once an insulator is chosen for further inspection, the UAV hovers while maintaining the given position and altitude. Otherwise, the onboard payload must be additionally stabilized, for instance, by using gyroscopes, an active vibration isolation system, or drives. Maintaining a given speed of a UAV helicopter is much more complicated in comparison with the fix-wing method because helicopter autopilots are unable simply reduce the throttle to decrease speed. Further, autopilots must estimate and control three times more of a dynamic state compared to fix-wing. These states are controlled by feedback loops which are not in use simultaneously. Autopilot enables feedback loops as required to control different aspects of flight.

Every single loop was adjusted during real terrain flights. The first rudder loop was adjusted as it controlled the most important part of helicopter, the tail boom. Due to the sensitivity of the tail rotor to wind, an additional external gyroscope was used to improve the tail rotor servo control task. This external yaw gyroscope was responsible for tail boom stabilization within ±0.5° (internal heading loop), while a second gyroscope acted as feedback sensor in an external heading loop responsible for controlling the changes of the helicopter course (heading). The rudder feedback loop was the most important and most challenging loop in terms of tuning because, if the rudder lost stability and started to oscillate or even increase the amplitude of the tail oscillation, it would have been difficult to recover the helicopter from such a state and may have caused a crash. An example heading loop performance during the fully autonomous flight through the predefined waypoints with a wind speed up to 10 m/s is presented in Figure 8.

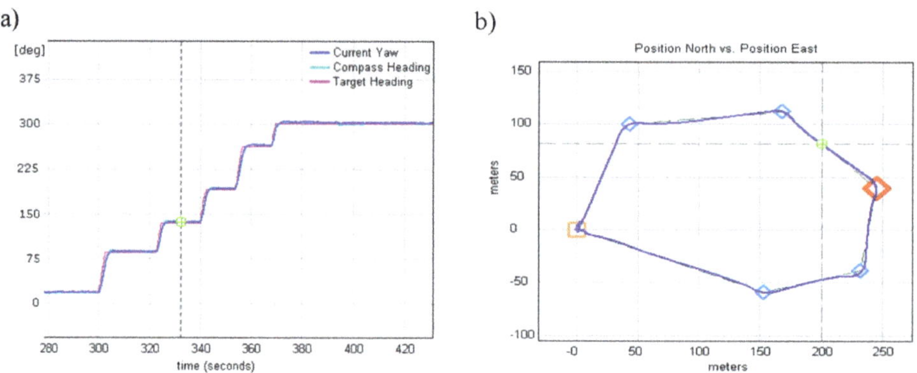

Figure 8. (**a**) Current and desired heading. (**b**) Predefined waypoints and achieved flight path.

Once tail boom stability was achieved and no rudder oscillations were seen in data logs, the pitch and roll control loops were tuned. Pitch or elevator feedback loop controlled the longitudinal cyclic pitch to minimize the difference between the actual and the desired pitch, while the roll or aileron feedback loop controlled the lateral cyclic pitch to minimize the difference between the actual and the desired roll. These two loops were also called the inner loops as they controlled helicopter attitude in any aspect of flight (take off, hover, flight, or landing), while the outer loops utilized them to achieve the given forward speed, for instance. To start the tuning of the inner loops, the inner loops were isolated from the outer loops to simplify the adjusting procedure. This was done by enabling the arcade modes in the MP2128Heli autopilot. The arcade modes [12] are a set of hybrid control modes which allow a pilot to control a helicopter's higher-level behavior without having to actually fly the helicopter. The MP2128Heli stabilizes the helicopter and manages the helicopter's flight controls based on the higher-level inputs provided by the pilot. There are three main CIC arcade modes. To find the best P, I, and D terms for the abovementioned pitch and roll loops, maximum desired (±20°) input signals were applied from the RC transmitter during flight during CIC attitude arcade mode. The helicopter reactions (current roll and pitch) for the given inputs, along with the P, I, and D, terms were presented to the ground control station and analyzed in real time during the flight. This way, base values of the PID terms without causing attitude oscillations were quickly found only in few flights. Examples of the obtained results are illustrated in Figure 9.

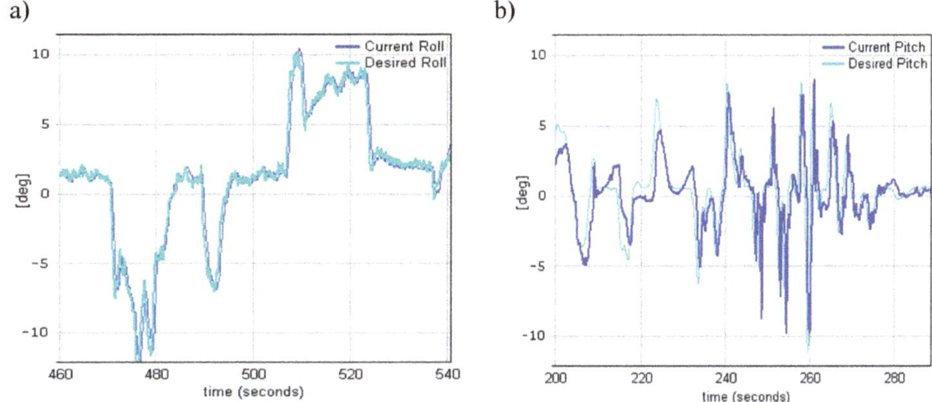

Figure 9. (**a**) Current and desired roll angle. (**b**) Current and desired pitch angle.

As can be seen, better performance was achieved for the loop which controlled the helicopter roll angle (rotation about the X axis). This was due to the nature of the helicopter mass distribution, which was much more concentrated around the center of gravity on the Y axis than the X axis, which in turn required the application of more forces to make a helicopter pitch rotation.

At this stage of tests, only one internal loop was left to tune, i.e., the altitude loop responsible for controlling the throttle in response to the Z velocity. This loop was adjusted in the same manner as the previous loops. The desired altitude was set either by the RC transmitter stick or the ground control station. The response of the helicopter altitude was observed and analyzed at the ground control station and the PID terms were adjusted in the real time. The final results of the achieved altitude loop performance are shown in Figure 10a. The error between the current and desired altitudes did not exceed 0.2 m RMS.

During the next test flights, two external control loops responsible for maintaining helicopter velocity were tuned. These loops were "external" because they utilized two already tuned roll and pitch control loops. The first outer loop was the hover pitch from the X velocity and controlled the longitudinal cyclic to minimize the difference between the actual and the desired hover X velocity. The second loop was the hover roll from the Y velocity and controlled the lateral cyclic to minimize the difference between the actual and the desired hover Y velocity. For testing purposes, a flight trajectory, as shown in Figure 10b, was defined. The trajectory consisted of a few tight turns and a few smooth turns, which are more complicated maneuvers from a helicopter dynamic point of view. The constant flight speed was defined as 10 m/s.

As seen in Figure 11a, the first part of the chart corresponding to flying from takeoff waypoints (point 0,0 in Figure 10b) to the first waypoint (point 0,100 in Figure 10b) was marked. During this part of the trajectory, two loops (X and Y body velocity control loops) were responsible for accelerating the helicopter from 0 to the given 10 m/s. Once 6.6 m/s was reached, small speed disturbances were visible as the helicopter entered the first turn. Similar speed disturbances were seen while the helicopter approached the second waypoint (Figure 11b). At this part of trajectory, the desired flight speed was accomplished. The worst speed disturbances were observed when the helicopter approached a tight turn (point 350, −25 in Figure 10b) flying at 10 m/s, followed quickly by the next tight turn. This case is illustrated in Figure 11a.

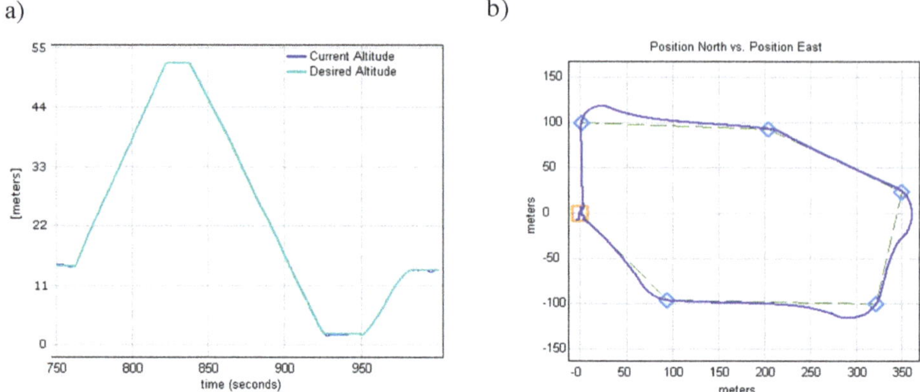

Figure 10. (**a**) Altitude loop performance. (**b**) Flight trajectory through waypoints.

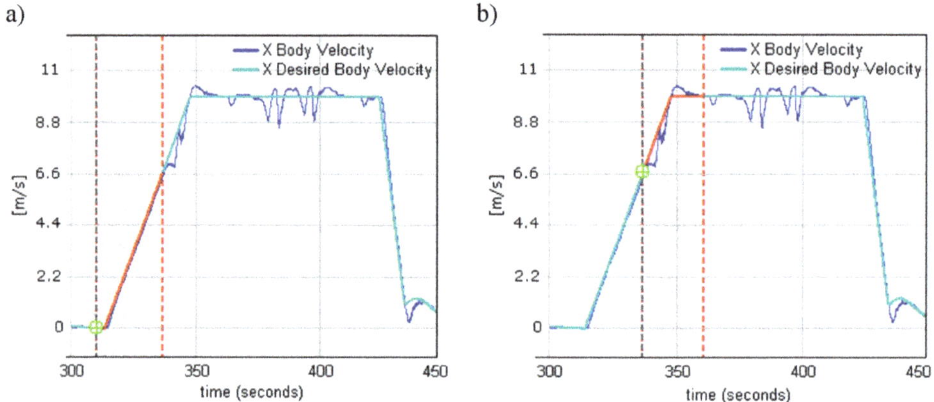

Figure 11. (**a**) X body velocity while approaching the 1st waypoint. (**b**) X body velocity while approaching the 2nd waypoint.

When fast helicopter deceleration was required, e.g., as determined by an anti-collision system, it was not possible to find PID terms that satisfied both the smooth velocity changes in forward flight and fast deceleration. This was due to a rotor head mechanical construction and the way it acted when the helicopter was flying forward and backward. Many tests were performed to find a compromise between these two requirements; however, when the helicopter was able to decelerate from, for instance, 10 m/s in 5 seconds, it was not able to perform smooth forward flight without pitching up and down (rotor head swashplate oscillation).

After tuning all of the internal and external loops, the fully autonomous test flight was accomplished, which aimed to ascend according to a given climb speed of 6 m/s to a desired altitude of 20 m and then hover until a new command was received. As the position did not change, this maneuver theoretically should have been done in the same relative X and Y positions. Figure 12b displays the north vs. east relative positions. The relative positions during the test did not go beyond a 2 m × 1 m rectangle (marked in red). However, the GPS receiver CEP error, which was 2.5 m, should be kept in mind. The GPS position drifted when the helicopter was still on the ground, which is marked in orange in Figure 12. After the inner and outer feedback loops were tuned, WABIK was ready to receive installation of the optoelectronic head and perform one of its remote sensing tasks, namely, human detection and recognition.

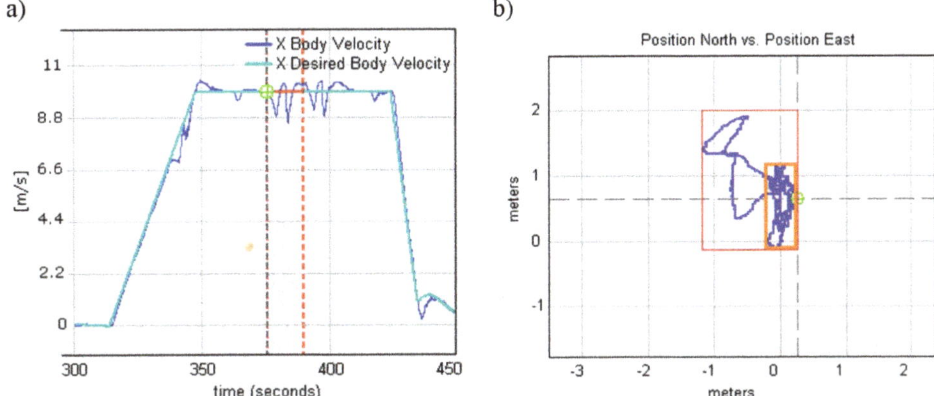

Figure 12. (**a**) X body velocity while approaching the 4th waypoint. (**b**) Relative position during autonomous takeoff.

4. Fields of Use

The developed monitoring system using vertical takeoff and landing unmanned aerial vehicles could be used to [15,19,20]:

Monitor gas transmission and distribution networks;

Patrol forests and act to recognize fires;

Monitor forest and agricultural degradation;

Function in air reconnaissance during floods and flooding;

Patrol border areas;

Surveil communication routes surveillance (railway traction and roads), particularly key transportation hubs;

Surveil special strategic importance land infrastructure (industrial complexes, power plants, airports, and seaports);

Monitor stadiums or outdoor events.

In the abovementioned situations, the use of unmanned aerial vehicles as discrete and non-invasive means of obtaining information seems to be expedient and the most reasonable solution. In regard to gas transmission network monitoring, the solution currently in development is aimed at assessing the technical condition of the infrastructure, monitoring the area in the immediate vicinity of the gas pipeline, and detecting possible damage. WABIK could be equipped with methane detectors (as shown in Figure 13) and, thanks to its small size and autonomous control system, it would be able to approach a gas pipeline at a distance that allows effective and reliable measurement.

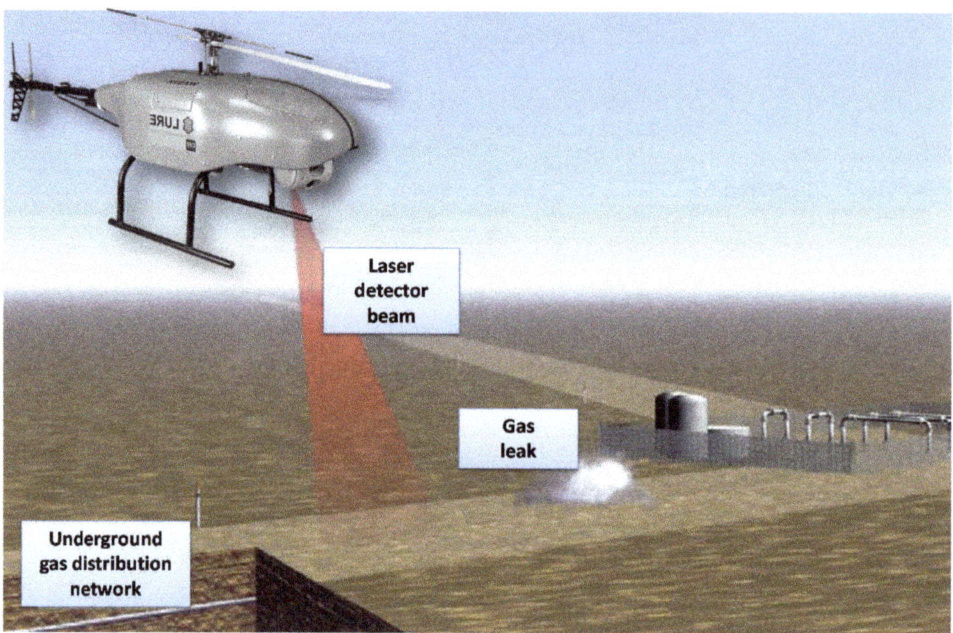

Figure 13. Remote sensing system in use to detect gas leaks from pipelines.

Thanks to its relatively big payload, WABIK is used in continuously developing remote sensing systems to obtain high-quality multispectral imaging. This special type of imaging is increasingly used in modern agriculture to obtain information regarding plant health and growth. Farm management decisions include information collected from remote sensing systems to increase crop productivity. Figure 14 shows an image of a Normalized Difference Vegetation Index (NDVI) distribution obtained using WABIK equipped with multispectral camera system. NDVI is a vegetation index used to quantify green vegetation. It normalizes green leaf scattering in the near infra-red wavelength and chlorophyll absorption in the red wavelength. NDVI is a measurement of the reflectivity of plants expressed as: the ratio of near-infrared reflectivity (NIR) minus red reflectivity (VIS) over NIR plus VIS:

$$NDVI = \frac{NIR - VIS}{NIR + VIS} \tag{1}$$

Figure 14 shows red circles, indicating corn field areas in worse condition, and white circles, indicating crops in very poor condition. Objects that are not plants are indicated blue circles. Therefore, NDVI is a simple way to measure general plant health.

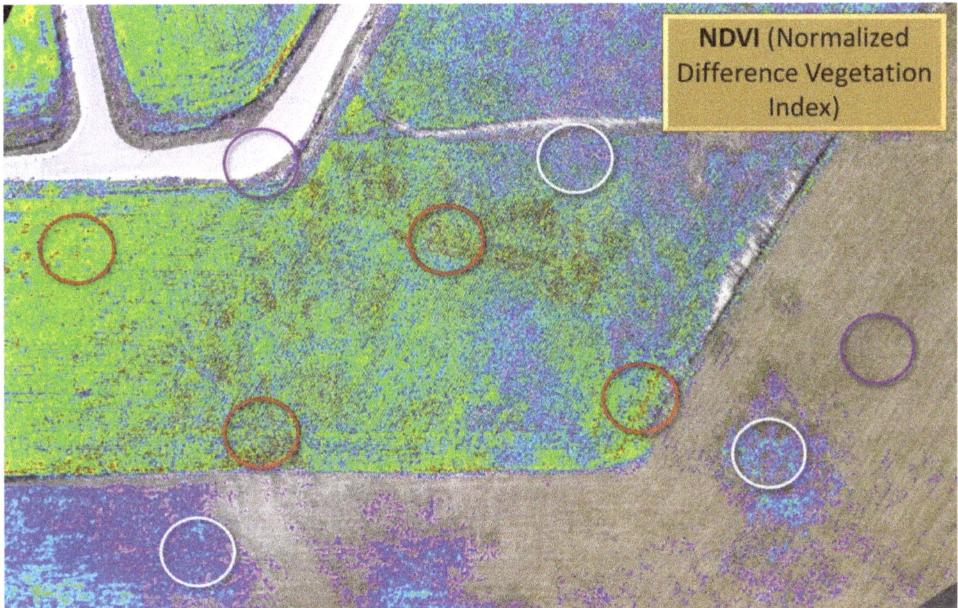

Figure 14. Remote sensing system in use to obtain high-quality multispectral imaging.

Measurements using the multispectral camera were carried out using recordings of the accurate aircraft position.

5. Discussion

The progress of miniaturization of electronics and the continuous development of robotics has allowed for the development of unmanned aerial vehicles, which can successfully perform tasks that, until now, were reserved for manned platforms. The use of UAV's in remote sensing systems allows us to significantly eliminate human errors. Until now, many people were involved in data acquisition, e.g., as pilots, observers, aircraft maintenance unit members, or data processing specialists. The use of unmanned aerial vehicles will automate this process, making it much cheaper and more reliable. Economically this is also justified, because the operation of manned helicopters is several times more expensive compared to the operation and maintenance of small unmanned helicopters.

The effective operation of the monitoring system directly depends on the functionality of its individual components. Therefore, during the development process of vertical takeoff and landing platform, particular emphasis was placed on ensuring its high functionality. Achieved in-flight WABIK stability approved by data analysis, as presented in Figures 9–11, allowed target observation by an onboard optoelectronic head, resulting in detection, recognition, and identification. WABIK can also integrate quickly and easily with many others sensors, like thermal imaging cameras, hyperspectral cameras, or a specialized system designed to monitor natural gas transmission and distribution networks. High-quality multispectral imaging can support agriculture (Figure 14) and can also be used during natural disasters for localization of water bodies. Current solutions to monitor gas transmission and distribution networks require the use of manned helicopters with specially trained pilots, who identify gas leaks based on changes in the environment of the pipeline. The use of an automated system based on developed UAVs could allow increased efficiency of tasks and reduced costs of entire operations.

The main advantages of helicopter-based remote sensing vehicles over electric multicopter vehicles include the much bigger payload capacity and flight time. In the category of helicopter-type mini-UAVs,

WABIK has two main competitors, namely, Aeroscout Scout B1-100 [21] from Switzerland, which has an 18 kg payload capacity, and YAMAHA Rmax [22] from the USA, which has an 16 kg payload capacity. Flight endurance, payload options, operational range, or payload capacity are all parameters that are quite similar among the models, but WABIK is distinct from the others in terms of price, as it is much cheaper compared to the other models.

Work is currently underway to adapt WABIK to the agriculture industry. A spraying system equipped with two electric atomizers capable of atomizing 50 mm particles is under design. Two additional spray tanks could be mounted next to fuel tanks, and instead of the optoelectronic head, a diaphragm sprayer pump with a driver integrated with ground control station software could be installed at the front. This way, the spraying system could be controlled from the same ground control station by the same operator.

Author Contributions: Conceptualization, A.O.; data curation, G.C.; funding acquisition, A.O.; investigation, Ł.K., R.R., G.C., M.R., M.M., and Ł.O.; methodology, R.R.; project administration, A.O.; resources, G.C., M.R., M.M., and Ł.O.; software, G.C.; supervision, A.O.; validation, G.C.; visualization, Ł.K.; writing—original draft, Ł.K. and G.C.; writing—review and editing, Ł.K. All authors have read and agreed to the published version of the manuscript.

Funding: Results that are presented in this paper were obtained during research works under the following projects: Modernizacja i budowa nowej infrastruktury naukowo badawczej Wojskowej Akademii Technicznej i Politechniki Warszawskiej na potrzeby wspólnych numeryczno doświadczalnych badań lotniczych silników turbinowych - POIG.02.02.00-14-022/09 – EU and NCBR funds co-financing; Wsparcie Klastra Konstrukcji i Technologii Lotniczych General Aviation – RPMA.01.06.00-14-012/12-00 – EU and national funds co-financing; Doświadczalno-numeryczne badania rozwojowe technologii badawczych z weryfikacją warunkującą osiąganie zakładanych celów techniczno-taktycznych powietrznych platform bezzałogowych – PBS WAT nr 940 – national funds co-financing.

Conflicts of Interest: The authors declare no conflict of interest. The funders had no role in the design of the study, in the collection, analyses, or interpretation of data, in the writing of the manuscript, or in the decision to publish the results.

References

1. Olejnik, A.; Kiszkowiak, Ł.; Rogólski, R.; Chmaj, G.; Radomski, M.; Majcher, M.; Omen, L. Precise Remote Sensing Using Unmanned Helicopter. In Proceedings of the 2019 IEEE 5th International Workshop on Metrology for AeroSpace (MetroAeroSpace), Torino, Italy, 19–21 June 2019; pp. 544–548.
2. Schowengerdt, R. *Remote Sensing: Models and Methods for Image Processing*, 3rd ed.; Academic Press: Cambridge, MA, USA, 2007; p. 2, ISBN 978-0-12-369407-2.
3. Goetz, S.J. Remote Sensing of Riparian Buffers: Past Progress and Future Prospects. *JAWRA J. Am. Water Resour. Assoc.* **2006**, *42*, 134–143. [CrossRef]
4. Sriharan, S.; Everitt, J.H.; Yang, C.; Fletcher, R.S. Mapping riparian and wetland weeds with high resolution satellite imagery; In Proceedings of the IEEE International Geoscience and Remote Sensing Symposium IGARSS 2008. Boston, MA, USA, 7–11 July 2008; Volume 1, pp. 476–479.
5. Ma, L.; Li, M.; Tong, L.; Wang, Y.; Cheng, L. Using unmanned aerial vehicle for remote sensing application. In Proceedings of the 2013 21st International Conference on Geoinformatics, Kaifeng, China, 20–22 June 2013; pp. 1–5.
6. Austin, R.G. Unmanned Multimode Helicopter. Google Patents US4163535A, 1979.
7. Vanderlip, E.G. Omni-Directional Vertical-Lift Helicopter Drone. Google Patents Patent number US3053480A, 1962.
8. Abhiram, D.; Ganguli, R.; Harursampath, D.; Friedmann, P.P. Robust Design of Small Unmanned Helicopter for Hover Performance Using Taguchi Method. *J. Aircr.* **2018**, *55*, 1–8. [CrossRef]
9. Godbolt, B.; Lynch, A.F. An unmanned helicopter control with partial small body force compensation: Experimental results. *Robotica* **2018**, *36*, 1436–1453. [CrossRef]
10. Micropilot. Available online: https://www.micropilot.com (accessed on 18 May 2019).
11. Brzozowski, B.; Daponte, P.; De Vito, L.; Lamonaca, F.; Picariello, F.; Pompetti, M.; Wojtowicz, K. A remote-controlled platform for UAS testing. *IEEE Aerosp. Electron. Syst. Mag.* **2018**, *33*, 48–56. [CrossRef]

12. Brzozowski, B.; Rochala, Z.; Wojtowicz, K. Overview of the research on state-of-the-art measurement sensors for UAV navigation. In Proceedings of the 2017 IEEE International Workshop on Metrology for AeroSpace (MetroAeroSpace), Padua, Italy, 21–23 June 2017; pp. 565–570.
13. Eisenbeiss, H. A mini Unmanned Aerial Vehicle (UAV): System overview and image acquisition. In Proceedings of the International Archives of Photogrammetry. Remote Sensing and Spatial Information Sciences, 36(5/W1), Pitsanulok, Thailand, 18–20 November 2004.
14. Olejnik, A. *Autonomiczne Bezzałogowe Statki Powietrzne Wyposażone w Środki Monitorowania i Nadzorowania Wspomagające Działania Policji i Straży Pożarnej*; WAT Internal Report; Military University of Technology: Warsaw, Poland, 2013.
15. Olejnik, A.; Rogólski, R.; Chmaj, G.; Kiszkowiak, Ł.; Kramarski, I. Public Security Forces Net-Centric Support System Based on Unmanned Aerial Vehicles. In *30th International Conference on Security Engineering - Protection Against Effects of Extraordinary Threats*; EKOMILITARIS 2016; Bel Studio Sp. z o.o.: Warsaw, Poland, 2016; ISBN 978-83-7798-195-5.
16. Aeroscout—Unmanned Aircraft Technology. Available online: https://www.aeroscout.ch (accessed on 21 May 2019).
17. Precision Agriculture. Available online: https://www.yamahamotorsports.com/motorsports/pages/precision-agriculture-rmax (accessed on 21 May 2019).
18. Chmaj, G. Inspection Flying Robot. Ph.D. Thesis, University of Science and Technology in Cracow, Kraków, Poland, 2010.
19. Kwang, H.K.; Jang, G.L. Adaptive Two-Stage EKF for INS-GPS Loosely Coupled System with Unknown Fault Bias. *J. Glob. Position. Syst.* **2006**, *5*, 62–69.
20. Olejnik, A.; Kołodziński, E.; Rogólski, R.; Kiszkowiak, Ł.; Chmaj, G.; Kramarski, I. *Monitoring and Surveillance Net-Centric System of the Selected Province Area*; Współczesne konteksty bezpieczeństwa, Wydawnictwo Wyższej Szkoły Gospodarki Euroregionalnej im; Alcide De Gasperi w Józefowie: Józefów, Poland, 2016; pp. 153–174. ISBN 978-83-62753-69-7.
21. Olejnik, A.; Danilecki, S.; Zalewski, P.; Łącki, T. Opracowanie projektu i budowa demonstratora technologii ultralekkiego samolotu jako elementu sieciocentrycznego systemu wsparcia rozpoznania i dowodzenia. *Mechanik* **2011**, *84*, 633–640.
22. Olejnik, A.; Danilecki, S.; Zalewski, P.; Łącki, T. *Opracowanie Systemu Sieciocentrycznego do Wykrywania i Rozpoznawania Zagrożeń Bezpieczeństwa w Dużych Skupiskach Ludzkich oraz Powodowanych Wybranymi Klęskami Żywiołowymi*; WAT Internal Report; Military University of Technology: Warsaw, Poland, 2013.

© 2020 by the authors. Licensee MDPI, Basel, Switzerland. This article is an open access article distributed under the terms and conditions of the Creative Commons Attribution (CC BY) license (http://creativecommons.org/licenses/by/4.0/).

Article

Integration and Investigation of Selected On-Board Devices for Development of the Newly Designed Miniature UAV †

Aleksander Olejnik, Robert Rogólski *, Łukasz Kiszkowiak and Michał Szcześniak

Faculty of Mechatronics and Aerospace, Military University of Technology, Warsaw 00-908, Poland; aleksander.olejnik@wat.edu.pl (A.O.); lukasz.kiszkowiak@wat.edu.pl (Ł.K.); michal.szczesniak@wat.edu.pl (M.S.)

* Correspondence: robert.rogolski@wat.edu.pl; Tel.: +48-261837374

† This is paper is an extended version of the paper: "Olejnik, A.; Rogólski, R.; Kiszkowiak, Ł.; Szcześniak, M.: Specific Problems of Selecting and Integrating Equipment Components in the Course of Developing a Technology Demonstrator for the mini-UAV" published in Proceedings of the 2019 IEEE International Workshop on Metrology for Aerospace, Torino, Italy–June 19–21, 2019; pp. 284–289; specified in the References as item 3.

Received: 31 December 2019; Accepted: 16 March 2020; Published: 21 March 2020

Abstract: The article is a development of the topic generally presented in the conference proceedings issued after the 2019 IEEE International Workshop on Metrology for Aerospace. In contrast to topics presented in the conference, the article describes in detail avionic equipment and on-board systems integration process and their in-flight adjustment in regard to the newly designed miniature unmanned aerial vehicle (mini-UAV). The mini-airplane was constructed and assembled in the course of the research project, the purpose of which was to show implementation of a totally new mini-UAV design. The intention of the work was to develop a new unmanned system including an originally constructed small airplane with elements purchased from open market. Such approach should allow to construct a new aerial unmanned system, which technologically would not be very advanced but should be easy to use and relatively inexpensive. The demonstrator mini-airplane has equipment typical for such an object, i.e., electric propulsion, autopilot system, camera head and parachute device for recovery. The key efforts in the project were taken to elaborate an original but easy to use system, to integrate subsystem elements and test them so to prove their functionality and reliability.

Keywords: miniature UAV; autopilot; on-board sensors; miniature propulsion

---

1. Introduction

Miniature unmanned aerial vehicles are very small reconnaissance flying objects of which the total weight according to widely known categorization documents is approximately between 5 and 30 kg, wherein stated definitions for this type of UAV are rather convergent both from the side of responsible civil and military organizations [1,2]. Fragment of the official classification presenting the position of miniature objects is shown in Table 1. Due to their small size (usually adequate for large flying models) and light payload, they are not a very expensive assortment of products offered by the aviation industry. Not very high technological requirements, general availability of construction, drive and equipment components, low price of both components and the final product–these all are reasons which make the use of mini-UAV more and more common and justified in various fields of activity. In military applications, a mini-plane equipped with a TV camera and silent electric drive is an excellent recognition instrument for the platoon/company level subdivision or an independent special group. There are many cameras dedicated to this type of mission. However, in addition to strictly military applications, mini-UAVs are widely used in civilian enterprises related to the activities of both state institutions and private enterprises.

Table 1. Extract of UAV categories defined by UVS-International, according to [1] or [2].

No	Category Name	Mass [kg]	Range [km]	Flight Altitude [m]	Endurance [h]
1	Micro	<5	<10	250	1
2	Mini	<25/30/150	<10	150/250/300	<2
3	Close Range	25–150	10–30	3000	2–4
4	Medium Range	50–250	30–70	3000	3–6
5	High Alt. Long Endurance	>250	>70	>3000	>6

A significantly crucial problem during the development of a new unmanned aerial system (UAS) is the integration of all devices mounted in a miniature airplane. To assure full system complexity and, in consequence, required functionality, this integration must be fulfilled at the following three levels: mechanical assembly, electric connectivity and achieving readiness for operation of each subsystem. The mini-UAV described here and previously presented more generally [3] was developed as a demonstrating product of some R&D projects carried out to prove usability and effectiveness of a small flying platform designed and constructed for the purposes of video monitoring and inspecting in close range and at low altitude. The primary goal was to design, construct and test a few technology demonstrators of mini-UAV obligatory equipped with the following devices or subsystems:

- electric propulsion, with an electric engine, pulling propeller and controller;
- automatic flight control system, ensuring a stable flight on a given route in autonomous mode (commercially offered autopilot MP 2128g was applied in the project);
- camera gimballed head, enabling day and night terrain recognition (in TV or thermal mode);
- communication subsystem, ensuring transmission of telemetry and image data;
- recovery system, with automatically ejected parachute;
- power supply system with electric sources and wiring.

UAS system components are subjects of numerous researches and tests described in wide bibliography, available in electronic or paper forms; exemplary items [4–10] are worth citing. In the presented project challenges, one can distinguish certain trends of activities determining the success of the finally developed system. These lines of action can be divided as follows:

- aerodynamic-and-structural airframe design and construction;
- selection of components and optimization of the propulsion;
- material and manufacturing technologies;
- architecture of avionic on-board subsystems, including devices of flight control and stabilization, navigation, recognition, communication, self-recovery and others;
- ground control station architecture;
- mechatronic integration and system validation tests;
- research and qualification tests;
- developing optimal operating and maintenance procedures.

Each action is taken at a specific stage of system development, and each stage defines specific technical or technological tasks. However, only meeting a set of requirements at all development levels together can contribute to achieving the final success, which is the development of a fully usable and commercially available product.

The problems presented in the paper were chosen as the most complicated and demanding from the range of all development activities undertaken towards developing a new system. Specific elements of the abovementioned equipment were being sequentially mounted on mini-airplane and then tested separately—at first during on-ground investigations and then during flight-tests. The target pre-prototype aircraft was equipped with all possible subsystems. It was fully completed and intended for use in the mode of supervised trial operation. Specific aspects of the UAV construction, equipment adaptation and its testing were described here. Particularly important measurements taken to verify operational readiness of deployed devices were carried out in the context of the integration of the autopilot system and the selection of the power unit components.

2. UAV Design

Mini-UAV Rybitwa (*eng. Tern*) of which CAD designs are presented in Figure 1 is a miniature made-of-composite airplane with traditional aerodynamic outline (high-wing monoplane with T-tailed boom). The lifting plane is assembled with 4 semi-monocoque segments and is equipped with ailerons and flaps. The central sections are connected to the fuselage by putting them over rigid carbon girder spar. There is a parachute recovery set inside the fuselage bay. The propulsion system consists of three-phase electric motor and two- or three-blade tractor propeller. The on-board avionic equipment (MP2128g autopilot and gimballed video sensor) are assembled inside the under-fuselage pod. The airframe configuration in a classic arrangement with a suspended slender container has been successfully used in such constructions as: Skylark (Elbit Systems), Skyblade (ST Aerospace) or experimental Kiwit (ATE Technology). Fundamental dimensions and performances achieved in case of the final UAV configuration shown in Figure 2 are the following:

Wingspan	3450 mm	V_{max}	130 km/h
Total length	1980 mm	V_{min}	35 km/h
Maximum take-off weight	9 kg	Service ceiling	200–400 m
Motor	electric three-phase 2.5 KW	Endurance	60 min
Propeller	22″ x 10″ (two-blade)	Start/recovery	hand-throw/parachute landing

Figure 1. Developing stages of the mini-UAV CAD model—the initial conception and the finally developed configuration.

Figure 2. Unmanned aerial system (UAS) components: mini-UAV with the video head and recovery parachute, portable Ground Control Station—transported in a container, communication terminal.

The recognition system distinguishes an air component, i.e., an unmanned aerial vehicle and a ground component comprising a ground control station (GCS) with an antenna terminal. Connectivity between telemetry data, image recognition and remote-control signals is provided between the components. The main components of GCS are portable industrial computer (with a joystick), antenna terminal, transceiver modules, portable source of electric supply. Conceptual view of a system division is presented in Figure 3.

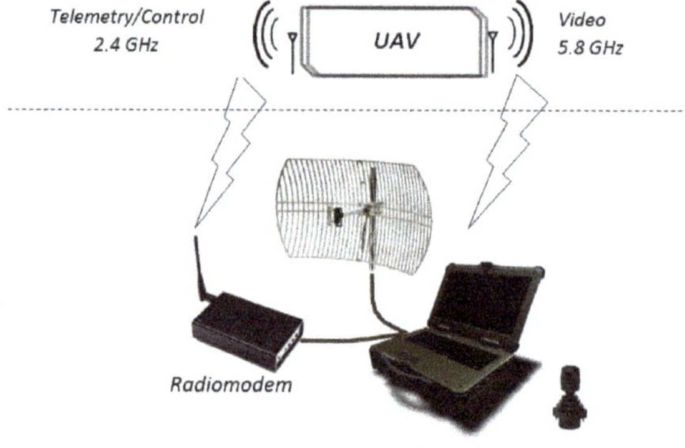

Figure 3. Conceptual view of the mini-UAS embracing the aerial component (mini-UAV) and parts of the on-ground component (GCS).

3. Autopilot Components and Control System Integration

The MP2128g autopilot—components of which are described in [11,12]—is dedicated to stabilizing and controlling UAVs of various types and sizes, i.e., to enable steady flights in fully autonomous mode. It is possible thanks to many on-board sensors included in the set supporting both start or landing and air cruise navigated in accordance up to 1000 waypoints. The device enables operating in different modes by maintaining such navigational parameters like altitude, airspeed, heading in reference to GPS/INS data and turn coordination. Using additionally added magnetometer, it is possible to estimate the wind correction vector especially needed at crosswind flight.

MP2128g was utilized in many UAV projects with success, which was described in papers like [13–15]. In those and many other papers, the challenge related to autopilot application is presented as a purely technological problem or as an original scientific task. The case of typically technological problem occurs, when the new UAV construction is ready and is to be equipped with a dedicated mission equipment, and at the same time, the priority commercial requirement is to deploy an available and reliable control and navigation system. In such a situation, the application of MP2128g is perhaps the best possible choice. The UAV equipped with that autopilot can be investigated in a virtual simulating environment [15], or directly in real flight tests [13]. On the other hand, the leading challenge could be the development of a quite new control system, based on alternative components and original self-elaborated software. For last over a dozen years many designs of own autopilot systems have been developed, which were being validated in the hardware-in-the-loop tests or in real flights [7,16,17]. However, even in those cases, using the commercial autopilot is rather recommended, as it enables to conduct comparison tests or measure flight parameters, which could be some reference data for the one identified by the newly applied set. The type of such dual research is presented in [14]. In this work, parameters recorded by the newly developed autopilot based on the PC/104 module were compared with the MP2000 autopilot parameters, while both devices were mounted simultaneously on the same flying platform.

Although MP2128g is the highly advanced solution, it is anyway not a plug-and-play device, which can be easily integrated with a UAV, especially if the vehicle is newly designed. The most common issues, which must be investigated and solved during integration process, include:

- Design of vibration isolation system;
- Components arrangement inside tight mini-airplane fuselage;
- Selection of radio frequency modems.

The main source of vibration is the engine and a propeller. Vibration generated by these two elements contributes to increasing errors of IMU, especially vibration increases gyroscope bias. An incorrectly selected vibration system may cause gyroscope saturation, which can lead to UAV crash. Since the airframe structure is relatively stiff and the imbalance of rotating masses in the electric drive is also small, autopilot vibro-isolation system was put together in the form of a passive system of four silicone cushions, on which the autopilot was mounted. The suspension, despite the small dimensions of the insulating supports used, turned out to be sufficiently rigid in relation to the deposited mass (50 g). During subsequent flight tests, no malfunctions were noted due to structure vibrations, and recorded flight accelerations did not exceed the value of 1 g. What is more important, gyroscope output was always stable. Figure 4 presents final isolator setup, on the left, and one of the initially tested ones, on the right. As can be seen on the left graph, gyroscope output was reduced almost twice comparing to the data on the right graph—these results were achieved only by changing isolators stiffness. Achieved results were also confirmed by calculating some statistic data, which shows that standard deviation for gyroscope data presented on the left is only 1.46 [deg/s] while standard deviation for gyroscope data presented on the right graph is up to 6.55 [deg/s].

Component arrangement inside tight mini-plane fuselage is not an easy task. Many components like GPS receiver, compass module, radio modem or autopilot include embedded microprocessors or memories which work at high frequency. In these conditions even wrongly led cables can cause interferences between these components. For example, GPS antenna cable placed on the top of GPS receiver module caused sudden increase of GPS jamming. Compass cable, through which SPI signal is sent, placed closed to GPS receiver also caused increase of GPS jamming. Radio modem placed too close to autopilot board caused all uncontrolled servos jerks. The same issue was observed when miniature diversity module was placed too close to PCB servo board. One of Rybitwa requirements was to send real-time video from onboard camera. This requirement caused additional radio modem installation onboard the UAV. Having three radio frequency modules onboard small UAV required installation of three antennas in a manner that did not cause mutual interference. Many tests with spectrum analyzer were done to find the best antenna location.

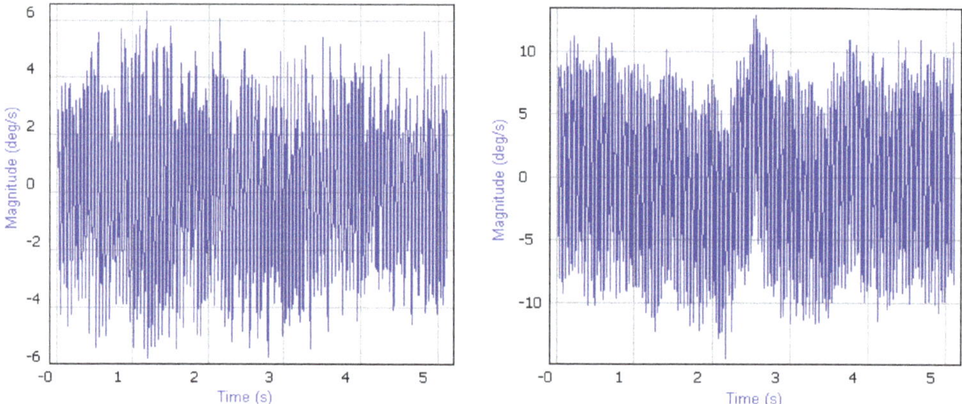

Figure 4. Gyroscope output during vibration tests.

Items of the autopilot set applied in the project are specified in the Figure 5. The primary ones of them are the following: main core (40 × 100 mm, 28 g), servo expansion board (40 × 39 mm, 18 g), ultrasonic AGL sensor (95 × 55 mm, 92 g), magnetometric sensor (43 × 13 mm, 29.5 g), GPS antenna (38 × 34 mm, 40 g), 2.4 GHz radio modem for telemetric transmission and miniature RC digital servomecanisms (Hitec HS-5125MG).

Figure 5. MP2128g autopilot main core and its primary peripheral elements.

Before adjusting autopilot to control UAV autonomously, it is necessary to check mini-airplane flight qualities in manually controlled operating, i.e., in RC mode. The flying platform should be capable to hold stable and controllable flight in any possible weather conditions. Only after verifying dynamic capabilities and performances and fixing accepted weights and balance shifts, the integration process can be initialized. In the state of free and stable flight, servos should be configured in such way to assure trim conditions with all control surfaces set in neutral positions (i.e., at zero deflection angles). Demonstration of a tendency towards static stability in the phase of manually controlled flights opens the prospect of testing success in the later autopilot testing phase. Control diagrams of RC mode and mixed (PIC-CIC) mode applied in subsequent flight-test stages are presented in Figures 6 and 7.

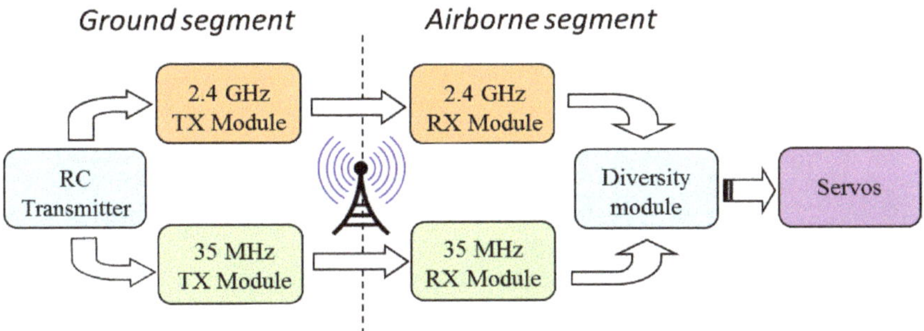

Figure 6. Block diagram of standard RC control system used in mini-airplane to test its flight qualities and to trim control surfaces—transmission redundancy was guaranteed by application two various transmission channels in frequencies: 33 MHz and 2,4 GHz; the diversity device was responsible for quick sampling, comparing and transceiving one of two signals with better quality.

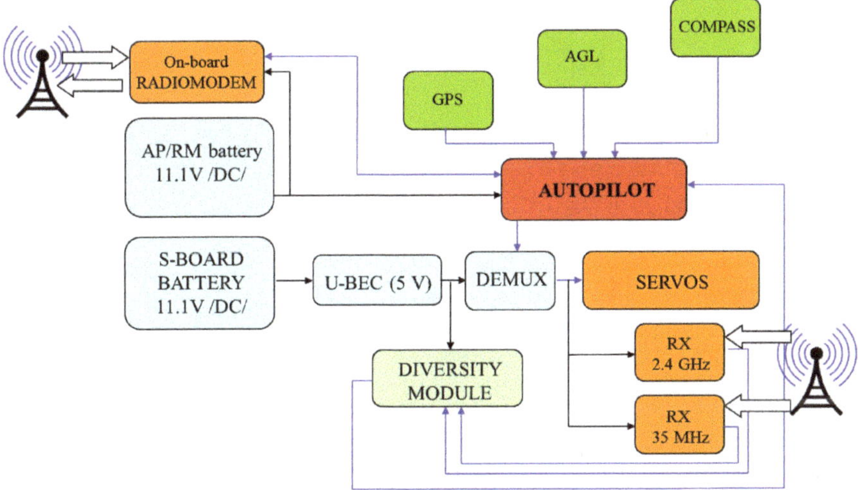

Figure 7. Block diagram of control system with MP2128g included–configuration allowing switch between manual mode (PIC–Pilot-in-Command) and automated mode (CIC–Computer-in-Command); in PIC mode remoted control is always supported by two independent transmissions.

After installing components and adjusting controls to hold conditions of aerodynamic equilibrium, there is a necessity to execute flights with autopilot working in flight recording mode. Having verified registered data and assuring that control system modified in accordance to a new architecture (Figure 7) acts properly, the new testing stage could be initialized. During the flight in semi-autonomous mode, the autopilot controls movement and position parameters in accordance with implemented control principles. Any possible state of flight (cruising, climbing, descending, turning or other maneuvers) is maintained by movements of control surfaces (indicated in Figure 8) as effect of combination of some feedback loops. In case of MP2128g, there are the nine basic control loops based on tuned PID controllers [11]:

- Aileron from roll–controls ailerons deflection so that the difference between the currently estimated roll value and the required value is as small as possible;
- Elevator from pitch—controls elevator deflection so that the difference between the current pitch value and the required pitch value is as small as possible;

- Rudder from $-y$ accelerometer—controls rudder deflection so that the difference between the current relative-to-y-axis acceleration value and the required–y acceleration is minimalized;
- Rudder from heading—controls rudder deflection so that the difference between the current heading angle value and the required heading value is as small as possible;
- Throttle from airspeed—controls throttle set so that the difference between the current airspeed value and the required airspeed value is as small as possible;
- Throttle from altitude—controls throttle set so that the difference between the current altitude value and the required altitude is as small as possible;
- Pitch from airspeed—controls elevator deflection (pitch as effect) so that the difference between the current airspeed value and the required airspeed is as small as possible;
- Roll from heading—controls roll (ailerons and rudder deflection as effect) so that the difference between the current heading angle value and the required heading is as small as possible;
- Pitch from descent—controls pitch (elevator deflection as effect) so that the difference between the current descent rate and the required descent rate is as small as possible.

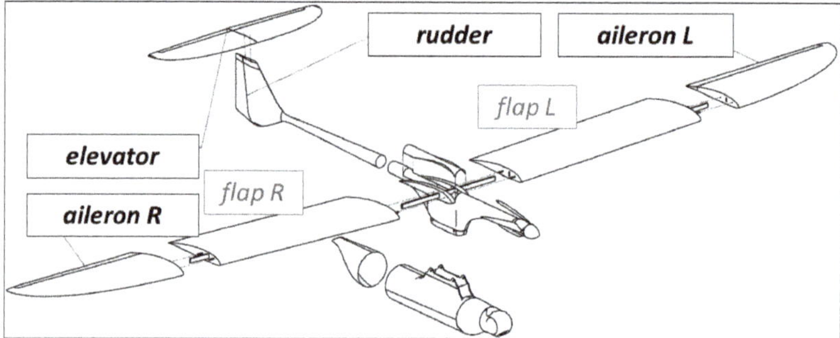

Figure 8. Control surfaces of the mini-UAV Rybitwa.

The system with active autopilot enables autonomous object control with the possibility of current correction of PID controller settings. However, it is still possible at any time to take manual control over the mini-plane by switching into radio receivers ensuring transmission of the RC signal.

System items were attached inside the mini-airplane in way shown in Figure 9. In case of applying magnetometer, the sensor should be separated from any sources of electromagnetic emissions such as high voltage wires or motor coils. The GPS antenna cannot be shielded. Both the autopilot core and radio modem require DC power supply from the range of 5 up to 27 V. It is rather necessary to apply separate power sources for autopilot and servomechanisms. Servos could consume large quantities of electricity in short time periods, which could interrupt autopilot operation by sudden voltage drop. General method and selected specified techniques useful by configuring system elements and adjusting controller gains were taken from works like [5,13,14,18–20].

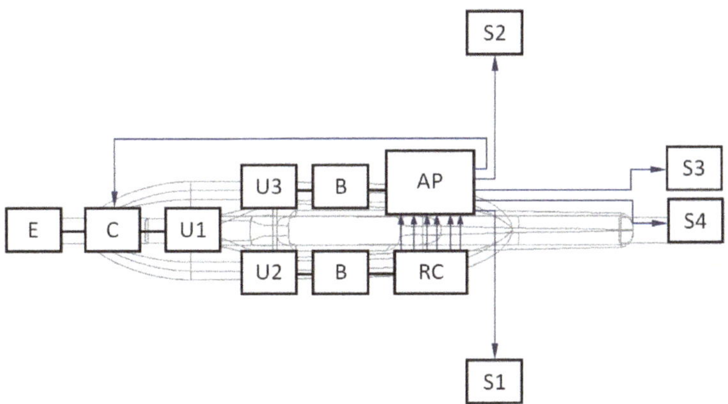

Figure 9. Block diagram of the avionics arrangement on the background of the fuselage chamber and the suspended pod: E, electric engine; C, electronic speed controller; U1, U2, U3, power sources (Li-Po type)—B, battery eliminator circuit (BEC)—AP, autopilot MP2128g—RC, receiver of remote radio-control signal—S1, S2, S3, S4, servomechanisms.

All electronic items should be securely fastened and properly connected with wires or with silicone pneumatic hoses. This applies strictly to pressure ports. Distribution of electronic components inside the under-fuselage pod is presented in Figure 10. Inside the storage, there is a carbon plate with installed telemetric radio-modem, autopilot core embedded on vibro-isolators (inside the glass-fiber box) and servoboard module. Li-Po battery packages are suspended under the plate. Inside the hemispherical rear chamber (on the left), there are radio-modem antenna and magnetometer installed. Inside the analogous front chamber (on the right), there are datalog DB-9 connector and the 2.4 GHz RC receiver. Both tip hemispherical chambers are attached to the cylindrical part with Velcro straps. Disassembling the plate with equipment is quite easy; therefore, recharging or replacing batteries is not a complicated service activity for users.

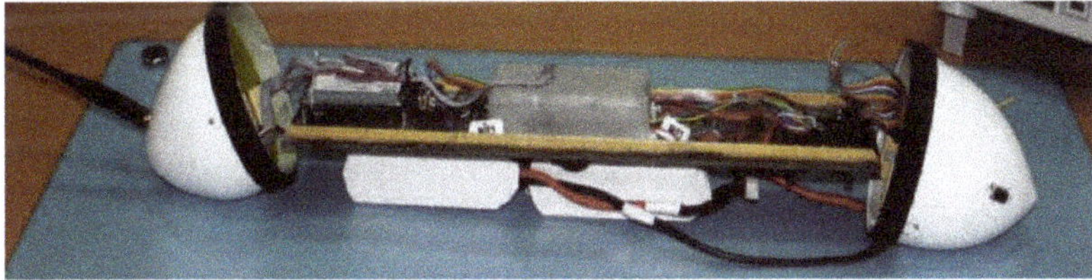

Figure 10. Mounting tray to be installed inside the pod with electronic devices and suspended battery packages; there are hemispherical chambers with additional elements on both sides of the container.

4. System Verification in Flight Tests

After installing autopilot set series of verification, flight tests begun. Strict safety procedures must be followed during both preparations and flight operations. Securing and supervising a separate flight zone is absolutely necessary. It is necessary to take into account the terrain, weather conditions and the possibility of collision with other flying objects. Especially hazardous are those ones flying illegally using prohibited frequencies or generating very high-power level, which can cause interferences with UAV equipment. Therefore, the doubled RC-transmission with two separate frequencies served by on-board miniature diversity module was deployed (marked in Figure 5, Figure 6) just to minimize the possibility of jamming the currently emitted control signal.

The task of tuning the autopilot is carried out during a series of flights with switching from the manual control mode (PIC) to the computer in command mode (CIC) during short flight sections. This is because autopilot controls UAV using 13 PID feedback loops, and each of them has to be adjusted separately in an iteration process. In the following flights, the autopilot commands are changed to allow all of the feedback loops to be tuned one after the other. After successful regulation of the loop, the drone should be able to perform on autonomous flights.

4.1. Adjusting Accessible Feedback Loops

As was mentioned in one of the previous sections, despite MP2128g autopilot being already integrated with many fix-wings, it is still not a plug and play device. MP2128g controls the UAV using a series of PID feedback loops. The term feedback loop refers to any mechanism that controls a system by adjusting the input of the system based on the measured output of the system; that is why for the newly designed UAV, which is Rybitwa, all 13 PID loops have to be adjusted in real flights. For instance, aileron from roll feedback loop controls the ailerons to minimize the difference between desired roll and actual roll; elevator from pitch feedback loop controls the elevator to minimize the difference between actual pitch and desired pitch, heading from crosstrack error feedback loop controls the desired heading to minimize the distance between UAV and the line defined by the previous waypoint and the next waypoint. These are only three of them, and each gain has the ability to define the gain schedule based on the current airspeed. What is more, output of some of the feedback loops feeds the input of other feedback loops, and autopilot can enable and disable feedback loops during the course of a flight in order to accomplish its user-defined mission. Figure 11 shows which feedback loops are enabled by autopilot during level flight.

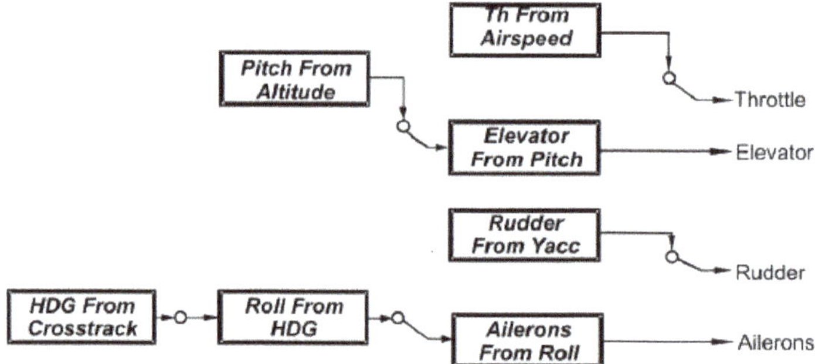

Figure 11. Scheme of enabled feedback loops during an exemplary level flight.

Among feedback loops, one can distinguish the two following sorts: inner loops and navigation loops. The inner loops are the primary stability loops and must be tuned before proceeding to navigation loops. Navigation loops are dependent upon ground speed, and they require gain scheduling because UAV is equipped with a slower update rate GPS. Typical test flight conducted under mixed control, that means switched between the PIC and CIC modes, is being started always at PIC mode. As was mentioned before, in the first stage, it is necessary to set gains firstly for priority loops, i.e., elevator from pitch, aileron from roll, and rudder from lateral acceleration. These are internal loops, the most important in terms of providing basic flight stability what is explained in References [5,11,13,20,21].

The pClimb command declared in file *.fly file displayed in Figure 12 below is dedicated to maintaining constant pitch angle while entering the targeted ceiling attempting to hold the roll and the sideslip at 0 values. With "climbpitch" parameter set to zero, the drone stops climbing, so the "pClimb" command would not allow to exceed the limit altitude of 9994 m. Parameter "thoverride" set to 0 enables the user to control the throttle manually. This procedure allows to adjust gains in flights with or into the wind for any speed ranges.

```
Metric
[recordHistory]=50
[thOverride]=1          // set manual throttle control in CIC mode
[climbPitch]=0          // temporarily set the pitch
takeoff
pClimb 9994             // initiate a pClimb at zero pitch, becouse climbPitch is set to zero
flyTo (0, 0)            // autopilot destination, just in case
repeat -1               // the pClimb altitude altitude is rached
```

Figure 12. Section of the *.fly file with feedback loop set for climbing.

Feedback loops are regulated by the PIC controllers' settings. Proper gain values can be modified in the *.vrs file after each flight. However, there is also some other method to adjust gains with the use the Horizon Software option called "Status Monitor". This tool enables the operator to observe flight parameters displayed on graphs in the real time and tune PID gains directly if necessary. The modified data can be transmitted to the autopilot via telemetric link, and the UAV stays in further control in accordance with actually tuned loops. However, that data is available only temporarily; therefore, in-flight gain changes must be copied from autopilot RAM to the *.vrs file just after landing.

In Figures 13–15 showing runs of flight trajectory and of specific parameters curves of PIC mode parameters are marked in red and sections referring to CIC mode are marked in blue. The CIC flight heading was set close to the wind direction to avoid possible wind affect onto the basic flight stability. When the drone was flying in the CIC mode, all angle values had to be visually tracked to assure the stabilization is properly maintained, i.e., without oscillations. At the same time, in real-time, P, I, D terms and current and desired roll and pitch signals are displayed and monitored on GCS console by the operator. If the plane's nose seems to oscillate up and down and then gain values for the loop, elevator deflection from pitch had to be decreased in real-time. Similar actions were taken in respect to other loops: ailerons from roll and yaw from lateral acceleration. To achieve better UAV turns without the aircraft nose dropping and adverse yaw, additional term called "feed forward" was set for three feedback loops responsible for control of the rudder based on Y accelerometer data, control of the elevator to minimize the difference between actual pitch and desired pitch and control of the throttle to maintain desired aircraft speed.

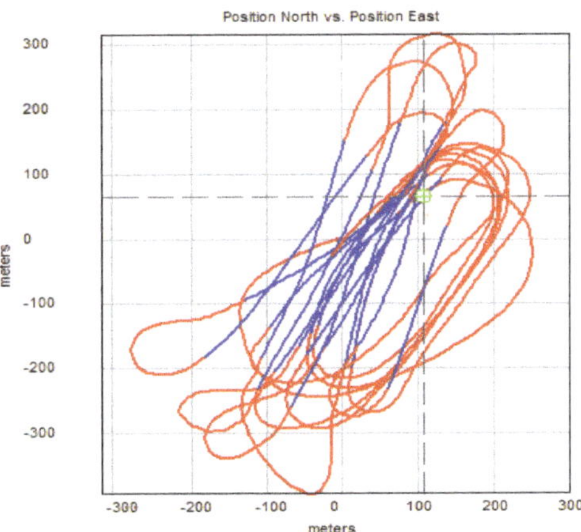

Figure 13. Flight path of mini-UAV received from logged coordinates—track sections performed in CIC mode are marked with blue.

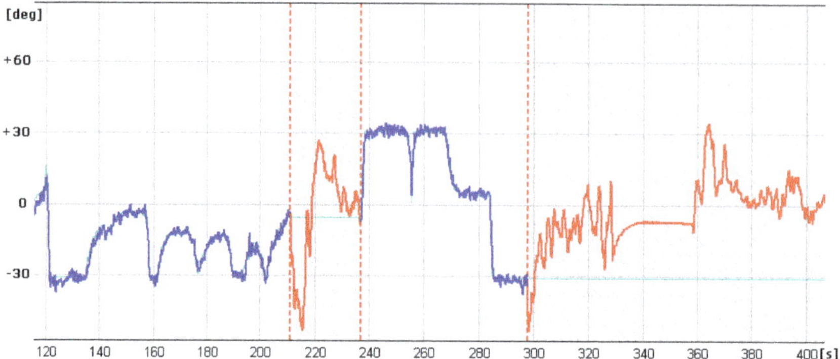

Figure 14. Curves of current roll angle and desired roll angle versus time.

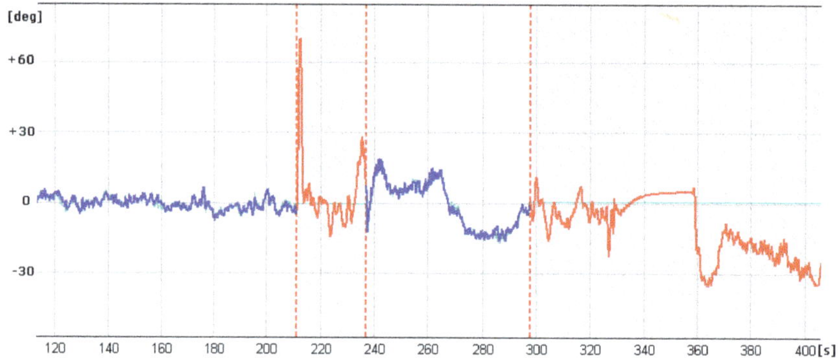

Figure 15. Curves of current pitch angle and desired pitch angle versus time.

It is generally noticeable that in the following CIC intervals, the regulation effect was being improved. High values of red curves in Figures 14 and 15 concern flight sections with turnings performed in PIC mode. During flight sections performed in CIC modes runs of current parameters are strictly close to the desired ones.

To estimate quality of loop adjustment, in CIC mode, error between current and desired values were calculated. For instance, for the pitch feedback loop (Figure 16), standard deviation for the time between 120 and 200 s is 0.84 degrees. This result should be considered as good especially if we take into account that there is a delay between current and desired pitch signals.

4.2. Testing System in Autonomous Flight

Once the inner loops were adjusted, navigation loops can be investigated to find the best P, I, D terms which allow fully autonomous flights. MP2128g autopilot ensures autonomy of the mission at any kind of flight stage. There are many types of possible takeoff and landing procedures, including winch, launchers, classic runway takeoffs, parachute recovery, deep stall or classic landings. In this study the UAV was started by the hand-throw and continued flight along the preprogrammed track autonomously [11,12].

Using the *.fly file included below, the UAV can make autonomous flight through 4 defined waypoints. The waypoint input format is longer than takeoff point in rectangular coordinate system. There are also three models added with preprogrammed conditions. They are set by the operator in the GCS. The safety procedure should be introduced to enable getting the UAV back to the takeoff area in case of losing RC transmission.

```
Metric
[recordHistory]=50
[elDrivesAlt]=1 // Throttle will control Airspeed
[thOverride]=0 // Autopilot controls throttle
takeoff
climb 100
waitclimb 70
flyto (-200,-200) // square pattern
flyto (200,-200)
flyto (200,200)
flyto (-200,200)
repeat -4


// Descend and fly around current location
definePattern 0
climb 50
flyTo (0, 0)
repeat -1

// Circle to the right - hover if using heli
definePattern 1
[rotatePattern]=[currentHeading]   //circle is located such that the UAV
                                   // begins flying on its cirucumfenece
circleRight 50           //and the pattern is rotated
                         //so that the UAVs heading is tangent to the circle
repeat -1

//RC Transmitter Lost - Fly towards home until RC returns
definePattern rcFailed
flyTo [home]
repeat -1
```

Figure 16. The content of exemplary *.fly file prepared for autonomous flight.

In Figure 17 below, flight path based on 4 arbitrary waypoints is shown. Trajectories crossing corner points of previously defined square field were circled during navigation loop adjustment. Each waypoint can be moved or deleted at any time during the flight, however, to ensure constant conditions for the procedure the constant trajectory was used to allow comparison between each adjustment stage.

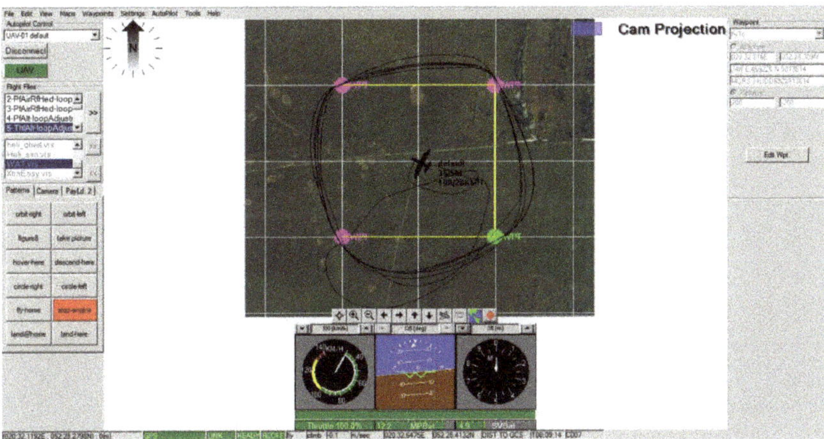

Figure 17. Exemplary view of the Horizon Software interface of the GCS with the path track referring to waypoints defined in previously set *.vrs file.

On the graphs below (Figures 18 and 19) selected parameter curves are shown—desired and measured values of altitude, heading and roll in the specified period of time during the left turn. Even in case of pretty strong turbulences, controlled values of roll and heading were very close to the desired ones. Estimated heading error occurred was never bigger than 10%. As can be seen in Figure 18, the heading loop was adjusted so that the current heading is slowly keep up the desired values. This desired behavior is because of the onboard electro-optical head. If the head is in tracking mode, sudden changes in heading can cause huge LOS error between target and pointing vectors. Such small UAV has payload capacity limitation, which affects the onboard payload. Limitation of take-off weight or power consumption affect the flight performance as it is in that case.

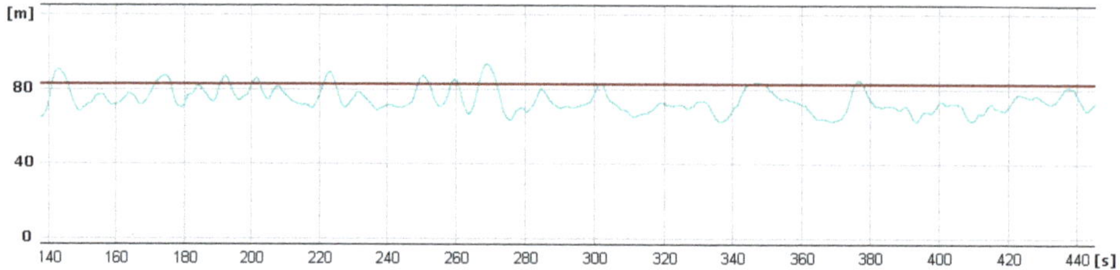

Figure 18. Currently estimated altitude (blue curve) and desired altitude (black horizontal line) in the autopilot control interval.

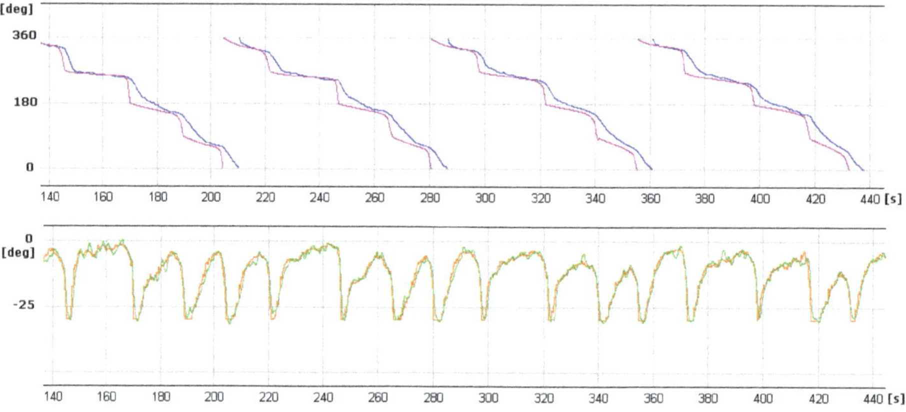

Figure 19. Currently estimated and desired runs of heading (top chart) and roll (bottom chart) angles.

5. Investigations of the Miniature Electric Propulsion Dedicated to UAV

In the aspect of selecting the components of the mini-airplane propulsion system, a series of measurement experiments were carried out. Some dependencies determined after measurements of propulsion qualities allowed to compare various configurations, which differed one from another because of component type. The research method is based on directions described in [9] and [22]. The aim of the study was to determine the following dependencies:

- Thrust vs. rotational speed T = f(n);
- Torque vs. rotational speed M = f(n);
- Effective power vs. rotational speed P_{eff} = f(n).

Appropriate measurement requires axial mounting of the engine on the measuring testbed. To perform the tests, some special equipment items are required, such as rotational speed controller, PWM pulse generator (JETIBOX), battery packs. In order to determine the abovementioned characteristics, the following measurements are to be performed:

- Thrust F [N];
- Force from the reaction moment R [N];
- Rotational speed of the propeller n [rpm];
- Battery voltage level U [V];
- Current consumed by engine I [A].

The test object was a three-phase GT5325-11 electric motor. The measuring system assembled on the testbed and illustrated in Figure 20 consists of the following items generally available on the market:

- Engine GT 5325-11 with APC 20 × 10 propeller (connected to the controller EMAX ESC-100A-HV);
- Miniature datalogger and clamp amperemeter (for measuring and storing current/voltage data);
- Force measurement track (for measuring F and R values);
- PWM pulse generator;
- Voltage supply source for engine (4/8 cells).

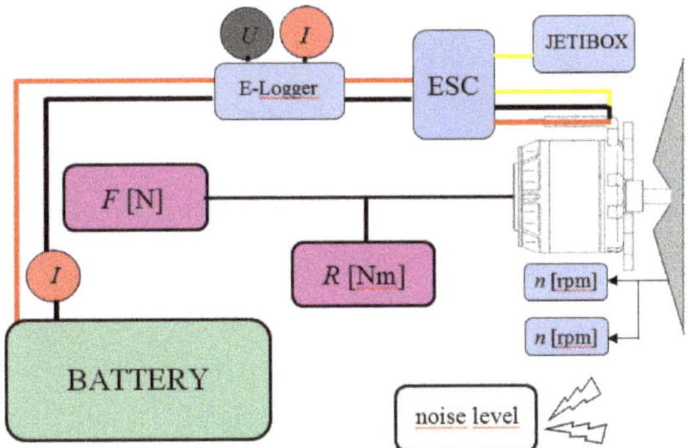

Figure 20. Scheme of measuring system installed on testbed for testing electric propulsion.

The tested engine was mounted on the testbed with the use of specially prepared steel mount. The tested motor is powered by three-phase AC voltage, which requires the use of a special speed controller for brushless motors. The engine speed is controlled by the EMAX ESC-100A HV regulator. The applied controller is adapted to supply from 4 to 10 Li-Pol cells. It is also characterized by a temporary maximum current up to 100 A; this value exceeds by 20% the maximum amount of current that can flow through the stator windings of the tested engine. In order to unambiguously determine the measurement points, a JETIBOX module was used to generate the PWM pulse, which could control the speed controller in the range from 1.024 ms to 2.047 ms. During the tests, the three different power configurations were applied:

(1) 4 s Li-Pol battery pack – 1 × 14.8 V (5300 mAh/25 C),
(2) 6 s Li-Pol battery pack – 1 × 22.2 V (5550 mAh/28 C),
(3) 8 s Li-Pol battery pack – 2 × 14.8 V (5300 mAh/25 C).

The conducted tests enable unambiguous determination of key characteristics for investigated UAV propulsion, i.e., torque, thrust, effective power and efficiency—each parameter versus rpm (revolution per minute rate). Their graphs are shown in Figures 21 and 22.

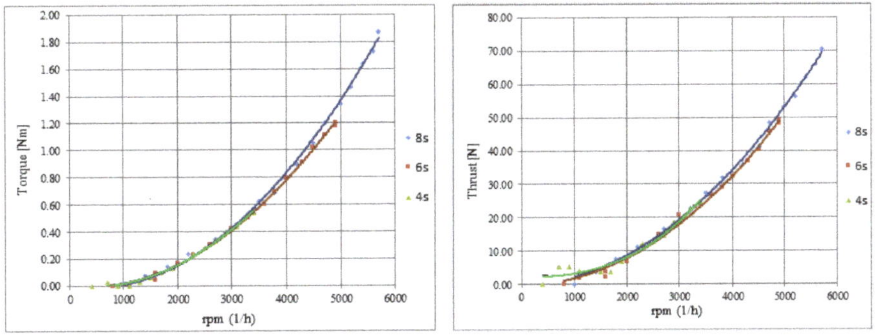

Figure 21. Graphs of torque and thrust versus rpm for three power sources (4 s/6 s/8 s).

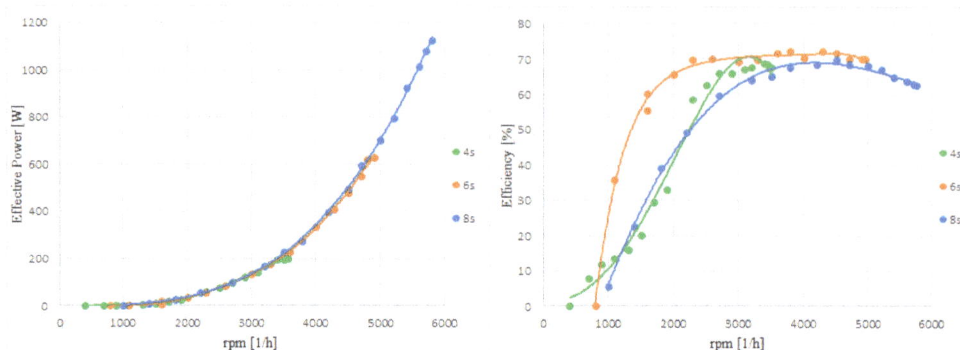

Figure 22. Graphs of effective power and efficiency (estimated as a ratio: P_{eff}/P_{elect}) versus rpm for three power sources (4 s/6 s/8 s).

The aim of the study was to select the most suitable power source for a specific engine and propeller, which means the pre-selected propulsion performance should be enough to assure required thrust (thrust-to-weight ratio is more adequate) and time of flight mission. In the course of the research, a series of measurements were carried out for each of the three available power packages of nominal voltage: 14.8 V (4 s pack.), 22.2 V (6 s), 29.6 V (8 s).

On the basis of the parameters obtained, a group of characteristics was prepared for various battery package configurations. It can be concluded that the mechanical efficiency was the highest for six power cells. Efficiency value is at the level about 70%, and it is sustained in the quite high range, i.e., between 2000 and 5000 rpm. The graph curves generated as an effect of approximation with polynomials of at least third degree do not deviate from the known theoretical runs for similar applications of miniature propulsions. The maximum recorded parameters occurred at the highest measured rpm rate gained at the voltage value of 33.6 V (achieved for 8 s package). Current value does not exceed the threshold 54 A, which allows to conclude the applied speed controller is absolutely sufficient taking into consideration that its maximum acceptable current value is 100 A. The nominal value of the maximum current flowing through the motor winding is about 30 A higher than the maximum recorded value. This means that it is possible to use a larger propeller, which could improve the performance of the tested power unit.

Analyzing thrust and current consumption values for applied voltage sources, it can be concluded that a package consisting of six cells (6 s) fits best to the developed mini-UAV at its maximum take-off weight of about 9 kg—the T/W (thrust to weight ratio) value is in that case about 0.54. Equally acceptable is the 4 s-package, which gave the maximum thrust equal to 24.4 N and in effect the T/W ratio equal to 0.27—value still acceptable for miniature and small UAV. On the other hand, the propulsion system powered with an 8 s-battery seems to be significantly oversized. This configuration is distinguished by the highest level and the largest useful range of the efficiency, but the T/W ratio obtained here is as much as 0.78—much more proper for a combat jet than for a quite slow-flying UAV. Such a high value of the ratio suggests that the propulsion produces too much thrust, which seems to be unprofitable when applied to the object under consideration.

It is necessary to note that the above described tests were carried out in static conditions, i.e., at zero cruising speed. It is known that when the speed of incoming airflow goes up (increased by the term of cruising speed), the efficiency of propellers decreases. Therefore, the above considerations do not reflect the actual description of the phenomena in the full operating range of UAV.

6. Concept of Simple Observation Video-Head for Training or Test Flights

For carrying out practice and video transmission tests during flight, a training observation video-head was constructed. The target recognition device is obviously the MicroPilot product: MP

Dayview/Nightview. However, to avoid the risk of damaging expensive equipment for the purpose of ad hoc flight tests, a much cheaper and simpler to use device has been developed. The head was equipped with the following components (shown in Figure 23):

- OEM block camera module—Sony FCB-IX11 AP;
- Executive servo device HS-5125MG (to drive camera in tilt rotation);
- On-screen display module—ETS OSD PRO (enabling on-screen projection of any subtitles);
- Video image transmission module—Immerson 5.8 GHz with 600 mW pure output power;
- RC receiver—Jeti Duplex R14 (for servo control).

Figure 23. Lateral views of the video-head thin-walled construction with electronic items installed.

Using the OSD module, it is possible to impose the flight parameters of the mini-airplane on the image transmitted to the ground. The module is directly connected to the TV signal output from the camera. The TV signal is processed by the head video-transmitter and is emitted to the GCS receiver.

The area observed by the head can be changed by tilting the optical axis of the optical sensor in the range of ~30°. This range of camera movement allows novice operators to perform preliminary training flights. The main emphasis in designing this device was concentrated on achieving the same mass in relation to the target head, elaborating simple construction and ensuring minimal possibilities to conduct entry training for flight operators.

Initially, one can estimate the area that the camera observes depending on the flight altitude of aircraft H. Using the geometrical relationships shown in the Figure 24, it is possible determine the depth of field of observation denoted as b. Two characteristic angular quantities resulting from the imaging scheme of the monitored area in the plane of the flight path should be taken into account, namely, τ, camera tilt angle; φ, camera view angle.

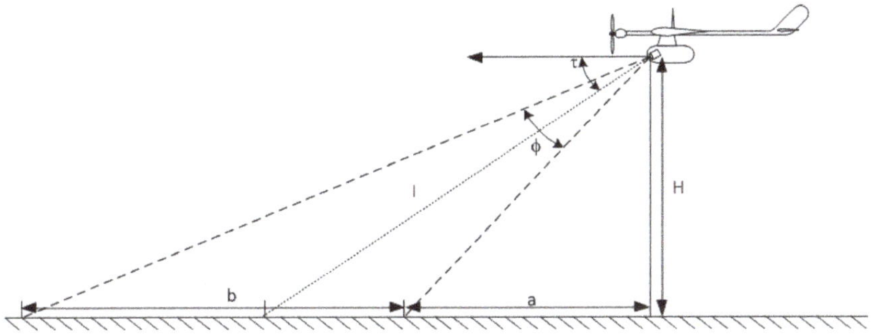

Figure 24. Scheme of geometric parameters for camera viewing in plane of flight trajectory.

The two horizontal distances shown in the scheme can easily be linked to the angles through the appropriate trigonometric relationships:

$$a + b = \frac{H}{tg\left(\tau - \frac{\varphi}{2}\right)} \tag{1}$$

$$a = \frac{H}{tg\left(\tau + \frac{\varphi}{2}\right)} \tag{2}$$

After linking both formulas, a function determining the depth of the observed area can be specified:

$$b = H\left[\frac{tg\left(\tau + \frac{\varphi}{2}\right) - tg\left(\tau - \frac{\varphi}{2}\right)}{tg\left(\tau + \frac{\varphi}{2}\right)tg\left(\tau - \frac{\varphi}{2}\right)}\right] \tag{3}$$

For the purposes of preliminary calculations, a reasonably estimated camera tilt angle $\tau = 40$ ° and a mission ceiling H = 200 m can be assumed. The angle of view is closely related to the focal length f and the height of the CCD matrix h, which demonstrates the simplified scheme of light projection in lateral view. Based on the scheme (Figure 25), one can determine the viewing angle according to the following formula:

$$\varphi = 2arctg\frac{h}{2f} \tag{4}$$

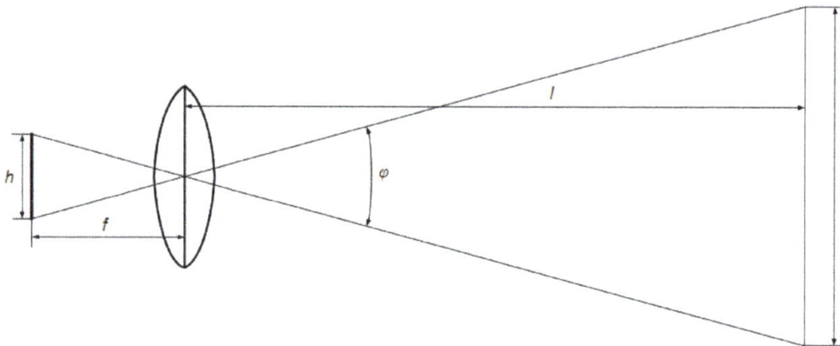

Figure 25. Simplified scheme of characteristic geometric parameters in side-view for relation between a CCD matrix, lens and an observed object.

After substituting Formula (4) into the initial Formula (3), a more detailed expression for the depth of field observation function can be obtained:

$$b = H\left[\frac{tg\left(\tau + arctg\frac{h}{2f}\right) - tg\left(\tau - arctg\frac{h}{2f}\right)}{tg\left(\tau + arctg\frac{h}{2f}\right)tg\left(\tau - arctg\frac{h}{2f}\right)}\right] \tag{5}$$

The Sony FCB-IX11 AP camera mounted in the head device is equipped with a CCD matrix having dimensions of 2.7/3.6 mm (height/width). The camera has a lens with a variable focal length in the range from f_{min} = 4.1 mm to f_{max} = 73.8 mm. Using the given relationships, it is possible to determine the approximate depth of visual penetration of the sensor. For the reasonably assumed mission altitude $H = 200$ and the given range of variation f, this parameter can vary from b_{min} = 150 m to b_{max} = 265 m.

7. Conclusions

In the paper, some crucial topics selected from the overall design-and-development process of a new concept mini-UAV are presented. In the whole prototyping process, there are at least five basic stages of system development, namely:

- defining overall conception and specification of technical requirements;
- elaborating technical design;
- manufacturing and integrating elements and components;
- investigating and validating devices or subsystems;
- test operating and correcting the whole system.

The problems described here appeared and were being solved during the stages of integrating components and investigating subsystems. The variety of undertaken tasks results from the nature of the research project, which was a typical technological-and-development activity aimed at developing, manufacturing and checking the properties of the unmanned system. Therefore, the performance and validation of the finished product characteristics required various, often unrelated tasks, whose success, however, determined the success of the entire undertaking. The research and engineering activities presented are certainly thematically diverse. However, the technological challenge undertaken here was in this case to develop and test a handy, inexpensive and relatively effective miniature unmanned system, which was to be built from ready-made elements commercially available. It was not about carrying out specialist tests and research focused on a single innovative device. The overarching goal was to build a system that, as serviceable and affordable, would have a chance for quick commercial implementation.

Key benefits gained from described work are the following:

- original UAV-construction demonstrating useful technological and aerodynamic qualities;
- effective adaptation of the MP2128g in terms of complex integration and successful setting all PID-loops;
- entry stage of mini-UAV propulsion system optimization, i.e., measuring values of thrust and effective power for selected battery configurations;
- elaborating self-made simple video-head useful for training and testing flights when high quality of the video imaging is rather secondary target.

The most useful and fully completed implementation seems to be the autopilot integration and successful tuning of its control loops. The development works on optimizing the mini-propulsion system and on constructing alternative versions of cheaper and easy-to-use video-head are being still carried out in the research team of the Institute of Aviation Technology of the MUT. The target mini-UAV version must be equipped with an avionics system fully integrated with the mission equipment, which will probably be an MP Dayview/Nightview camera. Preparation of the final prototype version requires additional operational tests; therefore, time and new financial outlays for the project are still necessary.

Author Contributions: Conceptualization and methodology, A.O. and R.R.; research logistics, integration, testing and validation, R.R., Ł.K.; data preparation, R.R., Ł.K., M.S.; paper writing (text, graphics, review and editing), A.O., R.R., M.S.; supervision, A.O. All authors have read and agreed to the published version of the manuscript.

Funding: The content of the study refers to the R&D project: Autonomous Unmanned Aerial Vehicles Equipped with Monitoring and Inspecting Means Destined for Aiding Police and Fire Department Operations. The project carried out in the years 2010–2013 was funded by the Polish government organization The National Centre for Research and Development (NCR&D). Furthermore, some aspects presented herein were continued in the framework of the statutory project: Experimental and Numerical Researches of New Technologies with Verification Allowing to Achieve Some Assumed Tactic and Technical Tasks in the Area of UAV, granted to FMA MUT by the Polish Ministry of Science and Higher Education for the years 2016–2018.

Conflicts of Interest: The authors declare no conflict of interest.

References

1. *Unmanned Aircraft Systems. The Global Perspective 2008/2009*; UVS International, Blyenburgh & Co.: Paris, France, 2008; p. 150.
2. *2010-2011 UAS Yearbook–UAS: The Global Perspective*, 8th ed.; UVS International, Blyenburgh & Co.: Paris, France, June 2010; pp. 61–62.
3. Olejnik, A.; Rogólski, R.; Kiszkowiak, Ł.; Szcześniak, M. Specific Problems of Selecting and Integrating Equipment Components in the Course of Developing a Technology Demonstrator for the mini-UAV. In Proceedings of the 2019 IEEE International Workshop on Metrology for Aerospace, Torino, Italy, 19–21 June 2019; pp. 284–289.
4. Austin, R. *Unmanned Aircraft Systems—UAVs Design, Development and Deployment. Aerospace Series*; John Wiley & Sons Ltd.: Hoboken, NJ, USA, 2010.
5. Gruszecki, J. *Unmanned Aerial Vehicles. Control and Navigation Systems /in Polish/*; Oficyna Wydawnicza PRz: Rzeszów, Poland, 2002.
6. Valavanis, K.P. (Ed.) Advances in Unmanned Aerial Vehicles—State of the Art and the Road to Autonomy. In *Intelligent Systems, Control and Automation: Science and Engineering*; Springer: Dordrecht, The Netherlands, 2007.
7. Lam, T.M. (Ed.) Aerial Vehicles. In-Tech: Croatia. Available online: www.intechweb.org, (accessed on 15 January 2009).
8. Shima, T.; Rasmussen, S. (Eds.) *UAV Cooperative Decision and Control—Challenges and Practical Approaches*; Society for Industrial and Applied Mathematics (SIAM): Philadelphia, PA, USA, 2009.
9. Kotvani, K. *Design and Development of Fully Composite Mini Unmanned Aerial Vehicle (Mini-UAV)*; Department of Aerospace Engineering, Indian Institute of Technology IIT: Kanpur, India, 2003.
10. Chung, P.H.; Ma, D.M.; Shiau, J.K. Design, Manufacturing and Flight Testing of an Experimental Flying Wing UAV. *Appl. Sci.* **2019**, *9*, 3043. [CrossRef]
11. MicroPilot. *Manuals for Installation, Operation and Use: Autopilot, Microhard Modems, Analog to Digital Converters, Electronic Compass, HORIZON, Vibration Isolation, Camera Software*; Micro-Pilot Inc.: Stony Mountain, MB, Canada, 2010.
12. Downing, D.; Ganley, S. MicroPilot Autopilot (Report). In *Proceedings of the Annual Montana Tech Electrical and General Engineering Symposium (Montana Tech Library)*; Montana Tech of The University of Montana: Butte, MT, USA, 2015.
13. Puscov, J. Flight System Implementation in UAV (diploma work). Examensarbete utfört vid Fysikinstitutionen, KTH, Stockholm, Sweden, 2002.
14. Sobieraj, W.; Homziuk, A.; Rochala, Z. Problems of Control of Mini Unmanned Aerial Vehicle (miniUAV). In *Scientific Letters of Rzeszow University of Technology. Mechanics 2007*; Rzeszow University of Technology: Rzeszow, Poland, 2007; pp. 31–42.
15. Grankvist, H. *Autopilot Design and Path Planning for a UAV. FOI-R—224—SE—Scientific Report*; FOI Defence Research Agency: Stockholm, Sweden, 2006.
16. Adiprawita, W.; Ahmad, A.S.; Sembiring, J. Harware in the Loop Simulation for Simple Low Cost Autonomous UAV (Unmanned Aerial Vehicle) Autopilot System Research and Development. In Proceedings of the International Conference on Electrical Engineering and Informatics, Bandung, Indonesia, 17–19 June 2007; pp. 218–221.
17. Fu, X.; Zhou, Z.; Xiong, W.; Guo, Q. MEMS-Based Low-Cost Flight Control System for Small UAVs. *Tsinghua Sci. Technol.* **2008**, *13*, 614–618. [CrossRef]
18. Brzozowski, B.; Rochala, Z.; Wojtowicz, K.; Gawełda, B.; Kaźmierczak, K. Miniature airflow probe for an unmanned aerial vehicle. In Proceedings of the 2016 IEEE Metrology for Aerospace (MetroAeroSpace), Florence, Italy, 22–23 June 2016; pp. 500–505.
19. Brzozowski, B.; Rochala, Z.; Wojtowicz, K. Overview of the research on state-of-the-art measurement sensors for UAV navigation. In Proceedings of the 2017 IEEE International Workshop on Metrology for AeroSpace (MetroAeroSpace), Padua, Italy, 21–23 June 2017; pp. 565–570.
20. Nelson, R.C. *Flight Stability and Automatic Control*, 2nd ed.; Mc Graw-Hill Book Company: New York, NY, USA, 1998.

21. Abzug, M.J.; Larrabee, E.E. *Airplane Stability and Control: A History of the Technologies that Made Aviation Possible*, 2nd ed.; Cambridge University Press: Cambridge, UK, 2005.
22. Kozakiewicz, A.; Zalewski, P. *Criteria for Selecting Propulsion System for Unmanned Aerial Vehicles*; Przegląd WLOP, Air Force and Air Defence Force Review; Polish Air Forces: Warsaw, Poland, 2002; in Polish.

© 2020 by the authors. Licensee MDPI, Basel, Switzerland. This article is an open access article distributed under the terms and conditions of the Creative Commons Attribution (CC BY) license (http://creativecommons.org/licenses/by/4.0/).

Article

Hybridized-GNSS Approaches to Train Positioning: Challenges and Open Issues on Uncertainty †

Susanna Spinsante [1,*] and Cosimo Stallo [2]

1 Dipartimento di Ingegneria dell'Informazione, Università Politecnica delle Marche, Via Brecce Bianche 12, 60131 Ancona, Italy
2 RadioLabs, Corso d'Italia, 00198 Rome, Italy; cosimo.stallo@radiolabs.it
* Correspondence: s.spinsante@staff.univpm.it; Tel.: +39-071-220-4102

† This paper is an extended version of our paper published in Spinsante, S.; Stallo, C. Uncertainty Issues in Hybridized-GNSS Approaches to Train Positioning. In Proceedings of the 2019 IEEE International Workshop on Metrology for Aerospace (MetroAeroSpace), Torino, Italy, 19–21 June 2019.

Received: 10 February 2020; Accepted: 25 March 2020; Published: 29 March 2020

Abstract: In recent years, the development of advanced systems and applications has propelled the adoption of autonomous railway traffic and train positioning, with several ongoing initiatives and experimental testbeds aimed at proving the suitability and reliability of the Global Navigation Satellite System signals and services, in this specific application domain. To satisfy the strict safety and accuracy requirements aimed at assuring the position solution's integrity, availability, accuracy and reliability, recent proposals suggest the hybridization of the Global Navigation Satellite System with other technologies. The integration with localization techniques that are expected to be available with the upcoming fifth generation mobile communication networks is among the most promising approaches. In this work, different approaches to the design of hybrid positioning solutions for the railway sector are examined, under the perspective of the uncertainty evaluation of the attained results and performance. In fact, the way the uncertainty associated to the positioning measurements performed by different studies is reported is often not consistent with the Guide to the Expression of Uncertainty in Measurement, and this makes it very difficult to fairly compare the different approaches in order to identify the best emerging solution. Under this perspective, the review provided by this work highlights a number of open issues that should drive future research activities in this field.

Keywords: GNSS; autonomous railway traffic; uncertainty; accuracy; positioning error; 5G

1. Introduction

Accurate radio-based positioning has recently boosted the exploitation of *location awareness*, i.e., the ability of a device to share its own physical location associated to a person or an object, in many different markets, spanning from Intelligent Traffic Systems (ITS) [1], to autonomous vehicles, Industry 4.0, and the improved management of communication networks [2].

So-called Location Determination Systems (LDSs) encompass the necessary technological components to enable location awareness. Among them, the Global Navigation Satellite System (GNSS), a constellation of satellites transmitting positioning and timing data from space to GNSS receivers, that use these data to determine their location [3], still is the only LDS able to provide position, velocity, and time (PVT) necessary for localization purposes, on a global scale and on a continuous time basis. By definition, GNSS provides a global coverage, ensured by the harmonization of the technical specifications (such as frequencies and bandwidth allocation) among the European Galileo, the NAVSTAR Global Positioning System (GPS) from USA, the Russian GLONASS, and the BeiDou Navigation Satellite System (BDS) from China [4]. As of January 2019, there were 115 available GNSS satellites, with 32 GPS satellites, 25 GLONASS satellites, 34 BDS satellites, and 24 Galileo satellites. The GNSS satellites' positions at 23:00 on 31 December 2019 is shown in Figure 1.

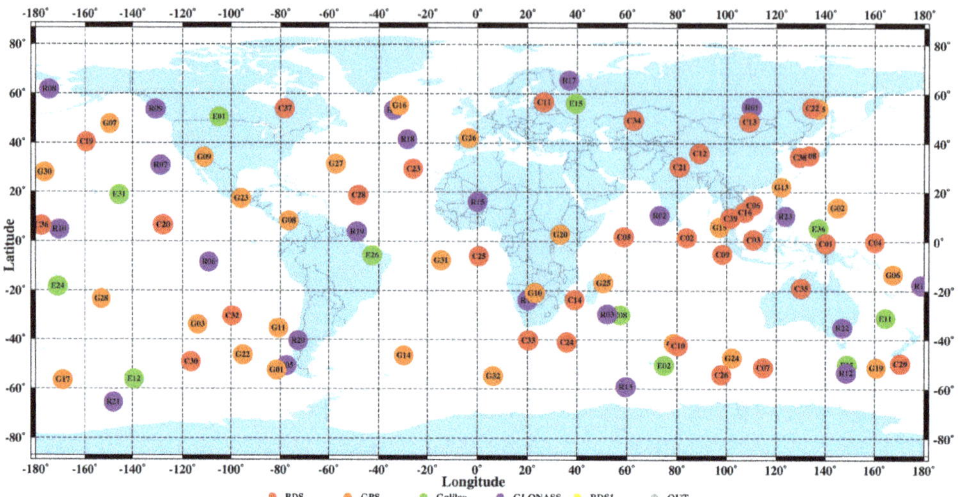

Figure 1. GNSS satellites' positions at 23:00 on 31 December 2019 (from http://www.igmas.org, retrieved on 7 February 2020).

Galileo is the European contribution to GNSS, which, differently from other systems, is under civilian control and has been designed in response to the diverse needs of different user communities. Galileo supports four services, namely Open Service, Commercial Service, Search and Rescue, and Public Regulated Service, with different levels of position information accuracy, robustness, authentication, and security. The fully operational system foresees 30 satellites emitting a coded radio signal. A Galileo receiver computes its position calculating latitude, longitude, and time based on signals received from four satellites. Galileo-enabled mobile terminals feature an accurate and ubiquitous GNSS localization, with an accuracy within 1 m for professional users and 5 m for the general public. Such a relevant accuracy performance from Galileo is made possible by a number of technical improvements, compared to GPS and GLONASS, such as a stronger signals' robustness to multipath, with a code phase error of ranging signals much lower than GPS, and the three orbital planes inclination enabling better Earth coverage at high latitudes. When all the 30 Galileo satellites are deployed and operational, six to eight satellites will always be visible from most locations, allowing positions and timing to be determined very accurately, down to a few centimeters. The reliability of Galileo services, and GNSS in general, is further increased thanks to the interoperability with GPS satellites.

Accurate and reliable positioning information provided by GNSS is nowadays exploited not only in aviation and autonomous cars or trains, but also in contemporary marine vessels, which are equipped with complex navigation and communication systems such as Electronic Chart Display & Information System (ECDIS), Automatic Identification System (AIS), Global Maritime Distress & Safety System (GMDSS), Integrated Navigation Systems (INS). GNSS PVT data are exploited by all of them, to assess the integrity of the positioning information, and mitigate interference, jamming, and spoofing in marine navigation [5]. In the majority of merchant ships, shipborne autonomous GPS receivers are still the primary means for positioning. More advanced GNSS receivers and satellite-based augmentation systems (SBAS) are still rare, but expected to become more common in the near future. Existing requirements, performance standards, and future concepts of maritime GNSS integrity are discussed in [6].

In autonomous cars and vehicles, several target Key Performance Indicators (KPIs) are set, which need to be strictly satisfied [1]. For example, a position accuracy < 20 cm is required by safety-critical automated driving [7]. Safety requirements are as relevant as those related to accuracy, for both autonomous trains and cars [8], and this has an impact on the requested integrity of the position solution. The per-hour Tolerable Hazard Rate (THR), defined as the occurrence rate the vehicle control systems fails to stop the vehicle at the desired location, or its speed exceeds the prescribed value, is used to quantify the safety requirements. In railway systems, an overall THR better than 10^{-9}/h or even 10^{-10}/h is mandatory, and this can be attained only through a cross-check with a GNSS-independent LDS, as shown by Lo et al. [9].

At present, the use of GNSS in European rail is primarily for non-Safety of Life (SoL) applications, such as asset management and passenger information services [10].

As a matter of fact, GNSS positioning still lacks the requested accuracy and continuous availability [11] in complex propagation environments such as high-speed moving trains, dense urban scenarios, tunnels, or multistory car parks, despite its global geographical coverage. In the aforementioned cases, augmentation of the vehicle localization capability by satellite-independent systems is necessary. The latest technological developments show that augmented GNSS, together with specific sensors, can help compliance to the stringent requirements associated to critical scenarios (railway LDSs or high-speed train positioning) for which the Comité Européen de Normalisation Électrotechnique (CENELEC) requires to fulfill the Safety Integrity Level (SIL) 4 [12].

In the Moving Block Signaling (MBS) system designed to efficiently improve the operation of railway lines, the moving authority of each train is calculated from the real-time positioning information of the preceding one [13]. A moving block (MB), defined as a virtual zone surrounding the train and dynamically determined from its position and velocity data, operates as a safety buffer inside which the train resides: other trains are forbidden to enter, and each train cannot leave its own safety buffer. To ensure safety of train traffic managed by the MB principle, a continuous, accurate, and reliable train positioning service is mandatory. It is actually a nontrivial work to develop a reliable MBS system for railways because of the requirements of highly accurate positioning, high capacity communications, and high performance separation control. Actually, there is no high-speed railway line operating in MBS mode in the world; instead, the Fixed Block Signaling (FBS) system mode is widely used because of its relatively high-safety property.

To increase the adoption of satellite-based positioning for railway signaling, the European GNSS (E-GNSS) Agency is working together with rail and space industry stakeholders. Their joint efforts led to the European Train Control System (ETCS), now being adopted both in Europe and beyond, as one of the components of the European Rail Traffic Management System (ERTMS). In the ETCS framework, nowadays the positioning of a train is based on balises, i.e., physical elements mounted at specific intervals along the railway track. Wherever possible, physical balises should be replaced by virtual ones (VBs), based on precise GNSS positioning, without any operational or safety implications on the ETCS. The main projects carried out to achieve an E-GNSS enabled ETCS, are presented in the related publicly available roadmap [14].

The discussion presented above shows that GNSS-based positioning still suffers several limitations with respect to the requirements set for reliable autonomous railway traffic, which could be tackled by hybridization techniques aimed at improving and augmenting the positioning data provided by GNSS. Among the proposed approaches, the integration of GNSS with localization techniques enabled by the fifth generation (5G) mobile communication is specifically targeted. Positioning will play a key role in the next 5G communication systems [15]: network pervasiveness due to densely deployed access nodes (ANs) will support precise location with enhanced availability, exploiting the network control links and signals in all kinds of environments, especially urban canyons, where most GNSS signals are blocked or affected by severe multipath. In those cases, unavailable or corrupted GNSS observables may be replaced by 5G observations added into GNSS solutions.

Several approaches have been proposed in the literature, but a direct comparison or evaluation among them is quite difficult to carry out. In fact, the overview of the recent state-of-the-art proposals for GNSS hybridization provided in this work, under the perspective of the related uncertainty issues and attained performance [16], shows that in many cases the declaration of the uncertainty in the positioning measurements attained by different studies is not consistent with the Guide to the Expression of Uncertainty in Measurement (GUM) [17]. It is consequently difficult to assess the different proposals and identify the best-performing one. The review carried out in this work aims to highlight such a limit in the currently available scientific and technical data resulting from different experiments and testbeds, in order to raise awareness about the need for standardized approaches that could enable a fair benchmarking. To the authors' best knowledge, such a kind of analysis is not yet found in the literature.

This work is organized as follows. Section 2 reviews the main measurement techniques used in wireless positioning systems, focusing on GNSS positioning. The most recent state-of-the-art proposals for hybridized-GNSS systems applied to train positioning are presented in Section 3, mentioning a recently appeared 5G-only approach too. The selected literature is analyzed under the uncertainty perspective and with respect to the results obtained in Section 4, while the limits of the methodologies adopted in the different studies are discussed with respect to the requirements prescribed in railway traffic in Section 5. Finally, Section 6 concludes the paper and provides some insights on future trends based on currently ongoing research initiatives.

2. Wireless Positioning Systems

Wireless positioning systems encompass either nodes with known locations and nodes whose position has to be determined. They are identified as *known nodes* and *unknown nodes*, respectively. The former may be at fixed locations, such as anchor nodes in local positioning systems, or not, as it happens in satellite positioning systems to provide a global coverage. The latter may be stationary or mobile nodes, as in navigation systems.

2.1. Measurement Techniques

The conventional approach to estimate the location of an unknown node relies on a two-steps method: first, measurements are derived from the received radio waves' characteristics; then, the position of the node is estimated by means of different calculations performed on the measurements taken. The last ones include, among others: (i) the difference in the power of the Received Signal Strength Indicator (RSSI), compared to the original signal strength; (ii) the Time of Arrival (ToA) and Time Difference of Arrival (TDoA); and (iii) the angle of Arrival (AoA) of the signal. A set of measurements of the same kind (AoA, RSSI, or ToA) is used by some position estimation approaches to establish the location of an unknown node; other solutions combine AoA information to distance estimates derived from RSSI or ToA.

Table 1 summarizes the main wireless positioning systems with their corresponding measurement techniques (elaborated from [18]). Although localization has long been considered an optional feature in the framework of standardization, implementation, and exploitation of existing cellular networks, the global cellular communication infrastructure deployed around the world can actually support positioning services. Several contributions have been provided by the research community to the development of positioning strategies within each generation of cellular technology, from the first one (1G) to the upcoming 5G. As appears in the table, different measurement techniques have been adopted in mobile cellular systems, according to the advent of different mobile networks generations.

The approximate location of a user's device, relative to the cells of the mobile communication network, is identified by a numerical parameter called Cell ID, in those cellular systems where local positioning is supported. For third-generation (3G) mobile terminals, the Enhanced Observed Time Difference (E-OTD) is a standard localization method, based on time difference measurements performed in the handset rather than the network, and a mechanism to pseudo-synchronize the network [19]. In 4G Long Term Evolution (LTE) mobile networks, the Observed Time Difference of Arrival (OTDoA) is used to determine the position of User Equipments (UEs) [20,21]. Multilateration is exploited in the Uplink-Time Difference of Arrival (U-TDoA) [22], based on the accurate measurement of the time it takes a signal to travel from a mobile phone to multiple sensitive receivers, called Location Measurement Units (LMUs). No specific chip onboard the UE is needed, as computations are performed by the network, thus this technique is available also to older generation mobile terminals.

Most of the cellular networks only provide basic localization methods and assistance data for GNSS, essentially because of the increased costs network operators should otherwise face. 3GPP LTE (4.5G) includes potential enhancements for positioning technologies which are both depending on Radio Access Technology (RAT), and RAT-independent, as summarized below [23]:

- OTDoA enhancement based on more density in the time domain, a new pattern and an extension of the Positioning Reference Signal bandwidth enabled by Carrier Aggregation, even combined with cell-specific reference signal and transmitted in unlicensed bands (LTE-U);
- Device-to-Device aided positioning: user equipments with a known position (acting as anchored nodes) could cooperate with target equipments to improve their location accuracy, exploiting the location of neighbour devices or using ranging and signal strength measurements;
- Multiple Input Multiple Output (MIMO): multiple antennas and beamforming techniques can improve the vertical positioning of the user equipment;
- WLAN/Bluetooth: ranging and signal strength from WLAN and Bluetooth networks could be combined with LTE-based measurements for positioning, exploiting inter-RAT functionalities;
- Terrestrial Beacon Systems: beacon and Positioning Reference signals could be transmitted with a dedicated infrastructure to enhance the network-supported positioning capabilities; and
- Barometer: barometric sensors embedded in user equipments can improve the accuracy of vertical positioning.

In Release 13 of TS 36.305 [24], the accuracy requirements for ranging or Reference Signal Time Difference (RSTD) measurements were updated, according to the use of received signals at the same carrier frequency (intra-frequency measurements) or at different carrier frequencies (inter-frequency measurements), as shown in Table II of [23]. As discussed in [25], 5G networks aim for sub-meter accuracy in 95% of served areas, comprising both indoor and urban environments, thus overcoming by at least one order of magnitude the accuracy level of the current state-of-the-art solutions, including GNSS in non-Line Of Sight (LOS).

Table 1. Wireless positioning systems and corresponding measurement techniques.

System	Measurement Technique
GNSS (e.g., Galileo, GPS)	TDoA
A-GNSS	TDoA
WLAN	AoA/RSSI/ToA, radio fingerprinting
Cellular	Cell ID/E-OTD/OTDoA/U-TDoA

2.2. GNSS Positioning

A GNSS receiver collects signals from the GNSS satellites that in a given moment are located in its field of view, as shown in Figure 2. By applying correlation techniques on the received signals, the receiver measures the so-called *pseudoranges* from each satellite. These measurements are affected by various sources of uncertainty, such as variations in the measured signals, due to multipath. Sub-metric position accuracy is attainable only by mitigating as much as possible the different sources of uncertainty. An additional fixed GNSS receiver with a precisely determined position is used in many techniques. Such a receiver acts as a *reference station*: it collects GNSS signals similar to a classical receiver, but, based on its exactly known position, the corrections to be applied on the pseudoranges may be computed. They allow obtaining sub-metric accuracy when the GNSS receiver is placed in an open-sky environment, not subject to strong multipath.

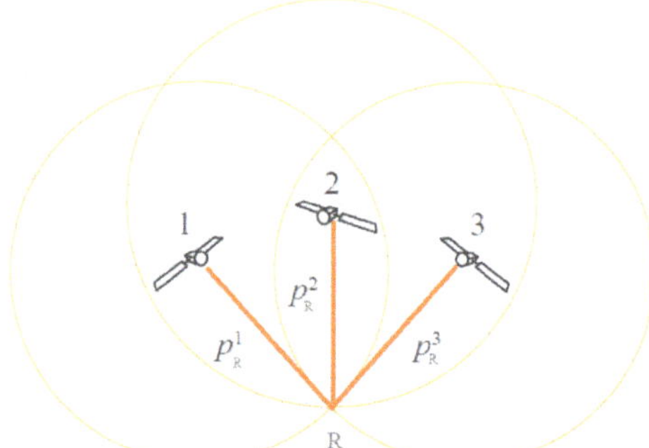

Figure 2. Receiver's pseudoranges (p_R^i, $i = 1, 2, 3$) from three satellites.

Multipath originates from the combination of LOS and Non-LOS (NLOS) signal components, due to reflections off surroundings before reaching the receiver's antenna. In the presence of multipath, the code tracking procedure performed by the receiver cannot correctly distinguish the actual correlation peak between the received signal and the locally generated replica at the receiver, with code phases errors up to tens of meters. In GNSS, the available redundant measurements due to the presence of many satellites in the receiver's field of view allow implementing a specific procedure to mitigate multipath, which is based on detecting and excluding the affected measurements. This procedure is known as de-weighting [26,27]: suitable algorithms are needed to detect and de-weight distorted signals, at the same time avoiding significant degradation in the measurement geometry. In fact, a poor geometry may even exacerbate the effects of remaining errors and worsen the ultimate position solution. The incidence of geometry degradation may be estimated by a metric called weighted Dilution of Precision (DOP).

Precise Point Positioning (PPP) [28] and Differential GNSS (DGNSS) [29] are the two coexisting methods to generate and provide range corrections. Then, two main DGNSS approaches exist: standard and Real Time Kinematic (RTK), depending on the type of measurement used. In standard DGNSS, the user's receiver computes its position by applying algorithms based on GNSS signals' code measurements from the reference station and the receiver itself. RTK algorithms, on the contrary, are based on carrier phase measurements. Decimeter level accuracy may be attained by PPP technique in kinematic mode; centimeter level or better is possible in static mode, thanks to precise orbit, clock, and error models. When GPS is the only constellation used for DGNSS (which is then called DGPS), accuracy is in the order of 1 m for users located in the range of few 10s km from the reference station, growing (i.e., worsening the quality of the positioning information) at the rate of 1 m every 150 km of separation. RTK DGNSS reaches centimeter-level positioning accuracy.

3. Hybridized GNSS for Train Positioning

When GNSS will be able to ensure the same safety level achieved by traditional systems, the European Railway Traffic Management System/European Train Control System (ERTMS/ETCS) will experience a disruptive innovation. To this aim, augmentation solutions, and integrity monitoring techniques are needed. An additional system, usually a terrestrial one, can cooperate to improve the positioning performance of a satellite navigation system, through augmentation. In Assisted GPS (A-GPS), i.e., GPS augmented with a cellular network, the burden of position computation is shifted from the receiver to the network, which makes the information available to the UEs within the cell.

In [30], two approaches to train positioning are compared, namely the DGNSS and the code double difference, both featuring an external augmentation to GNSS signals. Based on the assumption that the train is constrained to the track during its ride (so-called track constraint imposition), the train position is derived as the curvilinear abscissa of the track where the reference point of the train is lying on. The difference between the two approaches relies in the way the curvilinear abscissa is used to replace unknowns in the mathematical models used for positioning computation, both based on the extended Kalman filter (EKF) to solve the resulting equation set. The survey by Otegui et al. [31] analyzes the main train positioning solutions currently in use, looking at common parameters and criteria, such as applied sensors and algorithms, validation tests, results obtained. Sensors typically used in railway positioning include onboard Doppler radar, wheel sensor (called tachometer or odometer) and balise transponder. Recent studies have shown that Eddy current sensors can be used as speed sensors. The position information provided by fixed external balises, gathered by the train balise transponder, is currently exploited to compensate the long-term drift of the onboard odometry device. Ongoing research activities aim to replace the external balises with a Virtual Balise Reader (VBR) based on GNSS and embedded in the train itself, to periodically get positioning data sampled at higher frequency. A GNSS-based VBR would functionally operate as a physical balise, or a balise group. The last ones would still be necessary in situations where the VB cannot be available, such as in tunnels, or where GNSS provides poor performance. Environmental conditions and local effects [12] can increase the uncertainty of the GNSS-measured position, making it unacceptable with respect to user requirements [32].

It is well-known that *pseudoranges* measured by a GNSS receiver in a VBR are affected by various sources of uncertainty:

- Satellite clock: Even if the timing equipment of GNSS satellites is very precise and corrections are broadcasted in the GNSS signal, a small clock bias remains. In the downlink data, the satellite provides the user with an estimate of its clock offset, which gives a typical uncertainty of about ± 2 m, although this value can vary between different GNSS systems.
- Satellite ephemeris: A GNSS satellite broadcasts its own position within the signal, but, even with the corrections from the GNSS ground control system, small errors in the orbit can result in up to ± 2.5 m uncertainty.
- Ionosphere and troposphere: Ionospheric activity can delay signals in the GNSS frequency bands, causing a significant uncertainty in satellite position, typically much greater than ± 5 m. Variations in tropospheric delay are caused by the changing humidity, temperature, and atmospheric pressure in the troposphere, but they are very similar on a local scale, so that their effects can be compensated by DGNSS or RTK techniques.
- Multipath: GNSS signals are reflected and delayed by ground infrastructures and vegetation, resulting in a typical position uncertainty of ± 2 m.
- Receiver noise: It originates in the thermal noise generated by the receiver RF, providing a typical ± 0.1 m uncertainty.
- Receiver clock: Mass-market receivers contain oscillators cheaper than the ones on-board satellites. However, because all GNSS signals are similarly affected by the receiver clock, this error is easily mitigated by the receiver during the computation of the position.

The above sources of uncertainty are summarized in Table 2.

Table 2. Sources of uncertainty in GNSS pseudoranges measured by a receiver.

Source	Typical Uncertainty	Notes
Satellite clock	±2 m	Can vary between different GNSS systems
Satellite ephemeris	up to ±2.5 m	Even with corrections from GNSS ground control system
Ionospheric activity	>> ±5 m	Signals in GNSS bands may be significantly affected
Tropospheric activity	not relevant	Compensated by DGNSS or RTK
Multipath	±2 m	Caused by reflections and delay from ground infrastructures and vegetation
Receiver noise	±0.1 m	Thermal noise by receiver RF
Receiver clock	not relevant	Due to cheap oscillators onboard receivers, easily mitigated

PPP techniques may mitigate the first three contributions, while multipath cannot be compensated by a reference station, being a very local source of error. The receiver noise cannot be corrected either, being specific to each GNSS receiver. Hybridization of GNSS with one or several other localization techniques in an integrated navigation solution can overcome these issues: by adding devices or signals, position information can be provided on a continuous basis, even when satellite signals are not available, such as in tunnels. The fusion of GNSS and ten Degree Of Freedom (DOF) Inertial Measurement Unit (IMU) is evaluated in [33], for train positioning. The proposed method to calculate train velocity is based on the raw measurements provided by the GNSS receiver and the IMU, used to compute track features. They are then compared to a reference velocity obtained from tachometer and Doppler radar readings. Field measurements obtained in the study show that, even with a simple data fusion algorithm, the calculated velocity approaches the one returned by higher cost sensors, such as Doppler radars typically available onboard a train. However, in the case of a GNSS signal outage, additional information is required to guarantee the validity of the solution.

Due to environmental limits, service availability of satellite signals, which is essential for field applications of satellite positioning in train operation condition monitoring and safety control, cannot be guaranteed completely. Satellite positioning availability can be improved through multi-sensor information fusion, used to compensate signal shadowing in constrained operating areas, such as railway stations with canopy. To this aim, the so-called pseudolite (PL) technology makes it possible to enhance the GNSS availability: even when the true GNSS signals are completely blocked, thanks to a pseudolite signal similar to the one radiated by navigation satellites, the receiver can still perform the navigation calculation [34]. It is not a true hybridization approach, but it may be mentioned among the sensor fusion-based solutions. In [35], a method to optimize the pseudolite constellation design in railway stations with a constrained GNSS visibility condition is presented, considering both a degraded GNSS observing scenario, and a fully GNSS-denied operating scenario, the latter taking place when the train is moving in the canopy-covered track area of the station. The Performance Indication Factor (PIF) is used as a metric to assess the performance of the satellite constellation, considering both GNSS and PL, and aiming for its minimization. The paper shows how, in the degraded GNSS scenario, the PIF from the hybrid GNSS/PL mode is lower than that from the GNSS-alone mode. Such a result is attained involving pseudolites in the station area, joint an optimization strategy for their enhanced placement. The paper does not report results in terms of positioning accuracy, because performance is given in terms of PIF that is found to be, in some conditions, relatively high, thus showing that the adopted pseudolite constellation fails to provide the expected locating enhancement capabilities.

Minetto et al. [36] provided a comparison between the legacy Extended Kalman Filter (EKF) and a suboptimal Particle Filter (s-PF) within an integration paradigm aimed at the collaborative hybridization of independent heterogeneous measurements, for enhanced vehicular localization capabilities. The proposed framework relies on the exchange of navigation data and positioning solutions among GNSS receivers through an ad-hoc network, which supports the combination of GNSS observable measurements. In a non-collaborative framework, s-PF easily overcomes EKF performances at the cost of a higher computational cost, in a realistic scenario, just providing network connectivity among few GNSS receivers, allows a hybridized EKF implementation to reach and improve over s-PF outcomes. The paper shows that EKF can be enhanced by integrating auxiliary correlated information more efficiently than adding more particles in s-PF implementation, for which the attainable average accuracy improvement decreases from 23.02% in good visibility to 12.36% in poor visibility conditions. It is important to point out that all the results are obtained by simulations, and no field experiments in a real scenario have been performed.

The 5G-CHAMPION project [37] proposed a GNSS and 5G integrated positioning solution to overcome both insufficient satellite availability and multipath in harsh environments. Pseudoranges from GNSS are integrated by angular information provided by the 5G network, in order to lower the number of GNSS satellites required to compute a position. GNSS signals affected by multipath can be removed from computation, as the 5G Line of Sight (LoS) information provides extra equations to solve the positioning problem. In [38], the proposed GNSS/5G hybridization consists in adding 5G observations in the measurement vector of both single and PPP GNSS, again by exploiting the angular information from 5G. An accuracy in the order of 1 m or even below is expected for positioning supported by 5G networks; additionally, Werner et al. [39] showed that positioning algorithms can run at the network side, ensuring a highly energy efficient approach from the UE's perspective.

In the connected vehicle scenario, for which 5G will act as a fundamental enabler, technical solutions that facilitate obtaining and providing device location information with both high-accuracy and low power consumption at the UE are possible [40]. Simulation results indicate achievable localization accuracies below ±30 cm by using cellular ranging measurements with 50 and 100 MHz of system bandwidth. A hybrid positioning solution combining 5G and GNSS is presented in [41], based on a two-step approach to tightly couple the two technologies. The authors assumed a UE connected to a 5G base station and able to receive a number of GNSS satellites, thanks to an antenna array to serving both 5G and GNSS signals. In the theoretical study, the Fisher Information Matrix (FIM) of a hybrid 5G-GNSS localization estimation is derived, as well as the position and rotation error lower bound. Results are obtained through simulations only, and they indicate that beyond a certain value of the carrier-to-noise signal ratio (CNR), a practical 5G positioning can improve its accuracy through GNSS integration. The value of such a CNR threshold depends on the number of visible satellites. The authors stated that a specific positioning requirement of 50 cm can be achieved by a very favorable GNSS-only configuration (five visible satellites and high CNR), or, alternatively, by the hybrid solution with three visible GNSS satellites, lower GNSS CNR but good 5G signal-to-noise ratio.

The use of mmWave technology is one of 5G communications' key aspects: it enables high data rates and beamforming [42] for AoA positioning, through the use of a large number of elements on antenna arrays for massive multiple-input-multiple-output (MIMO) systems. In [43], high-speed train positioning using specific, beamformed 5G New Radio (NR) downlink synchronization signals (SS) is investigated, based on the sub-meter positioning accuracy requirements specified by the 3GPP Release 14. The Primary Synchronization Signal (PSS), the Secondary Synchronization Signal (SSS), and the Physical Broadcast Channel (PBCH) are transmitted by a network of fixed Remote Radio Heads (RRHs): this ensures very good built-in correlation properties, resulting in accurate ToA and Angle of Departure (AoD) measurements. However, the good results obtained from simulations do not yet satisfy mission-critical requirements, thus requiring an appropriate fusion with GNSS measurements. Despite being challenging to deploy and expensive, the envisioned architecture could be an effective replacement of the obsolete one currently in use, i.e., the international wireless communications system for railway communication and applications (GSM-R).

4. Results analysis

To qualify the result of a measurement activity, its associated uncertainty must be provided [17]. This section briefly recalls the results presented in the studies analyzed before, highlighting the way uncertainty has been expressed.

Two GNSS augmentation approaches tested in [30] provide similar results in terms of positioning *error*, compared to a ground truth position information obtained by means of an external RTK service. A car travelling on a highway instrumented with an ad-hoc augmentation network emulates a train ride. Either in differential or double difference mode, the differences between measured and ground truth position exhibit a Gaussian distribution, from which a ±1 m standard uncertainty at 99% confidentiality level arises. In this case, as prescribed by the GUM, the statistical distribution of the measurements collected in the experiments is determined and, based on that, the standard uncertainty with associated confidence is provided. The double difference mode declares four epochs of system unavailability, against the less conservative differential one. The former has the advantage of removing the clock issues from the navigation solution.

The performances of a local augmentation and integrity monitoring network were analyzed in a previous work by some of the authors [44], where a standard uncertainty of ±2 m for a Gaussian probability distribution of the differences between the measured and the ground truth positions is presented. Except for some cases in which the standard uncertainty increases to ±3 m, the declared confidence level is 99%. Having adopted the same modality to express either the experimental outcomes or the measurement uncertainty, based on a Gaussian distribution in [30,44], the results obtained may be fairly compared.

A real setting, with three there and back train rides between two stations, for a total covered distance of 54.53 km, has been arranged to test the fusion of GNSS and 10 DOF IMU measurements in providing train velocity data in [33]. The presence of tunnels along the railway determined GNSS signal outages, whose time duration was varied by modulating the train's velocity. Reference values of train velocity were obtained by synchronized tachometer and Doppler radar readings. According to the results detailed in the paper, the uncertainty in estimating the train velocity was within ±2 km/h with a 93.53% probability, and within ±1 km/h with 80.90% probability, when the drift in IMU-only measurements (due to GNSS outages) were excluded. Taking into account the GNSS outages, the resulting uncertainty was within ±2 km/h and ±1 km/h with a 88.90% and a 76.85% probability, respectively. In this study, the attained results are expressed in terms of estimated train velocity, and not in terms of train positioning error, as specified in the works analyzed above. As a consequence, the direct comparison among the different approaches is not possible. Additionally, the statistical distribution of the attained measurements is not specified, whereas it would have been important to characterize the influence of GNSS signal outages on the resulting uncertainty.

The fusion of GNSS signals with other sensory data was also discussed by Wu et al. [45], who evaluated the reliability of a GNSS-based train localization unit that exploits measurements from different sensors and a Petri Net to attain localization information. The train localization unit includes a GNSS receiver, an Inertial Navigation System (INS), and a data processing unit (DPU): the INS provides original localization determination, with position and velocity based on the output of internal gyroscope and accelerometer, while GNSS offers the pseudorange and pseudorange rate. A Kalman Filter (KF) corrects the localization determination and the output of the INS. The technical specifications of the INS and GNSS components are provided in the paper. The evaluation of the train localization unit is performed on field test data collected from the Beijing–Shenyang railway in August 2018. The reference system used simultaneously with the unit under test is a NovAtel SPAN-FSAS GNSS/INS receiver, featuring a centimeter-level positioning. The analysis of the experimental data shows that, if the localization unit uses INS alone for about 10 s, the positioning offset with respect to the reference system reaches about 20 m, which cannot satisfy some localization-based applications in railway. Adding GNSS data, the attained failure-rate of the localization is not low but there is room for further improvement. This study provides results in terms of positioning *offset* that can be related to the positioning *error* mentioned in [30,44]; however, details about the qualifying uncertainty are not provided.

In [38], the performances of hybrid GNSS and 5G positioning are assessed in a field test-bed with clear-sky, urban and canyon GNSS visibility conditions. The L1 band only is used to collect GNSS observations, as it mostly happens with the operational capabilities of current mass market receivers, whereas 5G observations are provided through a Transmission Control Protocol (TCP) stream. The 5G conditions are described by the relative position of the 5G base stations (BS) and the UE: 20 m to the north (N20); 20 m to the north and 20 m to the east (N20 E20); and 20 m to the north and 20 m to the southeast (N20 SE20). The same locations are also tested at a distance of 50 m. In both cases, the uncertainty of the BS-to-UE distance is not reported. A geo-referenced position in the open-sky environment provides the ground truth, whereas a precisely computed position is used for the urban and canyon settings. According to the way results are reported (Table 1 in [38]), the following outcomes hold: in the clear sky setting, the authors reported a best mean accuracy of 0.320 m at 99th percentile in (N20 E20) configuration with PPP and 5G; in urban scenario, best mean accuracy of 0.918 m is reported in (N20 E20) with 5G only; and, finally, in canyon scenario, the best mean accuracy declared is 0.7272 m in (N20 E20), using 5G only. 5G is especially important in the urban environments where GNSS signals are less available and the base stations are close enough to the UE.

High-speed train positioning using 5G only is simulated in [43], where a track of over 43 km is modeled, with variable train velocity of up to 400 km/h. Using channel models suitable for the considered scenario, and AoD and ToA measurements with the EKF-based tracking model, the authors declare a mean positioning accuracy of 0.66 m (below the 1 m 5G target), with 95% errors below 1.7 m, and 99% ones below 2.3 m. Separate PSS/SSS sequence identities are assigned to RRHs uniformly located at 500 m intervals along the upper side of the railroad, at a distance of 15 m from the railroad. This way, the same combination of PSS and SSS sequences cannot be heard in the same location from multiple RRHs, and the PSS sequences of adjacent RRHs are always different. Despite the promising outcomes, it is observed that, whenever the train acceleration changes, the tracking algorithm introduces a lag, which affects the tracking accuracy over a certain period. In the works by both Maymo-Camps et al. [38] and Talvitie et al. [43], the attained performances are expressed in terms of mean positioning accuracy; however, details about the statistical distribution of the measurements are not provided, and the coverage factor used to compute the accuracy is not specified as well. Additionally, despite being based on the fusion of different technologies, i.e., GNSS and 5G, the former study does not analyze the uncertainty propagation associated to the combination of different quantities to derive the position information.

5. Discussion

In the discussion of solutions based on GNSS measurements hybridization (i.e., fusion) with other measurements of different nature, it is important to recall the fundamental benefit motivating such approach, i.e., *reducing uncertainty* by combining information from multiple sources [46]. The results summarized in Table 3 motivate some considerations. First, performance comparison among different hybridization approaches is not straightforward. In fact, even if all the works considered aim for the same outcome, i.e., improving train positioning, it is evident how different studies evaluate their proposed approaches in diverse test-beds, obtaining different measures (either train positioning or velocity) whose uncertainty typically is not expressed according to the GUM. This way, it is not easy to say which hybridization approach performs better. Second, each study differs in theoretical foundation and performs different experiments or simulations, usually in environmental and operational settings that are in no way similar. Some studies rely on simulations only, others are supported by real-world experiments, but in conditions that are difficult to replicate, or cannot be fully controlled, e.g., outages, multipath, or fading effects due to the environment in which the train moves. Lastly, sensors used to test the proposed hybridization approaches are different from one study to another, and their own metrological characterization is usually missing.

Table 3. Summary of train positioning approaches and reported results.

Approach	Test Conditions	Reported Result
Augmented GNSS [30]	Emulated train ride and RTK ground truth	train positioning error < 2 m
Locally Augmented GNSS [44]	Static OBU and 2 RS located along the track: Leipzig (IGS), Zurich (EGNOS)	train positioning error in $[-2 \div 4]$ m, some spikes in $[-5 \div 6.5]$ m
GNSS and 10 DOF IMU fusion [33]	Field test: 3 there and back travels (54.53 km covered), variable time duration GNSS outages in tunnels, reference velocity by tachometer and Doppler radar readings	train velocity error (no GNSS outages) < 2 km/h, 93.53% prob. < 1 km/h, 80.90% prob. train velocity error (with GNSS outages) < 2 km/h, 88.90% prob. < 1 km/h, 76.85% prob.
GNSS & 5G hybridization [38]	Test-bed with GNSS visibility conditions: clear-sky, urban, canyon environments. 5G BS-UE locations: N20, (N20 E20), (N20 SE20), N50, (N50 E50), (N50 SE50). Single frequency GNSS (L1). Ground truth: geo-ref. position in open-sky; precisely computed position in urban and canyon environments	best mean positioning accuracy at 99th percentile: clear-sky: 0.320 m, in (N20 E20) PPP&5G urban: 0.918 m, in (N20 E20) 5G only canyon: 0.7272 m, in (N20 E20) 5G only
5G only [43]	Simulated track of over 43 km with variable train velocity of up to 400 km/h	mean positioning accuracy: 0.66 m 95% errors < 1.7 m, 99% errors < 2.3 m

According to the GUM, the measurement procedure shall be replicable and repeatable *under the same measurement conditions*. In real railway operation conditions, the positioning accuracy attained by the different studies shows strong variations; at the same time, the *dynamic* train localization accuracy is among one of the most relevant KPIs of the GNSS/hybridized-GNSS receiver, mandated by safety constraints in railway applications. As a consequence, the accuracy in both the along-track and cross-track directions of the train positioning should be evaluated under dynamic movement, possibly using bidirectional error models. However, measurements in which at least one of the quantities of interest is time-dependent, i.e., it changes on scales from picoseconds up to several minutes (called *dynamic measurements* [47]), pose additional challenges to the existing mathematical models and measurement methodologies. With the same motivations that steered the deployment of the *Galileo above* test bed (Galileo Application Center for Ground-Based Vehicles) in Germany in 2010 [48], different GNSS-hybridization approaches should be tested under identical, repeatable test conditions, which cannot happen if a pre-defined railway track is not used as a reference by all the experiments. In fact, the ISO 5725 [49] defines the repeatability limit as the value less than, or equal to which the absolute difference between two test results, obtained *under* the repeatability conditions, may be expected to be, with a probability of 95%. For measurements aimed at train localization by GNSS or hybridized-GNSS receivers, the experiment could be made twofold: static measurements at fixed antenna location, and dynamic measurements along a pre-defined railway track [50], with uncertainty expressed according to the GUM.

Considering the safety performances of the approaches analyzed in Table 3, the solutions presented in [30,44,51] report four epochs of system unavailability, and normal operation of the system for all the considered epochs, respectively. No details about the safety performances were provided by Otegui et al. [33] and Maymo-Camps et al. [38], whereas Talvitie et al. [43] reported a submeter accuracy with 75% availability, not enough for mission-critical use cases, such as autonomous trains. In [52], the authors studied and compared two fault detection methods for GNSS, based on an embedded odometer and on a single-axis accelerometer, respectively. The methods suffer from uncertainty on the accuracy of the track geometry information that is a necessary parameter for both of them, but this aspect is not further developed in the mentioned paper.

Table 4. PNT (including GNSS) Integrity requirements for rail [53].

Train LDS functionality	Position Accuracy (2σ)
VB detection (vital)	1 m - The PNT solution shall provide the train position with a horizontal accuracy lower than 1 m
VB detection (non vital)	5 m + 5% s - s is the distance travelled from the last calibration of the odometric device. The PNT solution shall provide the train position with a horizontal accuracy lower than 5 m
Track discrimination	1.9 m - The PNT solution shall provide the train position with a horizontal accuracy lower than 1.9 m

In the framework of the ERTMS evolution plan, GNSS is seen as a *game changer* innovation, with the potential to eliminate most of the track circuits and to increase the transport capacity, thus reducing the costs and contributing to a greener transport means. Moreover, the main goal of the introduction of the GNSS is the cost reduction without compromising the system safety. Considering that conventional LDS, such as those based on RF transponders (balises) deployed along the railway, are not economically sustainable for regional and freight lines, and that these lines constitute a big percentage of the European railways, the market slice addressed by such an innovation is quite huge. In fact, compliance with the safety requirements imposed by CENELEC norms is very challenging, so that adoption of an external AIMN (Augmentation and Integrity Monitoring Network) is highly recommended [9].

Concerning the availability, in the rail environment, the integration among the GNSS and a set of external sensors (e.g., mechanical odometers, inertial units or imaging sensors) can be exploited to increase system reliability and availability in GNSS denied areas (e.g., tunnels) as well as to improve system accuracy and integrity (e.g., by detecting and mitigating severe multipath or electromagnetic intentional and unintentional interferences).

From the operational point of view, uses of the train LDS include: (i) determination whether the train is in the proximity of those locations that may be trespassed only by authorized trains (this function is marked as VITAL in ERTMS); (ii) determination of the track on which the train is operating and its location on that track (function marked as NON-VITAL in the standard); (iii) detection of the eventual split of a train (train integrity assessment); and (iv) determination of the mutual distance among trains. Concerning Cases (i) and (ii), although different accuracy requirements to the LDS may be set to take into account the accuracy offered by today LDSs, the approach followed by the railway operators leading the introduction of the EGNSS in ERTMS is to define as target the same accuracy achieved by the conventional balise readers, as illustrated in Table 4.

This choice is the key to support full interoperability allowing trains equipped with next generation LDS to operate on lines equipped with physical balises, without the need of adopting different train control rules and procedures depending on the on-board equipment. Cases (iii) and (iv) are relevant for the upcoming ERTMS Level 3, so that the related requirements are still to be defined. Nevertheless, concerning performance, the driver is the decimeter accuracy required at least for the VITAL functions.

To reach the ambitious SIL-4 (Safety Integrity Level) target in terms of THR, the ERTMS has eliminated the need for control via light signals to drivers, eliminating the potential for human errors. The train is remotely controlled by the RBC (Radio Block Center), which periodically transmits to the train driver a proper authorization to proceed until the next location under speed supervision. If the train speed exceeds the authorized speed profile, the train is automatically stopped without human intervention. To satisfy the requirements reported in Table 4, data fusion of GNSS and other independent sensors is needed.

6. Conclusions and Future Perspectives

With the progresses in the deployment and near-term completion of Galileo, and thanks to the successful implementation and adoption of Europe's space-based positioning augmentation system, EGNOS, the basis for a more widespread integration of satellite-based support in the railway sector is now laid. The question to address in the near future is not about what type of possible support GNSS can provide to the railway sector, but how such a support may substantiate. The research studies and experiments discussed in this work have shown that GNSS is able to make rail systems less complex, cheaper, more reliable and responsive, while maintaining the highest standards of safety, but the roll-out of space technology in the railways sector needs to be accelerated, by defining the future system architecture and quickly moving the solutions from the laboratory or pilot deployment onto the track. Interoperability and standardization are the preconditions for a true industrial innovation in the railway sector, which could bring additional advantages such as decarbonizing the transport system and increased sustainability. Satellite technologies are key to increasing capacity and efficiency in the system by enhancing quality and consistency of train localization, but all new rail systems must be certified and still there is an open issue about who should be responsible for such a certification. A similar discussion was carried out in the aviation sector, where the introduction of space technologies had a disruptive role as is expected for the railway transport system. Within the European Shift2Rail Joint Undertaking [53], the project GATE4Rail (GNSS Automated Virtualized Test Environment for RAIL) [54] aims to implement a geo-distributed simulation and verification platform connecting GNSS centers of excellence and ERTMS/ETCS laboratories, to evaluate the GNSS performances in the railway environment with agreed methodologies and tools. This way, it will be possible to stress the global system in presence of very rare fault events instead of long and expensive tests on field, thus developing a standard methodology for representative test cases based on a common test process framework, for zero on-site testing, to minimize deployment cost and time.

In this work, the main GNSS and hybridized-GNSS based techniques for automated railway traffic and train positioning proposed in the literature are reviewed and analyzed, with a specific focus on the issues related to the expression of the position location uncertainty and the measurements accuracy declared by the different studies.

The performed review highlights a lack of repeatable and comparable results, typically attained with different experiments and test-beds, as well as a number of different ways to express uncertainty, not always in line with the Guide to the Expression of Uncertainty in Measurement. Additional challenges to the existing measurement models and methodologies are posed by the dynamic nature of the positioning systems, and by the fact that the proposed approaches rely on the same satellite-based localization systems that should represent the reference against which the attainable accuracy must be evaluated. These open issues should be taken into account and drive future measurement-related research in the field.

Author Contributions: Both the authors contributed equally to the development of this work, the analysis of the literature, the discussion, and the editing of the manuscript. All authors have read and agreed to the published version of the manuscript.

Funding: This research was internally funded by Università Politecnica delle Marche.

Conflicts of Interest: The authors declare no conflict of interest.

References

1. Chiocchio, S.; Cinque, E.; Persia, A.; Salvatori, P.; Stallo, C.; Salvitti, M.; Valentini, F.; Pratesi, M.; Rispoli, F.; Neri, A.; et al. A Comprehensive Framework for Next Generation of Cooperative ITSs. In Proceedings of the 2018 IEEE 4th International Forum on Research and Technology for Society and Industry (RTSI), Palermo, Italy, 10–13 September 2018, pp. 1–6. [CrossRef]
2. Taranto, R.D.; Muppirisetty, S.; Raulefs, R.; Slock, D.; Svensson, T.; Wymeersch, H. Location-Aware Communications for 5G Networks: How location information can improve scalability, latency, and robustness of 5G. *IEEE Signal Process. Mag.* **2014**, *31*, 102–112. [CrossRef]
3. Falone, R.; Stallo, C.; Gambi, E.; Spinsante, S. SDR GNSS receivers: A comparative overview of different approaches. In Proceedings of the 2014 IEEE Metrology for Aerospace (MetroAeroSpace), Benevento, Italy, 29–30 May 2014; pp. 326–331. [CrossRef]

4. European Global Navigation Satellite Systems Agency. What Is GNSS? Available online: https://www.gsa.europa.eu/european-gnss/what-gnss (accessed on 15 November 2019).
5. Zalewski, P.; Bilewski, M. GNSS Measurements Model in Ship Handling Simulators. *IEEE Access* **2019**, *7*, 76428–76437. [CrossRef]
6. Zalewski, P. GNSS Integrity Concepts for Maritime Users. In Proceedings of the 2019 European Navigation Conference (ENC) , Warsaw, Poland, 9–12 April 2019; pp. 1–10. [CrossRef]
7. *Report on Road User Needs and Requirements—Outcome of the European GNSS' User Consultation Platform*; Technical Report Issue/Revision: 1.0; European GNSS Agency: Prague, Czech Republic, 2018.
8. Daponte, P.; De Vito, L.; Mazzilli, G.; Picariello, F.; Rapuano, S. Power consumption analysis of a Wireless Sensor Network for road safety. In Proceedings of the 2016 IEEE International Instrumentation and Measurement Technology, Taipei, Taiwan, 23–26 May 2016; pp. 1–6. [CrossRef]
9. Lo, S.; Pullen, S.; Blanch, J.; Enge, P.; Neri, A.; Palma, V.; Salvitti, M.; Stallo, C. Projected performance of a baseline high integrity GNSS railway architecture under nominal and faulted conditions. In Proceedings of the 30th International Technical Meeting of the Satellite Division of the Institute of Navigation, Portland, OR, USA, 25–29 September 2017; Volume 4, pp. 2148–2171.
10. European Global Navigation Satellite Systems Agency. European Rail, Supported by European GNSS. Available online: https://www.gsc-europa.eu/news/european-rail-supported-by-european-gnss-3 (accessed on 15 November 2019).
11. Dovis, F.; Muhammad, B.; Cianca, E.; Ali, K. A Run-Time Method Based on Observable Data for the Quality Assessment of GNSS Positioning Solutions. *IEEE J. Sel. Areas Commun.* **2015**, *33*, 2357–2365. [CrossRef]
12. Stallo, C.; Neri, A.; Salvatori, P.; Coluccia, A.; Capua, R.; Olivieri, G.; Gattuso, L.; Bonenberg, L.; Moore, T.; Rispoli, F. GNSS-based location determination system architecture for railway performance assessment in presence of local effects. In Proceedings of the 2018 IEEE/ION Position, Location and Navigation Symposium (PLANS), Monterey, CA, USA 23–26 April 2018; pp. 374–381. [CrossRef]
13. Gao, S.; Dong, H.; Ning, B.; Zhang, Q. Cooperative Prescribed Performance Tracking Control for Multiple High-Speed Trains in Moving Block Signaling System. *IEEE Trans. Intell. Transp. Syst.* **2019**, *20*, 2740–2749. [CrossRef]
14. European Global Navigation Satellite Systems Agency. GNSS in Rail-Roadmap-European GNSS Agency. Available online: https://www.gsa.europa.eu/sites/default/files/rail-roadmap2018.pdf (accessed on 18 December 2019).
15. Roth, J.D.; Tummala, M.; McEachen, J.C. Fundamental Implications for Location Accuracy in Ultra-Dense 5G Cellular Networks. *IEEE Trans. Veh. Technol.* **2019**, *68*, 1784–1795. [CrossRef]
16. Spinsante, S.; Stallo, C. Issues on Uncertainty to Train Positioning in Hybridized-GNSS Approaches. In Proceedings of the 2019 IEEE 5th International Workshop on Metrology for AeroSpace (MetroAeroSpace), Torino, Italy, 19–21 June 2019; pp. 387–392. [CrossRef]
17. *Evaluation of Measurement Data—Guide to the Expression of Uncertainty in Measurement*; Technical Report; JCGM: Sèvres, France, 2008.
18. Oxley, A. Chapter 1—Positioning and Navigation Systems. In *Uncertainties in GPS Positioning*; Oxley, A., Ed.; Academic Press: Cambridge, MA, USA, 2017; pp. 1–18.
19. Charitanetra, S.; Noppanakeepong, S. Mobile positioning location using E-OTD method for GSM network. In Proceedings of the Student Conference on Research and Development, 2003, Putrajaya, Malaysia, 25–26 August 2003; pp. 319–324. [CrossRef]
20. Holma, H.; Toskala, A. LTE Release 8 and 9 Overview. In *LTE Advanced: 3GPP Solution for IMT-Advanced*; Wiley: Hoboken, NJ, USA, 2012. [CrossRef]
21. Holma, H.; Toskala, A. Release 11 and Outlook towards Release 12. In *LTE Advanced: 3GPP Solution for IMT-Advanced*; Wiley: Hoboken, NJ, USA, 2012. [CrossRef]
22. Mardeni, R.; Shahabi, P.; Riahimanesh, M. Mobile station localization in wireless cellular systems using UTDOA. In Proceedings of the 2012 International Conference on Microwave and Millimeter Wave Technology (ICMMT), Shenzhen, China, 5–8 May 2012; Volume 5, pp. 1–4. [CrossRef]
23. del Peral-Rosado, J.A.; Raulefs, R.; López-Salcedo, J.A.; Seco-Granados, G. Survey of Cellular Mobile Radio Localization Methods: From 1G to 5G. *IEEE Commun. Surv. Tutor.* **2018**, *20*, 1124–1148. [CrossRef]
24. *Stage 2 Functional Specification of UE Positioning in E-UTRAN*; Release 13, V13.0.0; Technical Report 3GPP TS 36.305; Sophia Antipolis: Cannes, France, 2016.

25. Koivisto, M.; Hakkarainen, A.; Costa, M.; Kela, P.; Leppanen, K.; Valkama, M. High-Efficiency Device Positioning and Location-Aware Communications in Dense 5G Networks. *IEEE Commun. Mag.* **2017**, *55*, 188–195. [CrossRef]
26. Pirsiavash, A.; Broumandan, A.; Lachapelle, G.; O'Keefe, K. Detection and De-weighting of Multipath-affected Measurements in a GPS/Galileo Combined Solution. In Proceedings of the 2019 European Navigation Conference (ENC), Warsaw, Poland, 9–12 April 2019; pp. 1–11. [CrossRef]
27. Pirsiavash, A.; Broumandan, A.; Lachapelle, G.; O'Keefe, K. GNSS Code Multipath Mitigation by Cascading Measurement Monitoring Techniques. *Sensors* **2018**, *18*. [CrossRef] [PubMed]
28. Zumberge, J.; Heflin, M.; Jefferson, D.; Watkins, M.; Webb, F. Precise point positioning for the efficient and robust analysis of GPS data from large networks. *J. Geophys. Res. Solid Earth* **1997**, *102*, 5005–5017. [CrossRef]
29. Grewal, M.; Weill, L.; Andrews, A. *Global Positioning Systems, Inertial Navigation, and Integration*; Wiley: Hoboken, NJ, USA, 2007. [CrossRef]
30. Salvatori, P.; Stallo, C.; Coluccia, A.; Rispoli, F.; Neri, A. Differential GNSS and Double Difference Approaches Comparison for High Integrity Railway Location Determination System. In Proceedings of the 2018 5th IEEE International Workshop on Metrology for AeroSpace (MetroAeroSpace), Rome, Italy, 20–22 June 2018; pp. 1–5. [CrossRef]
31. Otegui, J.; Bahillo, A.; Lopetegi, I.; Díez, L.E. A Survey of Train Positioning Solutions. *IEEE Sens. J.* **2017**, *17*, 6788–6797. [CrossRef]
32. Beugin, J.; Legrand, C.; Marais, J.; Berbineau, M.; El-Koursi, E. Safety Appraisal of GNSS-Based Localization Systems Used in Train Spacing Control. *IEEE Access* **2018**, *6*, 9898–9916. [CrossRef]
33. Otegui, J.; Bahillo, A.; Lopetegi, I.; Díez, L.E. Evaluation of Experimental GNSS and 10-DOF MEMS IMU Measurements for Train Positioning. *IEEE Trans. Instrum. Meas.* **2019**, *68*, 269–279. [CrossRef]
34. Gan, X.; Yu, B.; Heng, Z.; Huang, L.; Li, Y. Indoor combination positioning technology of Pseudolites and PDR. In Proceedings of the 2018 Ubiquitous Positioning, Indoor Navigation and Location-Based Services (UPINLBS), Wuhan, China, 22–23 March 2018; pp. 1–7. [CrossRef]
35. Zhao, X.; Liu, J.; Cai, B.; Lu, D.; Wang, J. Research on Optimized Pseudolite Constellation Design under Constrained GNSS Environment in Railway Stations. In Proceedings of the 2019 IEEE Intelligent Transportation Systems Conference (ITSC), Auckland, New Zealand, 27–30 October 2019; pp. 3475–3481. [CrossRef]
36. Minetto, A.; Falco, G.; Dovis, F. On the Trade-Off between Computational Complexity and Collaborative GNSS Hybridization. In Proceedings of the 2019 IEEE 90th Vehicular Technology Conference (VTC2019-Fall), Honolulu, HI, USA, 22–25 September 2019; pp. 1–5. [CrossRef]
37. Mueck, M.; Strinati, E.C.; Kim, I.; Clemente, A.; Dore, J.; Domenico, A.D.; Kim, T.; Choi, T.; Chung, H.K.; Destino, G.; et al. 5G CHAMPION - Rolling out 5G in 2018. In Proceedings of the 2016 IEEE Globecom Workshops (GC Wkshps), Washington, DC, USA, 4–8 December 2016; pp. 1–6. [CrossRef]
38. Maymo-Camps, R.; Vautherin, B.; Saloranta, J.; Crapart, R. 5G positioning and hybridization with GNSS observations. In Proceedings of the International Technical Symposium on Navigation and Timing (ITSNT), Toulouse, France, 13–16 November 2018. .
39. Werner, J.; Costa, M.; Hakkarainen, A.; Leppänen, K.; Valkama, M. Joint User Node Positioning and Clock Offset Estimation in 5G Ultra-Dense Networks. In Proceedings of the 2015 IEEE Global Communications Conference (GLOBECOM), San Diego, CA, USA, 6–10 December 2015; pp. 1–7. [CrossRef]
40. Rastorgueva-Foi, E.; Costa, M.; Koivisto, M.; Leppänen, K.; Valkama, M. User Positioning in mmW 5G Networks Using Beam-RSRP Measurements and Kalman Filtering. In Proceedings of the 2018 21st International Conference on Information Fusion (FUSION), Cambridge, UK, 10–13 July 2018; pp. 1–7. [CrossRef]
41. Destino, G.; Saloranta, J.; Seco-Granados, G.; Wymeersch, H. Performance Analysis of Hybrid 5G-GNSS Localization. In Proceedings of the 2018 52nd Asilomar Conference on Signals, Systems, and Computers, Pacific Grove, CA, USA, 28–31 October 2018; pp. 8–12. [CrossRef]
42. Tariq, S.; Psychoudakis, D.; Eliezer, O.; Khan, F. A new approach to antenna beamforming for millimeter-wave fifth generation (5G) systems. In Proceedings of the 2018 Texas Symposium on Wireless and Microwave Circuits and Systems (WMCS), Waco, TA, USA, 5–6 April 2018; pp. 1–5. [CrossRef]

43. Talvitie, J.; Levanen, T.; Koivisto, M.; Pajukoski, K.; Renfors, M.; Valkama, M. Positioning of high-speed trains using 5G new radio synchronization signals. In Proceedings of the 2018 IEEE Wireless Communications and Networking Conference (WCNC), Barcelona, Spain, 15–18 April 2018; pp. 1–6. [CrossRef]
44. Palma, V.; Salvatori, P.; Stallo, C.; Coluccia, A.; Neri, A.; Rispoli, F. Performance evaluation in terms of accuracy positioning of local augmentation and integrity monitoring network for railway sector. In Proceedings of the 2014 IEEE Metrology for Aerospace (MetroAeroSpace), Benevento, Italy, 29–30 May 2014; pp. 394–398. [CrossRef]
45. Wu, B.; Cai, B.; Lu, D. An Advanced Reliability Evaluation Method For GNSS-based Train Localization Unit. In Proceedings of the 2019 International Conference on Electromagnetics in Advanced Applications (ICEAA), Granada, Spain, 9–13 September 2019; pp. 0558–0562. [CrossRef]
46. Hall, D.L.; Llinas, J. An introduction to multisensor data fusion. *Proc. IEEE* **1997**, *85*, 6–23. [CrossRef]
47. Eichstädt, S.; Elster, C.; Smith, I.; Esward, T. Evaluation of dynamic measurement uncertainty—An open-source software package to bridge theory and practice. *J. Sensors Sens. Syst.* **2017**, *6*, 97–105. [CrossRef]
48. Hoelper, C.; Pölöskey, M. Galileo above — A terrestrial Galileo test environment for vehicular applications: Automotive & rail. In Proceedings of the 2010 5th ESA Workshop on Satellite Navigation Technologies and European Workshop on GNSS Signals and Signal Processing (NAVITEC), Noordwijk, The Netherlands, 8–10 December 2010, pp. 1–5. [CrossRef]
49. Deldossi, L.; Zappa, D. ISO 5725 and GUM: comparison and comments. *Accredit. Qual. Assur.* **2009**, *14*, 159–166. [CrossRef]
50. Lu, D.; Spiegel, D.; Becker, U.; Cai, B.; Wang, J.; Liu, J.; Liu, X. Repeatability test method of GNSS for safe train localisation in real and simulated environments. In Proceedings of the 2016 IEEE/ION Position, Location and Navigation Symposium (PLANS), Savannah, Georgia, USA, 11–16 April 2016; pp. 687–692. [CrossRef]
51. Stallo, C.; Neri, A.; Salvatori, P.; Capua, R.; Rispoli, F. GNSS Integrity Monitoring for Rail Applications: Two-Tiers Method. *IEEE Trans. Aerosp. Electron. Syst.* **2019**, *55*, 1850–1863. [CrossRef]
52. Heekwon, N.; Vezinet, J.; Milner, C. Simplified GNSS Fusion-based Train Positioning System and its Diagnosis. In Proceedings of the 32nd International Technical Meeting of the Satellite Division of The Institute of Navigation Miami, Miami, FL, USA, 16–20 September 2019; pp. 3033–3044. [CrossRef]
53. European Commission-S2R. Shift2Rail Joint Undertaking. Available online: https://shift2rail.org/ (accessed on 24 March 2020).
54. European Commission. Gate4Rail—GNSS Automated Virtualized Test Environment for Rail. Available online: https://cordis.europa.eu/project/id/826324 (accessed on 24 March 2020).

© 2020 by the authors. Licensee MDPI, Basel, Switzerland. This article is an open access article distributed under the terms and conditions of the Creative Commons Attribution (CC BY) license (http://creativecommons.org/licenses/by/4.0/).

Article

Application of GNSS/INS and an Optical Sensor for Determining Airplane Takeoff and Landing Performance on a Grassy Airfield †

Jaroslaw Pytka [1,*], Piotr Budzyński [1], Jerzy Józwik [1], Joanna Michałowska [2], Arkadiusz Tofil [1], Tomasz Łyszczyk [1] and Dariusz Błażejczak [3]

1 Faculty of Mechanical Engineering, Lublin University of Technology, 20-618 Lublin, Poland; p.budzynski@pollub.pl (P.B.); j.jozwik@pollub.pl (J.J.); atofil@pwsz.chelm.pl (A.T.); tomasz.lyszczyk@gmail.com (T.Ł.)
2 The State School of Higher Education, The Institute of Technical Sciences and Aviation, 22-100 Chełm, Poland; jmichalowska@pwsz.chelm.pl
3 Department of Construction and Usage of Technical Devices, West Pomeranian University of Technology in Szczecin, 70-310 Szczecin, Poland; Dariusz.Blazejczak@zut.edu.pl
* Correspondence: j.pytka@pollub.pl; Tel.: +48-515-155-341

† This paper is an extended version of conference paper Pytka, J.; Budzyński, P.; Józwik, J.; Łyszczyk, T.; Laskowski, J.; Gnapowski, E. Measurement of Takeoff and Landing Ground Roll of Airplane on Grassy Runway. In Proceedings of the 6th IEEE Workshop Metro Aerospace, Torino, Italy, 19–21 June 2019.

Received: 8 November 2019; Accepted: 9 December 2019; Published: 12 December 2019

Abstract: The performance of a PZL 104 Wilga 35A airplane was determined and analyzed in this work. Takeoff and landing distances were determined by means of two different methods: one which utilized a Global Navigation Satellite System/Inertial Navigation System (GNSS/INS) sensor and another in which airplane ground speed was measured with the use of an optical non-contact sensor. Based on the airfield measurements, takeoff and landing distances as well as rolling resistance coefficients were determined for the used airplane on a grassy runway at the Radawiec airfield, located near Lublin, southeast Poland. The study was part of the "GARFIELD" project that is expected to deliver an online information system on grassy airfield conditions. It was concluded that both sensors were suitable for the aimed research. The results obtained in this study showed the effects of high grass upon the takeoff and landing performances of the test airplane. Also, the two methods were compared against each other, and the final results were compared to calculations of ground distances by means of the chosen analytical models.

Keywords: airfield performance; landing and takeoff distances; rolling resistance coefficient; GNSS/INS sensor; optical sensor; measurement; grassy airfield

1. Introduction

1.1. Grassy Airfields

General aviation (GA) airplanes often operate on aerodromes with unprepared, grassy runways. Some examples of such operations are agricultural flights, firefighting, humanitarian flights in the third world countries, search and rescue flights, and sport and leisure aviation. Airfield performance, in terms of ground roll distances, is strongly affected by weather conditions. In order to ensure safe and efficient takeoffs and landing, it is necessary to analyze the performance of an airplane, and this is mainly affected by wheel–soil interactions. This study is an extended version of the paper entitled "Measurement of Takeoff and Landing Ground Roll of Airplane on Grassy Runway", presented during the 2019 IEEE International Workshop Metrology for Aerospace in Torino, Italy [1].

Grassy runways are one example of a natural terrain, and its mechanical strength is very sensitive to changes in moisture content [2–4]. A dry grassy surface is hard, and a wheel does not sink down. Wetting the grassy surface results in a decrease of strength which may have bad effects on airfield performance [5].

The two most important issues are (1) a decrease in bearing capacity which results in a reduction of gross weight and (2) an increase in the rolling resistance of the wheels. A solution by means of an instrumental method for the first problem was the introduction of the California Bearing Ratio (CBR) which is a measure describing mechanical characterization of homogenous soil. Four different categories of flexible surfaces were introduced by the International Civil Aviation Organization (ICAO): high, medium, low, and very low. These measures were quantified by means of the CBR values 15, 10, 6, and 3, respectively [6].

The problem of increased rolling resistance is critical for safe and economical operation of aircrafts on grassy airfields. Rolling resistance, F_{RR}, determines the dynamics of the longitudinal motion of wheeled vehicles including aircrafts. Generally, the higher the F_{RR}, the longer the takeoff distance (landing distance decreases, but the problem of low CBR may result in a nose-down incident). The rolling resistance is often mixed-up with braking friction. This is false, since the braking friction is, in fact, the coefficient of friction between the wheels and a surface. Besides, knowing the actual rolling resistance is critical in the determination of the airfield performance of an airplane [7–10].

The aim of this study was to measure the airfield performance of a light GA airplane operating on a grassy airfield under mixed surface conditions. A comparison of two sensor systems was also conducted. In addition, we compared analytical models that enable the calculation of the takeoff and landing ground distances.

The difficulty of measuring airfield performances of an airplane during takeoff or landing on grassy airfields lies mainly in the fact that the tire–soil interactions on a deformable grass surface depends on many factors such as soil moisture, grass length, humidity, and the weather. The high dynamics of changes in these factors means that traditional models of forecasting the aircraft's ground performance give vague results, differing from the real ones. On the other hand, measurements in real conditions using full-size objects (aircraft) are burdened with errors whose impact is difficult to estimate. Hence, there is importance in this undertaken research for the larger research community and for practitioners in general aviation. The subject matter is rarely discussed in the literature, and there are no proven and credible solutions. The cited methods of testing the takeoff and landing distances of an airplane are either not very accurate or complicated and difficult to access, especially in the case of grassy airfields. Therefore, the search for a new method or adaptation of existing solutions may contribute to a more rational and effective use of grassy airports, improvements in safety, and extension of the network of connections in air communication.

1.2. Methods for Determination of Rolling Resistance for Aircraft Tires: A Review

A revision of methods for rolling resistance determination was done by van Es [7] who developed a method for the determination of an aircraft's wheel rolling resistance in snow as well as by Shoop [9] who determined a method to estimate the rolling resistance of an airfield to predict takeoff distance.

Van Es assumed that high-speed compaction of a compressible material (e.g., soil, snow, grass) causes a resistance related to the speed with which the material is compressed. It was suggested that this resistance is related to the increase in kinetic energy of the particles from zero up to the aircraft ground speed. The resulting method assumes two terms: drag due to the compression and drag due to the motion. This method has limitations, one of them was that it is not recommended for wet snow or wet soil [7].

Shoop concluded that the total rolling resistance of an aircraft on loose soil is due to the internal mechanical resistance (e.g., landing gears, wheels suspensions, bearings), aircraft aerodynamic drag, low-speed compaction of the soil, and high-speed resistance due to the soil drag on the wheels and spray drag on the body of the aircraft. Soil compaction term was determined with an instrumented vehicle, and the speed-related term was estimated in real size flight tests in which a C-17 transport aircraft was used [9].

Rowe and Hegedus [10] have dealt with rolling resistance on extremely wet soils and proposed an equation in which the resistance of a wheel is due to the viscosity and static pressure and to velocity pressure (dynamic resistance).

Crenshaw [11] developed an equation for rolling resistance due to the fact of velocity and dynamic sinkage which is a function of tire diameter, deflection, section height, cone index, and other empirical constants. These factors can be determined from soil data and from measurements on an instrumented aircraft.

Hovland [12] developed a theory of inertial forces in moving soil and their effect on tractive forces. It was possible to predict at what velocity severe and immobilizing sinkage of the wheels will begin for an airplane landing on a soft surface. As an example, a theoretical relationship between soil lift force and aircraft ground speed for a Cessna 150 landing on soft playa or marsh was shown.

Gibbesch [13] presented a simulation method for calculation of tractive forces on wheels of an aircraft on soft soil. This method assumes the effect of both elasticity and viscosity on contact pressures in wheel–soil systems, and the soils were described by means of a rheological model. The method was applied to determine the vertical forces acting on the multi-wheeled gears of a transport aircraft.

1.3. Takeoff and Landing Ground Roll Distance Determination Methods

Takeoff and landing belong to the most critical maneuvers of airplanes, especially when operated on unpaved, short grassy runways. From the point of view of a pilot, a safe takeoff ground roll must be finished well before the runway ends, mainly because of obstacles clearance. It is a similar situation for landing, where the approach should take into account a safe clearance above obstacles. This situation is shown in Figure 1. A pilot preparing for takeoff or planning a flight must know the ground performance of her/his airplane related to the available runway length. Because of environmental factors (e.g., grass length, soil moisture content, wind component, density altitude), an actual takeoff or landing ground roll for a given airplane may differ very much. This is a major reason why takeoff and landing performance should be measured and analyzed.

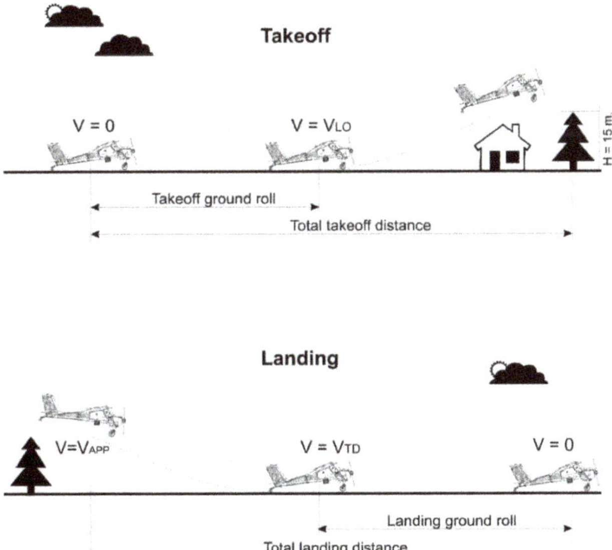

Figure 1. A schematic of takeoff and landing. V_{LO}—liftoff speed, V_{APP}—approach speed, V_{TD}—touchdown speed.

Several measurement methods for airplane ground roll during takeoff or landing have been developed through the years [14]. Some of them are briefly reported below.

The sighting bar method uses one or two sighting bars located at a known distance from the runway. By the use of these bars, a stopwatch, runway observers, and hand-recorded flight data, takeoff and landing distances data may be obtained. In the triangulation method, a camera and a scale located near the camera, parallel to the airfield direction, are used. The scale has two wires installed so that the gap between the scale and the wires ("runway wire" and "50 ft wire") is set to coincide with the runway centre line and the 50 ft (or 15 m) screen at the mid-point of the runway. The device is set level and parallel with the runway.

The movie theodolite method utilizes a movie camera that is used to record other data, such as time and azimuth, in addition to filming the airplane. In onboard theodolite method, a camera is mounted on the airplane to obtain three-axis position information. Runway lights and other objects on or along the runway of known size are used with photogrammetric techniques and perspective geometry to obtain airplane position and altitude.

Methods based on electronics include Del Notre Transponder, laser altimeter, and global positioning system (GPS) sensors. The Del Notre Transponder measures horizontal distance and, when combined with a radio altimeter for height information, provides all the distance and height information necessary for determining takeoff and landing distances. The Trisponder consists of a distance measuring unit (DMU), a master transponder, and associated antennae, cables, and power sources.

The laser method uses a laser landing altimeter. This unit measures distance using a modulated laser beam with centimeter accuracy.

A GPS sensor is used to determine aircraft longitudinal and vertical position together with time coordinates during a takeoff or landing. The weakness of this method is that the moment of liftoff is difficult to be determined with required accuracy.

A wheel dynamometer, also used in the automotive, can be either a rotating sensor, embedded in a roadwheel or a stator sensor, in the wheel hub. A critical point is bearing the wheel, since the relatively large size of the stator sensor usually leads to unconventional solutions and major changes in wheel rim construction. On the other hand, with rotating dynamometers one has to consider the need of transferring data signals from the rotating sensor which can be done by means of a slip ring or a radio telemetry system. Also, a disadvantage of rotating wheel dynamometers is that four of the six components measured are dependent on the wheel position (the sensor rotates and changes its orientation related to acting forces). This requires recalculation of the measured data with respect to the wheel position which has to be determined independently. For our dynamometer system, we chose the stator sensor as a simpler solution. Since the space available for the sensing element core and the transducers was very tight (approximately 150 mm in diameter × 200 mm width), it was impossible to fit all the needed transducers into the stator sensor, and the M_Y sensor was placed off the stator sensor and embedded into the brake arm (see the following section).

2. Materials and Methods

2.1. GNSS/INS Sensor

The GNSS/INS (global navigation satellite system/inertial navigation system) is a system based on GPS, upgraded by adding a ground reference station that is wirelessly connected with the vehicle in motion and improves the performance of the basic system. The referencing system recalculates the outgoing data and, being stationary, assures higher accuracy of the readings. The inertial navigation system (INS) consists of a set of laser fiber gyroscopes and accelerometers that enable the calculation of the kinematics of a running (flying) vehicle, based on the measured linear and angular accelerations. The results of those measurements, in the form of time courses, are then integrated in order to obtain velocities (1st integral after time) and distances (2nd integral after time).

The system used in this study was the OXFORD RT3002 (Oxford Technical Solutions Ltd., Oxfordshire, UK) which consisted of a main unit installed in the aircraft with a 12 V power supply, a remote base station, and a portable computer for data acquisition and storage. The system had to be initiated before tests by moving with a constant velocity of approximately 10 km/h, and it was performed in a ground vehicle. Initiating the system in the airplane was not possible because of the high vibration level. The acquisition time of the system was 10 ms, and the positioning resolution was 20 mm. The navigation unit enabled measurement of the 39 kinematical parameters of the aircraft's motion. Figure 2 shows the installation of the GNSS/INS unit in the test airplane.

Figure 2. Installation of the global navigation satellite system/inertial navigation system (GNSS/INS) unit in the test airplane.

2.2. The CORREVIT L400 Optical Sensor

The CORREVIT optical sensor is a type of sensor used typically in the automotive industry or in research. It consists of a light emitter that illuminates the surface, and the scattered light is analyzed in the optical receiver. Since the motion of an object (i.e., vehicle, airplane) causes changes in the light wavelength, relative velocity can be determined with accuracy of 0.1 km/h and the range of measurements is 0–250 km/h. The main subsystem was the light emitter/receiver which was installed on a vehicle so that the distance between the emitting lamp and the road or airfield surface was kept in a required range. Changing this distance, for example, due to the fact of vehicle hopping, results in missing data. Therefore, for this study, we used a special version of the CORREVIT sensor, the L400 (Kistler Group, Winterthur, Switzerland), which features a high sensor with a surface base distance of 400 mm as well as an extended range of tolerant changes in this distance: ±130 mm. This is of critical importance when the sensor has to be used on an airplane operation on a grassy airfield. Another issue was the installation of the sensor on the test airplane. The emitting/receiving unit had to be installed outside the vehicle with a close proximity to the ground surface, but since common installation racks are for road vehicles, we had to construct an untypical installation system. This system, shown in Figure 3, ensured safe mounting of the sensor. The complete CORREVIT system consisted of additional electronics and a 12 V power supply that were placed in the airplane together with a SONY DAT digital tape data recorder (Sony Corporation, Minato, Tokyo, Japan) which was used for saving the gathered data. The use of the DAT recorder instead of a Notebook computer was due to the much higher reliability of data recording.

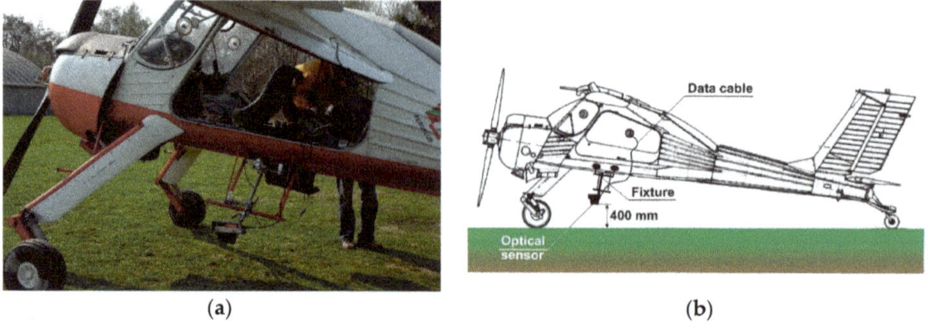

(a) (b)

Figure 3. Installation of the CORREVIT L400 sensor on the test airplane. (**a**) a photograph showing the test airplane with the Correvit optical sensor, (**b**) a schematic of the installation.

2.3. The Test Airplane

A four seat, single engine, multipurpose STOL (short takeoff and landing) aircraft PZL 104 Wilga (PZL Warszawa Okęcie, Warszawa, Poland) was used for the tests (see Figure 4). The airplane is a high wing monoplane, powered by the AI 14R 192 kW radial 7 cylinder engine (PZL Kalisz, Kalisz, Poland). It has a non-retractable, tail-dragger-type landing gear with main wheels with brakes, and low-pressure tires of 500 × 200 mm in size. The main legs are rocker type with oleo-pneumatic shock absorbers. They are castored and have a positive rake angle of 18°; the axle offset is 400 mm. The airplane is 8.10 m in length with a wing span of 11.2 m and wing area of 15.5 m^2. The empty mass is 900 kg. In the flight tests, there were four persons on board and the takeoff weight with 125 liters of fuel was 1150 kg.

Figure 4. The airplane used for the flight tests: the PZL 104 Wilga with the CORREVIT optical sensor.

2.4. Airfield Conditions and Procedures

The measuring systems were maintained by two persons on board. The tests were conducted on a sport airfield in Radawiec, near Lublin, Poland. Tests on low (10 cm) and mid-high (20 cm) grasses were conducted. For both grass conditions, a total of 10 flights were performed, 5 of them with flaps in takeoff position and 5 with no flaps during takeoff. On landings, flaps were extended to normal landing position for all 10 flights.

2.5. Data Reduction Methods

Data from the GNSS/INS system contained numerous kinematics measures of aircraft motion, but only a few of them were used and evaluated: (1) height (altitude) above the ground, (2) ground speed, and (3) takeoff or landing ground roll. Of high importance for the precision of the analysis was to determine the time moment when the aircraft lifts off and when it touches down. This was done on the altitude graphs. Example altitude graphs are shown in Figure 5. Here, we had the time points of the lift off and touch down. Knowing the time coordinates, it was possible to analyze the mentioned measures (i.e., altitude, speed, and ground roll), and typical sets of graphs for takeoff and landing are shown in Figure 5. The numerical data were collected for further analysis (integrating the velocity courses).

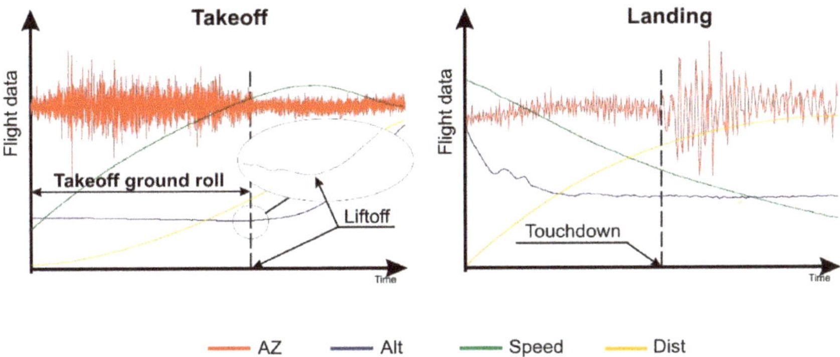

Figure 5. Sample results of the measurements and the idea of determining the ground roll distance.

The data from the optical sensor was analyzed regarding two additional inputs:

- Manually given by a test engineer who was observing the undercarriage wheel of the test airplane. This input was given by simply clicking the microphone, connected to the DAT recorder, and marking the time point when the airplane lifted off;
- Instrumentally obtained by analyzing the outgoing signal when it vanished due to increasing altitude.

3. Results

3.1. Airplane Ground Speed Profiles

These results are from the optical sensor. Figure 6 shows two sample airplane ground speed courses versus time. A comparison can be made between the two airfield conditions: short, wet grass and long, dry grass. We can see that the difference was in the early stage of the takeoff, when the airplane accelerated intensively. The difference was in the course of the ground speed: for the short, wet grass, this course had a less intensive increase in the first few seconds. This difference was quite surprising. It may be caused by a greater rolling resistance from the uneven pavement which is more pronounced when the grass is low. Long grass compensates for surface irregularities, and then the increase in speed during takeoff is more intensive.

Based on the determined ground speed profiles, it was possible to calculate the rolling resistance coefficient. In order to determine rolling resistance, the captured speed data were analyzed by solving a differential equation of the airplane's horizontal motion as below:

$$\frac{dV}{dt} = \frac{g}{W}[T - D - k_{RR}(W - L)] \quad (1)$$

where V = aircraft ground speed, W = aircraft weight, T = thrust, D = aerodynamic drag, L = aerodynamic lift, g = gravity acceleration, 9.81 m/s^2, and k_{RR} = rolling resistance coefficient.

This equation includes both aerodynamic forces, lift and drag, as well as the propeller thrust. The aerodynamic forces were calculated with the classic equations. The aerodynamic coefficients were determined from experimental data based on the assumptions:

- in the after-liftoff flare, the aerodynamic lift force is equal to aircraft's weight with respect to climb angle:

$$L = \frac{1}{2}\rho V^2 S C_L = Q\cos\beta \tag{2}$$

- the aerodynamic drag is equal to the propeller thrust force minus the longitudinal acceleration of the entire aircraft (determined from speed data for the after-liftoff flare):

$$D = \frac{1}{2}\rho V^2 S C_D = T - \frac{Q}{g}a_x \tag{3}$$

where C_D, C_L = aerodynamic coefficients, S = reference area, and ρ = air density.

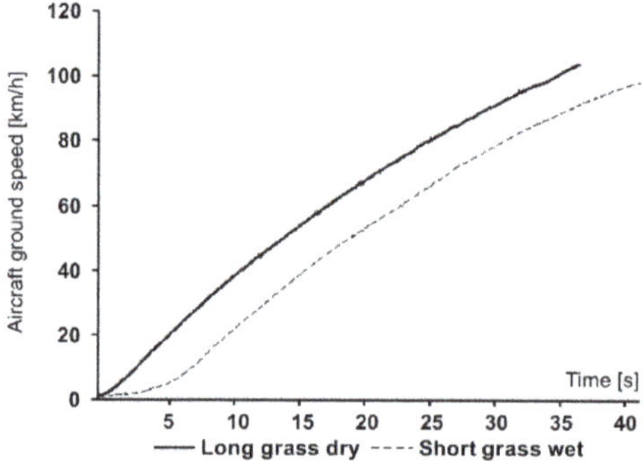

Figure 6. Airplane ground speed profiles for two surface conditions.

Aerodynamic data for the airplane were provided by the producer, but in order to determine true engine–propeller thrust, we performed a test in which the thrust was measured with the use of a load cell at takeoff RPM settings of the airplane engine.

Having all parameters values, we performed data integration. Using the Euler method, the takeoff speed course was divided into several small-time intervals (i.e., 10), assuming the acceleration was constant during each interval, and the rolling resistance coefficient k_{RR} was obtained for the end of each time interval (Figure 7). The complete algorithm for determining the k_{RR} from the speed courses is included in the work by Pytka [8]. It is worth noting that the value of rolling resistance coefficient is not constant in the takeoff speed range, and, additionally, the course of this coefficient value decreases with the area of momentary increase in the middle range of the takeoff speed. The difference between the two surfaces—short grass and long grass—was significant, although the nature of the course remained similar. The shape of the course was interesting and may result from rheological effects occurring in plant matter (in the grass). Namely, the effect of deformation speed caused an increase in the elasticity of the grass material which was manifested by lower losses associated with the absorption of deformation energy. It happened at higher speeds, and, at low speed, we observed the opposite. In the middle range, on the other hand, the k_{RR} value was influenced by aerodynamic forces that relieved the chassis wheels.

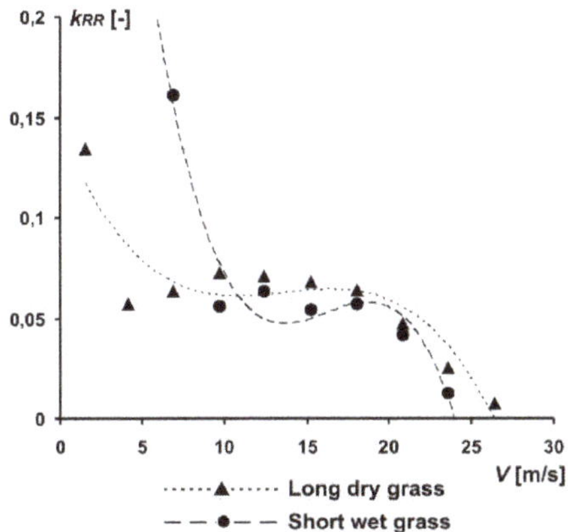

Figure 7. Rolling resistance coefficient for the test airplane wheels on a grassy runway.

3.2. Distances of Takeoff and Landing of the Wilga Airplane on a Grassy Runway

During the first experiment, there was a 5 m/s wind and the aircraft was flown facing the wind. The results of the takeoff and landing distances are presented graphically in Figure 8. There averaged values of distance with error bars are shown.

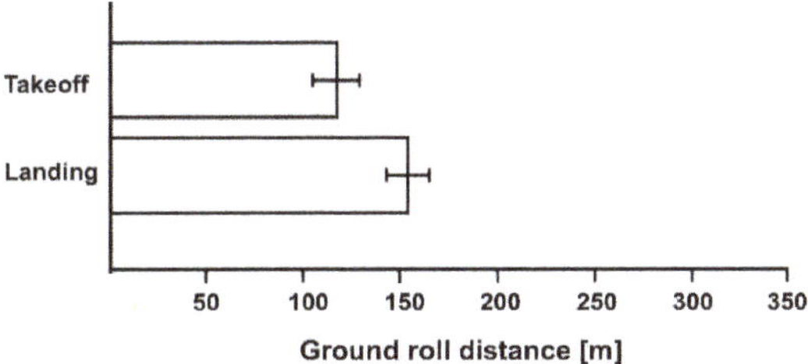

Figure 8. Takeoff and landing distances on short wet grass. Headwind = 5 m/s.

In this experiment, the aircraft was taking off with normally deployed flaps (21°). On the second day, the wind was 0–1 m/s, and the results were much different. We performed tests with deployed flaps as well as with no flaps. The averaged values are presented together with error bars in Figure 9. The takeoff and landing distances were significantly longer. The effect of grass length was the probable cause, although we were unable to analyze the effect of wind precisely. Besides, based on the comments by the pilots, this airplane was not as sensitive to headwinds during takeoffs; this is the effect of a powerful engine with shaft reduction and high diameter propeller.

Averaged values of takeoff and touchdown speed (V_{TO} and V_{TD}) for the two modes of operation (flaps and no flaps) are presented graphically together with error bars in Figure 10. Note that the values for the no-flaps mode were higher; moreover, in this mode, the difference between V_{TO} and V_{TD} was higher. We tried the "clean configuration", inspired by one piloting handbook which suggested it as a remedy for shortening the takeoff distance. This suggestion is probably right for a hard-paved runway, but, based on the results of our flight tests, not for the natural, soft surfaces of a grassy airfield. The touchdown speed was almost identical for the two configurations—flaps and no-flaps; this is because the landing was always performed with extended flaps (the landing configuration for this aircraft was 44° flaps).

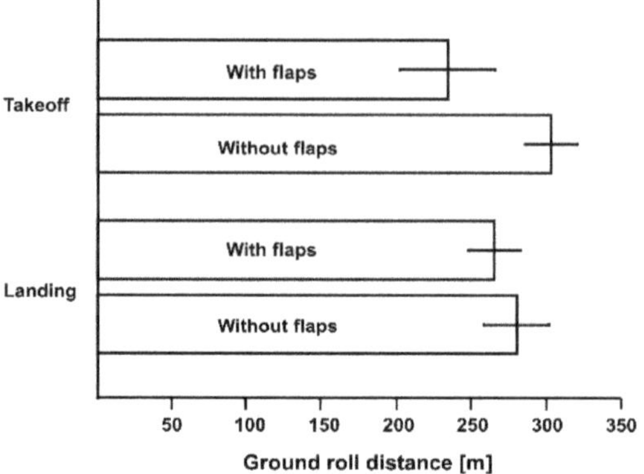

Figure 9. Takeoff and landing distances for the aircraft with and without flaps (wind 0–1 m/s). Long, dry grass.

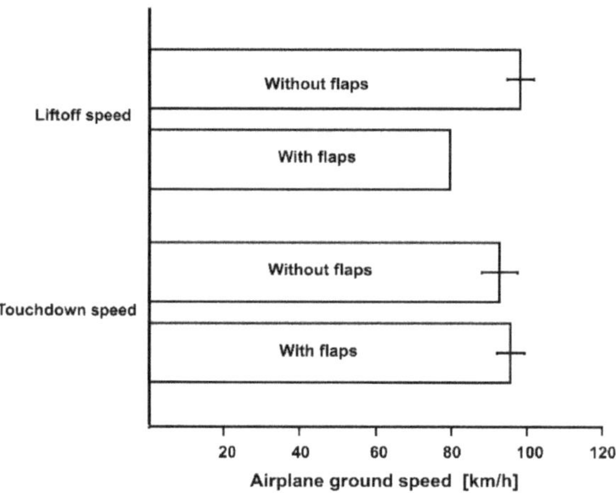

Figure 10. Takeoff and landing speed for the two operation modes (i.e., flaps and no flaps). Long, dry grass.

4. Comparison of the Two Sensors

Two sensors were enabled to measure data that was required to determine the ground roll distances of airplane takeoff and landing. However, the methods differed from each other in some details. It will be discussed below.

4.1. Accuracy and Functionality

One of the most important features of a measuring method is its accuracy. The accuracy of the method depends generally on two important factors:

- Accuracy of measurement of the basic parameter;
- Human factor—manual determination of the lift off or touchdown.

The accuracy of the measurement of basic parameters can be determined by means of known methods or can be read from a measuring device's data sheet. In the case of the optical sensor, it is known that its accuracy within the whole measuring range (0–250 km/h) is 0.01 km/h. From that, we could calculate the optical sensor's accuracy of distance determination. A^{OM}, taking into account the data sampling rate of 1 ms:

$$A^{OS} = (0.01 \text{ km/h} / 3.6) \times 0.001 \text{ s} = 0.0027 \text{ m} \tag{4}$$

On the other hand, the GNSS/INS method is based on acceleration measurement, and its accuracy is 0.01 m/s^2. The acquisition time was 10 ms. The GNSS/INS method's accuracy of distance determination A^{DGPS} is as below:

$$A^{GNSS} = (0.001 \text{ m/s}^2 / 0.01) \times 0.01 \text{ s} = 0.001 \text{ m} \tag{5}$$

Here, we should point out that the acquisition time could be set for 1 ms and, thus, the accuracy would be one order higher, but it was impractical, since the data files would be enormous (approximately 100–150 MB for one flight test). Besides, the accuracy of the both measuring methods was high enough. For a reliable verification of the GARFIELD models, an accuracy of approximately 1 m was enough.

However, the second factor affecting the total accuracy of the methods may be of critical importance. This was because of the manual determination of the liftoff or touchdown moment. Simply, the test engineer observed the landing gear wheel during the takeoff or landing and sent an impulse to the data recorder. This factor for the optical sensor method can be set for 1 s, based on a general human's time of reaction. So, the human factor for the optical method is as follow:

$$HF^{OS} = 1 \text{ s} \times 27 \text{ m/s} = 27 \text{ m} \tag{6}$$

On the other hand, this factor for the GNSS/INS sensor depends upon the method of data reduction which is described in the second section of this paper. When analyzing the time courses of the acceleration, it is noticeable that the amplitude of acceleration suddenly drops—this is the moment when the airplane lifts off. But the exact time point is difficult to determine. A method of simple discrimination has been applied, and it was assumed that the liftoff point is where the amplitude reaches 50% of its value, taken from the ground roll reading. A time between the 50% and low amplitude is set as a determinant for the accuracy (Figure 11). Therefore, the human factor for the GNSS/INS method is as follow:

$$HF^{GNSS} = t^{50\%} \times 27 \text{ m/s} = 5.4 \text{ m} \tag{7}$$

where a typical value of the $t^{50\%}$ = approximately 0.1 s.

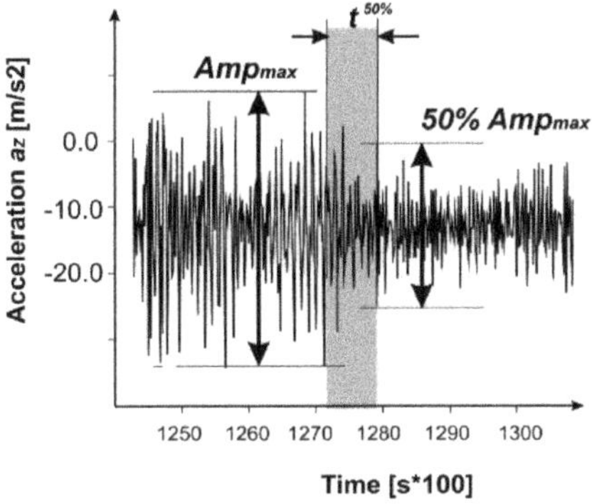

Figure 11. A schematic of the $t^{50\%}$ time discrimination method.

From the above, it can be seen that human factors may lower the accuracy of the methods significantly. Both human factors HF^{OS} and HF^{GNSS} can make the methods' accuracies unacceptable for GARFIELD modeling purposes.

4.2. Installation and Operation

The installation of the CORREVIT optical sensor was difficult. The light emitter/receiver unit was installed outside the cabin at a certain position above the ground. This required some improvisation by constructing the fixture. Another issue was that the unit was left outside for the flight test, exposed to possible collisions with things inducted by the propeller. The operation, on the other hand, was very simple, since the sensor sends linear signal and we used a DAT recorder which was very comfortable for the onboard test engineer.

The GNSS/INS sensor required to be installed by fixing in any location in the cabin. Its operation was rather difficult, since the measurement was controlled from a PC notebook computer which had to be on during the whole test flight. The onboard test engineer had to operate data acquisition software in a vibrating cabin, and there was one case when the data was lost due to the problems with operation of the computer during the flight.

4.3. Cost and Other Issues

Both sensor systems are expensive, although the GNSS/INS sensor is significantly more expensive compared to the CORREVIT optical sensor. The CORREVIT sensor that we used belongs to the Lublin University of Technology, where it is used for automotive applications; thus, we had uses it at zero cost. On the other hand, the GNSS/INS sensor that we used belonged to one military institute, and we had to pay for renting it for our tests. When analyzing the costs, it is important that, for both sensors, a professional should also be hired to operate it.

Table 1 includes a summary of the sensors comparison. The authors aimed to develop a measuring system based on a low-cost GNSS receiver. Such a system should have enough accuracy for very short baselines, such as the takeoff and landing distances, analyzed in this work [15].

Table 1. Comparison chart for the two sensors used in the study.

Functionality	Accuracy *	Installation and Operation	Data Handling **	Cost
		CORREVIT^R Optical Sensor		
Full functionality in the described research	0.0027 m	Difficult and time consuming	Easy and reliable	High
		GNSS/INS Sensor		
Full functionality in the described research	0.001 m	Easy and quick installation, but difficult data gathering	Difficult, the data can be missed	High

* The accuracy of the method depends partially on error prone human factors, (see text). The given values describe the accuracy of the measurement of basic parameters. ** Data handling for a complete measuring system, i.e., with a data acquisition device.

5. Verification of Measurements with Calculated Values of Takeoff Ground Roll Distances

The next step in this research was to compare the obtained experimental results of the takeoff ground roll distances with the calculated results. The literature describes several equations with different degrees of simplifications and assumptions that allow for a calculation of takeoff and landing distance [16,17]. We chose three different approaches: a simple equation used by designers of homebuilt aircrafts, a general equation for airfield performance, and an equation for GA aircrafts for design projects.

5.1. Homebuilt Designer Equation

This calculation method was presented by Jarząb [18] and was proposed for amateur builders and designers of airplanes. It is important that homebuilt aircrafts are mostly designed to operate on grassy runways with poor surfaces. Their design utilizes tubes and fabric structures for fuselages and wooden wings with fabric covering. The overall aerodynamics of such airplanes can be poorer than factory-built, so the effective takeoff performances are critical factor. The proposed equation can be used for calculations of preliminary ground performance during the advanced stages of a design as well as in the flight-testing process. It consists of two equations: one for determining the rate of climb and the second one for calculating the final ground roll. The rate of climb equation is:

$$w = \frac{57\frac{R^2}{S}}{\left(\frac{Q}{N}\right)^2 + \frac{Q}{S}} \tag{8}$$

where w—rate of climb, R—wingspan, S—wing surface, Q—aircraft mass, N—engine power.

The final ground roll equation is:

$$L = K\frac{\left(\frac{Q}{S}\right)^2}{w^2} \tag{9}$$

where L—ground roll and K—an empirical coefficient.

This approach is very simple and assumes the airplane is powered with a low HP engine and typical landing gear with low pressure tires. No environmental impact is regarded in the equation by Jarząb, so the effect of grass or soil deformability is neglected. The K-coefficient could probably be used to compensate for those effects; however, no further information is provided in the cited publication.

5.2. Filippone Equation

The second approach chosen for our comparison was presented by Filippone [19] and assumes the variable rolling resistance coefficient, k_{RR}, which can be either measured experimentally or, for the purposes of this analysis, taken from the literature. However, the value of k_{RR} is assumed to be constant which is in opposition to the results obtained by the present authors (Figure 7). This approach can be useful when applied to takeoffs and landings on grassy airfields with unpaved runways. The following is the final equation for ground roll distance:

$$S_G = \frac{mV^2}{2\frac{\eta P}{V} - \frac{1}{4}\rho AC_D V^2 - f_t mg} \tag{10}$$

where m—aircraft mass, V—liftoff speed, η—propeller efficiency, P—engine power, A—wing surface, C_D—aerodynamic drag coefficient, ρ—air density, f_t—rolling resistance coefficient, and g—gravitational acceleration.

For the calculations we assumed a propeller efficiency of 0.5, takeoff speed of 27 m/s, drag coefficient of 0.04, and standard air density. Other parameters were obtained experimentally, for example, the aircraft was weighted with 4 persons on board, and the engine/propeller thrust was measured. Details are included in the publication by Pytka [8].

5.3. Rapid Takeoff Distance Equation

The third calculation method was defined by Gudmundsson [20]. This approach is the most general and requires detailed knowledge about aircraft parameters, coefficient values, and weather conditions during takeoff. This method requires calculating the acceleration of the aircraft first and then the final ground roll distance which is very similar to the approach by Raymer [16]. The basis equation can be adopted to both turbine and piston engine powered, typical GA aircrafts of a total mass below 5700 kg. However, this approach can be simplified to one equation when applied to piston-powered light airplanes:

$$S_G = \frac{V^2 W}{\frac{\eta_P(550P)}{V}g + \frac{1}{2}g\rho V^2 S(\mu C_L - C_D) - 2g\mu W} \tag{11}$$

where V—speed at liftoff, W—aircraft weight, η_p—propeller efficiency, P—engine power, S—wing surface, g—gravitational acceleration, ρ—air density, C_L—aerodynamic lift coefficient at takeoff, C_D—aerodynamic drag coefficient at takeoff, and μ—ground friction coefficient.

5.4. Results Comparison

Figure 12 shows a simple comparison of all calculated and measured results. The esults obtained from both an amateur constructors' method (211 m) and the Filippone equation (213 m) were surprisingly close to the measured average (234 m). Theoretical ground roll distance calculated using the Gundmundsson equation (304 m) was further from the experimental results. This substantial difference is suspected to be caused by the nature of the Gundmundsson equation which is suitable rather for faster GA airplanes, turboprops or high HP piston machines that operate from rigid surfaces. Nevertheless, even the simplest of the compared methods proved itself to be effective when tested against real measured values.

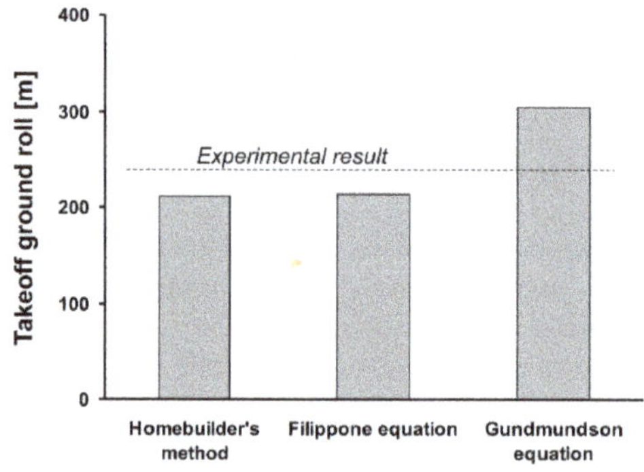

Figure 12. Measured versus calculated ground roll comparison.

6. Conclusions

Flight test experiments on airfield performance for takeoff and landing of a light airplane on a grassy airfield were performed. Two different sensor systems were used in the tests: the GNSS/INS sensor and the CORREVIT optical sensor. The results showed that the effect of grass length was significant for ground velocity time course and the distances for takeoff and landing ground roll. Also, wing flaps had a measurable effect on ground roll, either by takeoff or by landing. The obtained values for the rolling resistance coefficient of the airplane's undercarriage wheels changed along with the ground speed range.

Three different analytical models were used to calculate the ground roll of a takeoff, and the results obtained in the flight tests were compared to the calculations. All three calculation methods proved itself to be effective when tested against real measured values.

A comparison of the two sensor systems used in the tests was performed. It was concluded that both sensors were able to measure the data to determine the required parameters of the ground distance of takeoff and landing. However, both methods have drawbacks, and the intention of the authors is to seek another method. The major prerequisite is that the optimal method should be easier in handling for reliable use in multiple tests.

The data obtained in this study will be used for further analysis, and a general application of these results is to identify and verify a wheel-grass model which is a basic analytical tool of the GARFIELD information system.

Author Contributions: Conceptualization, J.P., P.B.; Methodology, J.P., P.B.; Software, P.B., T.Ł., Validation, J.M., Formal Analysis, A.T.; Investigation, T.Ł.; Resources, J.J.; Data Curation, J.P., T.Ł.; Writing-Original Draft

Preparation, J.P.; Writing-Review and Editing, P.B. and T.Ł., Visualization, J.P.; Supervision, A.T. and D.B.; Funding Acquisition, P.B. All authors provided critical feedback and collaborated in the research.

Funding: The project/research was financed in the framework of the project Lublin University of Technology—Regional Excellence Initiative, funded by the Polish Ministry of Science and Higher Education (contract no. 030/RID/2018/19). The CORREVIT is a trademark owned by the Kistler Group, Switzerland.

Conflicts of Interest: The authors declare no conflict of interest.

References

1. Pytka, J.; Józwik, J.; Łyszczyk, T.; Budzyński, P.; Laskowski, J. Gnapowski Measurement of Takeoff and Landing Ground Roll of Airplane on Grassy Runway. In Proceedings of the 2019 IEEE 5th International Workshop on Metrology for AeroSpace (MetroAeroSpace), Torino, Italy, 19–21 June 2019.
2. Dąbrowski, J.; Pytka, J.; Tarkowski, P.; Zając, M. Advantages of all-season versus snow tyres for off-road traction and soil stresses. *J. Terramech.* **2006**, *43*, 163–175. [CrossRef]
3. Pytka, J. A semiempirical model of a wheel-soil system. *Int. J. Automot. Techn.* **2010**, *11*, 681–690. [CrossRef]
4. Pytka, J.; Józwik, J.; Budzyński, P.; Łyszczyk, T.; Tofil, A.; Gnapowski, E.; Laskowski, J. Wheel dynamometer system for aircraft landing gear testing. *Measurement* **2019**, *148*. [CrossRef]
5. Pytka, J.; Budzyński, P.; Kamiński, M.; Łyszczyk, T.; Józwik, J. Application of the TDR Moisture Sensor for Terramechanical Research. *Sensors* **2019**, *19*, 2116. [CrossRef] [PubMed]
6. Shoop, S.A.; Diemand, D.; Wieder, W.L.; Mason, G.L.; Seman, P.M. *Predicting California Bearing Ratio from Trafficability Cone Index Values*; CRREL Techn Rep TR-08-17; Engineer Research and Development Center Hanover Nh Cold Regions Research and Engineering Lab: Hanover, NH, USA, 2008.
7. van Es, G.W.H. Method for predicting the rolling resistance of aircraft tires in dry snow. *J. Aircr.* **1999**, *36*, 762–768. [CrossRef]
8. Pytka, J. Identification of Rolling Resistance Coefficients for Aircraft Tires on Unsurfaced Airfields. *J. Aircr.* **2014**, *51*, 353–360. [CrossRef]
9. Shoop, S.A. Estimating rolling friction of loose till for aircraft takeoff on dirt runways. In Proceedings of the 13th International Conference of the ISTVS, Munich, Germany, 14–17 September 1999; pp. 421–427.
10. Rowe, R.S.; Hegedus, E. *Drag Coefficients of Locomotion over Viscous Soil*; Report No.54; Department of the Army, Ordnance Tank—Automotive Command, Land Locomotion Laboratory: Centerline, MI, USA, July 1959.
11. Crenshaw, B. Soil—Wheel interaction at high speed. *J. Terramech.* **1972**, *8*, 71–88. [CrossRef]
12. Howland, H.J. Soil inertia in wheel-soil interaction. *J. Terramech.* **1973**, *10*, 47–65. [CrossRef]
13. Gibbesch, A. Reifen-Boden Interaktion von Flugzeugen auf Nachgeibigen Landebahnen beihohen Geschwindigkeiten. In Proceedings of the Luft-Und Raumfahrt Kongress DLR-Deutsches Zentrum Fur Luft-undRaumfahrt, München, Germany, 17–20 November 2003; pp. 2079–2085.
14. Pytka, J.; Józwik, J.; Łyszczyk, T.; Gnapowski, E. Method for Determination of Airplane Takeoff and Landing Distance. In Proceedings of the 2018 5th IEEE International Workshop on Metrology for AeroSpace (MetroAeroSpace), Rome, Italy, 20–22 June 2018.
15. Albéri, M.; Baldoncini, M.; Bottardi, C.; Chiarelli, E.; Fiorentini, G.; Raptis, K.; Strati, V. Accuracy of flight altitude measured with low-cost GNSS, radar and barometer sensors: Implications for airborne radiometric surveys. *Sensors* **2017**, *17*, 1889. [CrossRef] [PubMed]
16. Raymer, D. *Aircraft Design. A Conceptual Approach*; American Institute of Aeronautics and Astronautics, Inc.: Washington, DC, USA, 1992.
17. Stinton, D. *Flying Qualities and Flight Testing of the Aeroplane*; Blackwell Science: Oxford, UK, 1998.
18. Jarząb, K. Obliczanie zdolności startowej. *Skrzyd. Pol.* **1986**, *1808*, 12.
19. Filippone, A. *Flight Performance of Fixed and Rotary Wing Aircraft*; AIAA Education Series; AIAA: Washington, DC, USA, 2006.
20. Gudmundsson, S. *General Aviation Aircraft Design. Applied Methods and Procedures*; Elsevier: Amsterdam, The Netherlands, 2016.

© 2019 by the authors. Licensee MDPI, Basel, Switzerland. This article is an open access article distributed under the terms and conditions of the Creative Commons Attribution (CC BY) license (http://creativecommons.org/licenses/by/4.0/).

Article

Improved Multi-GNSS PPP Software for Upgrading the DEMETRA Project Time Monitoring Service

Wei Huang [1,3,*], Pascale Defraigne [2], Giovanna Signorile [3] and Ilaria Sesia [3]

1 Politecnico di Torino, 10129 Torino, Italy
2 Observatoire Royal de Belgique, 1180 Brussels, Belgium; pascale.defraigne@oma.be
3 Istituto Nazionale di Ricerca Metrologica, 10135 Torino, Italy; g.signorile@inrim.it (G.S.); i.sesia@inrim.it (I.S.)
* Correspondence: wei.huang@polito.it

Received: 30 August 2019; Accepted: 9 October 2019; Published: 11 October 2019

Abstract: The H2020 DEMETRA project provides short latency clock monitoring services to the time users using the Atomium precise point positioning (PPP) software developed by the Royal Observatory of Belgium. In this paper, three recent updates of the current Atomium software are introduced: adding Galileo signals in the PPP computation; the option to constrain the receiver clock; PPP with integer ambiguity resolution. The advantages of these updates are demonstrated: Combining the Galileo and global positioning system (GPS) signals for PPP time transfer will further improve the frequency stability inside the computation batch; PPP with receiver clock constraint is not only used to reduce the short-term noise of the clock measurements but can also be used for some specific applications to a keep continuous clock solution in the computation batch or retrieve correct clock measurements from extremely noisy environments; the integer PPP allows a continuous clock solution, and improves the mid-term and long-term stability of the frequency transfer compared to the current PPP frequency transfer techniques.

Keywords: precise point positioning; GNSS; time transfer; frequency transfer; Galileo; constrained PPP; integer PPP; DEMETRA; time service

1. Introduction

The global navigation satellite systems (GNSS) include the United States of America's global positioning system (GPS), Russian's global navigation satellite system (GLONASS), Europe's Galileo and China's BeiDou navigation satellite system. They are well known for their positioning services with global coverage. With the GNSS precise point positioning (PPP) techniques, the positioning accuracy can reach centimeter-level accuracy [1]. Another important function that GNSS provides to the user is the timing service, based on the accurate and reliable GNSS time references, e.g., GPS system time (GPST), and their predicted difference to the Coordinated Universal Time (UTC) [2].

GNSS are widely used for accurate time and frequency dissemination, synchronization, and remote clock comparison (called time transfer in time metrology). In the industrial domain, GNSS time synchronization is widely used in the telecom network, energy grid, and financial system [3] and has proved its advantages in various specific applications, e.g., time synchronization of the intelligent transportation systems [4] and spacecraft [5]. In the scientific domain, clock comparison with GNSS is essential for the computation of International Atomic Time (TAI), with more than 50% of time laboratories contributing to TAI and UTC equipped by GNSS receivers for their time link comparisons [6]. GNSS timing is also the primary choice for some scientific activities in which all the stations from a given network observing the same event need to be highly time-synchronized, e.g., seismic monitoring [7], deep space tracking [8], and neutrinos speed measurement [9].

Many users rely on GNSS code measurements for the time synchronization purpose, this method has a statistical uncertainty of several nanoseconds over one day [10]. Better performance of GNSS time transfer is achieved by PPP techniques, which use both dual-frequency carrier phase and code measurements from the GNSS signals and the more precise satellite products instead of the broadcast navigation message for the analysis. The statistical uncertainty of this method can reach 100 ps over one day [6,11].

DEMETRA (*DEM*onstrator of *E*GNSS services based on *T*ime *R*eference *A*rchitecture) [12] is a project funded by the European Union in the Horizon 2020 program coordinated by the Italian National Metrology Institute (INRIM) that involved 16 European partners, including the Royal Observatory of Belgium (ORB). The DEMETRA project aims at providing the end-users in both industrial and scientific domains with improved and new time services, some of them being based on the European GNSS. Nine different timing services with improved accuracy or availability and any requisite as certification or redundancy were designed, developed, and tested. Most of the time, services are currently still operative through their infrastructures maintained at the INRIM premises. As an example, INRIM implemented the fiber link between the financial district in Milan and INRIM for UTC time distribution service [13].

One particular service in DEMETRA called "Time Monitoring and Steering", provides the user, in quasi-real-time, the time difference between the user atomic clock and the reference time scale of DEMETRA from an hourly PPP solution. This PPP solution is based on the data from the user geodetic GNSS receiver transmitted hourly to the DEMETRA platform. With the data provided by the service, the user can make sure its clock is highly synchronized to the DEMETRA reference and continue being alerted to any possible non stationarities on its clock, namely phase or frequency jumps. Meanwhile, the service offers the necessary quality monitoring data (such as satellite sky plot, multipath, cycle slips, etc.) visible directly on the personal area of the DEMETRA web page. An example of part of the user personal page is attached in the Appendix A, and all the details can be found on the homepage of the DEMETRA project [14]. In addition, some parameters are sent to the users to steer their clock to UTC if needed. Because these steering parameters are computed based on the monitoring data, in this paper, we focus on the algorithms of the time monitoring part of the service.

The core software in the DEMETRA timing monitoring service is called Atomium, which is developed by ORB [11]. Atomium is a GPS+GLONASS PPP software dedicated to GNSS time transfer and geodetic positioning. It uses the least-square method to estimate the following parameters: receiver clock error at each epoch; user antenna position for the whole daily data batch; tropospheric zenith wet delay at defined interval, and float phase ambiguities. The software uses satellite products made available from International GNSS service (IGS) and estimates the clock errors as "IGS time-user clock". On the DEMETRA side, the computation "IGS time-DEMETRA reference" is also carried out with PPP, for the same time period. Differentiating these two quantities, we obtain estimates of the user clock with respect to the DEMETRA reference. The user clock can be either the internal clock of the GNSS receiver or an external clock (e.g., hydrogen maser, cesium, rubidium atomic clock) used to drive the GNSS receiver at the user side. For convenience, all these clocks are called user clock or receiver clock, in general, in this paper. The DEMETRA reference time is chosen to be the UTC(IT) time scale that feeds the GNSS receiver at the INRIM side. UTC(IT) is the local realization of UTC made by the INRIM and is based on a hydrogen maser steered on UTC.

Atomium PPP has been dedicated to precise time and frequency transfer for more than 10 years [15]. Currently, the PPP time transfer is mainly limited by the time discontinuity at the daily batch boundary, which is caused by the averaging of the noisy code measurements in the daily batches [16]. A common way to avoid the daily boundary discontinuities is to extend the estimation batch to multi-day (e.g., with PPP software developed by Natural Resources Canada (NRCAN) [17]). However, the boundaries still exist at the end of the multi-day batches. In addition, this method may cause the long-term artificial variation of the clock estimation, introducing additional random walk noise inside the batch [18]. Moreover, while being the state-of-the-art GNSS technique for modern high-quality clock comparison, short-term stability of the PPP is currently limited by the thermal noise in the carrier phase data. Further developments are, therefore, still needed to improve the stability of the PPP clock solution and mitigate the boundary issue. In this paper we present three updated versions of the Atomium PPP software that we have integrated into DEMETRA for improving the time monitoring part of the service: (i) Atomium Galileo PPP; (ii) Atomium constrained PPP; (iii) Atomium integer PPP. The basic ideas of the proposed new algorithms have been presented in [19], while they have not been included in any proceedings paper. In the current work, the mathematic principles of these improved PPP algorithms are described in detail, and results based on experimental data are presented enabling the evaluation of the improvements of the new methods and to define possible application scenarios.

Atomium Galileo PPP includes Galileo measurements in the PPP computation, in addition to the GPS and GLONASS ones in the old version. Though the Galileo satellite system is still not fully operational for the moment, comparable performance to GPS in time transfer with code-only measurements has already been demonstrated in [20]. This paper presents the impact of using GPS+Galileo data in Atomium PPP for time and frequency transfer.

Second, Atomium constrained PPP will be presented. This approach further constrains the receiver clock with the estimated frequency offset and drift, to retrieve the short-term stability of the measurements on the clocks with good quality (e.g., hydrogen maser), by restricting the impact of thermal noise on the receiver clock solutions [21]. Applying a stochastic model in kinematic PPP has shown improvement in the kinematic positioning [22], and frequency stability also improves in real-time PPP time transfer with a receiver clock between-epoch constrained model as described in [23]. In this paper, a simple and well-defined receiver clock model is introduced and tested in the post-processing mode. Furthermore, the constrained PPP is also shown to be able to keep the continuity of the clock solution in case of satellite tracking loss or to retrieve the clock solution from very noisy measurements.

Finally, Atomium integer PPP will be presented. This approach fixes the carrier phase ambiguities as integer numbers in the least square computation. The old version of Atomium PPP in DEMETRA treats the carrier phase ambiguities as float values, ignoring the fact that these ambiguities should be integers. It was reported in [24] that using the carrier phase measurements with integer ambiguity resolution will improve the stability of GNSS time transfer and also eliminate the clock jumps or random walk introduced by float ambiguity solutions. In [25], a new method for fixing integer ambiguity on un-differenced phase measurements was described, and it leads to a PPP with centimeter-level precision. Since the GPS satellite integer products started to be available at the IGS [26], it has become possible for the single receiver user to perform the PPP time transfer with integer ambiguity resolution. Due to the features of the techniques, the independent solution of integer PPP can be only used for frequency transfer, while it cannot be used for time transfer. However, it shows that integer PPP can also perform excellently in long term time transfer, mitigating the problem of time discontinuity at the daily batch boundary if its solution can be calibrated by another time reference [27].

The mathematic details of the three upgrades of Atomium PPP are presented in Section 2, and the associated clock solutions, compared to the current PPP performances, are given in Section 3.

2. Updated Atomium PPP Algorithms

2.1. Atomium Galileo PPP

As for GPS, the P(Y) code and carrier phase signals ionosphere-free combination on L1 and L2 band are used in the old version of Atomium PPP. The code and carrier phase ionosphere-free measurements of Galileo on E1 and E5a band are added in the new Atomium Galileo PPP computation. The post-processed products of GNSS satellite clocks and orbits are selected for the PPP computation to provide the clock solution with the highest possible stability. To include Galileo, the PPP uses the post-processed satellite products from the IGS in the frame of the MGEX (Multi-GNSS Experiment), which contains products for all the GNSS constellations [28].

To deal with both GPS and Galileo signals in PPP to generate a coherent receiver clock solution, an additional parameter called the inter-system bias (ISB) between GPS and Galileo measurements must be estimated in addition to the antenna position, the receiver clock, and the troposphere delay as the outputs of the least square computation. This bias is the sum of both the differences between the hardware delays in the receiving chain for GPS and Galileo signals and the bias between the time references for the GPS and Galileo satellite clocks in the MGEX products. This ISB is considered as a constant value during the estimation in one daily data batch, based on the assumption that the receiver delay is stable enough during one day, and the same handling scheme is used in the MGEX analysis center [29].

Note that the observed distance between satellite and receiver are measured between the phase centers of the antennas at both ends, while the satellite positions provided by IGS are at the center of mass, they need to be corrected to the phase center positions. The file of IGS phase center corrections for the GNSS antennas at receiver and satellite is utilized. It contains the mean phase center offset and phase center variation corrections and is available on the IGS FTP [30] in ANTEX format (current version is igs14.atx).

The phase center variations applied to GPS and GLONASS satellite antennas are nadir-dependent, while the ones to the Galileo satellite antennas are both azimuth- and nadir-dependent. The azimuth angle of the satellite antenna is estimated in the satellite-fixed coordinate system [31].

The phase center variations for the GNSS receiver antennas are elevation- and azimuth-dependent. However, few antennas are currently calibrated for Galileo signals in the IGS ANTEX file. In this paper, we use the same values in the receiver antenna phase center corrections on GPS L1 and L2 frequencies for the corrections on Galileo E1 and E5a, separately.

Finally, the outputs of Galileo PPP are receiver antenna GPS/Galileo mean phase center position; receiver antenna ARP (antenna reference point) position; receiver clock errors to the MGEX time; troposphere zenith wet delay; ISB.

2.2. Atomium Constrained PPP

The Atomium constrained PPP is based on a 2-step process. The first step is a classic PPP. The instantaneous frequency of the PPP clock solution "receiver clock—IGS/MGEX time" is then computed at each epoch as

$$frq_i = \frac{clock_{i+1} - clock_i}{t_{i+1} - t_i}, \tag{1}$$

where $clock_i$ is the PPP clock solution at epoch t_i. Then, from these instantaneous frequencies, we estimate the parameters $y0$ and d of a linear model.

$$frq_i = y0 + d * t_i \tag{2}$$

with $y0$ the frequency offset and d the frequency drift. The constraint between the clock solutions at adjacent epochs can then be expressed as

$$\begin{aligned} delt_i &= clock_{i+1} - clock_i \\ &= (y0 + d * t_i) * (t_{i+1} - t_i). \end{aligned} \tag{3}$$

The estimation variance of the constraint can be predicted as

$$\sigma_i^2 = (ADev(1s))^2 * (t_{i+1} - t_i), \tag{4}$$

where $ADev(1s)$ is the user input Allan Deviation of the receiver clock at 1 s, which can be found in the clock specification or can be estimated from the clock solution $clock_i$. In this paper, we use the $ADev(1s)$

provided in the specification document of the clocks, e.g., 2×10^{-13} for a hydrogen maser, 5×10^{-12} for a cesium clock. The weight of the clock constraint $delt_i$ can be expressed as

$$W_i = \frac{\sigma^2}{\sigma_i{}^2}, \tag{5}$$

where σ^2 is the a posteriori estimation variance on the clock solution at epoch t_i from the least square in the classic PPP.

The second step of the approach is then to compute a new PPP, with the additional constraints on the clock solution inserted in the least square computation of PPP. This constraint model is built based on the assumption that the receiver clock is mainly affected by white frequency noise during the analysis period.

In addition, constrained PPP offers the option for outlier recovery. If the recovery mode is on, it assumes that the clock solution is very noisy, the software will remove the frequency outliers before estimating the linear frequency model. With this refined clock constraint, the clock outliers will be pulled back to a normal noise level in PPP computation. If the recovery mode is off, it assumes there are some intentional changes causing a clock jump. In that case, the clock will not be constrained around those jumps, allowing the true clock jump to appear in the clock solution. This recovery mode is on by default, and if it is known there are some physical changes during the day of computation, from user information, it is then switched off.

2.3. Atomium Integer PPP

To fix the integer ambiguities of the carrier phase measurements for improved PPP frequency transfer, the GRG satellite products [26] are used in the Atomium integer PPP software instead of the IGS/MGEX products. The GRG products are generated at the CNES (Centre National d'Etudes Spatiales) and CLS (Collecte Localisation Satellites) IGS analysis center. These products include the satellite clock, satellite orbit, earth rotation parameters (ERP), and satellite wide-lane biases (wsb). Since the wsb files for Galileo are not regularly available yet, only GPS signals are included in the current integer PPP software.

The computation takes place in three main steps: 1. Resolve the wide-lane (WL) integer ambiguity; 2. Resolve the narrow-lane (NL) integer ambiguity; 3. Compute the final solution.

1. In the first step, the Melbourne–Wubbena (MW) combination is built at each epoch using phase and code measurements for each GPS satellite:

$$\text{MW} = (\alpha_{WL} L1 - \beta_{WL} L2) - (\alpha_{NL} P1 + \beta_{NL} P2), \tag{6}$$

where (L1, L2) and (P1, P2) represent the carrier phase measurements and code measurements on the L1 and L2 band separately, $(\alpha_{WL}, \beta_{WL})$ and $(\alpha_{NL}, \beta_{NL})$ are the coefficients in front of the related measurements for the MW combination. Equation (6) can be further expressed as

$$\frac{\text{MW}}{\lambda_{WL}} = N_{WL} + WRB - WSB, \tag{7}$$

where λ_{WL} is the WL wavelength (in meter), N_{WL} is the integer ambiguity of WL combination (in cycle), WRB is the WL receiver bias, and WSB is the WL satellite bias provided by GRG products.

With the least square method, the real-valued N_{WL} and WRB are estimated by each satellite track and each epoch separately. To avoid the singularity of WRB estimation, an absolute constraint on WRB is set for the first day estimation. Then the GNSS bootstrapping method [32] is adopted to fix the N_{WL} to a closest integer cycle per track, and WRBs are the remaining fractional cycles in the MW combination at each epoch.

In this step, the N_{WL} is fixed only when its success rate of fixing [32] is over 90% and its real-valued ambiguity is not too far from an integer value (smaller than 0.25 WL cycle), only the track with fixed

N_{WL} is used for the computation in the next step. To obtain coherent clock solutions for multiple days, the mean of the solved WRB in the previous day is chosen as the constraint on the WRB estimation of the next day. So, the estimated WRBs do not have to be always a fractional cycle of WL as it is evolving, but it has to be continuous to avoid the receiver clock misalignment at different days caused by the integer cycle estimation error of WRB.

2. In step 2, the GPS ionosphere free (IF) code and carrier phase combination can be first expressed as

$$P_{IF} = \rho + T^R + \text{zpd} + \text{e} + \varepsilon_P, \tag{8}$$

$$L_{IF} = \rho + T^R + \text{zpd} + \lambda_{NL}(N_1 + (\lambda_{WL}/\lambda_2)N_{WL}) + \text{W} + \text{e} + \varepsilon_L, \tag{9}$$

where P_{IF} and L_{IF} are the code and carrier phase ionosphere free combination, respectively, ρ is the distance between satellite and receiver phase center, T^R is the receiver clock error, zpd is the troposphere zenith wet delay, W is the wind-up effect to be corrected, e is the other common errors between code and phase measurements (e.g., troposphere dry component delay, relativistic effect, satellite clock error), ε_P and ε_L are the code and phase noises, respectively, λ_{NL} is the NL wavelength, N_1 is the integer ambiguity to be resolved, and integer N_{WL} has already been fixed in step 1.

Since $\lambda_{NL} = 17\lambda_{IF} = 10.7$ cm, it is much easier to resolve the integer ambiguity. The code and carrier phase IF measurements are put in the least square to compute the real-valued N_1 per track, and the integer N_1 is fixed again by the GNSS bootstrapping method. The N_1 ambiguities are also partially fixed as in step 1, the measurements with un-solved ambiguities are excluded from the next computation.

3. In the last step, since the integer ambiguities of the carrier phase IF combination have already been resolved, the carrier phase IF measurement is not ambiguous anymore and is rewritten again as

$$L_{IF} - \lambda_{NL}\left(N_1 + \left(\frac{\lambda_{WL}}{\lambda_2}\right)N_{WL}\right) - \text{W} - \text{e} = \rho + T^R + \text{zpd} + \varepsilon_L, \tag{10}$$

where the variables on the left side are already solved, the ones on the right side are to be solved, including receiver position (in ρ), receiver clock error, and troposphere error. The wind-up corrections are computed from step 2, together with the ambiguity fixing.

At last, the carrier phase IF measurements are put in the least square without code IF measurements to solve the final solutions.

If the WRB estimation is coherent during the whole period and the ambiguities are resolved correctly, the values of the discontinuities at the daily boundaries for the integer PPP clock solutions are the integer multiple of the GPS λ_{NL}. Hence, the clock solutions for multiple days can be easily aligned with minimal error introduced. Attention also needs to be paid to the discontinuity inside the daily batch. If it is caused by any a hardware change in the receiver system, the integer feature of discontinuity is lost because the WRB may endure an integer cycle estimation error in Equation (7) after the discontinuity, and bias the solution in Equation (10) in terms of an integer multiple of $\lambda_{NL}(\lambda_{WL}/\lambda_2)N_{WL}$.

3. Processing and Results

In this section, the performances of the updated PPP software are evaluated in several cases of remote clock comparison. To focus on the performance of the software themselves without the effect of the quality of the compared clocks, two kinds of experiments were built: 1. Common clock difference (CCD), where the two GNSS stations are driven by the same external clock; 2. "PPP–OPT", which means the difference of clock comparison results from PPP and optical fiber link.

The purpose of both experiments was to cancel the effect of the clocks themselves. The CCD results were assumed to be around zero if both GNSS stations were perfectly calibrated since it is the common clock that was compared. Two CCD experiments were set: 1-month common clock comparison using stations GR01 and GR02 in INRIM during the modified Julian date (MJD) 58591–58636 and another comparison using stations BRUX and RTBS in ORB during MJD 58583–58621. "PPP–OPT" removes the clock component by subtracting the clock comparison results from two individual techniques, in which the optical fiber link results are expected to be more stable and accurate than the GNSS results. The astrogeodynamic observatory of the space research center (AOS) and the central office of measures (GUM) provide a continuous optical link comparison between their atomic clocks [33]. Meanwhile, the clocks are also compared through their GNSS stations AO_4 and GUM4 using NRCAN PPP (both measurements can be downloaded from the BIPM FTP ftp://tai.bipm.org/TimeLink/LKC/ [6]). Two periods of GNSS observation data and available measurements from both the optical link and NRCAN PPP at AOS and GUM were collected for the "PPP–OPT" experiment: MJD 58417–58447 and MJD 58537–58572. For the stations in the CCD experiments, the satellite MGEX and GRG products from CNES/CLS IGS analysis center were used, and for station AO_4 and GUM4, which only receive GPS signals, the IGS rapid products were used for the study of constrained PPP, and the GRG products from CNES/CLS were used for the integer PPP clock comparison.

Note that the accuracy of the GNSS station calibration is not within the scope of this paper, only the stability and precision of the clock solution are studied here. Actually, these calibration values are normally constant values during clock comparison, which introduces only an offset to the GNSS clock output from the software but does not affect the stability. In this section, all the clock comparison results were shifted towards zero to ease the comparison between different solutions.

3.1. Atomium Galileo PPP

For both the CCD experiments in ORB and INRIM, the common clock was compared through their two GNSS stations with the Atomium PPP using GPS signals and GPS + Galileo signals. As shown in Figure 1, the two solutions in ORB show a good coherence, however, due to the presence of daily boundary jump, the performances of the GPS-only solution and GPS + Galileo solution are hard to compare.

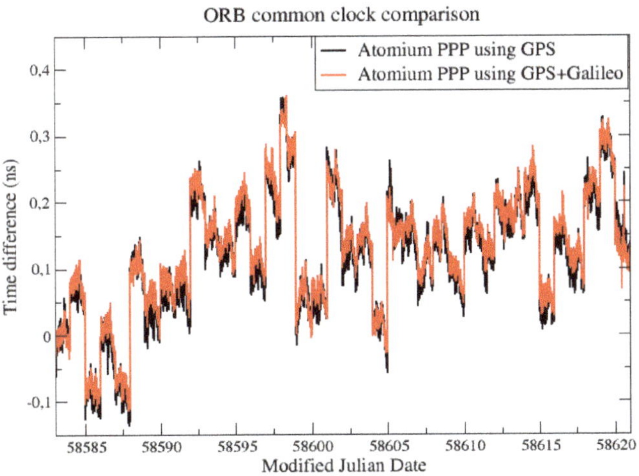

Figure 1. Common clock difference (CCD) Galileo results between BRUX and RTBS.

For a better comparison, the daily batch boundaries were minimized by aligning the daily results using 2nd order extrapolation at the border [6]. This method will cause an accumulated error in the long-term, so only 3 days of data from Figure 1 were aligned and compared, as displayed in Figure 2.

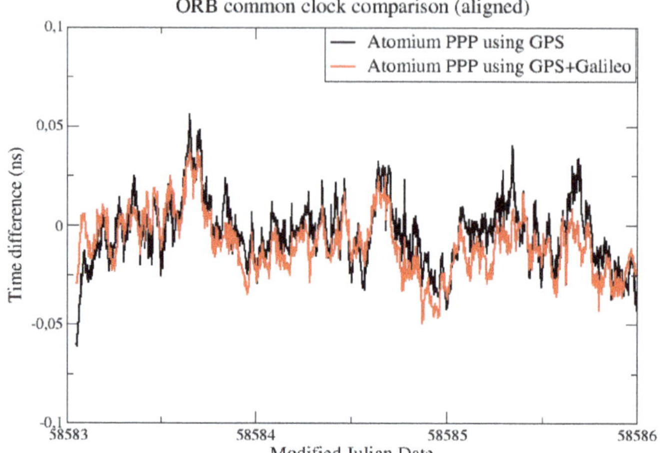

Figure 2. 3 days aligned CCD Galileo results from Figure 1.

From Figure 2, it is observed that the daily PPP clock comparison noise was generally within 0.1 ns, and with GPS + Galileo signals, the PPP obtained slightly more precise results, and a similar conclusion can be made from the CCD experiments in INRIM. To better elaborate the improvement by adding Galileo signals, they were compared in the form of Allan DEViation (ADEV). The ADEVs for the CCD results in ORB and INRIM are plotted in Figures 3 and 4.

The ADEVs of GPS + Galileo PPP results were generally lower than the GPS-only ones, as shown in Figures 3 and 4. The improvement in the frequency stability at medium and short averaging times was generally between 10% and 25%. However, due to the discontinuity at the batch boundary, as shown in Figure 1, it is hard to compare the long-term stability.

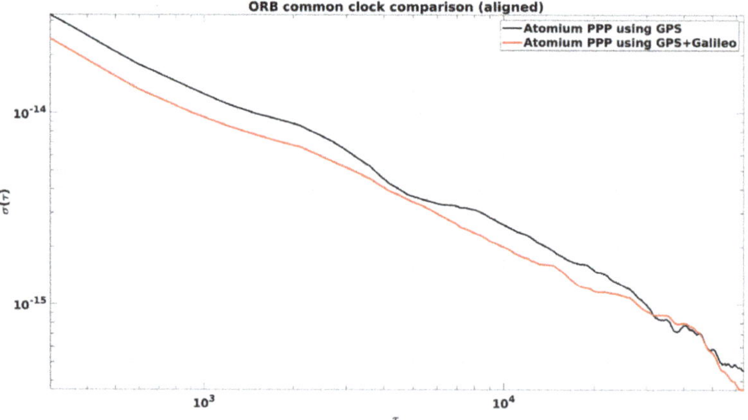

Figure 3. Allan DEViations (ADEVs) for the CCD global positioning system (GPS)-only and GPS + Galileo precise point positioning (PPP) results in the Observatory of Belgium (ORB) from Figure 2.

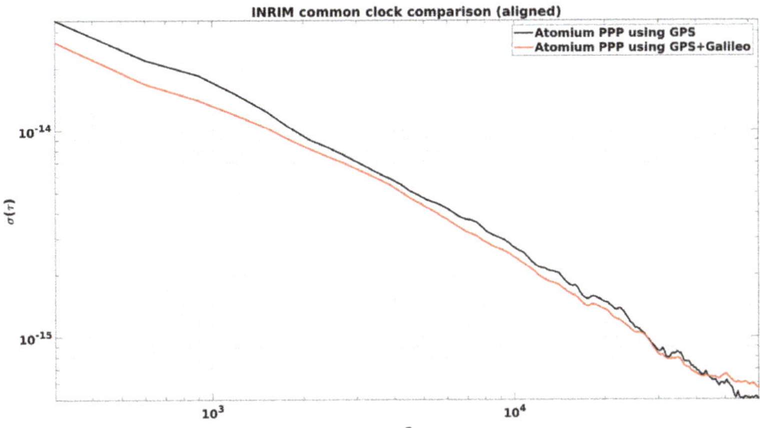

Figure 4. ADEVs for the CCD GPS-only and GPS + Galileo PPP results in the Italian National Metrology Institute (INRIM).

3.2. Atomium Constrained PPP

The constraints were applied to GPS-only PPP and GPS + Galileo PPP results separately, and the results were compared with the original clock results in Figure 5. Both INRIM and ORB have their hydrogen maser clock for the comparison in this experiment, and the values of $ADev(1s)$ were both set to 2×10^{-13}. As seen in Figure 5, the constrained PPP can reduce the noise of the clock solution inside the batch and keep the long-term stability as for the original PPP. The noise reduction for the constrained PPP was more visible in an aligned solution using 3 days' clock data, as shown in Figure 6.

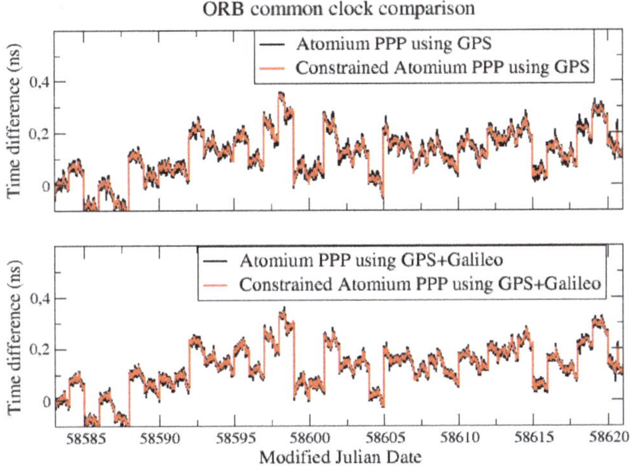

Figure 5. CCD constrained results in ORB.

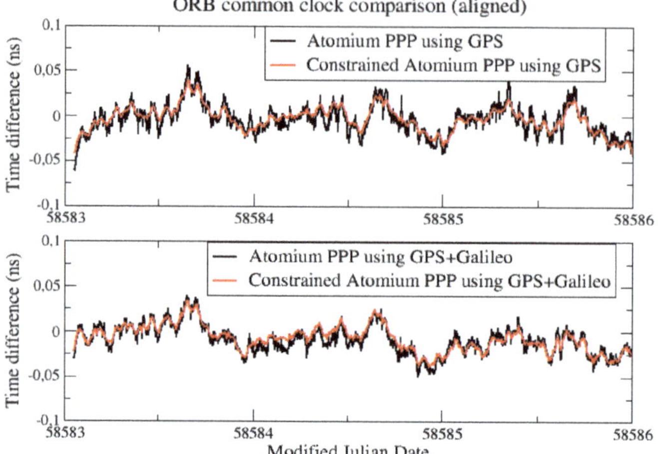

Figure 6. Aligned CCD constrained results in ORB.

The ADEVs of constrained PPP, using the aligned 3-day results, were compared with the classical PPP ones in Figures 7 and 8 for both the CCD experiments in ORB and INRIM.

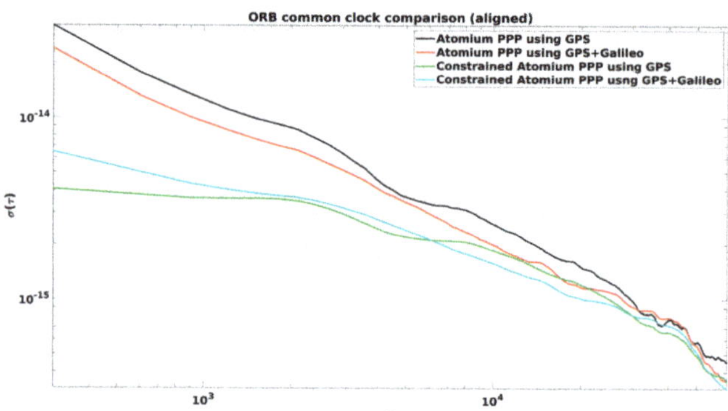

Figure 7. ADEVs of aligned CCD results in ORB.

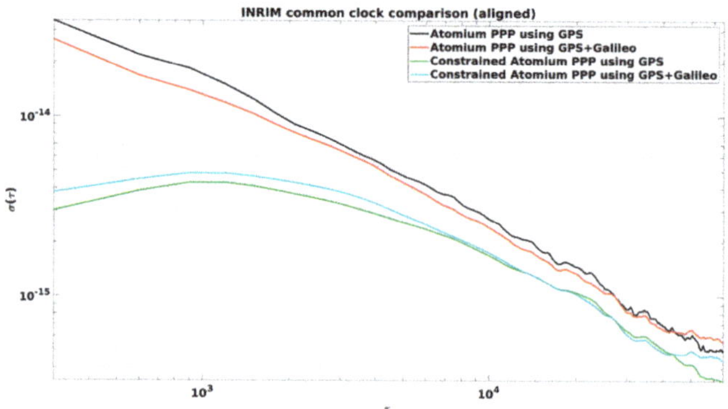

Figure 8. ADEVs of aligned CCD results in INRIM.

It is very clear that constrained PPP improves clock frequency stability (see Figures 7 and 8), especially at short averaging times, as reported in Table 1. It can be observed in both Figures that the best short-term stability was provided by GPS-only constrained PPP, rather than the GPS + Galileo constrained PPP. It is most probably because the GPS + Galileo PPP has more satellite measurements at each epoch, and the relative weight of the constraint gets smaller for GPS+Galileo than for GPS-only.

Table 1. The stability improvements when applying constraint model to global positioning system (GPS) precise point positioning (PPP) (black to the green line in Figures 7 and 8) and GPS + Galileo PPP (red to the light blue line in Figures 7 and 8), respectively.

Averaging Time	Improvement of Stability	
	GPS	GPS + Galileo
5 min	89.3%	79.4%
1 h	50.0%	36.7%
3 h	32.3%	23.1%
12 h	21.1%	17.7%

The "PPP—OPT" experiments were also built here to test the constrained PPP. Since both AO_4 and GUM4 stations receive only GPS signals, GPS + Galileo PPP results were not available for the following "PPP—OPT" experiments. The GPS-only results were compared in Figure 9.

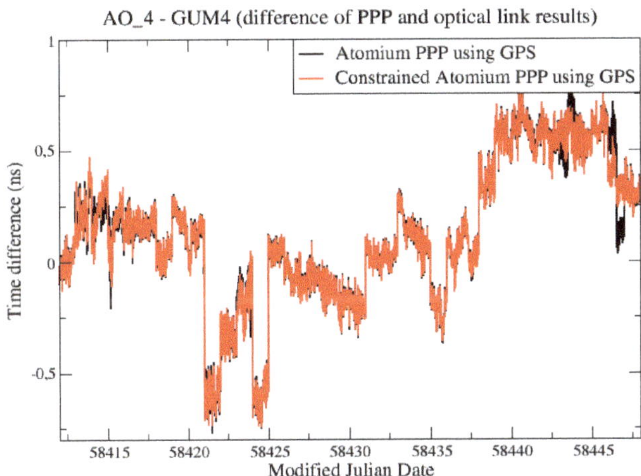

Figure 9. "PPP—OPT" experiments to evaluate constrained PPP.

It was found (Figure 9) that the constrained PPP did not reduce the noise obviously as in the former CCD experiments. The reason is that, while the AO_4 is driven by a hydrogen maser, GUM4 is driven by a caesium clock. As a result of its lower ADev(1s), the caesium clock will be constrained in PPP with a much smaller weight than a hydrogen maser, as we can see from Equations (4) and (5). Although the constraint improved the stability dramatically at the AO_4 side, the caesium clock estimation at the GUM4 side was not improved effectively by the constraint, as shown in Figure 10. The noise of caesium clock dominated the stability of the two clock comparison so that the final constrained PPP solution did not show any improvement.

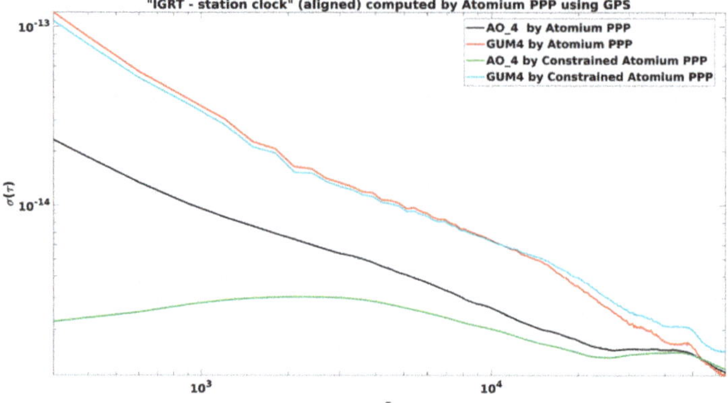

Figure 10. ADEVs in "PPP—OPT" experiments to evaluate constrained PPP. IGRT stands for the reference time of International global navigation satellite systems service (IGS) rapid products that are used in the PPP computation.

In addition to the normal function of constrained PPP, the applications of constrained PPP in different scenarios were also studied in this paper:

3.2.1. The Signal Tracking Loss

There was a notable difference between the constrained PPP and non-constrained PPP results in the last test periods of the former "PPP—OPT" experiment, as can be seen in Figure 9. These differences were zoomed in and plotted again in Figure 11.

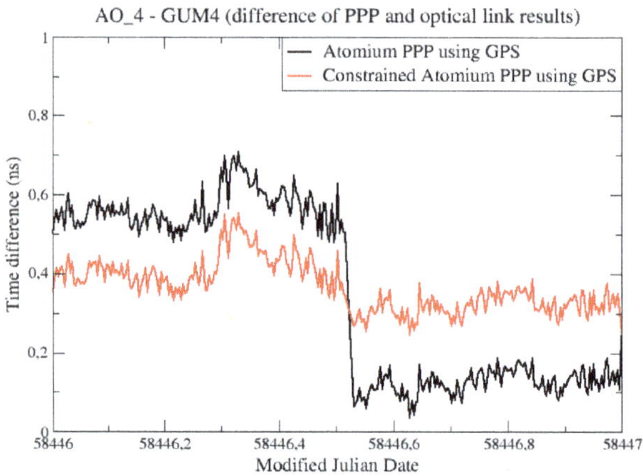

Figure 11. "PPP—OPT" experiments to evaluate constrained PPP.

The jump that happens in the middle of the MJD 58446 was not a daily batch jump, but because all the satellite tracks restarted so that new ambiguities were determined for all the satellites, at that epoch in the PPP computation for station AO_4, and the jump happened just as at the batch boundary. This jump was overcome by the constrained PPP, in which the receiver clock measurements were relatively constrained at all epochs.

3.2.2. For an Extreme Noisy Environment

Station GUM4 was observed to have very noisy receiver measurements for around 2 months, and it returned to normal after some receiver modification. The basic idea was constraining the measurements' noise by constrained PPP, as illustrated in Figure 12.

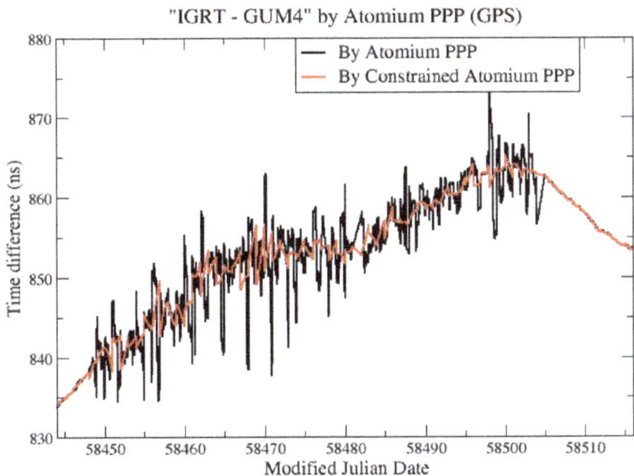

Figure 12. "IGRT—GUM4" from constrained and non-constrained PPP.

Since the cesium clock estimation at GUM4 was not affected much by the constraint model as mentioned before, a larger weight of the clock constraint at GUM4 was set in this experiment to further restrict the abnormal measurements' noise (the same weight as for the hydrogen maser was chosen). The constrained PPP reduced the noise but still not at a satisfactory level, which can be seen in Figure 12. This is because the frequency offset and drift over each day were estimated with larger uncertainties due to the very noisy measurements.

To further retrieve the correct clock solution from the noisy environment, we needed a more accurate frequency measurement of the clock. For this purpose, we estimated the averaged frequency deviation "IGRT—GUM4" ($Frq_{\text{"IGRT—GUM4"}}$) as the difference of $Frq_{\text{"IGRT—AO\_4"}}$ and $Frq_{\text{"AO\_4—GUM4"}}$. $Frq_{\text{"IGRT—AO\_4"}}$ was computed from the clock solution IGRT—AO_4 through Atomium PPP, and $Frq_{\text{"AO\_4—GUM4"}}$ was computed from independent clock comparison results of AO_4—GUM4. Then the constraints were computed from the frequency data and applied to Atomium PPP to estimate the constrained "IGRT—GUM4". In the present case, we used the optical link for AO_4—GUM4. This link is even more stable than the PPP, so there is no real advantage of using it to get a constrained PPP solution. However, in some other links, the only available independent link might be a two-Way satellite time and frequency transfer (TWSTFT) [34], which provides only 24 data points per day, and therefore, does not provide short-term stability. In that case, using the TWSTFT to get the frequency deviation of the link will be very helpful to get a constrained PPP, and hence, a frequency transfer with very good short-term frequency stability.

The results of constrained PPP at GUM4 using estimated constraints are displayed in the red line in Figure 13. To further mitigate the daily boundary effect, and also to verify the correctness of the constrained clock results inside the batch, the multi-day constrained PPP computations were also performed: For each daily batch, in addition to the relative constraints, the receiver clock at the first epoch was further absolutely constrained based on the estimated clock at the last epoch of the previous batch and the increment predicted from the frequency model of the previous batch. The corresponding multi-day constrained PPP results are plotted in Figure 13 in the green line. It is observed from Figure 13 that the constrained PPP using the estimated constraints can highly reduce the clock measurements' noise, and the multi-day constrained PPP can further provide continuous clock solutions by reducing the size of the daily boundary jump. However, the multi-day constrained method should be used carefully since it may also introduce an artificial variation up to hundreds of ps during 1-month clock measurements. Hence, it is recommended to use this multi-day constrained PPP in extremely noisy environments only to restrict the large daily boundary jump, which is the main factor influencing the stability of the constrained PPP clock solution, as can be seen in Figure 13.

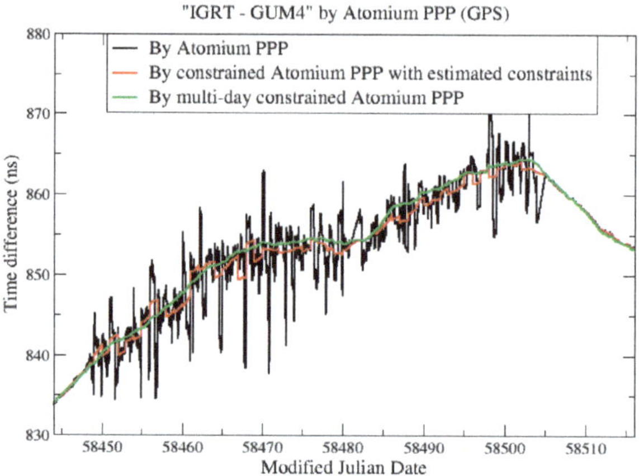

Figure 13. "IGRT—GUM4" from PPP and constrained PPP.

The correct frequency data are expected to be obtained from external products, e.g., TWSTFT. Though it is also possible to directly use the phase data from external products if it exits, the calibration error of the external measurements will also be introduced. Meanwhile, the estimated frequency data is not affected by these calibration errors and can be used by the target clock without any bias.

3.3. Atomium Integer PPP

As described before, to check if there was any integer cycle estimation error of N_{WL} in the integer PPP results, the continuity of the estimated WRB needs to be ensured. Figure 14 shows the estimated WRB at the first step of integer PPP for the station BRUX and RTBS in the CCD experiments, and no integer cycle error was found during the period of estimation.

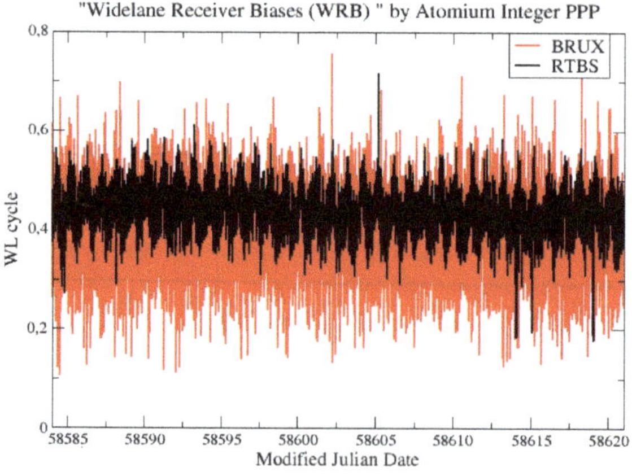

Figure 14. Wide-lane receiver bias (WRB) at station BRUX and RTBS estimated by integer PPP.

As shown in Figure 15, the common clock comparison results using integer PPP also showed some discontinuities at the daily batch boundaries. However, these discontinuities were only integer multiples of the GPS λ_{NL} (are 1 or -1 λ_{NL} in Figure 15), and the continuous clock solution could be easily obtained after alignment by adding integer multiples of λ_{NL}.

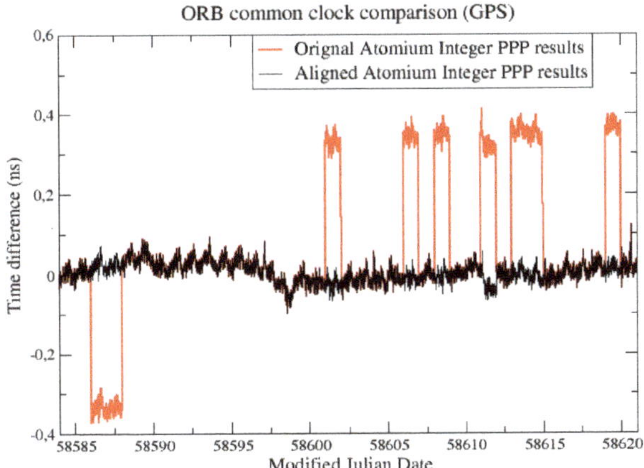

Figure 15. CCD experiments at ORB with integer PPP.

This continuous integer PPP CCD result was compared to the original Atomium PPP result with daily boundary jumps and the aligned original Atomium PPP result in terms of phase offset and ADEVs, as illustrated in Figures 16 and 17, separately. Compared to the original Atomium PPP results, integer PPP clearly improved the frequency stability from several hours to longer averaging times, and the frequency accuracy of 1×10^{-16} could be reached by 4 days averaging. Aligning the daily Atomium PPP results using 2nd order extrapolation to remove the daily boundary jumps could largely improve the frequency stability as shown in Figure 17, however it caused a drift in the time solution, as shown in Figure 16 (where the aligned data in green line accumulate an error of about 0.6 ns in a month). It is, therefore, not recommended to align the daily time solutions from float ambiguities PPP over a long time span to not affect the time accuracy.

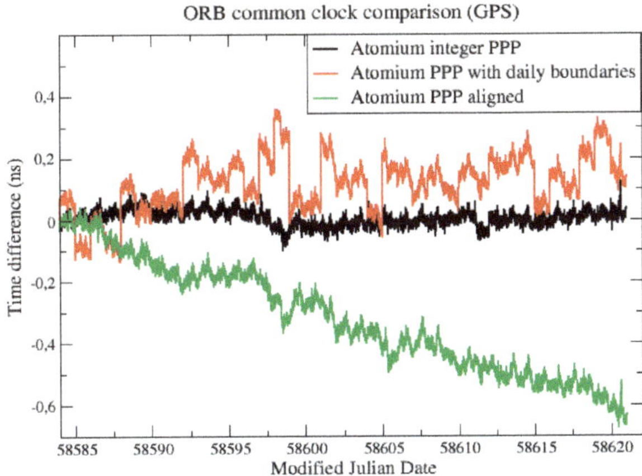

Figure 16. CCD results in ORB using Atomium integer PPP and original Atomium PPP.

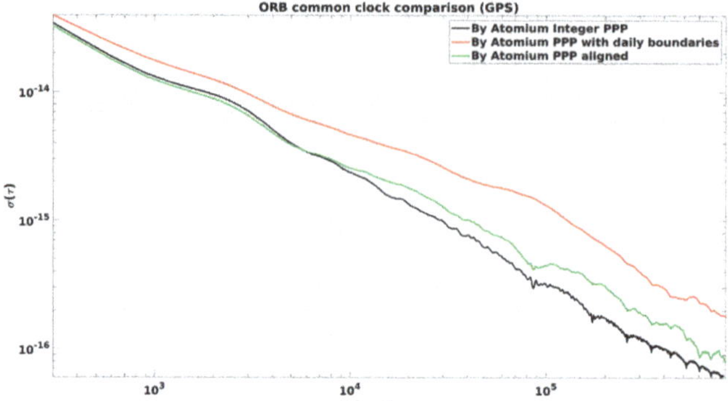

Figure 17. ADEVs between integer PPP and PPP CCD results in ORB.

The improvements of frequency stability at different averaging times by using Atomium integer PPP instead of Atomium PPP (corresponding to the black and red lines in Figure 17) are recorded in Table 2. It is shown that the largest improvement happened at the averaging time of one day, which was exactly the occurrence rate of the daily boundary jump.

Table 2. The stability improvements applying integer ambiguity resolution to Atomium PPP (GPS).

Averaging Time	Improvement of Stability
5 min	13.3%
1 h	31.2%
3 h	50.2%
12 h	63.3%
1 day	80%
5 days	65.3%

Figure 18 compares the "PPP—OPT" results from integer PPP with the one from NRCAN PPP, which was based on a multi-day batch to avoid the daily jumps within the batch. The NRCAN PPP results showed a slight divergence and reach 0.4 ns at the end of the one-month batch.

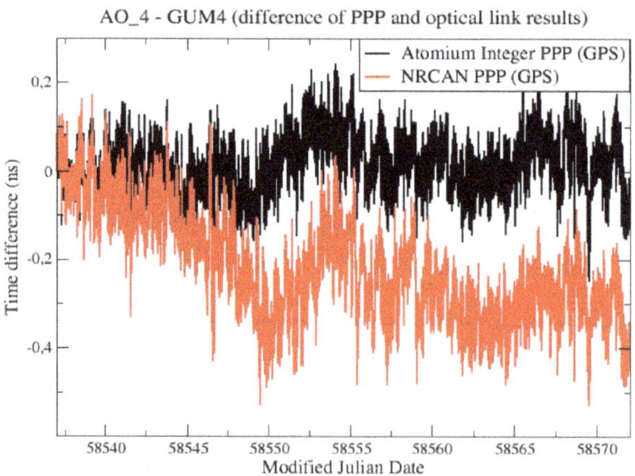

Figure 18. Comparison of integer PPP and NRCAN PPP results.

The ADEVs of the clock measurements in Figure 18 are further compared in Figure 19. A similar conclusion can be made as from the CCD experiments: The integer PPP improved the clock frequency stability from an average of 2.5 h to longer times compared to the NRCAN PPP, and at 10 days averaging, the stability of 1×10^{-16} was acquired by using integer PPP.

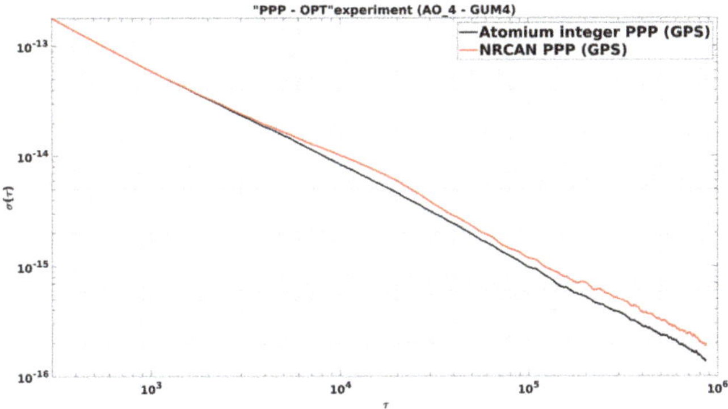

Figure 19. ADEVs between integer PPP and NRCAN PPP results.

4. Conclusions

Three updated versions of Atomium PPP software were presented in this paper: Atomium PPP with GPS + Galileo measurements; Atomium PPP with receiver clock constrained model; Atomium PPP with integer ambiguity resolution. Two kinds of experiments were built to test the performance of their clock solutions. Using Galileo in addition to GPS in PPP computation reduced the noise of the clock solution inside the daily batches, while the long-term stability was hardly improved due to the presence of daily jumps. The basic function of constrained PPP is to reduce the short-term noise of the clock measurements on good quality clocks as hydrogen masers, by restraining the short-term thermal noise. In addition, two more application scenarios for constrained PPP were introduced in this paper. First, constrained PPP can keep providing continuous clock solution during the period when the receiver experiences short-term loss of lock. Second, constrained PPP can be used to retrieve true clock solutions from very noisy measurements as long as the related clock frequency data can be estimated from external products, e.g., from TWSTFT or optical link measurements, without introducing their calibration errors at the phase measurements. Finally, integer PPP shows an advantage to the current float ambiguity PPP techniques on the clock frequency transfer stabilities at mid-term to long-term averaging times; a frequency stability of 1×10^{-16} can be reached at an averaging of 4 days to 10 days, depending on the time link. It is expected that when the Galileo integer ambiguity satellite products are regularly available, the integer PPP with GPS + Galileo will further improve the current GNSS clock frequency transfer capacity.

Currently, all the updated Atomium PPP software have been integrated into the DEMETRA server to provide the daily solutions for users who need post-processing services with better performances. GPS and Galileo daily quality monitoring data are both available for the users. And the test and the integration of the Atomium Galileo PPP hourly solution is in progress.

Author Contributions: Conceptualization, P.D. and W.H.; methodology, P.D. and W.H.; software, W.H.; validation, P.D., G.S. and I.S.; formal analysis, W.H.; investigation W.H., P.D., G.S. and I.S.; resources, P.D., G.S. and I.S.; data curation, W.H.; writing—original draft preparation, W.H.; writing—review and editing, P.D., G.S. and I.S.; visualization, W.H.; supervision, P.D., G.S. and I.S.; project administration, G.S. and I.S.

Funding: This research received no external funding.

Acknowledgments: The authors would like to thank Patrizia Tavella, the director of the Time Department at the International Bureau of Weights and Measures (BIPM), for formerly coordinating the DEMETRA project and her sustained guidance during PPP software updating. The authors also acknowledge the IGS/MGEX analysis centers for the free availability of their satellite products.

Conflicts of Interest: The authors declare no conflict of interest.

Appendix A

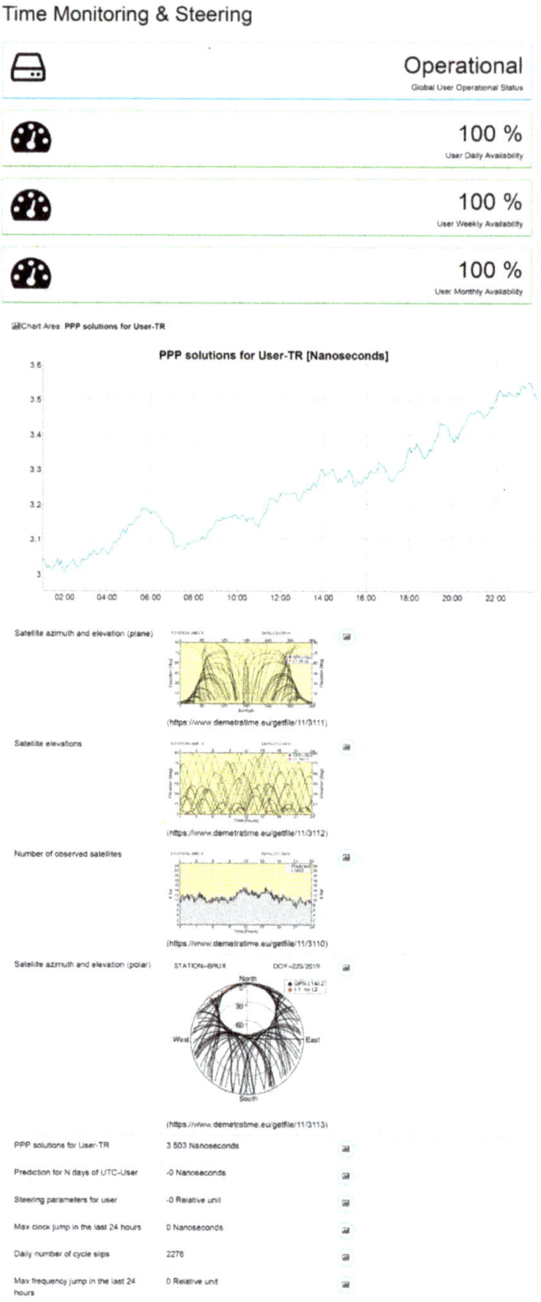

Time Monitoring & Steering

Operational
Global User Operational Status

100 %
User Daily Availability

100 %
User Weekly Availability

100 %
User Monthly Availability

Chart Area. PPP solutions for User-TR

PPP solutions for User-TR [Nanoseconds]

Satellite azimuth and elevation (plane)
(https://www.demetratime.eu/getfile/11/3111)

Satellite elevations
(https://www.demetratime.eu/getfile/11/3112)

Number of observed satellites
(https://www.demetratime.eu/getfile/11/3110)

Satellite azimuth and elevation (polar)
(https://www.demetratime.eu/getfile/11/3113)

PPP solutions for User-TR	3.503 Nanoseconds
Prediction for N days of UTC-User	-0 Nanoseconds
Steering parameters for user	-0 Relative unit
Max clock jump in the last 24 hours	0 Nanoseconds
Daily number of cycle slips	2276
Max frequency jump in the last 24 hours	0 Relative unit

References

1. Lachapelle, G.; Petovello, M.G.; Gao, Y.; Garin, L. Precise point positioning and its challenges, aided-GNSS and signal tracking. *Inside GNSS* **2006**, *1*, 16–18.
2. Defraigne, P. GNSS time and frequency transfer. In *Springer Handbook of Global Navigation Satellite Systems*; Teunissen, P.J., Montenbruck, O., Eds.; Springer Handbooks: Cham, Switzerland, 2017; pp. 1187–1206.
3. European GNSS Agency (GSA). *GNSS Market Report*; GSA: Washington, DC, USA, 2017; pp. 82–89.
4. Hasan, K.F.; Feng, Y.M.; Tian, Y.C. GNSS time synchronization in vehicular ad-hoc networks: benefits and feasibility. *IEEE Trans. Intell. Transp. Syst.* **2018**, *19*, 3915–3924. [CrossRef]
5. Grelier, T.; Garcia, A.; Peragin, E.; Lestarquit, L.; Harr, J.; Seguela, D.; Issler, J.L.; Thevenet, J.B.; Perriault, N.; Mehlen, C.; et al. GNSS in space part 1: Formation flying radio frequency missions, techniques, and technology. *Inside GNSS* **2008**, *November/December*, 40–46.
6. Petit, G.; Kanj, A.; Loyer, S.; Delporte, J.; Mercier, F.; Perosanz, F. 1×10^{-16} frequency transfer by GPS PPP with integer ambiguity resolution. *Metrologia* **2015**, *52*, 301–309. [CrossRef]
7. Pallares, O.; Shariat-Panahi, S.; del Rio, J.; Manuel, A. Time synchronization of a commercial seismometer through IEEE-1588. In Proceedings of the Fourth International Workshop on Marine Technology, Cádiz, Spain, 22–23 September 2011.
8. Clements, P.A. Global positioning system timing receivers in the DSN. *TDA Prog. Rep.* **1981**, *November and December*, 42–67.
9. Caccianiga, B.; Cavalcante, P.; Cerretto, G.; Esteban, H.; Korga, G.; Misiaszek, M.; Orsini, M.; Pallavicini, M.; Pettiti, V.; Plantard, C.; et al. GPS-based CERN-LNGS time link for Borexino. *J. Instrum.* **2012**, *7*, P08028. [CrossRef]
10. Radiocommunication Bureau. *ITU Handbook: Satellite Time and Frequency Transfer and Dissemination*; Radiocommunication Bureau: Geneva, Switzerland, 2010.
11. Defraigne, P.; Bruyninx, C.; Guyennon, N. PPP and Phase-only GPS time and frequency transfer. In Proceedings of the 2007 IEEE International Frequency Control Symposium Joint with the 21st European Frequency and Time Forum, Geneva, Switzerland, 29 May–1 June 2007; pp. 904–908.
12. Defraigne, P.; Tavella, P.; Sesia, I.; Cerretto, G.; Signorile, G.; Calonico, D.; Costa, R.; Clivati, C.; Cantoni, E.; de Stefano, C.; et al. Demonstrator of Time Services based on European GNSS Signals: The H2020 DEMETRA Project. In *Precise Time and Time Interval Meeting*; ION PTTI: Monterey, CA, USA, 2017; pp. 127–137.
13. Calonico, D.; Clivati, C.; Mura, A.; Tampellini, A.; Gertosio, M.; Levi, F. Time and frequency distribution over fibre for geodesy, seismology and industry. In Proceedings of the 2018 IEEE International Symposium on Precision Clock Synchronization for Measurement, Control, and Communication (ISPCS), Geneva, Switzerland, 30 September–5 October 2018.
14. DEMETRA homepage. Available online: www.demetratime.eu (accessed on 10 October 2019).
15. Defraigne, P.; Guyennon, N. *Atomium: A New Tool for Time and Frequency Transfer in PPP Mode*; TAI Working Group: Paris, France, 2006.
16. Defraigne, P.; Bruyninx, C. On the link between GPS pseudorange noise and day-boundary discontinuities in geodetic time transfer solutions. *GPS Solut.* **2007**, *11*, 239–249. [CrossRef]
17. Kouba, J.; Héroux, P. Precise point positioning using IGS orbit and clock products. *GPS Solut.* **2001**, *5*, 12–28. [CrossRef]
18. Rey, J.; Senior, K. Geodetic techniques for time and frequency comparisons using GPS phase and code measurements. *Metrologia* **2005**, *42*, 215. [CrossRef]
19. Huang, W.; Defraigne, P.; Signorile, G.; Sesia, I. New precise point positioning software for upgrade of the time monitoring and steering service of the H2020 DEMETRA project. In Proceedings of the 2019 IEEE International Workshop on Metrology for Aerospace, Turin, Italy, 19–21 June 2019.
20. Defraigne, P.; Verhasselt, K. Multi-GNSS time transfer with CGGTTS-V2E. In Proceedings of the 2018 European Frequency and Time Forum (EFTF), Turin, Italy, 10–12 April 2018; pp. 270–275.
21. Cerretto, G.; Lahaye, F.; Tavella, P.; Vitrano, S. Precise Point Positioning: Implementation of the constrained clock model and analysis of its effects in T/F transfer. In Proceedings of the EFTF-2010 24th European Frequency and Time Forum, Noordwijk, The Netherlands, 13–16 April 2010.
22. Wang, K.; Rothacher, M. Stochastic modeling of high-stability ground clocks in GPS analysis. *J. Geod.* **2013**, *87*, 427–437. [CrossRef]

23. Ge, Y.; Zhou, F.; Liu, T.; Qin, W.; Wang, S.; Yang, X. Enhancing real-time precise point positioning time and frequency transfer with receiver clock modeling. *GPS Solut.* **2019**, *23*, 20. [CrossRef]
24. Delporte, J.; Mercier, F.; Laurichesse, D.; Galy, O. GPS carrier-phase time transfer using single-difference integer ambiguity resolution. *Int. J. Navig. Obs.* **2008**, *2008*, 273785. [CrossRef]
25. Laurichesse, D.; Mercier, F. Integer ambiguity resolution on undifferenced GPS phase measurements and its application to PPP. In Proceedings of the ION GNSS 2007 20th International Technical Meeting of the Satellite Division, Fort Worth, TX, USA, 25–28 September 2007; pp. 839–848.
26. Loyer, S.; Perosanz, F. Zero-difference GPS ambiguity resolution at CNES–CLS IGS analysis center. *J. Geod.* **2012**, *86*, 991–1003. [CrossRef]
27. Leute, J.; Gérard, P.; Exertier, P. High accuracy continuous time transfer with GPS IPPP and T2L2. In Proceedings of the 2018 European Frequency and Time Forum (EFTF), Turin, Italy, 10–12 April 2018; pp. 249–252.
28. The Multi-GNSS Experiment (MGEX). Available online: http://mgex.igs.org/ (accessed on 15 August 2019).
29. Zhou, F.; Dong, D.; Li, P.; Li, X.; Schuh, H. Influence of stochastic modeling for inter-system biases on multi-GNSS undifferenced and uncombined precise point positioning. *GPS Solut.* **2019**, *23*, 59. [CrossRef]
30. International GNSS Service (IGS) The Antenna Exchange format (ANTEX) file, version 1.4. Available online: ftp://www.igs.org/pub/station/general/igs14.atx (accessed on 16 August 2019).
31. ANTEX Format Description, September 2010. Available online: https://kb.igs.org/hc/en-us/articles/216104678-ANTEX-format-description (accessed on 16 August 2019).
32. Teunissen, P.J.G. GNSS ambiguity bootstrapping: Theory and application. In Proceedings of the International Symposium on Kinematic Systems in Geodesy, Geomatics and Navigation, Banff, AB, Canada, 5–8 June 2001; pp. 246–254.
33. Śliwczyński, Ł.; Krehlik, P.; Czubla, A.; Buczek, Ł.; Lipiński, M. Dissemination of time and RF frequency via a stabilized fibre optic link over a distance of 420 km. *Metrologia* **2013**, *50*, 133. [CrossRef]
34. Kirchner, D. Two-way satellite time and frequency transfer (TWSTFT): Principle, implementation, and current performance. In *the Review of Radio Sciences 1996-1999*; Ross Stone, W., Ed.; Oxford University Press: Oxford, UK, 1999; pp. 27–44.

© 2019 by the authors. Licensee MDPI, Basel, Switzerland. This article is an open access article distributed under the terms and conditions of the Creative Commons Attribution (CC BY) license (http://creativecommons.org/licenses/by/4.0/).

Article

"MicroMED" Optical Particle Counter: From Design to Flight Model †

Diego Scaccabarozzi [1,*], Bortolino Saggin [1], Riccardo Somaschini [1], Marianna Magni [1], Pietro Valnegri [1], Francesca Esposito [2], Cesare Molfese [2], Fabio Cozzolino [2] and Giuseppe Mongelluzzo [2]

1 Department of Mechanical Engineering, Politecnico di Milano, 23900 Lecco, Italy; bortolino.saggin@polimi.it (B.S.); riccardo.somaschini@polimi.it (R.S.); marianna.magni@polimi.it (M.M.); pietro.valnegri@polimi.it (P.V.)

2 INAF-Astronomical Observatory of Capodimonte, Salita Moiariello 16, 80131 Naples, Italy; francesca.esposito@inaf.it (F.E.); cesare.molfese@inaf.it (C.M.); fabio.cozzolino@inaf.it (F.C.); giuseppe.mongelluzzo@inaf.it (G.M.)

* Correspondence: diego.scaccabarozzi@polimi.it

† This paper is an expanded version of "Design validation of MicroMED, a particle analyzer for ExoMars 2020" published in the Proceedings of the 2019 IEEE 5th International Workshop on Metrology for AeroSpace (MetroAeroSpace), Torino, Italy, 19–21 June 2019.

Received: 18 December 2019; Accepted: 19 January 2020; Published: 22 January 2020

Abstract: MicroMED (Micro Martian Environmental Dust Systematic Analyzer (MEDUSA)) instrument was selected for the ExoMars 2020 mission to study the airborne dust on the red planet through in situ measurements of the size distribution and concentration. This characterization has never been done before and would have a strong impact on the understanding of Martian climate and Aeolian processes on Mars. The MicroMED is an optical particle counter that exploits the measured intensity of light scattered by dust particles when crossing a collimated laser beam. The measurement technique is well established for laboratory and ground applications but in order to be mounted on the Dust Suite payload within the framework of ExoMars 2020 mission, the instrument must be compatible with harsh mechanical and thermal environments and the tight mass budget of the mission payload. This work summarizes the thermo-mechanical design of the instrument, the manufacturing of the flight model and its successful qualification in expected thermal and mechanical environments.

Keywords: MicroMED; thermo-mechanical design; Dust Suite; ExoMars 2020; flight model; qualification testing

---

1. Introduction

MicroMED (Micro Martian Environmental Dust systematic analyzer) was developed by a consortium of Italian research institutes and Spain's space agency on the basis of the heritage of MEDUSA (Martian Environmental Dust Systematic Analyzer), a particle counter designed for the Humboldt payload [1,2]. MicroMED's goal is measuring the abundance and size distribution of atmospheric dust close to the surface of the red planet, where dust lifting takes place. Such kinds of measurements have not been performed before on Mars, even though dust distribution in the Martian atmosphere has gathered so much attention in the scientific community, as reported in many studies exploiting indirect measurement methods applied to data collected by various experiments [3–8]. The expected impact of the MicroMED sensor will be the improvement of knowledge and understanding of the planet's climate, since it is well recognized that airborne dust and regional or global dust storms cause changes in the atmospheric thermal structure of Mars.

MicroMED will be mounted on the Dust Suite of the surface platform onboard the ExoMars 2020 ESA and Roscosmos mission. The Dust Suite payload is composed of five sensors developed to image the landing site, monitor the long-term climate, and perform investigation and studies of the Aeolian processes on Mars. The instrument working principle is typical of optical particle counters (i.e., an optical system measures the light scattered by the dust particles passing through a collimated laser blade); from the light profile levels and duration information about the dust grain size and speed can be obtained.

In order to be approved for launch with the mission, the instrument must prove compatibility with harsh mechanical and thermal environments, due mainly to the launcher and the descent module dynamic excitations foreseen in the mission profile and predicted temperature scenarios on Mars. Moreover, a limited mass allocation of 500 g including electronics was made available for this instrument. The feasibility design of the MicroMED has already been reported in previous studies [9–13], evidencing the possibility from analysis to fulfill the design requirements of the mission with the available resources.

In this work, the detailed thermo-mechanical final design of the instrument is reported along with the experimental activity carried out in expected mechanical and thermal environmental conditions to validate the developed numerical models of the instrument. Moreover, the tests required to qualify the MicroMED Flight Model prior to integration on the ExoMars landing platform are presented.

2. Methods

2.1. Instrument Layout

MicroMED layout is made by different subassemblies. View of the main parts and components is provided in Figure 1.

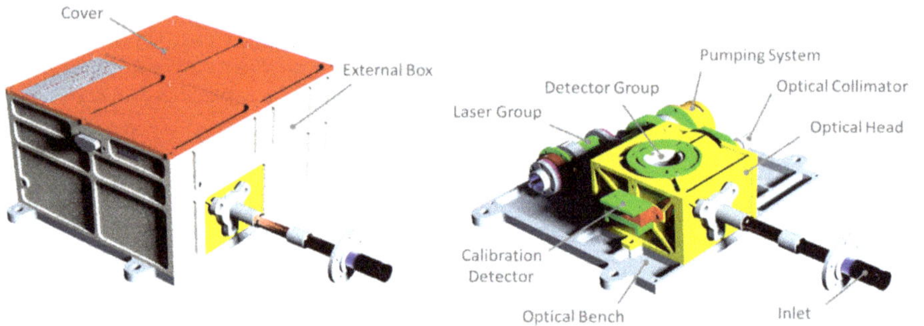

Figure 1. View of the MicroMED (Micro Martian Environmental Dust systematic analyzer) Flight Model layout three-dimensional (3D) CAD model.

The optical head (OH) is the main part of the instrument that holds science and reference detectors, the optical collimator (OC), the instrument fluid dynamic subsystem (inlet and outlet parts), and differential pressure sensors and temperature sensors. The selected detectors are Si PIN photodiodes (S5106 type @Hamamatsu), providing a sensing area of 5 × 5 mm^2 and compatibility with a low pressure environment. One of the two detectors is used to measure scattered light from analyzed particles, whereas the second one (used for calibration) tracks potential degradations of the laser and of the main detector as the consequence of aging and radiations. More details about the instrument's optical layout and performances are found in [9].

The selected pressure sensor is a differential MEMS pressure from Amphenol Nova Sensors (NPH-8-002.5-GH MEMS type). Its objective is to measure the differential pressure between the OH and the environment since, combining the measured pressure difference with the pressure loss characteristic curve of the instrument fluidic system, the measurement of the flowrate can be performed, allowing the assessment of the atmospheric dust density (i.e., the number of particles per unit of volume in the atmosphere). The pressure sensor is an off-shelf component, designed to operate in Earth's environment, but to be adopted for the mission, it underwent full performance characterization in the expected temperature range, in low pressure conditions, and with the expected dose of radiation during the mission. The performed characterization proved the suitability of the selected sensor for the intended application [14].

The pumping system (PS) is based on a space-designed rotary vane pump with an embedded brushless motor; it must provide the differential pressure required for the atmospheric gas and particles to flow through instrument sampling volume. Design feasibility of the PS concept was already reported by Scaccabarozzi and colleagues [15], but its full characterization in the representative environment was performed prior to integration in the MicroMED Flight Model. A testing campaign was performed with a specifically designed instrument to measure the flowrate of gasses in low pressure conditions [16]. The obtained calibration curves of the PS highlighted proved compatibility with the required fluid dynamic performances predicted by CFD analysis [17].

The OC concentrates laser light, conveyed by an optic fiber in the center of the OH, generating a sampling volume of $1 \times 1 \times 0.24$ mm^3 where the scattering of the particles occurs. The scattered light is focused by a parabolic mirror on the science photodiode. Thanks to the optical and CFD design, MicroMED is able to directly measure both the size distribution and concentration of dust grains suspended in the Martian atmosphere, exploring the particle size range from 0.4 μm to 20 μm.

Pt1000 temperature sensors are installed on the sensitive components of the MicroMED (i.e., LG, PS, OH) and on the electronics as well, in order to monitor the temperature during operative and non-operative phases and assure safety of the installed components. Moreover, one temperature sensor is located on the tip of the instrument' inlet, to provide an estimate of the Martian atmosphere temperature. The latter is expected to be different from the MicroMED interface one, and its knowledge is fundamental to accurately evaluate and control the flowrate within the instrument's fluid dynamic circuit. The control of flowrate is needed, especially to achieve the required dust speed on the sampling volume where science measurements are performed.

The main instrument components are mounted on the optical bench (OB), which is the interface with the Dust Suite platform. Finally, the external box (EB) holds the electronics group, mechanically shielding the instrument's components from external dust or contamination. The electronics allows conditioning and controlling of the subsystems and sensors and acquisition of the instrument housekeeping. The electronics is mounted on the instrument top and the EB provides an EMI shielding.

2.2. Thermo-Mechanical Design Requirements

Thermo-mechanical design requirements have been derived from the Surface Platform Information Package document [18]. The expected mechanical environment requires instrument survival with sweep sine and random excitations, whose profiles are summarized in Tables 1 and 2.

Table 1. Power sweep sine excitation.

Frequency (Hz)	Level [1] (g)
20	1
40	1.5
80	2
160	4
320	7
640	9
2000	9

[1] "g" has the meaning of standard acceleration gravity (i.e., about 9.81 m/s^2).

Table 2. Random levels.

Frequency (Hz)	Level [2] (g^2/Hz)
20	0.02
50	0.02
100	0.02
200	0.05
500	0.05
1000	0.025
2000	0.013

[2] "g" has the meaning of standard acceleration gravity (i.e., about 9.81 m/s^2).

Based on the launch and landing levels and gained experience from previous studies [19–22], the dynamic requirement was first set to a natural frequency larger than 150 Hz and a quasi-static loading to be used for the design phase of 1000 m/s^2. Moreover, size and mass allocations for MicroMED development were limited to a maximum envelope of 148 × 200 × 70 mm^3 and 500 g mass, respectively.

Temperatures at the Dust Suite Platform interface are between −50° C to 40 °C or between −20 °C and 40 °C for the storage and working phases, respectively. It should be noted that the MicroMED requires sampling of the Martian atmosphere through the inlet. Thus, the thermal design must account for the Martian atmosphere temperature variation and it has to be verified that the instrument does not cause large heat dissipation to the environment through its sampling inlet. Moreover, in order to ease the unit thermal control, the MicroMED is covered with thermal insulator (about 9 cm thickness) with protrusion of the inlet in the Martian atmosphere of at least 1 cm. The expected Martian temperature profiles in hot and cold working scenarios [18] are shown in Figure 2.

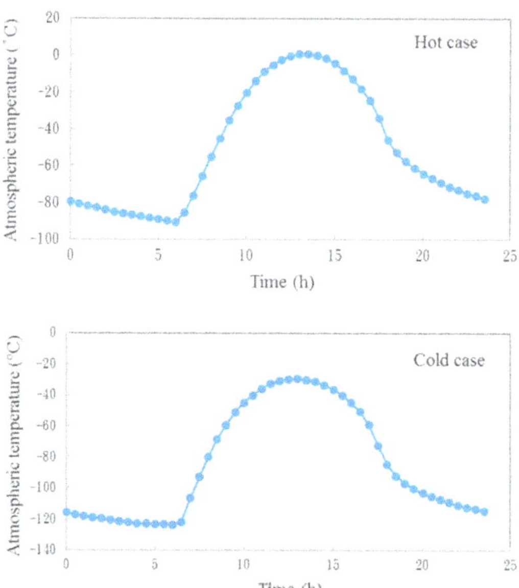

Figure 2. Hot (top) and cold (bottom) cases predict atmospheric temperatures for Mars.

The temperature trends shown in Figure 2 are used in thermal analysis to evaluate temperatures on sensitive elements of the MicroMED.

2.3. Flight Model Thermo-Mechanical Design

In order to complete the mechanical design of the MicroMED Flight Model, the finite element (FE) approach is used. Feasibility design of the MicroMED was already assessed in a previous study [9], showing the fulfillment of the requirement on the mass budget by means of mechanics based on ribbed geometry, which was optimized both for the OB and EB. Hereafter, the main results of the detailed design are presented.

Different FE models of the MicroMED assembly were developed and related analyses were performed to verify the mechanical resistance vs. quasi-static loading and instrument dynamic behavior. The assembly FE model comprises 59,885 tetrahedral and triangular elements and 17,274 nodes. The main subassemblies (i.e., OB, PS, OH, EB) and other components (inlet, outlet, and detectors) were modeled as lumped masses. This was mandatory in order to manage the FE model without jeopardizing the results' accuracy. The lumped masses were connected to their supporting structures by weighted links and the OB was fixed simulating the bolt's connection. The main material for the mechanical structures was aluminum alloy Al7075-T6, whereas the inlet tip, which is directly immersed in the Martian atmosphere, was designed with Vespel SP22. The latter is a well-known plastic material with space heritage. Details about the materials used in the FE model are given in Table 3.

Table 3. Mechanical and thermal properties of the materials.

Material	Density (kg/m^3)	Modulus of Elasticity (MPa)	Poisson's Coefficient	Tensile Yield Strength (MPa)	CTE (10^{-5}/K)	Thermal Conductivity (W/(mK))
Al 7075	2810	71,700	0.33	503	2.36	130
Al 6061	2700	68,900	0.33	55.2	2.34	180
Vespel SP22	1650	4830	0.4	51.7	3.8	0.89

In order to evaluate mechanical resistance in the expected temperature environment, a FE model was developed to investigate instrument behavior for the worst-case temperature scenario (i.e., mounting at room temperature and achievement minimum non-operational temperature). The model comprises the same features previously described with the addition of an aluminum plate 6061, to simulate Dust Suite mounting interface.

Results of first three modes of vibration and Von Mises stress along three main axes with quasi-static loading are provided in Figures 3 and 4, respectively. Computed temperature distribution on the MicroMED for the thermo-elastic analysis is shown in Figure 5.

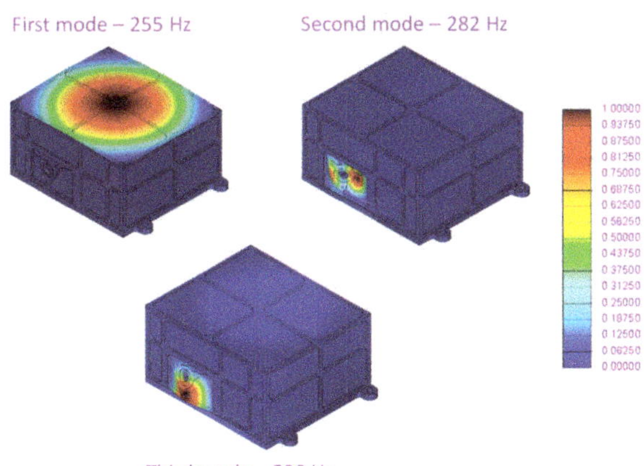

Figure 3. Computed non-dimensional displacement from modal analyses.

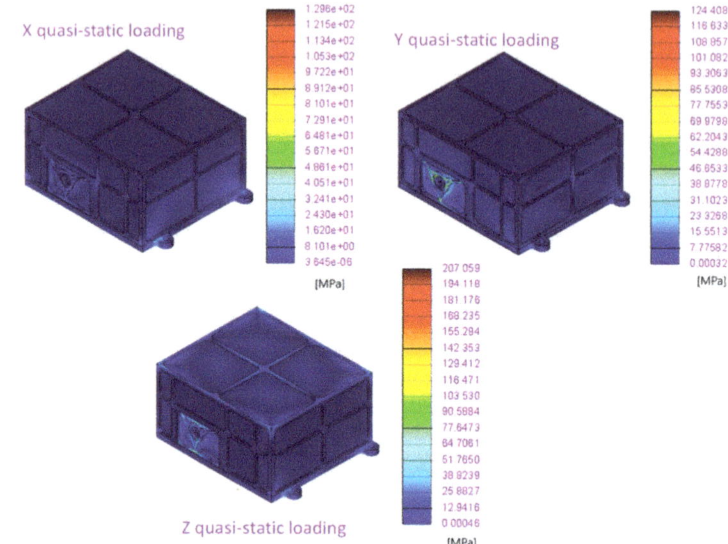

Figure 4. MicroMED Flight Model, quasi-static analyses results.

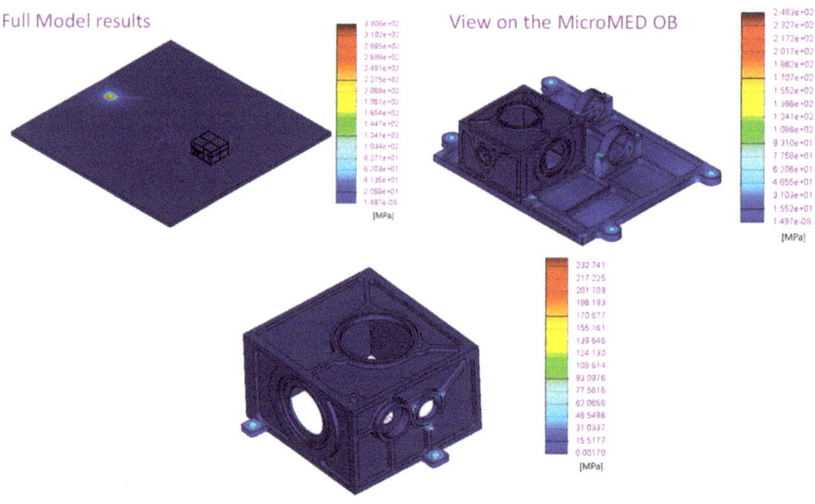

Figure 5. MicroMED Flight Model, thermo-elastic analyses results: computed Von Mises stresses on the MicroMED optical head (OH) and optical bench (OB), non-operative case, environmental temperature at −50 °C.

Table 4 summarizes the results of the mechanical resistance with quasi-static loading and thermo-elastic analyses, providing the computed margin of safety (MOS), which according to the ESA standard [23] is defined as:

$$MOS = 1 - \frac{\sigma_C}{\frac{\sigma_{lim}}{\eta_{lim}}} \quad (1)$$

where σ_c is the computed Von Mises stress, σ_{lim} is the maximum admissible stress, and η_{lim} is the related safety factor with respect to the limit condition. As for Equation (1), the MOS shall be positive in order to validate the mechanical design.

Table 4. Von Mises stresses for mechanical analyses.

Analysis	Maximum Von Mises Stress (MPa)	Margin of Safety (MOS)
Quasi-static X axis	129	0.62
Quasi-static Y axis	125	0.63
Quasi-static Z axis	207	0.38
Thermo-elastic cold case non-operative condition	248	0.26

2.4. MicroMED Thermal Model

A thermal model of the MicroMED was developed to determine the temperature ranges in various conditions for the main instrument parts. The unit was therefore modeled with shells representing the main structures, electronics board, and power sources (i.e., the PS and LG subassemblies). A convective link with the only exposed surface (i.e., the instrument inlet) was set with a convective coefficient of 0.2 W/(m^{2}K) and 2 W/(m^{2}K) for the hot and cold cases, respectively. Infrared emissivity of the developed model is shown in Figure 6. Thermo-optical properties used for the surfaces of the model are summarized in Table 5.

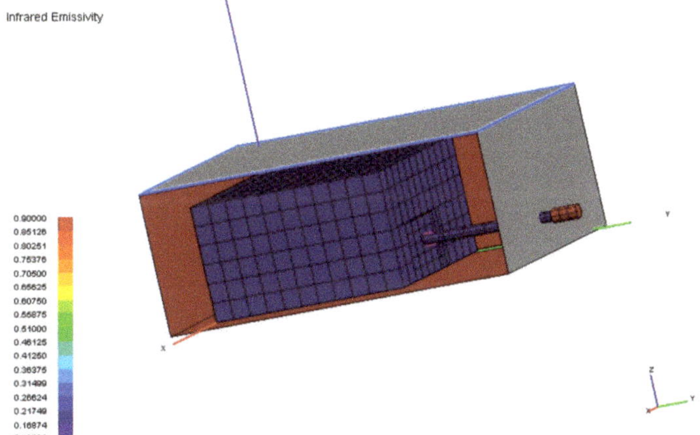

Figure 6. MicroMED thermal model, emissivity distribution over the geometrical model.

Table 5. Thermo-optical properties [3].

Part	Material	Coating	α_s	ρ_s	ε_h	ρ_h
MicroMED	Al 7075	Surtec 650	ns	ns	0.12	0.88
Electronics	PCB	PCB	ns	ns	0.8	0.2
OH	Al 7075	Aeroglaze Z306	0.95	0.05	0.9	0.1

[3] "s" means specular, "h" means hemispherical, "ns" means not significant.

As already explained, the thermal control architecture is based on a thermally conductive mounting on the Dust Suite platform as the temperature range specified at the interface compatible with the thermal requirements of all MicroMED components. The heat generated internally is therefore mostly damped through the mounting interface. Power dissipated by electronics ranges between 1.7 W and 2 W in operative hot and cold cases, respectively. The ON state duration for the electronics is about 300 s (time for each cycle) for the hot case and 126 s for the cold case. Power budgets for the PS and for the LG in both the operative cold and hot cases are 2.5 W and 0.4 W, respectively. Cycle time for both the LG and PS is set to 100 s.

The applied conditions for environmental temperatures are shown in Figure 2. The results of the thermal simulations are provided in Figures 7 and 8.

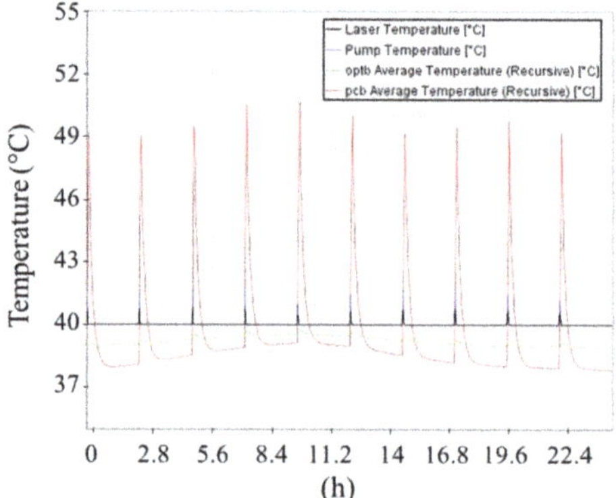

Figure 7. Hot operational case: temperature of critical components during second solar day of Mars operations.

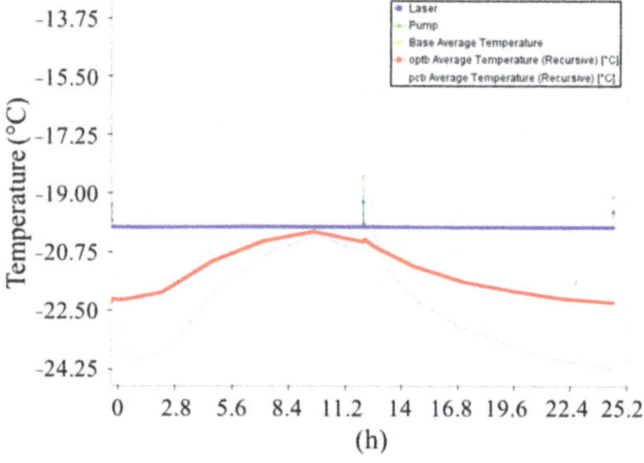

Figure 8. Cold operational case: daily cycle after two operational days.

Computed power exchanges at the Dust Suite interface are 1.7 W and 2.9 W in hot and cold operative scenarios, respectively.

2.5. Discussion

Performed FE analyses allowed verification of the MicroMED resistance within expected mechanical and thermal environments. It should be noted that the MOS for static and thermo-elastic simulations is larger than the minimum required. The modal analyses showed that MicroMED dynamic behavior is compliant to design requirements, highlighting first resonance about 60% over the minimum admissible value by design, providing a large safety margin with respect to the expected increased compliance due to real mounting and fixation.

The thermal analyses showed compatibility between the requirements for MicroMED's components and the predicted temperatures. The laser was the only part whose predicted operating temperature exceeded the manufacturer-specified maximum temperature; this result was unavoidable because the max operating temperature for the mounting interface is coincident with the laser operating one. Nevertheless, the predicted laser temperature increase was below 1 °C, a value that is considered not relevant for the reduction of the component lifetime. Moreover, refined thermal analyses with a more detailed geometrical model of the instrument provide evidence that temperature uniformity is generally achieved on the OH, with maximum temperature difference of about 0.3 °C for both operational cases.

Finally, the temperature of the MicroMED inlet tip is always different from the atmospheric one because of both the low convective exchange with the Martian atmosphere and the limited thermal insulation between the sensor and instrument body. Provided that the inlet tip temperature is the closest approximation of the atmospheric one, a correction procedure shall be developed in case the atmospheric temperature cannot be retrieved from other instruments of the atmospheric package sampling volume.

3. Manufacturing and Integration

MicroMED main components were manufactured and checked before instrument integration. Main tasks were focused on geometry, mass, design tolerances, and interference checking. After manufacturing, coating (SurTec 650 Tri-chrome) was applied on all the mechanical parts. This finishing is required for any flight mechanics to avoid passivation and stress corrosion of the sensitive 7075T6 alloy.

In Figure 9 some examples of manufactured components and pre-assembled elements are shown.

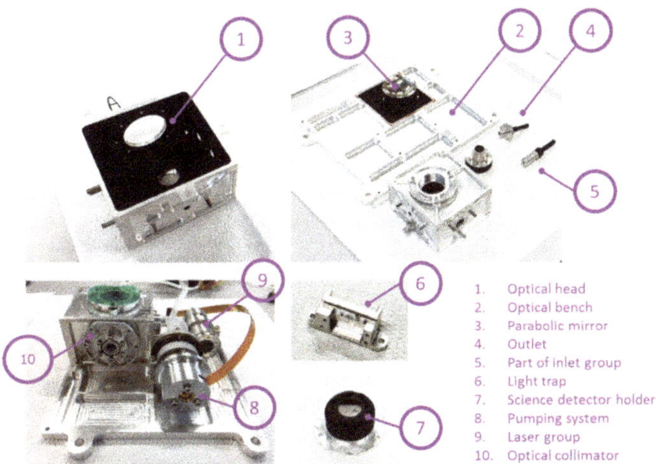

Figure 9. View of the manufactured components, preassembling activities.

Developed FE models were validated by means of experimental modal analyses on the manufactured main components.

In order to perform qualification testing on the MicroMED Flight Model, a protection box was designed and manufactured. The box provides the mechanical and thermal interfaces for the tested item, assuring fulfillment of the planetary protection requirements, even when the MicroMED must be transported and mounted on the Dust Suite interface. Thus, the MicroMED was integrated in a clean environment at the INAF-OAC (Astronomical Observatory of Capodimonte) clean room and mounted inside the box prior to qualification. In order to measure acceleration levels and temperature during the mechanical and thermal tests, different sensors were applied:

1. three piezoelectric accelerometers (Endevco 27A11 model, Brüel & Kjaer, nominal sensitivity 10 mV/g) placed on the top of the MicroMED Flight Model to sense acceleration along three testing directions (X, Y, and Z); and
2. four temperature sensors (Pt 1000, class A, IEC 751-95) were used: one was mounted on the MicroMED mounting interface (identified as the temperature reference point (TRP)), and the other three were positioned on the cover, on the OH, and on the inlet in order to have temperature distribution over the instrument during thermo-vacuum phases.

Views of the sensors positioning for mechanical and thermal tests and of the instrument during integration are shown in Figure 10. The box allows measurements of the acceleration, temperatures, and instrument data (temperatures, detectors' output, PS performances) by means of interface connectors located on a wall of the container.

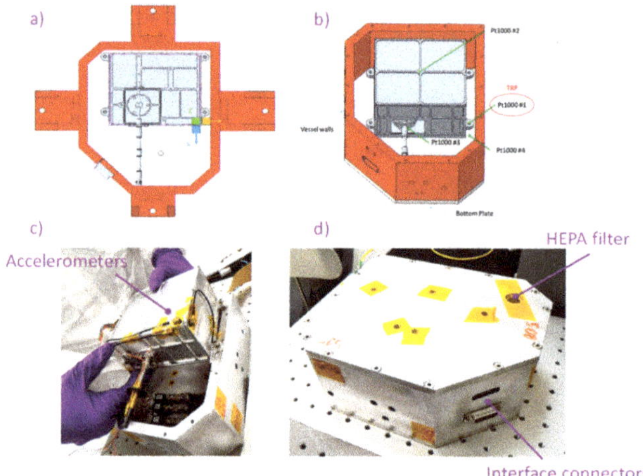

Figure 10. (**a**) Accelerometers positioning on the MicroMED; (**b**) temperature detectors mounting positions; (**c**) view of the MicroMED during integration within the vessel; (**d**) vessel after integration.

Stability of instrument's parameters, like the voltage of the reference detector and the PS rotational speed and power consumption, were identified as success criteria for the qualification testing.

4. Qualification Testing

4.1. Mechanical Testing

Facility for mechanical testing is based on an LDS V830 SPA-16 electro-dynamic shaker, with LMS SCADAS III controller for sine, random, and shock excitation. The shaker provides 8.9 kN peak force, maximum sine acceleration of 734.5 m/s^2, and nominal bandwidth up to 2250 Hz. The test procedure required testing of the unit along the three axes X, Y and Z, identified in accordance with the MicroMED mounting on the Dust Suite, with the acceleration levels and profiles summarized in Tables 1 and 2. Before and after each power test (sweep sine or random), a resonance search test was performed. The latter was a sweep sine test at low acceleration amplitude (5 m/s^2) from 20 Hz to 2000 Hz. This test measures, by means of installed accelerometers, the instrument resonances and tracks any change in the MicroMED dynamic behavior. In order to control applied acceleration at the mounting interface and monitor box accelerations, a triaxial (256B21 model, PCB Piezotronics,) and two monoaxial (4397 model, B&K) accelerometers were used. Accelerometers were calibrated with Brüel & Kjaer Calibration Exciter, type 4294. Figure 11 shows installation of the tested unit on the shaker for qualification along the X direction and related acceleration levels for power sweep and random are shown in Figure 12. Figure 13 provides measured frequency response functions during resonance search tests for the complete qualification.

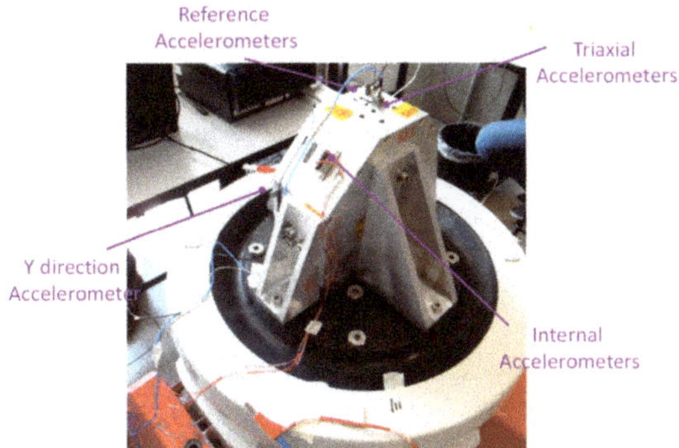

Figure 11. Testing unit mounting, qualification along X axis.

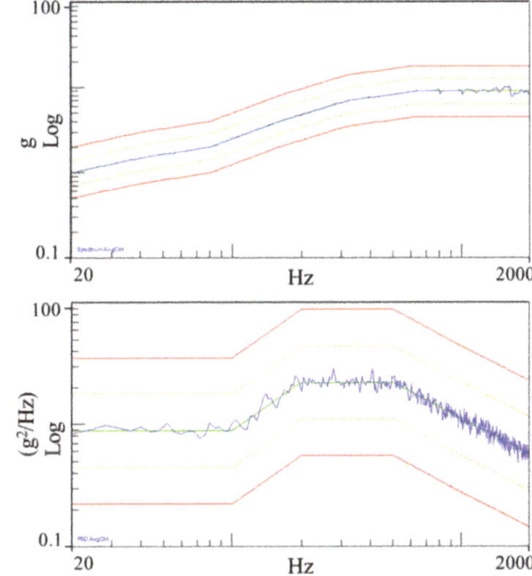

Figure 12. Applied sweep sine (top) and random (bottom) acceleration levels. Blue trend is the acceleration level measured at the interface whereas green trend shows reference profile. Control acceleration levels at 3 dB and 6 dB are shown in orange and red colors.

4.2. Thermo-Vacuum Testing

Cylindrical thermo-vacuum (TV) chamber (300 mm in diameter and 400 mm in length) was used for qualification testing. Both ends of the cylinder are removable flanges and inside the chamber a LiN2 cooled plate is used as a heat sink. Film heaters were used for fine thermal control. The TV chamber comprises a pressure gauge in the range from 1 bar to 10^{-4} Pa (Cathode Gauge LIC40239 gauge, Varian), a CC2C Varian conditioning unit, and connection with 37972A Keysight conditioning unit for temperature reading. Varian DS 402 rotary vane pump and T-Station 85 turbomolecular pumping station were used to achieve the required low pressure level in the vacuum chamber. In addition to temperature sensors on MicroMED parts (read by the instrument electronics) and the four ones installed on the MicroMED during integration on the protection box, additional sensors (named from T5 to T10) were installed on the testing unit and cold interface (T11) of the chamber. Figure 14 provides views of the box installation inside the vacuum chamber and of the temperature sensors' positioning.

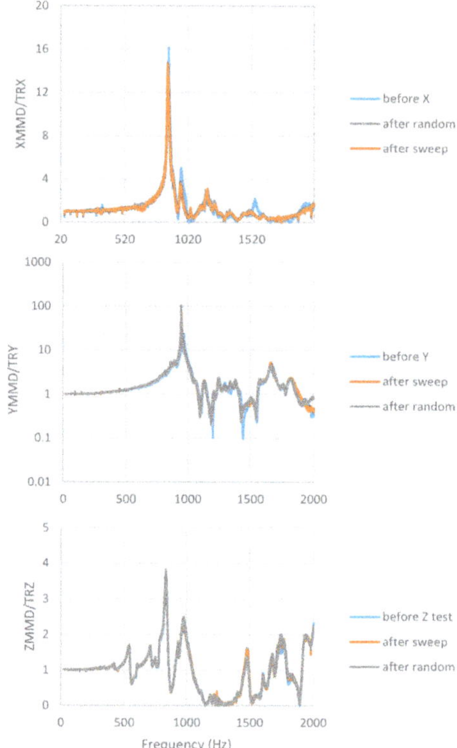

Figure 13. Measured frequency response functions (between internal accelerometer and reference one) for X (top), Y (middle), and Z (bottom) directions for the mechanical qualification.

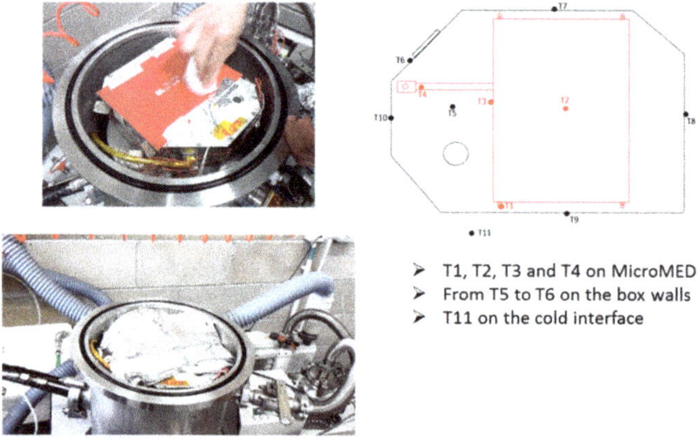

Figure 14. (Left) View of the mounting of the test unit inside the chamber and (right) scheme of the additional temperature sensors for the thermo-vacuum qualification.

Qualification requires temperature stabilization at defined test points (variation of less than 1 °C per hour) and its maintenance for at least 2 h dwell time. Thermo-vacuum levels and cycles for qualification are summarized in Table 6.

Table 6. Maximum and minimum temperatures for the qualification.

Cycle Number	Maximum Temperature (°C)	Minimum Temperature (°C)
I	60	−50
II	50	−30
III	50	−30
IV	35	−15

Some check points were defined during vacuum testing (at safe temperatures for the instrument's components) to monitor performance of the MicroMED (i.e., in particular the outputs of the detectors, straight light measurements and PS performances). Measured temperatures at the TRP are provided in Figure 15 as well as the indication of performed cycles.

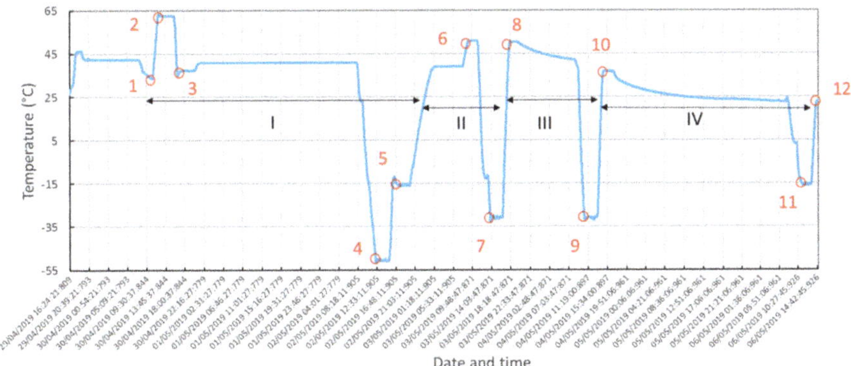

Figure 15. Thermo-vacuum cycles, measured temperature at the TRP.

Figure 16 shows overlap of the temperature readings from the sensors installed on the MicroMED during integration. Figure 17 shows measured temperatures from the additional sensors installed on the test unit.

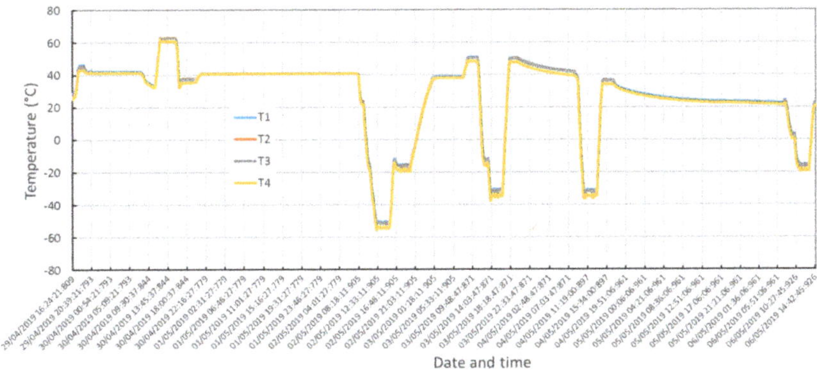

Figure 16. Thermo-vacuum qualification, measured temperatures, T1–T4 sensors.

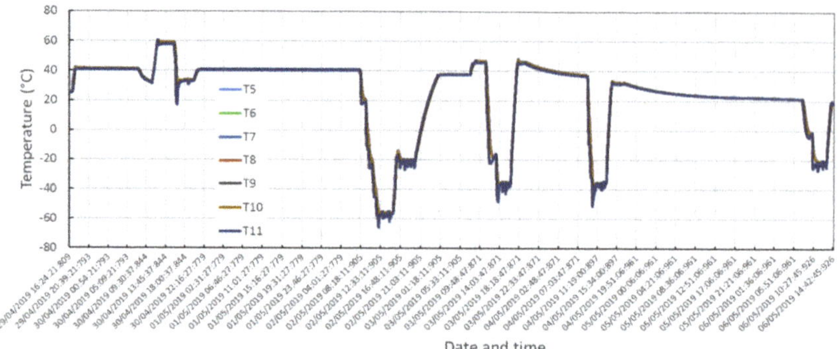

Figure 17. Thermo-vacuum qualification, measured temperatures, T5–T11 sensors.

4.3. Discussion

Qualification in mechanical environment of the MicroMED Flight Model was successfully completed. Mechanical testing evidenced survival of the instrument under expected mechanical loads. Functional tests performed after each direction did not show deviations from the nominal behavior of the instrument providing indirect proof that no mechanical changes occurred during the testing. Moreover, changes in measured resonances before and after power vibration tests (sine and random) were found to be within the allowed limits (less than 5 % of the reference value) justified by normal settlement of the mechanical structure bolted junctions.

Qualification in the thermal environment was successfully performed. No criticalities were detected during testing as shown by temperature plots in Figures 15–17. The optical group and the LG survived storage at maximum and minimum temperature predicted for the mission's scenarios. Moreover, checking the instrument's performance and housekeeping during thermal cycling evidenced compatibility of the measured straight light levels and PS power consumption with the reference values before qualification, validating the thermo-mechanical design and the workmanship of integration activities.

The main result of the successful qualification was that MicroMED Flight Model was ready to be delivered to the Russian team of IKI (Space Research Institute of the Russian Academy of Sciences) for the integration activity on the Dust Suite package.

5. Conclusions

This work provides validation of the thermo-mechanical design of MicroMED, a miniaturized particle analyzer for the ExoMars 2020. Modal and quasi-static analyses performed with finite element model allowed verification of the dynamic behavior and mechanical strength of the instrument, required to be compliant with the expected mechanical and thermal scenarios. The thermo-elastic and thermal analyses provided the generated stresses and the expected temperature distribution in different operational and non-operational conditions, showing that no criticalities were foreseen for the most sensitive components (i.e., the laser group and the pumping system). Moreover, predicted thermo-elastic behavior demonstrated no critical conditions within the qualification temperature range, spanning from −50 °C to 60 °C. The MicroMED Proto-Flight Model was manufactured and integrated for qualification testing. Qualification campaign was successfully completed both in mechanical and thermal environments, proving MicroMED's compliance with all conditions throughout all mission phases. Moreover, thanks to the stability of MicroMED's measured performances during the qualification, compliance with the operative environments of the sensor's optical and thermo-mechanical designs was demonstrated as well.

Author Contributions: Conceptualization, D.S., B.S., R.S., M.M., P.V., F.E., C.M., F.C., and G.M.; methodology, D.S., B.S., R.S., M.M., and P.V.; investigation D.S., B.S., R.S., M.M., P.V., F.E., C.M., F.C., and G.M.; data curation,

D.S., B.S., R.S., M.M., and P.V.; writing—original draft preparation, D.S.; writing—review and editing, D.S., B.S., R.S., M.M., P.V., F.E., C.M., F.C., and G.M.; supervision, F.E., C.M., B.S., and D.S.; project administration, F.E., C.M., and B.S.; funding acquisition, F.E. All authors have read and agree to the published version of the manuscript.

Funding: This work has been supported by ASI (contract's grant number: 2016/41/H.0).

Conflicts of Interest: The authors declare no conflicts of interest. The data used in this paper can be accessed upon personal request to the first author (diego.scaccabarozzi@polimi.it).

References

1. Esposito, F.; Colangeli, L.; della Corte, V.; Molfese, C.; Palumbo, P.J.; Ventura, S.; Merrison, J.P. MEDUSA: Observation of atmospheric dust and water vapor close to the surface of Mars. *Int. J. Mars Sci. Explor.* **2011**, *6*, 1–12. [CrossRef]
2. Colangeli, L.; Lopez-Moreno, J.J.; Nørnberg, P.; della Corte, V.; Esposito, F.; Epifani, E.M.; Merrison, J.; Molfese, C.; Palumbo, P.; Rodriguez-Gomez, J.F.; et al. MEDUSA: The ExoMars experiment for in-situ monitoring of dust and water vapour. *Planet. Space Sci.* **2009**, *57*, 1043–1049. [CrossRef]
3. Toon, O.B.; Pollack, J.B.; Sagan, C. Physical properties of the particles composing the Martian dust storm of 1971–1972. *Icarus* **1977**, *30*, 663–696. [CrossRef]
4. Colburn, D.S.; Pollack, J.B.; Haberle, R.M. Diurnal variations in optical depth at Mars. *Icarus* **1989**, *79*, 159–189. [CrossRef]
5. Horne, D.; Smith, M.D. Mars Global Surveyor Thermal Emission Spectrometer (TES) observations of variations in atmospheric dust optical depth over cold surfaces. *Icarus* **2009**, *200*, 118–128. [CrossRef]
6. Griffin, D.W.; Kellogg, C.A.; Shinn, E.A. Dust in the wind: Long range transport of dust in the atmosphere and its implications for global public and ecosystem health. *Glob. Chang. Hum. Health* **2001**, *2*, 20–33. [CrossRef]
7. Wolkenberg, P.; Giuranna, M.; Grassi, D.; Aronica, A.; Aoki, S.; Scaccabarozzi, D.; Saggin, B. Characterization of dust activity on Mars from MY27 to MY32 by PFS-MEX observations. *Icarus* **2018**, *310*, 32–47. [CrossRef]
8. Wolff, M.J.; Smith, M.D.; Clancy, R.T.; Spanovich, N.; Whitney, B.A.; Lemmon, M.T.; Bandfield, J.L.; Banfield, D.; Ghosh, A.; Landis, G.; et al. Constraints on dust aerosols from the Mars Exploration Rovers using MGS overflights and Mini-TES. *J. Geophys. Res.* **2006**, 112. [CrossRef]
9. Scaccabarozzi, D.; Saggin, B.; Pagliara, C.; Magni, M.; Tarabini, M.; Esposito, F.; Molfese, C.; Cozzolino, F.; Cortecchia, F.; Dolnikov, G.; et al. MicroMED, design of a particle analyzer for Mars. *Measurement* **2018**, *122*, 466–472. [CrossRef]
10. Mongelluzzo, G.; Esposito, F.; Cozzolino, F.; Molfese, C.; Silvestro, S.; Franzese, G.; Popa, C.I.; Lubieniecki, M.; Cortecchia, F.; Saggin, B.; et al. CFD analysis and optimization of the sensor "MicroMED" for the ExoMars 2020 mission. *Measurement* **2019**, *147*, 106824. [CrossRef]
11. Mongelluzzo, G.; Esposito, F.; Cozzolino, F.; Saccabarozzi, D.; Saggin, B. Optimization of the sensor "MicroMED" for the ExoMars 2020 mission: The Flight Model design. In Proceedings of the IEEE Metrology for Aerospace (MetroAeroSpace), Turin, Italy, 19–21 June 2019; pp. 749–753. [CrossRef]
12. Mongelluzzo, G.; Esposito, F.; Cozzolino, F.; Molfese, C.; Silvestro, S.; Popa, C.I.; Dall'ora, M.; Lubieniecki, M.; Cortecchia, F.; Saggin, B.; et al. Optimization of the fluid dynamic design of the Dust Suite-MicroMED sensor for the ExoMars 2020 mission. In Proceedings of the IEEE Metrology for Aerospace (MetroAeroSpace), Rome, Italy, 20–22 June 2018; pp. 134–139. [CrossRef]
13. Scaccabarozzi, D.; Saggin, B.; Pagliara, C.; Magni, M.; Tarabini, M.; Esposito, F.; Molfese, C.; Cortecchia, F.; Cozzolino, F.; Kuznetsov, I. Thermo-mechanical design of a particle analyzer for Mars. In Proceedings of the 2017 IEEE International Workshop on Metrology for AeroSpace (MetroAeroSpace), Padua, Italy, 21–23 June 2017; pp. 234–238. [CrossRef]
14. Saggin, B.; Scaccabarozzi, D.; Valiesfahani, A.; Valnegri, P.; Somaschini, R.; Esposito, F.; Molfese, C.; Cozzolino, F. Qualification of MEMS differential pressure sensors in Martian-like environment. In Proceedings of the 2019 IEEE 5th International Workshop on Metrology for AeroSpace (MetroAeroSpace), Torino, Italy, 19–21 June 2019; pp. 490–494. [CrossRef]
15. Scaccabarozzi, D.; Tarabini, M.; Almasio, L.; Saggin, B.; Esposito, F.; Cozzolino, F. Characterization of a pumping system in Martian-like environment. In Proceedings of the 2014 IEEE Metrology for Aerospace (MetroAeroSpace), Benevento, Italy, 29–30 May 2014; pp. 150–154. [CrossRef]

16. Saggin, B.; Scaccabarozzi, D.; Pagliara, C.; Valiesfahani, A.; Esposito, F.; Molfese, C.; Cortecchia, F. Design of a Flowrate Measurement System for Low-Pressure Gases. In Proceedings of the 2018 5th IEEE International Workshop on Metrology for AeroSpace (MetroAeroSpace), Rome, Italy, 20–22 June 2018; pp. 279–283. [CrossRef]
17. Mongelluzzo, G.; Esposito, F.; Cozzolino, F.; Franzese, G.; Ruggeri, A.C.; Porto, C.; Molfese, C.; Scaccabarozzi, D.; Saggin, B. Design and CFD Analysis of the Fluid Dynamic Sampling System of the "MicroMED" Optical Particle Counter. *Sensors* **2019**, *19*, 5037. [CrossRef] [PubMed]
18. Russian Academy of Sciences. *ExoMars 2018 Surface Platform Experiment Proposal Information Package*; Space Research Institute (IKI): Moscow, Russia, 2015.
19. Scaccabarozzi, D.; Saggin, B.; Alberti, E. Design and testing of a roto-translational shutter mechanism for planetary operation. *Acta Astronaut.* **2014**, *93*, 207–216. [CrossRef]
20. Saggin, B.; Tarabini, M.; Scaccabarozzi, D. Infrared optical element mounting techniques for wide temperature ranges. *Appl. Opt.* **2010**, *49*, 542–548. [CrossRef] [PubMed]
21. Scaccabarozzi, D.; Saggin, B.; Tarabini, M.; Palomba, E.; Longobardo, A.; Zampetti, E. Thermo-mechanical design and testing of a microbalance for space applications. *Adv. Space Res.* **2014**, *54*, 2386–2397. [CrossRef]
22. Shatalina, I.; Saggin, B.; Scaccabarozzi, D.; Panzeri, R.; Bellucci, G. MicroMIMA FTS: Design of spectrometer for Mars atmosphere investigation. In Proceedings of the Remote Sensing of Clouds and the Atmosphere XVIII and Optics in Atmospheric Propagation and Adaptive Systems XVI, Dresden, Germany, 25–26 September 2013. [CrossRef]
23. European Cooperation for Space Standardisation. ECSS-E-30 Part 2A. In *Space Engineering Mechanical—Part 2: Structural*; Requirements & Standards Division: Noordwijk, The Netherlands, 2000.

© 2020 by the authors. Licensee MDPI, Basel, Switzerland. This article is an open access article distributed under the terms and conditions of the Creative Commons Attribution (CC BY) license (http://creativecommons.org/licenses/by/4.0/).

Article

Design and CFD Analysis of the Fluid Dynamic Sampling System of the "MicroMED" Optical Particle Counter †

Giuseppe Mongelluzzo [1,2,*], Francesca Esposito [1], Fabio Cozzolino [1], Gabriele Franzese [1], Alan Cosimo Ruggeri [1], Carmen Porto [1], Cesare Molfese [1], Diego Scaccabarozzi [3] and Bortolino Saggin [3]

1 INAF—Astronomical Observatory of Capodimonte, Salita Moiariello 16, 80131 Naples, Italy; francesca.esposito@inaf.it (F.E.); fabio.cozzolino@inaf.it (F.C.); gabriele.franzese@inaf.it (G.F.); alan.ruggeri@inaf.it (A.C.R.); carmen.porto@inaf.it (C.P.); cesare.molfese@inaf.it (C.M.)
2 Department of Industrial Engineering, University of Naples "Federico II", Piazzale Tecchio 80, 80125 Naples, Italy
3 Department of Mechanical Engineering, Polo Territoriale di Lecco, Politecnico di Milano, Via Gaetano Previati 1c, 23900 Lecco, Italy; diego.scaccabarozzi@polimi.it (D.S.); bortolino.saggin@polimi.it (B.S.)
* Correspondence: giuseppe.mongelluzzo@inaf.it or giuseppe.mongelluzzo@unina.it; Tel.: +39-08155-754-65

† This paper is an expanded version of "Optimization of the sensor "MicroMED" for the ExoMars 2020 mission: The Flight Model design" published in the Proceedings of the 2019 IEEE 5th International Workshop on Metrology for AeroSpace (MetroAeroSpace), Torino, Italy, 19–21 June 2019.

Received: 20 October 2019; Accepted: 16 November 2019; Published: 19 November 2019

Abstract: MicroMED is an optical particle counter that will be part of the ExoMars 2020 mission. Its goal is to provide the first ever in situ measurements of both size distribution and concentration of airborne Martian dust. The instrument samples Martian air, and it is based on an optical system that illuminates the sucked fluid by means of a collimated laser beam and detects embedded dust particles through their scattered light. By analyzing the scattered light profile, it is possible to obtain information about the dust grain size and speed. To do that, MicroMED's fluid dynamic design should allow dust grains to cross the laser-illuminated sensing volume. The instrument's Elegant Breadboard was previously developed and tested, and Computational Fluid Dynamic (CFD) analysis enabled determining its criticalities. The present work describes how the design criticalities were solved by means of a CFD simulation campaign. At the same time, it was possible to experimentally validate the results of the analysis. The updated design was then implemented to MicroMED's Flight Model.

Keywords: MicroMED instrument; ExoMars 2020 mission; CFD; Mars

---

1. Introduction

The optical particle counter MicroMED (Figure 1) [1–5] is conceived to provide the first ever in situ measurements of airborne dust in Martian atmosphere. The sensor will be part of the upcoming ExoMars 2020 mission, and it is a miniaturized version of the sensor MEDUSA [6,7], previously developed at the INAF (Istituto Nazionale di AstroFisica) Astronomical Observatory of Capodimonte (OAC) in Naples, Italy, where the characterization of dust in Earth and planetary atmospheres has been the main focus of the research activities for years [8–14]. MicroMED will be able to directly determine both the size distribution and concentration of dust grains suspended in Martian atmosphere (in the 0.4–20 μm diameter range), which is a measurement that has only been performed indirectly so far, using the light scattering characteristics of the aerosol [15]. Such measurement could have a huge impact on our understanding of Martian climate, on the mechanism of saltation and dust lifting on Mars, as well as phenomena such as dust devils and dust storms [16]. The size distribution of suspended dust is indeed a key input parameter for the mesoscale climatic models. Given that dust absorbs solar radiation, the size distribution alters the quantity of solar radiation that is able to reach the Martian surface, thus influencing the values of atmospheric temperature. Moreover, the dimension of suspended dust grains is directly related to the wind speed present on the surface, since the higher the wind speed, the larger the grains lifted [17].

The instrument analyzes dust by means of an optical system, including an optical collimating system, a laser diode emitting a laser beam and a light trap that avoids reflections and is able to detect the light scattered by the dust grains, a parabolic mirror that is able to focus the scattered light on a photodiode, and the instrument electronics that process the received input and characterize it. In order to allow the detection, the fluid with embedded particles has to cross a 1 mm^2 sensing spot where the laser beam is focused. To do so, a proper fluid dynamic system was developed. Such a system is made of a sampling head that is exposed to outer atmosphere and able to collect dust grains, an inlet duct that conveys the fluid toward the sensing volume, an outlet duct that allows the expulsion of the sucked fluid after the optical sampling, and a pump that generates a pressure difference between the inlet and outlet sections of MicroMED, triggering the flow. While crossing the sensing volume, dust grains scatter light differently depending on their size and speed, so the amplitude and duration of the signals are directly related to those characteristics of the dust grains. The instrument's Elegant Breadboard was realized, and tests showed that the instrument could not detect large dust grains (15–20 μm in diameter) with high efficiency, especially in the presence of wind [1,2]. These results highlighted the need for a detailed fluid dynamic analysis of the instrument design in order to solve its criticalities. In those studies, MicroMED's Elegant Breadboard design was analyzed by means of the CFD code Fluent®, which allowed the identification of the causes of the observed low efficiency. Such causes were especially present for cold instrument conditions (the minimum operative temperature is around 253 K, given that MicroMED will be under a thermal cover limiting its temperature range to 253–313 K). Such studies also determined the optimum operating conditions of the pump needed to maximize MicroMED's efficiency. The analysis highlighted the causes of the reduction of the instrument's efficiency. A couple of undesired phenomena were individuated and will be described in the following sections. Then, the CFD analysis was enhanced in order to find a solution to such issues and improve the fluid dynamic design of MicroMED.

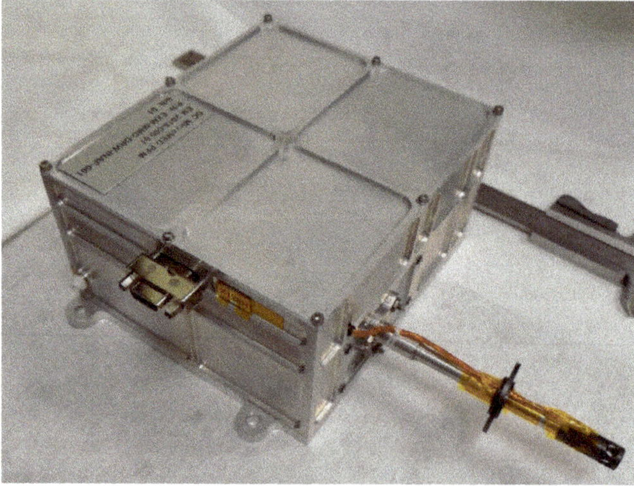

Figure 1. MicroMED's proto Flight Model.

MicroMED's design was updated to what is now the Flight Model design in order to correct those issues. The results for both CFD runs and laboratory tests show that a relevant improvement of the instrument's efficiency has been obtained.

2. Methods

The analysis was performed by means of the version 18.1 of the CFD solver "Ansys Fluent®". CFD numerical simulations are often used to study the movement of dust and aerosols in atmosphere [18–20]. As is well known, CFD methods are based on the conservation equations of mass, momentum, and energy for the flow (the equations are reported and discussed in Appendix A), which are valid only if the continuum hypothesis is valid. To verify that, the Knudsen number (Kn) for MicroMED was calculated (Kn is the ratio of the gas mean free path to the instrument characteristic length), and it was verified that $Kn < 0.1$ (the regime is considered continuum if $Kn < 0.1$, while in transitional regime if $0.1 < Kn < 50$). For our applications, Kn was indeed 0.007, allowing the use of conventional Navier–Stokes based solvers such as Ansys Fluent.

The simulation campaign was performed following a logic similar to the previous analyses [1,2]. The analysis is focused on the main fluid dynamic parameters and on the "sampling efficiency" parameter, which is the ratio of the number of dust grains that cross the laser-illuminated spot to the total number of dust grains that cross the instrument's inlet holes. This parameter allows evaluating the quality of the fluid dynamic design. Simulations considered different conditions in terms of ambient temperature and pressure in the range expected on Mars. Runs for five different values of ambient temperature between 190 and 280 K and for three different ambient pressure values (between 6 and 8 mbar) were performed. Seven different instrument temperatures inside the possible temperature range (253–313 K) were considered. Given that 95.3% of Mars' atmosphere is made of carbon dioxide, runs were performed with CO_2 as the fluid. The suction of fluid was simulated by means of the pressure difference generated by the pump between the inlet and outlet sections of MicroMED. For the Elegant Breadboard, the inlet–outlet Δp was simulated between 250 and 500 Pa. In the present work, simulations considered a Δp in the 100–300 Pa range, which is in accordance with the experimental results obtained by tests performed on the pump of MicroMED's Flight Model. Most simulations were performed with a simple model not considering surface roughness. Simulations considering a mean surface roughness of 10–20 μm were indeed performed, and the results showed that the instrument's sampling efficiency is barely influenced by surface roughness (variations always under 2% and most times under 1%). The analysis showed that the regime could be considered laminar similarly to what was obtained for the Elegant Breadboard design [1], allowing the use of laminar model for simulations. Indeed, the Reynolds number for the present application is always under 1000 given both the extremely low density of Mars atmosphere ($1.6–1.8 \times 10^{-2}$ kg/m^3) and the small characteristic dimensions of MicroMED (order of magnitude of millimeters). This, coupled with a high Knudsen number related to the particles' diameter (Kn_P ranges from 0.67 to 33.45), highlights the need for a correction factor in the drag law of the grains. In particular, the Cunningham correction factor for drag law was introduced [21]. The flow can be considered compressible similarly to what happened for previous works [1]. Dust grains in the sampling range have a Stokes number that ranges from 2×10^{-4} to 0.54, meaning that there could be a different behavior between large (15–20 μm) and small (0.4–1 μm) dust grains, with small dust grains more likely to follow the fluid streamlines along their entire path through the instrument. In CFD simulations, dust grains were simulated as spherical. Injections of dust grains of 16 different dimensions in the instrument's sampling range were simultaneously simulated. The interaction among dust grains and the effect of magnetic and electrical forces on grains were calculated, and given the small effect on the overall results (the maximum contribution of such forces to dust grains speed is in the order of 10^{-5} m/s), they have not been considered during simulations. The CFD model was already validated in previous works [1], showing how the model prediction matched test results with good accuracy.

3. Undesired Phenomena

Previous CFD analyses showed a couple of criticalities in the Elegant Breadboard design, causing a reduction of the sampling efficiency for both large and small dust grains, with different extents and causes. These phenomena are described hereafter.

3.1. Collisions on the Inlet Walls

MicroMED's sampling head is exposed to Martian atmosphere. A pump, connected to the outlet section of the instrument, generates a pressure difference with respect to the outside of MicroMED, triggering the suction of fluid. Then, an inlet duct conveys the fluid toward the sensing volume. The fluid drags the suspended particles, which follow the streamlines along the inlet duct. Given the particular geometry of the Elegant Breadboard's inlet head, fluid streams coming from opposite holes of the sampling head tend to cross (see Figure 2). The sharpness of the bending depends on the dust grains inertia: the lower the inertia, the sharper the deflection. This is an important aspect, as large dust grains are a lot more likely to hit the duct walls. Given the aforementioned assumption that the walls' mean surface roughness is of the same order of magnitude of the particles' diameter (in the 10–20 μm range), this may cause adhesion of the particles to the walls, preventing their detection. The Elegant Breadboard's duct shape promoted such phenomenon. Thus, the Flight Model design was modified to avoid the crossing of the trajectories and to linearize the fluid streamlines during the suction, also helping the laminarity of the flow, which is a design parameter (since it improves the instrument efficiency). In Figure 2, it is possible to see the dust grains' trajectories inside the sampling head and inlet duct, showing the phenomenon just described.

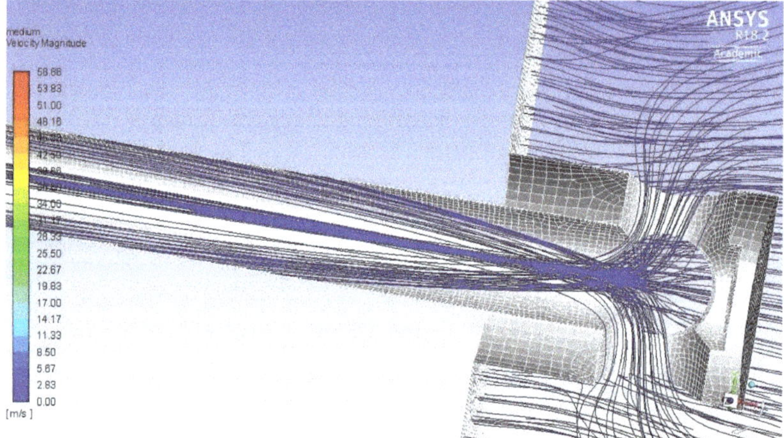

Figure 2. Particles trajectories inside the Elegant Breadboard's inlet head and duct.

3.2. Deflection of Dust Grains' Trajectories

Inside MicroMED's optical head and at the end of the inlet duct, a 4 mm gap is present, which is needed for the optical scan of the flow. The final section of the inlet duct has a 1 mm internal diameter. When the fluid reaches such gap, it expands, possibly deflecting the particles' trajectories. There is indeed the chance that some of the dust grains follow the streamlines and cross the sensing plane outside the 1 mm^2 laser illuminated spot, preventing their detection. This behavior is especially possible for small dust grains given their low Stokes number. Such phenomenon could alter the efficiency of MicroMED's Elegant Breadboard depending on the environmental conditions. Therefore, the geometry update was aimed at having good performances in every possible operating condition. Figure 3 shows the possible undesired behavior of particles.

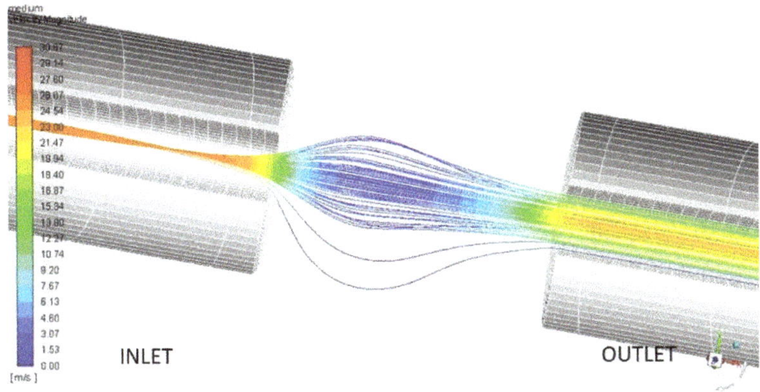

Figure 3. Dust grains behavior inside the sensing section of MicroMED's Elegant Breadboard.

4. Geometry Update

The fluid dynamic analysis of MicroMED's Elegant Breadboard highlighted the need to change the shape of the inlet head. Then, the inlet walls' thickness was reduced so that the small inlet cylindrical ducts that conveyed the fluid toward the main inlet duct disappeared. This variation helped reduce the curvature radius of the inhaled particles' trajectories. The sampling head internal shape was also slightly modified and made more conical with respect to the mostly cylindrical shape of the Elegant Breadboard. This variation sharpened the particles deflection, allowing a less complex inlet duct. For the Flight Model, indeed, the inlet duct has a simply conical shape compared to the conical-then-cylindrical-then-conical shape present in the previous design (see Figure 4). These variations led to a big improvement of the duct capability to direct dust grains toward the sensing spot, as will be shown hereafter. Figure 4 shows the geometry variations adopted for the inlet head and duct.

The outlet duct of MicroMED also had to be changed because of volumetric constraints. The variations made the outlet more compact and short (32.5 mm against 88.5 mm of the previous version), and the internal duct was designed as simply conical instead of a combination of two cylindrical ducts. Figure 5 shows such variations. Section 5 will detail the effects of all these variations on MicroMED's efficiency.

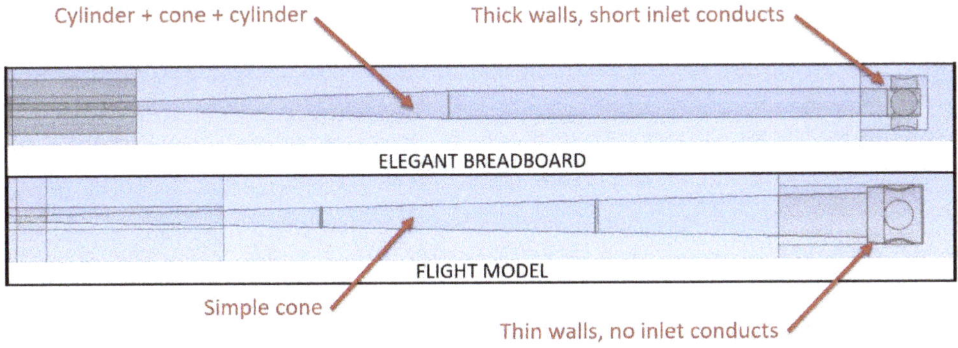

Figure 4. Geometry variations of MicroMED's sampling head and inlet duct (drawings not to scale).

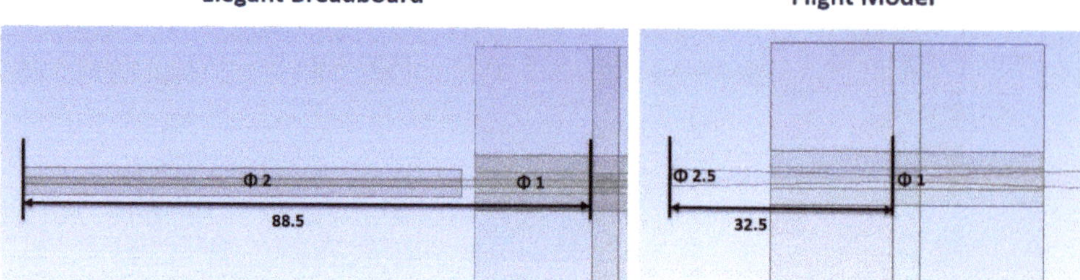

Figure 5. Geometry variations of MicroMED's outlet duct (dimensions in mm).

5. Results

Results are here reported as a comparison with the Elegant Breadboard's status, showing the improvements obtained. The effect of environmental parameters on MicroMED's efficiency was previously analyzed [1] for the Elegant Breadboard. The present analysis shows similar effects for the Flight Model. The instrument temperature is the most influential parameter both on the sampling efficiency and in the evaluation of the volumetric flow rate, which is needed in order to determine the dust concentration in the sample of gas inhaled. The analysis performed on the Elegant Breadboard [1] showed that other parameters can influence MicroMED's behavior. Moreover, the optimum conditions for tests had to be deduced, since good efficiency was not guaranteed in any environmental conditions. The Flight Model design provides improvements of the efficiency for every size and basically guarantees good efficiency for every possible environmental condition. The analysis performed on the Flight Model also shows how optimum results can be obtained with a Δp generated by the pump in the 200–300 Pa range. The Elegant Breadboard needed at least 300 Pa Δp in order to work, so the current design allows more flexibility in the choice of the operating conditions and a reduction of the power consumption of the instrument. The following sections describe the results obtained for such an updated design.

5.1. Sampling Efficiency of Large Dust Grains

As shown in Figures 6 and 7, CFD runs predict an important improvement of the instrument's ability to detect large (15–20 μm diameter) dust grains. When the instrument is "hot" (T_i = 313 K, the maximum allowed temperature under the thermal cover), MicroMED's Breadboard was already sufficiently efficient for such dust sizes; however, the updated design provides improvements as large as 10–14%. When the instrument is "cold" (T_i = 253 K, the minimum temperature allowed), the improvement is clear. The impacts of dust grains on the walls completely disappear, so that 100% of the large dust grains can be correctly detected by MicroMED's optical system compared to roughly 30% obtained with the Elegant Breadboard design (see Figure 6).

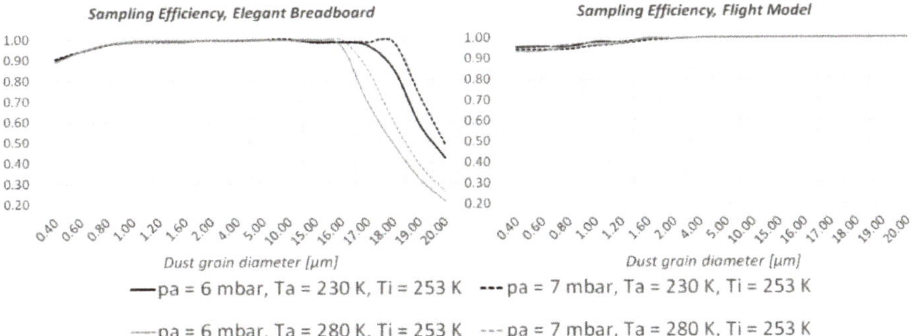

Figure 6. Comparison between the sampling efficiencies of MicroMED's Elegant Breadboard and Flight Model for "cold instrument conditions" (T_i = 253 K).

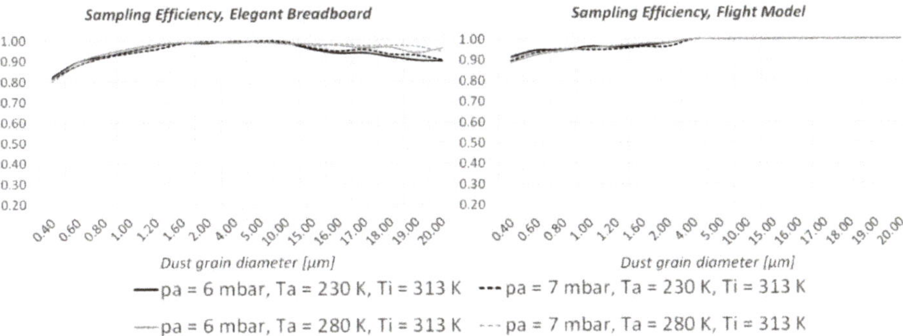

Figure 7. Comparison between the sampling efficiencies of MicroMED's Elegant Breadboard and Flight Model for "hot instrument conditions" (T_i = 313 K).

The instrument's ability to detect large dust grains was confirmed by tests performed at the INAF Astronomical Observatory of Capodimonte Laboratory. In such a laboratory, a Martian chamber and a clean room are installed, enabling the reproduction of Martian conditions in terms of pressure and atmospheric composition, thus reducing the amount of atmospheric dust that could alter the measurements and keep the instrument sterile in accordance to the planetary protection constraints. Moreover, the ATS (Autonomous Thermal Simulator) system installed in the Capodimonte laboratory allowed to perform tests at different instrument temperatures, simulating the possible different conditions foreseen at the lander level during the mission. During such tests, the instrument appeared to show a good ability to detect large dust grains, as Figure 8 shows, even though the results of the analysis are still preliminary, so they are only mentioned. The test showed in Figure 8 was made injecting in the Martian chamber monodispersed SiO_2 19.7 μm spherical calibrated particles (for the test setup, see Appendix B).

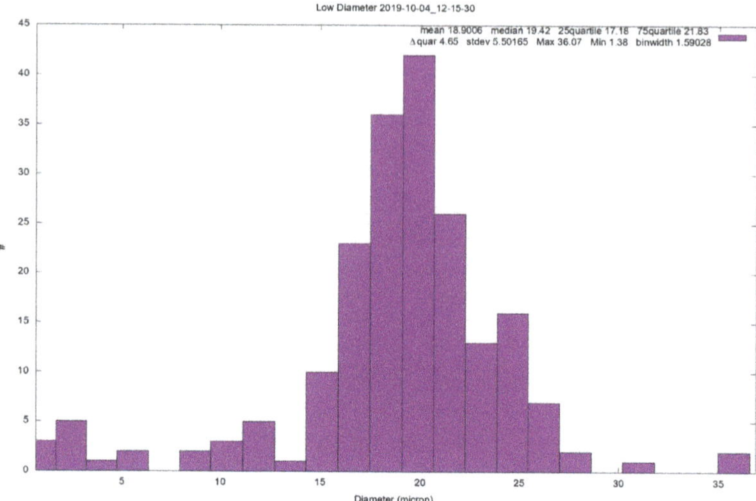

Figure 8. Example of test performed at the Capodimonte laboratory injecting 19.7-μm calibrated SiO_2 particles.

5.2. Dust Grains' Position in the Sensing Plane and Evaluation of Dust Grain Size Distribution

Particles trajectories along MicroMED were analyzed in order to understand not only if they cross the laser-illuminated spot, but also how far from the center of the spot they do so. Indeed, given that laser light is more uniform and intense at the center of the laser spot, errors in the determination of the dust grain size are smaller when particles cross the spot close to the center. For this reason, position histograms were derived from CFD simulations, determining the quality of the fluid dynamic design. The Elegant Breadboard already guaranteed good results in such an aspect, and the geometry variations adopted for the new design could potentially make the particles more "spread" through the sensing plane. Indeed, in some cases, the streamlines' deflection is smoother for the new design; therefore, the dust grains are more distributed inside the sampling spot rather than more concentrated. The comparison was made for two different working conditions, given that the optimum operating condition for the Elegant Breadboard version is related to an inlet–outlet pressure difference of 300 Pa while the Flight Model already works efficiently at 200 Pa, which was the operating condition considered in this paper. The results show that the new design does not alter the chances of a grain crossing the sampling spot in the proximity of the center (roughly 91% of all dust grains inhaled are within 400 μm, and 86% are within 350 μm from the center of the spot; these numbers pretty close to the ones obtained for the Breadboard design, see Figure 9). There are some cases where the efficiency for small dust grains could slightly decrease (by less than 3%) because of the dynamics previously described, but the geometry variation provides a definite improvement of the overall percentage of particles that are now detectable. For large dust grains (see Figure 6), there is a definite improvement (sampling efficiency 70% higher in some cases, as already stated). Moreover, the new design provides efficiency over 89% for all the small grains (diameter < 1 μm) for every possible environmental condition, differently to the Elegant Breadboard that had cases of efficiency dropping below 80%.

The new design also gave tangible improvements in terms of the instrument's ability to evaluate dust grain concentration. Figure 10 shows that the Flight Model is able to describe the size distribution of the particles inhaled with much better accuracy with respect to the previous design.

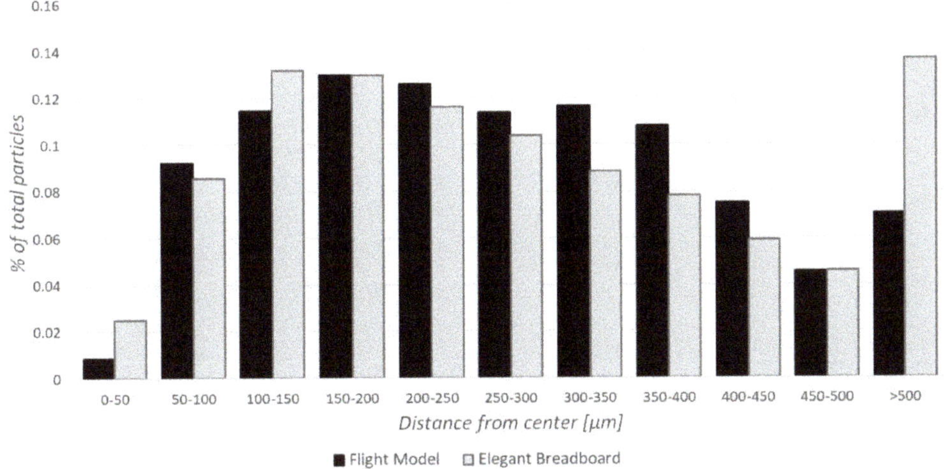

Figure 9. Example of position histogram comparing the distance of crossing particles from the center of the laser spot for both the Elegant Breadboard (GRAY) and the Flight Model (BLACK) design.

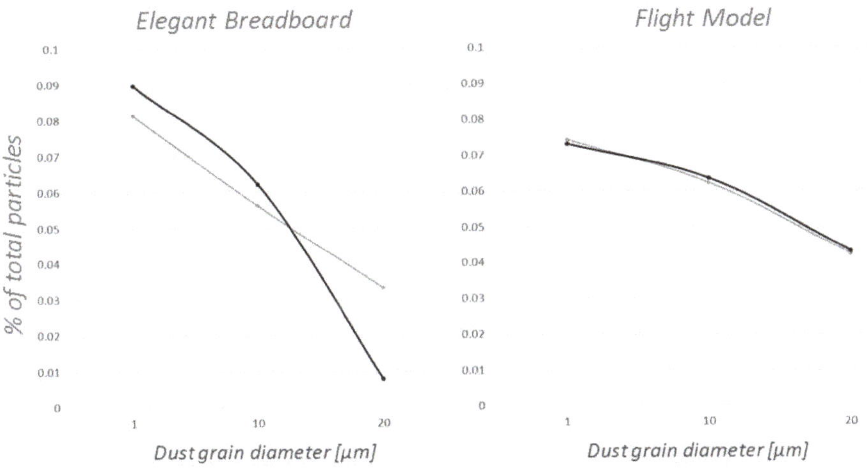

Figure 10. Comparison between the Elegant Breadboard's and the Flight Model's ability to correctly deduce size distribution (GRAY: real distribution, BLACK: measured distribution).

5.3. Results in Presence of Wind

A CFD analysis to predict MicroMED's behavior in windy environments was also performed, given that sustained wind is present in most occurrences on Mars. A CFD model was developed to simulate windy conditions, changing boundary conditions from the traditional CFD model used in this paper in order to generate a wind (of set speed) that passes over MicroMED's sampling head. Dust grains could only be simulated as spherical, which is expected to make simulated particles more stable than real particles, so the results obtained could underestimate the instrument's ability to detect particles while in the presence of wind (more stable dust grains are less probable to be deflected inside MicroMED). Indeed, while CFD analysis of the Elegant Breadboard stated that MicroMED could be unable to detect dust grains starting from a wind speed of 2 m/s, a test campaign performed at the Aarhus Wind Tunnel Simulator (AWTS) facility at Aarhus University in Aarhus, Denmark [22] showed that small dust grains are well detected, while the detection of large dust grains is related to a threshold value. For every size, there is a threshold value of wind speed after which MicroMED is unable to detect grains of such size: the larger the size of dust grains, the smaller the speed threshold value. The preliminary results of such a test campaign seem to show that the instrument was able to see dust grains for wind speeds up to around 10 m/s (see Appendix C for the description of test setup at the AWTS facility).

The results of CFD simulations on the Flight Model's geometry, compared with those on the Elegant Breadboard, showed clear improvements in the ability to detect small dust grains and a moderate improvement in the ability to detect larger grains. Figure 11 shows a comparison between the two analyzed geometries, highlighting the better overall efficiency obtained. However, according to CFD analysis, efficiency is still low. Tests at the AWTS were also performed on MicroMED's Flight Model, and the preliminary results show that the improvement was probably bigger than what was predicted by the CFD. It was found that MicroMED's updated design not only improves the instrument's efficiency in the detection of large dust grains, but it also provides good efficiency. As Figure 12 shows, MicroMED was able to detect 20-μm diameter particles even at the highest possible wind speed for the facility (15 m/s), which confirms that the CFD model is conservative, and that MicroMED's Flight Model better detects dust grains also in the presence of wind. Since the analysis of the measured signals is still ongoing, the figure reports data in terms of signal intensity (as measured by the instrument's detector) and not in size; however, the run relative to such a figure was performed injecting only 20.07-μm calibrated spherical particles with a wind speed of 15 m/s, so the signals detected are for sure 20.07-μm dust grains. The figure was still reported to show that hundreds of samples can be obtained in such tests. Such results confirm that the CFD model should be improved to predict MicroMED's behavior in windy conditions with better accuracy. Moreover, the Flight Model's ability to work properly with lower pump rpm speeds could help, as it could be possible to increase the pump speed if necessary. However, the analysis of data obtained in such a test campaign (plenty of tests were performed with 10 different monodispersed spherical sizes and with JSC-1 non-spherical Martian simulant, as well as tests with other broad distribution samples) is still preliminary; therefore, they are only mentioned in this work.

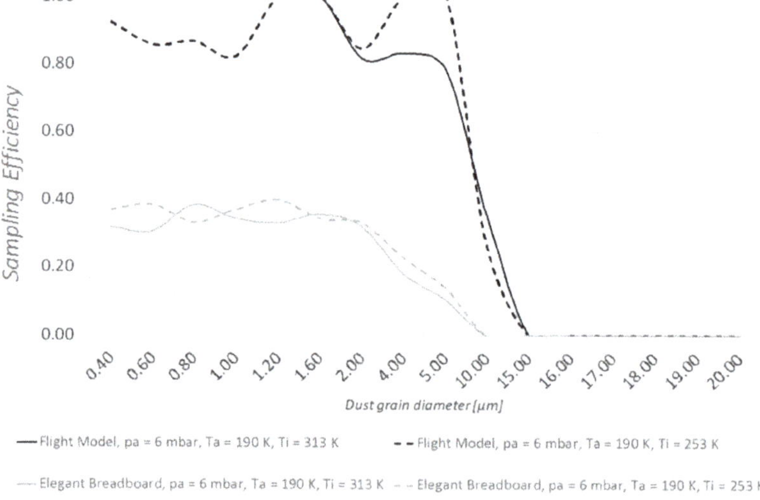

Figure 11. Comparison between Flight Model and Elegant Breadboard in terms of sampling efficiency, wind speed 2 m/s.

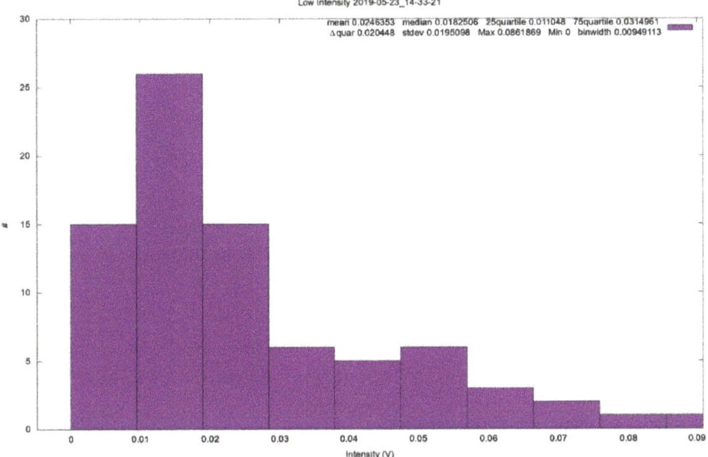

Figure 12. Example of test performed at AWTS in Aarhus, Denmark. The plot is reported in terms of signal intensity as the analysis is not ready; however, tests were performed with 20.07-µm particles only.

6. Conclusions

MicroMED's Flight Model design was developed by means of a CFD simulation campaign aimed at the improvement of the instrument's ability to detect dust grains, especially large ones (15–20 µm in diameter), in every possible environmental condition that the instrument could face during the ExoMars 2020 mission. This is important since the actual operating conditions of the instrument while on Mars are unpredictable. The analysis shows that the updated fluid dynamic design improves the detection of dust grains. The key result is a huge improvement in the ability to suck and detect large dust grains, avoiding hits on the walls and obtaining more accurate results in the measurement of size distribution curves for the samples. Optimum operating conditions can be obtained for an extended range of pump-generated Δp, which allows more flexibility in the choice of operating conditions in relation to the environment. The analysis results were confirmed by tests performed at the Astronomical Observatory's Laboratory in Naples, Italy, where abundant samples of large dust grains were detected by the instrument. The analysis was also extended to windy operating conditions. The model showed to be extremely conservative in predicting the outcome of tests. However, CFD results predicted an efficiency improvement, which was confirmed by tests performed at the AWTS facility in Aarhus, Denmark. Indeed, tests showed that MicroMED's Flight Model is able to detect significant amounts of large dust grains even at high wind speed (speeds until 15 m/s were tested), confirming that an improvement was obtained, as predicted by the CFD model.

Author Contributions: Conceptualization, G.M., F.E., C.M., F.C., D.S. and B.S.; methodology, G.M., F.E., F.C., D.S., B.S. and G.F.; software, G.M. and G.F.; validation, G.M., F.E., F.C., D.S., B.S. and G.F.; formal analysis, G.M., G.F., A.C.R. and C.P.; data curation, G.F., A.C.R., C.P. and G.M.; writing—original draft preparation, G.M.; writing—review and editing, G.M., F.E., C.M., F.C., D.S., B.S., G.F., A.C.R., C.P.; supervision, F.E., C.M., B.S.; project administration, F.E., C.M.; funding acquisition, F.E.

Funding: This work has been supported by ASI (contract's grant number: 2016/41/H.0). The instrument development was funded and coordinated by ASI under the scientific leadership of INAF-OAC, Naples, Italy. The test campaign at the AWTS Facility was funded by the Europlanet 2020 project. The data used in this paper can be accessed upon personal request to the first author (giuseppe.mongelluzzo@inaf.it).

Acknowledgments: Special thanks go to the Europlanet 2020 project and to the University of Aarhus, Denmark, which allowed performing tests that deeply helped the analysis. Trips at the AWTS facility allowed obtaining key results. The team led by J. Merrison was helpful and willing, and allowed optimizing our time at the facility.

Conflicts of Interest: The authors declare no conflict of interest.

Appendix A

This appendix shows the equations used by ANSYS Fluent to solve the flow for the present application. Given that laminar flow is present, conservation equations of mass and momentum are reported for such a regime. The equations are the following:

$$\frac{\partial \rho}{\partial t} + \nabla\cdot\left(\rho \vec{v}\right) = S_m$$

$$\frac{\partial\left(\rho \vec{v}\right)}{\partial t} + \nabla\cdot\left(\rho \vec{v}\,\vec{v}\right) = -\nabla p + \nabla\cdot\left(\overline{\overline{\tau}}\right) + \rho \vec{g} + \vec{F}$$

where ρ is density, $\vec{v}$ is the fluid velocity vector, S_m is the mass added to the continuous phase from the dispersed second phase (dust), p is the static pressure, $\overline{\overline{\tau}}$ is the stress tensor, and $\rho\vec{g}$ and $\vec{F}$ are the gravitational body force and the external body force (e.g., that arise from interaction with the dispersed phase). Such equations are valid also for compressible flows. An energy equation has to be added because thermal effects are also considered in simulations. The energy equation is

$$\frac{\partial(\rho E)}{\partial t} + \nabla\cdot\left(\vec{v}\,(\rho E + p)\right) = \nabla\cdot\left(k_{eff}\nabla T - \sum_j h_j \vec{J}_j + \left(\overline{\overline{\tau}}_{eff}\cdot\vec{v}\right)\right) + S_h$$

where k_{eff} is the effective conductivity ($k_{eff} = k + k_t$, where k_t is the turbulent thermal conductivity that is not active for the laminar model) and, $\vec{J}_j$ is the diffusion flux of species (no flux here). The first three terms of the right-hand side of the equation represent energy transfer due to conduction, species diffusion, and viscous dissipation, respectively. S_h includes effects due to chemical reaction and other contributions that are not present in this work. In these equations, the stress tensor $\overline{\overline{\tau}}$ is expressed as:

$$\overline{\overline{\tau}} = \mu\left[(\nabla\vec{v} + \nabla\vec{v}^{T}) - \frac{2}{3}\nabla\cdot\vec{v}I\right]$$

where μ is molecular viscosity, I is the unit tensor, and the second term on the right-hand side is the effect of volume dilation.

Appendix B

This appendix describes the procedure used for tests performed at the INAF–OAC laboratories of Capodimonte in Naples, Italy.

Tests were performed inside the INAF–OAC clean room, which includes a vacuum chamber. All the components inside the clean room have to be sterilized to avoid contamination of the instrument in accordance with the planetary protection constraints.

The tests simulated the data acquisition of MicroMED under Martian conditions. For this reason, an atmosphere with a pressure between 6 and 7 mbar was created inside the vacuum chamber.

Temperature conditions were set and controlled using the Autonomous Thermal Simulator (ATS) system, which is a custom thermal system produced by TransTech® that allowed varying the instrument temperature in the 253–313 K temperature range.

The electrical and grounding setup can be accessed upon request to the authors of the present paper.

Tests were performed with calibrated SiO_2 particles made by Microparticles® of different diameters (1, 2, 4, 6, 8, 10, 15, 16, 18, and 20 μm) and with a JSC-1 Martian simulant (which is made of non-spherical particles).

As for the injection of particles, a Medikron® atomizer containing a weighted sample of grains was used. The atomizer was filled with ethanol, mounted on a pulsing vortex mixer (to prevent the agglomerating of particles), and connected with a CO_2 tank by means of a tube, so that the injection of

CO_2 allowed the exit from the atomizer of a mushroom cloud of ethanol and dust. Ethanol tends to evaporate because of the low pressure, leaving only silica particles available to MicroMED for suction.

Papers describing the experiments in detail and giving an in depth analysis of the results are forthcoming.

Appendix C

This appendix describes the procedure used for tests performed at the AWTS facility at the University of Aarhus situated in Aarhus, Denmark. The test campaign was funded by the Europlanet 2020 project and allowed one week of detailed testing of the instrument under various conditions.

Tests were performed inside the facility's vacuum chamber. The chamber allowed reproducing Martian conditions in terms of pressure, atmospheric composition, and wind speed. Indeed, a fan was present inside the vacuum chamber, allowing wind speeds varying from 0.5 to 15 m/s.

Temperature conditions were controlled by means of a base plate.

Electrical and grounding setup can be accessed upon request to the authors of the present paper.

The injection of particles was made by means of a valve. First, 1 mg of grains (monodispersed calibrated SiO_2 particles made by Microparticles® or a broad distribution sample of non-spherical grains depending on the test) was positioned in a small duct connected to the valve. The opening of the valve caused suction of the dust grains because of the pressure difference between the inside and the outside of the chamber (6–8 mbar inside, 1 bar outside). Then, the flow generated by the fan moved the particles toward MicroMED, reproducing the dust-embedded wind expected on Mars.

Tests were performed with calibrated, spherical particles of different diameters (1, 2, 4, 6, 8, 10, 15, 16, 18, and 20 μm), with a JSC-1 Martian simulant (which is made of non-spherical particles) and with other samples of broad distribution. Papers describing in detail the results of the test campaign are currently under development.

References

1. Mongelluzzo, G.; Esposito, F.; Cozzolino, F.; Molfese, C.; Silvestro, S.; Franzese, G.; Popa, C.I.; Lubieniecki, M.; Cortecchia, F.; Saggin, B.; et al. CFD analysis and optimization of the sensor "MicroMED" for the ExoMars 2020 mission. *Measurement* **2019**, *147*, 106824. [CrossRef]
2. Mongelluzzo, G.; Esposito, F.; Cozzolino, F.; Molfese, C.; Silvestro, S.; Popa, C.I.; Dall'Ora, M.; Lubieniecki, M.; Cortecchia, F.; Saggin, B.; et al. Optimization of the fluid dynamic design of the Dust Suite-MicroMED sensor for the ExoMars 2020 mission. In Proceedings of the IEEE Metrology for Aerospace (MetroAeroSpace), Rome, Italy, 20–22 June 2018; pp. 134–139. [CrossRef]
3. Mongelluzzo, G.; Esposito, F.; Cozzolino, F.; Saccabarozzi, D.; Saggin, B. Optimization of the sensor "MicroMED" for the ExoMars 2020 mission: The Flight Model design. In Proceedings of the IEEE Metrology for Aerospace (MetroAeroSpace), Turin, Italy, 19–21 June 2019; pp. 749–753. [CrossRef]
4. Scaccabarozzi, D.; Saggin, B.; Pagliara, C.; Magni, M.; Tarabini, M.; Esposito, F.; Molfese, C.; Cozzolino, F.; Cortecchia, F.; Dolnikov, G.; et al. MicroMED, design of a particle analyzer for Mars. *Measurement* **2018**, *122*, 466–472. [CrossRef]
5. Scaccabarozzi, D.; Saggin, B.; Somaschini, R.; Magni, M.; Valnegri, P.; Valiesfahani, A.; Tarabini, M.; Esposito, F.; Molfese, C.; Molfese, C.; et al. Design validation of MicroMED, a particle analyzer for ExoMars 2020. In Proceedings of the IEEE Metrology for Aerospace (MetroAerospace), Turin, Italy, 19–21 June 2019. [CrossRef]
6. Esposito, F.; Colangeli, L.; della Corte, V.; Molfese, C.; Palumbo, P.J.; Ventura, S.; Merrison, J.P. MEDUSA: Observation of atmospheric dust and water vapor close to the surface of Mars. *Int. J. Mars Sci. Explor.* **2011**, *6*, 1–12.
7. Colangeli, L.; Lopez-Moreno, J.J.; Nørnberg, P.; della Corte, V.; Esposito, F.; Epifani, E.M.; Merrison, J.; Molfese, C.; Palumbo, P.; Rodriguez-Gomez, J.F.; et al. MEDUSA: The ExoMars experiment for in-situ monitoring of dust and water vapour. *Planet. Space Sci.* **2009**, *57*, 1043–1049. [CrossRef]
8. Esposito, F.; Debei, S.; Bettanini, C.; Molfese, C.; Rodríguez, I.A.; Colombatti, G.; Harri, A.-M.; Montmessin, F.; Wilson, C.; Aboudan, A.; et al. The DREAMS Experiment Onboard the Schiaparelli Module of the ExoMars 2016 Mission: Design, Performances and Expected Results. *Space Sci. Rev.* **2018**, *214*, 103. [CrossRef]

9. Bettanini, C.; Esposito, F.; Debei, S.; Molfese, C.; Rodríguez, I.A.; Colombatti, G.; Harri, A.; Montmessin, F.; Wilson, C.; Aboudan, A.; et al. The DREAMS experiment on the ExoMars 2016 mission for the study of Martian environment during the dust storm season. In Proceedings of the IEEE Metrology for Aerospace (MetroAeroSpace), Benevento, Italy, 29–30 May 2014; pp. 167–173.
10. Bettanini, C.; Esposito, F.; Debei, S.; Molfese, C.; Colombatti, G.; Aboudan, A.; Brucato, J.R.; Cortecchia, F.; di Achille, G.; Guizzo, G.P.; et al. The DREAMS experiment flown on the ExoMars 2016 mission for the study of Martian environment during the dust storm season. *Measurement* **2018**, *122*, 484–493. [CrossRef]
11. Colangeli, L.; Moreno, J.J.L.; Palumbo, P.; Rodriguez, J.; Bussoletti, E.; della Corte, V.; Esposito, F.; Herranz, M.; Jerónimo, J.M.; Lopez-Jimenez, A.; et al. GIADA: The Grain Impact Analyzer and Dust Accumulator for the Rosetta space mission. *Adv. Space Res.* **2007**, *39*, 446–450. [CrossRef]
12. Rotundi, A.; Sierks, H.; Della Corte, V.; Fulle, M.; Gutierrez, P.J.; Lara, L.; Barbieri, C.; Lamy, P.L.; Rodrigo, R.; Koschny, D.; et al. Dust measurements in the coma of comet 67P/Churyumov-Gerasimenko inbound to the sun. *Science* **2015**, *347*, aaa3905. [CrossRef] [PubMed]
13. Della Corte, V.; Rotundi, A.; Fulle, M.; Ivanovski, S.; Green, S.F.; Rietmeijer, F.J.M.; Colangeli, L.; Palumbo, P.; Sordini, R.; Ferrari, M.; et al. 67P/C-G inner coma dust properties from 2.2 au inbound to 2.0 au outbound to the Sun. *Mon. Not. R. Astron. Soc.* **2016**, *462* (Suppl. S1), S210–S219. [CrossRef]
14. Esposito, F.; Colangeli, L.; della Corte, V.; Palumbo, P. Physical aspect of an "impact sensor" for the detection of cometary dust momentum onboard the "Rosetta" space mission. *Adv. Space Res.* **2002**, *29*, 1159–1163. [CrossRef]
15. Wolff, M.J.; Smith, M.D.; Clancy, R.T.; Spanovich, N.; Whitney, B.A.; Lemmon, M.T.; Bandfield, J.L.; Banfield, D.; Ghosh, A.; Landis, G.; et al. Constraints on dust aerosols from the Mars Exploration Rovers using MGS overflights and Mini-TES. *J. Geophys. Res.* **2006**, *112*. [CrossRef]
16. Kok, J.F.; Parteli, E.J.R.; Michaels, T.I.; Karam, D.B. The physics of wind-blown sand and dust. *Rep. Prog. Phys.* **2012**, *75*, 1–72. [CrossRef] [PubMed]
17. Iversen, J.D.; Greeley, R.; Pollack, J.B. Windblown dust on earth, Mars and Venus. *J. Atmos. Sci.* **1976**, *33*, 2425–2429. [CrossRef]
18. Doronzo, D.M.; de Tullio, M.D.; Dellino, P.; Pascazio, G. Numerical simulation of pyroclastic density currents using locally refined Cartesian grids. *Comput. Fluids* **2011**, *44*, 56–67. [CrossRef]
19. Doronzo, D.M.; Khalaf, E.A.; Dellino, P.; de Tullio, M.D.; Dioguardi, F.; Gurioli, L.; Mele, D.; Pascazio, G.; Sulpizio, R. Local impact of dust storms around a suburban building in arid and semi-arid regions: Numerical simulation examples from Dubai and Riyadh, Arabian Peninsula. *Arab. J. Geosci.* **2014**. [CrossRef]
20. Almeida, M.P.; Andrade, J.S., Jr.; Herrmann, H.J. Aeolian transport layer. *Phys. Rev. Lett.* **2006**, *96*, 018001. [CrossRef] [PubMed]
21. Cunningham, E. On the Velocity of Steady Fall of Spherical Particles through Fluid Medium. *Proc. R. Soc. Lond. Ser. A Contain. Pap. Math. Phys. Character* **1910**, *83*, 357–365. [CrossRef]
22. Holstein-Rathlou, C.; Merrison, J.; Iversen, J.J.; Jakobsen, A.B.; Nicolajsen, R.; Nørnberg, P.; Rasmussen, K.; Merlone, A.; Lopardo, G.; Hudson, T.; et al. An Environmental Wind Tunnel Facility for Testing Meteorological Sensor Systems. *J. Atmos. Ocean. Technol.* **2014**, *31*, 447–457. [CrossRef]

© 2019 by the authors. Licensee MDPI, Basel, Switzerland. This article is an open access article distributed under the terms and conditions of the Creative Commons Attribution (CC BY) license (http://creativecommons.org/licenses/by/4.0/).

Article

Characterization of Back-Scattering and Multipath in a Suburban Area after the Calibration of an X-Band Commercial Radar

Gaspare Galati [1], Gabriele Pavan [1,*] and Christoph Wasserzier [2]

[1] Department of Electronic Engineering, Tor Vergata University of Rome and CNIT—National Inter-University Consortium for Telecommunications, RU of Rome, via del Politecnico 1, 00133 Rome, Italy; gaspare.galati@uniroma2.it

[2] Fraunhofer Institute for High Frequency Physics and Radar Techniques FHR, Fraunhoferstrasse 20, 53343 Wachtberg, Germany; christoph.wasserzier@fhr.fraunhofer.de or christoph.wasserzier@uniroma2.it

* Correspondence: gabriele.pavan@uniroma2.it

Received: 23 December 2019; Accepted: 11 January 2020; Published: 14 January 2020

Abstract: The increasing interest in the radar detection of low-elevation and small-size targets in complicated ground environments (such as urban, suburban, and mixed country areas) calls for a precise quantification of the radar detection capabilities in those areas. Hence, a set of procedures is devised and tested, both theoretically and experimentally, using a commercial X-band radar, to (i) calibrate the radar sensor (with an online evaluation of its losses) using standard scatterers, (ii) measure the multipath effect and compensate for it, and (iii) create "true radar cross section" maps of the area of interest for both point and distributed clutter. The above methods and the related field results are aimed at future qualification procedures and practical usage of small, cheap, and easily moveable radars for the detection of low-observable air targets, such as unmanned air vehicles/systems (UAV/UAS), in difficult ground areas. A significant set of experimental results as discussed in the paper confirms the great relevance of multipath in ground-based radar detection, with the need for correcting measures.

Keywords: radar cross-sections; multipath; radar calibration

---

1. Introduction

Unwanted echoes represent a significant problem in many radar applications, such as the surveillance of airports, harbors, and inhabited areas. These echoes, globally referred to as "clutter" [1], can be generated by both continuous (i.e., distributed) or discrete sets of scatterers. Discrete sources of clutter are, for example, road/rail infrastructures, buildings, water towers, and large lampposts. On the other hand, grass, vegetation, terrain, etc. represent distributed clutter sources, whose echoes are measured as the radar cross section per unit area. Clutter may be orders of magnitude more intense than the useful targets of interest, strongly affecting their visibility. Moreover, the dynamic range of the clutter echoes and their probability distribution widely vary both in time (at a given resolution cell) and in space (i.e., from adjacent radar resolution cells at the same time). These effects strongly depend on the radar installation in the operational area.

Ground clutter models are widely studied, see for instance [1–3], but, in spite of the available literature and databases, dedicated measurements are needed in order to obtain the clutter levels in a given environment, especially when both discrete and distributed sources are present.

In order to generate valuable clutter maps, it is necessary to operate with a well calibrated radar sensor, that is, setting of a precise relationship between the received power and the radar cross section (RCS) in each resolution cell. The classical way to calibrate a radar sensor by measuring the various terms that made up the radar equation [4] (i.e., antenna gain, transmitted power, losses, etc.) may be lengthy and cumbersome. In the literature [5,6], a more cost/effective approach is investigated based on reconstructing the relationship between the RCS of known, standard objects (metallic spheres and corner reflectors) and the received power. This relationship is a linear law above the noise plafond and below the receiver saturation zone.

After calibration, realistic clutter maps can be generated [7–10]. They may be used, for example, to design the siting of X-band surveillance radars [11] for the surveillance of airports (surface movements radar and apron control) and for the detection of unauthorized unmanned aerial vehicles (UAVs also known as drones) in urban and suburban areas. The possibility for the UAVs to carry on payloads makes their use both very interesting and potentially dangerous. The Federal Aviation Administration (FAA) regulations in the USA regarding the use of drones in commercial applications are clearly defined, while hobby or recreation usages are encouraged to follow the safety guidelines [12]. The problem of drones' control over a limited area is strongly rising up. The first idea is to, in some way, reproduce the surveillance system in Air Traffic Control (ATC) which uses dedicated radar sensors. However, they are very expensive and not easily deployable: typically, operating at S-band (2–3 GHz), these radars are more suitable for detecting big objects, like an airplane, rather than small drones. On the other hand, drones might be detected by cheaper commercial X-band (9.2–9.5 GHz) marine radars, either with a "solid state" transmitter or with a magnetron transmitter, which are very portable and easily deployable as compared with the large ATC surveillance radars.

The RCS of drones has recently been investigated. In [13], the numerical and experimental results are shown for a quadcopter drone (DJI Phantom2) at 10 GHz. The measurements of drones, carried out in an anechoic chamber, have shown RCS variations from -30 dBm^2 to -10 dBm^2. In [14], experimental results, at a range of up to 2 km, with an average altitude (above ground level) of 30 m, are reported using a commercial drone (DJI-Phantom4) and a ubiquitous FMCW radar system working at 8.75 GHz. The authors concluded that the drone echoes show the statistical behavior of Swerling I targets with an average RCS close to -20 dBm^2. In [15], the radar signals in both the range and Doppler domains are analyzed for a quadcopter drone (DJI Phantom2), as well as three different birds (Hooded Vulture, Eurasian Eagle Owl, and Barn Owl). The authors, using an S-band coherent multistatic radar, found that the micro-drone signatures were comparable (in RCS) to those of birds. In [16], X-band measurements of flying sea birds have been carried out in order to investigate the potential of extracting the features for target recognition from micro-Doppler signatures.

In spring 2017, at Tor Vergata University in Rome, the preliminary experimental results were obtained using a commercial X-band marine radar to detect drones with the aim to compare the drone's echo to those coming from known reflectors, in order to derive the radar visibility of the drone [17]. However, in this experimental research, it was soon clear that back-scattering and multipath, because of both continuous and discrete sources (often found in a suburban area), constitute a serious problem for small aircraft detection. The preliminary results, regarding the multipath analysis as a result of the land surface using an X-band radar, are shown in [18]. The relevant evaluations are based on the measured RCS variations of a calibrated target taken during the NATO SET-225 (Sensors and Electronic Technology Group-225) field trials in June 2018 at the Fraunhofer Institute FHR in Wachtberg, Germany.

During 2018 and in spring/summer 2019, more trials were carried out at Tor Vergata University. Some of the acquired data will be described in the following so as to highlight the problems related to the back-scattering and to the multipath in the frame of the detection of small targets.

This work is an extended version of the paper "Environmental Effects on Ground-based Radar Measurements" in Proceeding of the Sixth IEEE Workshop Metro Aerospace, Torino, Italy, June 2019 [19]. The paper is organized as follows. Section 2 recalls the system calibration procedure, including the estimation of the total radar losses from the antenna to the receiver. Section 3, considering the effects of terrain, describes the reflection coefficient and the multipath with its possible compensation. Sections 4 and 5 show the evaluation of some aspects of multipath in the Tor Vergata area and in the Wachtberg area using calibrated targets. Using opportunity targets, in Section 6, the non-stationary nature (in time and in space) of the multipath is highlighted in a suburban environment. Final considerations and perspectives are reported in Section 7. Appendix A recalls the multipath.

2. Radar System Calibration and Estimation Losses

For research purposes, the Radar Business Unit (RBU) of the Navico Company has supplied Tor Vergata University of Rome, on loan to use, an International Maritime Organization (IMO) compliant marine radar (model Simrad SRT). The main parameters of this set are shown in Table 1.

Table 1. Main parameters of the Navico/Simrad SRT radar.

Power transmitter (cavity magnetron): $P_t = 12$ kW (nominal) Operating frequency: $f = 9410$ MHz ($\lambda = 0.032$ m)
Transmitted pulse length: Short range mode 50 ns, medium range mode 250 ns, and long range mode 800 ns
Pulse repetition frequency (PRF): Short range mode 3000 Hz, medium range mode 1500 Hz, and long range mode 750 Hz
Antenna type: 6–feet (1.8 m) H-polarized slotted waveguide Maximum gain: $G_{dB} = 29$ dB Elevation beam-width: 25° Horizontal beam-width: 1.35°
Revolution speed: 24 – 48 RPM (revolutions per minute), selectable
Receiver: logarithmic with 95 dB dynamic range Noise figure: $F = 5$ dB (nominal)
Analog-to-digital-converter (ADC): 8 bit @ 50 Msample/s

Figure 1 shows the antenna and the Tx/Rx unit on the top of the Information Engineering building at Tor Vergata University in Rome. The radar is placed in the west corner of the upper terrace and is installed on a lifter capable of increasing the height of the radar antenna by up to 1 m. Arbitrary step positions are made possible for data acquisition, so as to evaluate the multipath effects, as explained in the following.

Figure 1. The Simrad/Navico SRT Radar and its lifter on the top of the Information Engineering building (Tor Vergata University).

All of the trials shown in this paper have been performed in the short range mode (see Table 1) with a pulse length of 50 ns (bandwidth B = 20 MHz) and PRF = 3000 Hz. In this set-up, the power of the receiver noise $P_{noise} = Fk_BTB$ (F is the noise figure, k_B is the Boltzmann's constant, and $T = 290$ K is the reference temperature) is equal to −96 dBm. The range resolution is 7.5 m, with a sampling interval of 3 m. Because of the horizontal beam-width of 1.35°, the nominal azimuth resolution is 12 m and 24 m at distances of 500 and 1000 m, respectively.

Downstream of the analog to digital conversion, for each antenna revolution, the acquired digital signals, coded in 256 levels, hereafter called ADC_{level} (0–255), are recorded in a (Range, Azimuth) matrix with 2048 rows (range-bins), corresponding to a maximum range of 6144 m (3 m × 2048), and a number of columns (azimuth-bins) nominally equal to 7500 at the rotating speed of 24 RPM and a PRF equal to 3000 Hz. The (Range, Azimuth) matrix allows us to display the plan position indicator (radar PPI).

The received power, expressed in dB above one milliwatt (dBm) is estimated by a linear relationship, as follows:

$$P(\text{dBm}) = \alpha \cdot ADC_{level} - \beta \tag{1}$$

where the constant α in (dB/level) is associated to each measured ADC level, while the constant β relates to the noise floor level (P_{noise}). These constants are obtained by radar calibration, without the need for dedicated measurements on the radar set.

The radar calibration is fully described in the literature [5]. The main steps are summarized as follows:

i. Three octahedral corner reflectors (CRs) having scaled dimensions of the side a (large $a = 45$ cm, medium $a = 22.5$ cm, and small $a = 11.25$ cm), and a coated sphere with a radius of $r = 20$ cm, are put in a suitable position (characterized with terrain masking by suited natural fences) with both land clutter and multipath effects practically absent.
ii. Four independent measurements of the ADC_{level}, obtained by averaging eight measurements for each target, are recorded.
iii. The theoretical RCS (in dBm^2) of CR's are normalized to the RCS of the sphere.
iv. From step (ii), on the plane (ADC_{level}, RCS_{norm}), four points are plotted.
v. The slope of the regression line among these points estimates the parameter α, while the parameter β is estimated imposing on the straight line to pass on the mean noise point.

The measurements did confirm that in the chosen configuration, the multipath factor [4] is very close to the unit, thus allowing us to estimate, with an accuracy of the order of 1%, the parameters α and β of Equation (1). It resulted $\alpha = 0.3792$ (dBm/level) and $\beta = 105.48$ (dBm). More details on this procedure are presented in [5].

The received power, P_r, is theoretically evaluated by the Radar Equation (2), including the overall system losses (L) up to the analog-to-digital-converter (ADC), as follows:

$$P_r = \frac{P_t G^2 \lambda^2}{(4\pi)^3} \cdot \frac{\sigma}{R^4} \cdot \frac{\eta^4}{L} \tag{2}$$

For a given distance, R, of the target, in the linear range of the receiver response, P_r is proportional to the RCS of the target (σ) and to the ratio $\frac{\eta^4}{L}$, where L denotes the losses and η^4 is the contribution due to multipath, analyzed in Appendix A. Inverting Equation (2) and using Equation (1) for the P_r expressed in dB, the RCS, $\sigma_{\text{dBm}^2} = 10 \cdot log_{10}(\sigma)$, of a single target is written as follows:

$$\sigma_{dBm^2} = \alpha \cdot ADC_{level} - \beta + L_{\text{dB}} - \eta^4_{\text{dB}} + 4 \cdot R_{\text{dB}} - C_{\text{dB}} \tag{3}$$

where $C_{\text{dB}} = 10 \cdot log_{10}\left[\frac{P_t G^2 \lambda^2}{(4\pi)^3}\right]$ includes the various radar parameters and constants. Using $\alpha = 0.3792$ (dBm/level) and $\beta = 105.48$ (dBm; under the hypothesis that in operational conditions, such as the protection by a fence, $\eta^4_{dB} \cong 0$), the losses L_{dB} in Equation (3) can be estimated as the value that minimizes the mean square error between the theoretical and the measured radar cross sections of the dedicated targets (four in this study). This resulted in $L_{dB} = -3.92$ dB. This value is compatible with the manufacturer data, kindly supplied by Navico RBU.

In general, it is not easy to evaluate, in one shot, radar losses with a residual error below 1 dB in typical conditions, and, to the best of the authors' knowledge, these overall black-box measurements of the radar receiver and overall losses, using standard reflectors and a fence, are novel.

3. Evaluation of Multipath and of Reflection Coefficient

The previous discussion is related to those very particular conditions (e.g., rays blocked by a fence) in which the multipath on the vertical plane is negligible. Conversely, in the presence of a multipath [1], that is a normal situation in surface radars, the calibration by standard reflectors requires evaluating the multipath factor η^4 in the particular geometry and propagations conditions. The multipath factor is the ratio between the amplitude of the electro-magnetic field on the target and the amplitude in the (theoretical) case of the absence of multipath. This factor depends on the complex reflection coefficient of the surface $\Gamma = \rho \cdot e^{\psi_r}$, and on the phase shift ψ_Δ, related to the path difference Δ between the direct and the reflected signals.

Following the developments and the notations in Appendix A, η^4 is expressed as follows:

$$\eta^4 = \left[1 + 2\rho \cdot \cos(\psi_r + \psi_\Delta) + \rho^2\right]^2 \tag{4}$$

For a given polarization, the parameters ρ and ψ_r depend on the electromagnetic properties of the surface, that is, the real and imaginary part of its dielectric constant $\epsilon = \epsilon' - j\epsilon''$, the frequency, the polarization, and on the grazing angle ψ_g. For the horizontal polarization, $\Gamma_H = \frac{\sin(\psi_g) - \sqrt{\epsilon - \cos^2(\psi_g)}}{\sin(\psi_g) + \sqrt{\epsilon - \cos^2(\psi_g)}}$.

In addition, considering the roughness of the surface, the global reflection coefficient is $\Gamma_{total} = \Gamma \cdot \rho_{rough}$, where $0 < \rho_{rough} < 1$ depend on the grazing angle, ψ_g, and on the r.m.s. roughness in the vertical direction σ_h.

For flat soil (clay) with a moisture content of 35 g/cm^3, we may use (for the horizontal polarization of the SRT) values of $\epsilon' = 14.8$ and $\epsilon'' = 6.7$ (valid at 10.0 GHz [20]) as a good approximation of the 9.41 GHz data. Including the contribution of the land roughness (σ_h), Figure 2 shows the amplitude, ρ, of the reflection coefficient for $\sigma_h = 0.05,\ 010,\ 0.15,$ and 0.20 m. For low grazing angles (in the order of 1 or 2 degrees), this coefficient is within $0.05 < \rho < 0.9$.

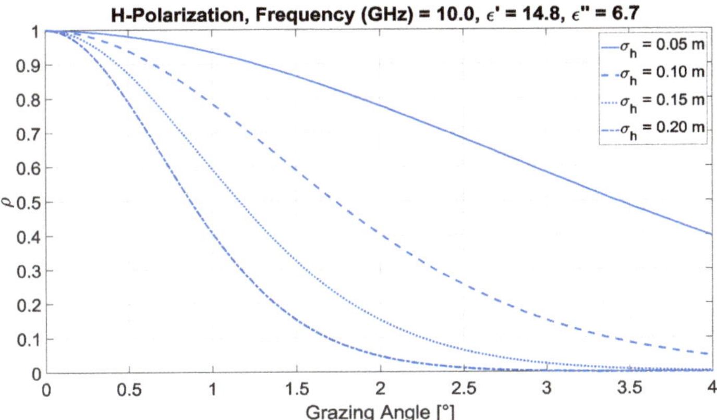

Figure 2. Modulus of the reflection coefficient ρ (X band, horizontal polarization) for soil (clay) with moisture of 35 g/cm^3, low grazing angles, and four values of σ_h.

In the horizontal polarization, the phase of the reflection coefficient for low grazing angles can be supposed to be equal to π, as follows: for grazing angles between 0.5° and 3°, its values range from 179.95° to 179.65° (H-polarization, $f = 10.0$ GHz, soil moisture 35 g/cm^3, $\epsilon' = 14.8$, and $\epsilon'' = 6.7$), that is, $\Gamma = -\rho$, is not affected by the σ_h variation. The roughness contribution to the reflection coefficient is real.

In Appendix A, Table A1 reports the maximum variation in the "apparent" RCS of the target due to multipath. For example, considering $\rho \cong 0.5$ (corresponding to 1.75° grazing angle and $\sigma_h = 0.1$ m), a radar target distance D = 1000 m and 30 m target height, Figure 3 shows a sample of the variation of the multipath factor with the radar antenna height, h_a, varying from 22.0 m to 23.1 m. The distance and the height of the target correspond to a certain target (lamppost L4, see Section 6) and the antenna height to the installation on the roof of the Information Engineering building at Tor Vergata University. The latter height can be controlled by varying the vertical radar position of the antenna by a mechanical elevator, as shown in Figure 1. With this set-up, we were able to reconstruct curves, such as in Figure 3, in order to estimate (with a best fit procedure) the modulus ρ of the reflection coefficient at the given experimental conditions.

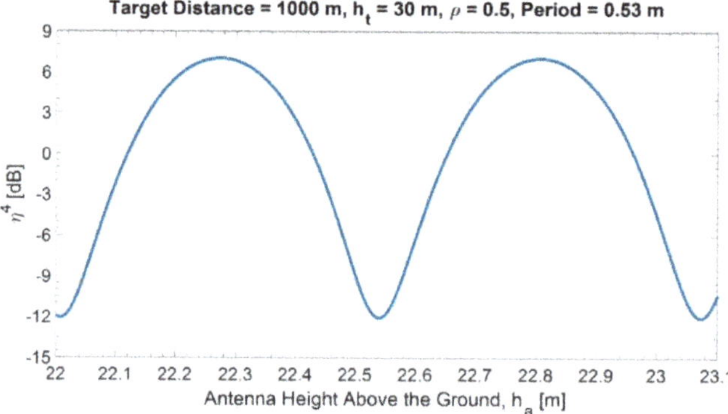

Figure 3. Multipath factor η^4 vs. the antenna height (target distance D = 1000 m and target height of $h_t = 30$ m). The period $\frac{\lambda D}{2h_t}$ is equal to 0.53 m.

From curves such as the one in Figure 3, the maximum value η_{max} and the minimum value η_{min} of the multipath factor are easily estimated. As η_{max} is equal to $1+\rho$, and η_{min} is equal to $1-\rho$, the modulus of the reflection coefficient is readily estimated, as follows:

$$\rho = \frac{\eta_{max} - \eta_{min}}{\eta_{max} + \eta_{min}} \tag{5}$$

In practice, the effect of the multipath in the RCS measurements can be corrected, as described in the following procedure.

(a) Set the target under test (distance D from the radar and the height of its main scatter h_t).
(b) Evaluate the period of the factor η^4 versus the antenna height h_a. It is $\frac{\lambda D}{2h_t}$ (Appendix A).
(c) Define the step Δh_a to obtain a significant number (e.g., ten) of positions inside the period.
(d) Record the received power, proportional to η^4, for each position, and draw the best-fit curve.
(e) Find the maximum and the minimum values of η^4.
(f) Use Equation (5) to obtain ρ.
(g) Use Equation (A7) to calibrate the received power.

4. Multipath Evaluation in the Tor Vergata Area

The first experiments were carried out in 2018 with a fixed radar installation for various heights of the same target (a corner reflector on the top of a pole, see left part of Figure 4).

Figure 4. R and T denote the positions of the radar on the roof of the Information building (41°51′6.40″ N, 12°37′0.50″ E) and of the target (41°51′6.40″ N, 12°37′0.50″ E), respectively.

Subsequently, in 2019, the radar elevator, shown in Figure 1, allowed us to vary the antenna height (in addition to the variation of the corner reflector height). The aim was to quantify the effect of the multipath on the RCS measurements, and to evaluate the reflection coefficient of the terrain (land with vegetation). For this goal, it was decided to keep the reflection point fixed on the ground, in order to be free from unwanted variations of the reflecting surface.

On 16 May 2019, a trihedral CR with an RCS of 7.2 dBm^2 has been posed (as a target) on a tripod at a distance of 690 m from the radar (see Figure 4). The weather conditions were cloudy sky with sunshine, $T = 16$ °C, and no wind.

Considering the starting position of the measurements (antenna height $h_a = 23$ m and the target height $h_t = 4.6$ m), the reflection point on the ground is 575 m away from the radar. To keep the distance of the reflection point unchanged, we varied the height of the antenna and the target, and being $h_a/h_t = 5$, it was necessary to vary the h_a and h_t (Equation (A1)) according to the values shown in Table 2. Considering the maximum ranges of elevation for the target and for the antenna, five positions were deemed sufficient.

Table 2. Five positions during the trials on 16 May 2019. The last column shows the radar cross section (RCS) estimate for the medium corner reflector (CR).

Position	Δ_a [cm]	Δ_t [cm]	Time	RCS [dBm^2]
1	0	0	10 : 45	12.1
2	−25	−5	10 : 55	9.1
3	−50	−10	11 : 07	8.3
4	−75	−15	11 : 20	6.0
5	−95	−19	11 : 35	6.0

Table 2 reports, for each position, the time of the acquisition and the measured RCS of the corner reflector. Remember that the theoretical value of the RCS of this CR is 7.2 dBm^2.

For the five positions, the mean grazing angle is 2.24°. Using Equation (1) with $\alpha = 0.3792$ and $\beta = 105.48$, the received power has been evaluated. Hence, the multipath factor has been estimated using the average of the five received power values for each position. Using Equation (5), the amplitude of the reflection coefficient results in $\rho = 0.17$. This value is compatible with vegetated land, with the height of the vegetation in the order of 10 to 15 cm, which is consistent with the observed height of the grass on the meadow. Figure 5 shows the theoretical multipath factor and the measured one.

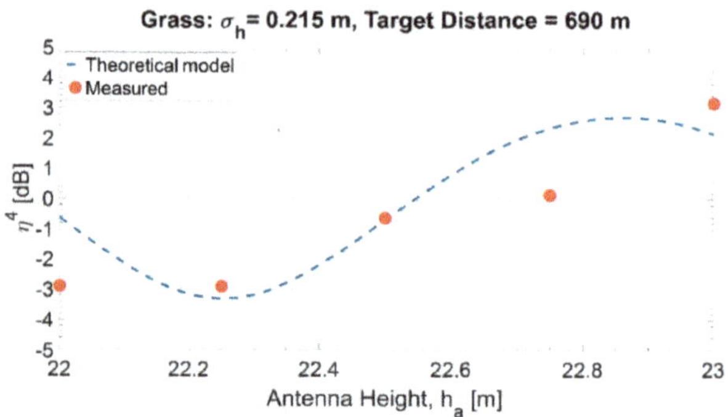

Figure 5. Theoretical and measured multipath factor with a constant position of the reflection point.

The differences between the theoretical model and the measurements of this early experiment have been attributed to various unfavorable elements of the experimental area, including the following: (i) the land not being flat, (ii) the vegetation not being uniform, and (iii) the variation of the grazing angle versus h_a.

In a more regular environment, such as the one described in Section 5, the multipath analysis has shown a better agreement with the theory [19].

The ensuing experiments were performed, and their results are presented in the following. One experiment, conducted at Fraunhofer FHR, investigates the influence of variations in the vegetation on the multipath effect, as presented in Section 5. A second set of experiments, performed over a period of more than one month, and presented in Section 6, was conducted at Tor Vergata, aiming for an investigation of the temporal influences on the multipath effect.

5. Multipath Effects of Different Vegetations Observed in the Area of Wachtberg

Trials were conducted in June 2018, with a continuous emission (CE) noise radar at the Fraunhofer FHR Institute in Wachtberg, Germany. They addressed different aspects of noise radar, and were carried out by members of the NATO research task group (RTG) SET-225. The noise radar (NR) used during these experiments was developed by FHR, and is referred to as FHR-NR; it operates in the X-Band, more precisely in the standard maritime navigation band ranging from 9200 MHz to 9500 MHz, and thus, the same frequency band as the Simrad SRT radar, which allows for a direct comparison of the multipath effects observed in Tor Vergata with the ones at the FHR outpost in Wachtberg.

During the multipath trials, the operational signal bandwidth was $B = 50$ MHz, corresponding to a range resolution not less than 3 m. The transmitted signal was a random waveform with a Gaussian distribution and less than 2 W of effective radiated power.

Dedicated experiments were conducted to characterize the effect of different kinds of vegetation on the multipath. Figure 6 shows the setup of the experiments with two positions of the same reference radar target. FHR-NR was installed inside a shelter at a small facility in Wachtberg, surrounded by fields. On one side of the shelter, there was a field of sweetcorn, with an approximate height of 2 m. On the other side, there was grassland of approximately 50 cm in height. The first position of the standard target was located in the sweetcorn field on a block of concrete at a 186 m distance to the radar (path A). The second target position had a radar distance of 159 m, and was located close to the end of the grass field (path B).

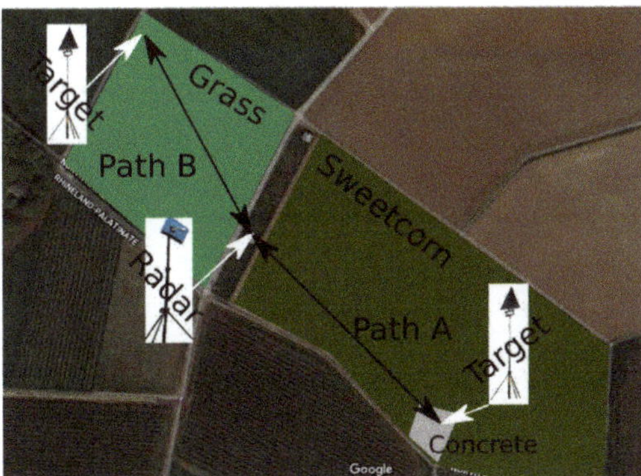

Figure 6. Illustration (from Google Earth) of the experimental setup in the fields of Wachtberg. Two different propagation paths were measured with the same radar and the same standard target. Sweetcorn was grown in the field of propagation path A (186 m), whereas grass was grown on the field of propagation in path B (159 m).

The standard target, used for both experiments, was a large trihedral corner reflector (CR) with a radar cross section not less than 1000 m^2. This corner reflector was mounted on a tripod, fully extended during all of the experiments at a constant height of 4.1 m. The height of the radar transmitting and receiving antenna system, made up by two standard horn antennas on a second tripod, as shown in Figure 7, varied in intervals of 10 cm from 2.6 m (path A), respectively 2.2 m (path B), up to a maximum height of 3.6 m.

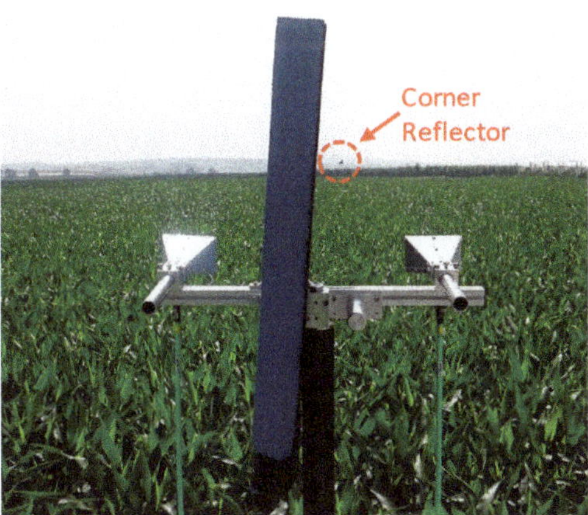

Figure 7. The two radar antennas were mounted on a tripod with variable heights. A plate of absorbing material was installed between the antennas for isolation purposes. This picture shows the propagation over sweetcorn along path A.

A detailed analysis of this experiment can be found in [19]. The main points are summarized here and compared with the previously discussed results from the experiments performed at the Tor Vergata University of Rome.

Figure 8 shows the measured multipath effect on the RCS measurement of the standard target.

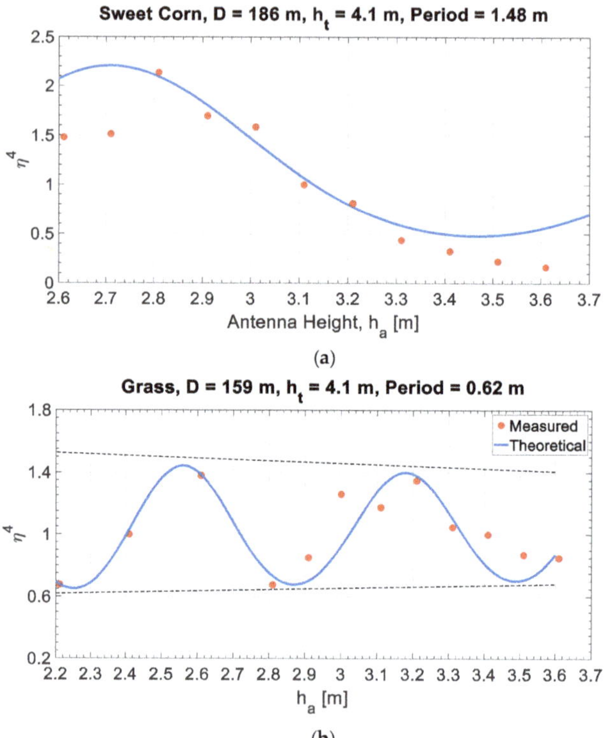

Figure 8. Measurement results. The period of the multipath effect is lower in propagation over grass (**b**) than over sweetcorn (**a**), due to the different height of the two kinds of vegetation.

In both graphs of Figure 8 the periodic multipath effect is clearly seen. The measured values (red dots) are in good agreement with the theoretical values drawn as a blue line. The influence on multipath of the different kinds of vegetation can be easily read from the graphs. The amplitude of the oscillation is higher in the case of propagation over sweetcorn while an oscillation with lower amplitude but higher frequency is observed over grassland.

These results represent a single snapshot of a single day without any significant variation in temperature, humidity and more; anyway, this single comparison between two different kinds of vegetation results in significant observations. Long-term measurements are presented in the ensuing section describing the variations of the multipath effect over time.

6. Long-Term Measurements of Radar Cross Section of Man-Made Targets

In the frame of the 2018 experiments at Tor Vergata campus and their 2019 follow-on, fixed identical and very large lampposts (30 m high, here referred to as L1, L2, L3, and L4) at different distances from the radar (798, 888, 570, and 1008 m, respectively) were considered as opportunity targets. Figure 9 displays an overview of the area, showing the position of the radar and of the lampposts.

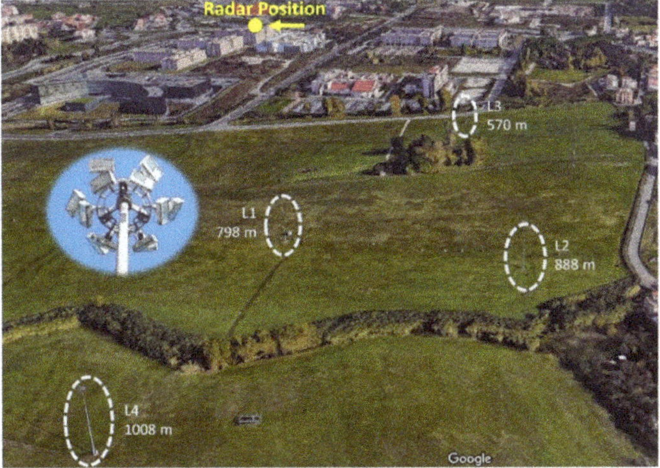

Figure 9. Overview (from Google Earth) of the trials environment at the Tor Vergata University area. The four opportunity targets (lampposts L1, L2, L3, and L4) are highlighted.

Each lamppost has six pairs of large lamps arranged on a metallic ring, as shown in the inset of Figure 9. As the top of each lamppost is similar to a single, small, but strong radar scatterer at 30 m above ground level, the radar echo from these targets are candidates to be significantly affected by multipaths. Using Equation (3) and the (Range, Azimuth) matrix of the ADC levels, an RCS map of the environment has been generated, as shown in Figure 10. In these measurement results, two significant target echoes are created by the compound of the Literature Faculty and the long commercial building in via Schiavonetti.

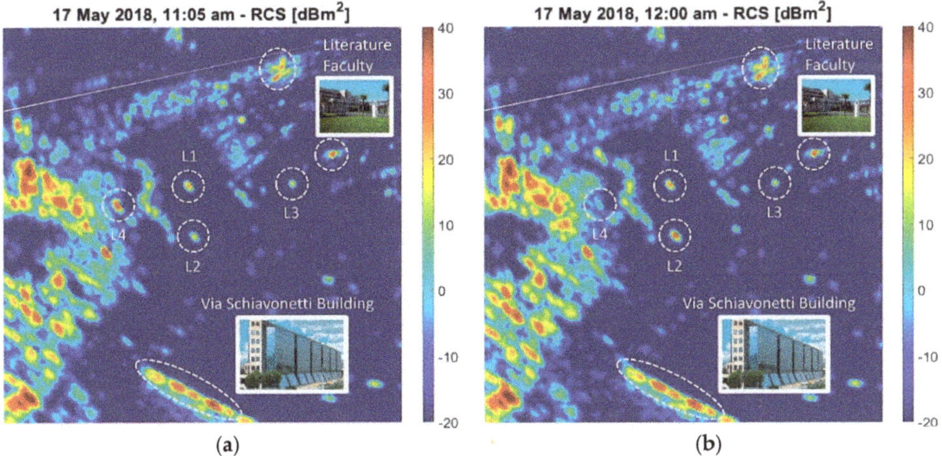

Figure 10. Measured RCS, 17 May 2018: (**a**) 11:05 and (**b**) 12:00.

Unexpectedly, during the trials on 17 May 2018, the estimated values of the RCS, in two close times of this day (11:05 and 12:00) have shown a considerable difference for lampposts L2 and L4. For L2, the measured RCS changed from 19.1 dBm^2 to 32.4 dBm^2, while for L4, it changed from 33.8 dBm^2 to 8.1 dBm^2 (see Figure 10a,b, where the RCS in dBm^2 is shown as a false color in an area including the four lampposts).

The rationale for these significant variations can only be attributed to the propagation effects, in particular, to some time-varying multipath. In fact, in some experiments, we found that the reflecting surface on the ground is characterized by many time-varying scatterers (such as vegetation, foliage of trees, tall grass, etc.), and we know that the multipath can significantly affect the measurements of the RCS.

The sensitivity of the measured RCS w.r.t. the height of the target (see Appendix A) increases significantly when this height (in wavelengths) increases. Hence, for a tall scatterer such as a building (e.g., in Figure 10, the building at Via Schiavonetti, and the ones of the Faculty of Literature), the multipath effects average along the elevation, and, thus, the measured RCS is stable; the opposite is true for a point target, such as a small flying object or, neglecting the contribution by its pole, a lamppost at a fixed height. Therefore, for large and tall targets and for most buildings, the measured RCS has steady values as the multipath effect vanishes ($\eta_{dB}^4 \cong 0$ dB).

To better understand this point, more measurements of the four lampposts were carried out in spring 2019, varying the antenna height. Five 25 cm-step positions, from a maximum elevation (position 1) to a minimum elevation (position 5), were used for radar data acquisition. Figure 11 shows the reflection points (see point M in Figure A1) on the ground for lampposts L1, L2, and L4.

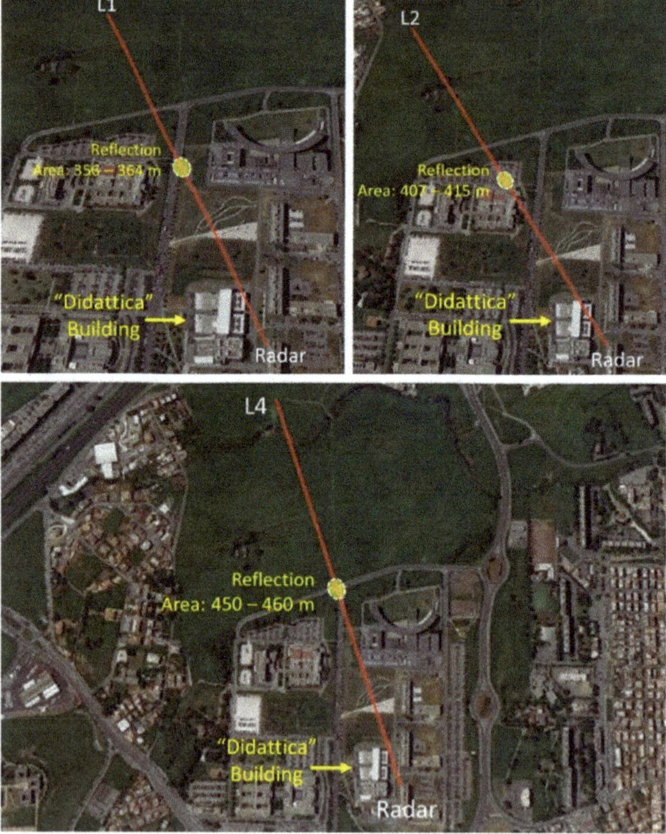

Figure 11. Reflection points on the ground for L1, L2, and L4.

For lampposts L1 and L4, the reflection points shifted on the road, by 7–8 m, due to the variation of the antenna height. Hence, the reflection points' positions may move from vegetated land into a road, possibly with some traffic. For the lamppost L2, the reflection point is on the parking lot of the Faculty of Literature, and the multipath may change rapidly if cars or buses move or appear/disappear during the acquisition. This explains the noticed, strong variations in RCS. The pole of the lamppost L3 is partially masked by the roof of the "Didattica" building (see Figure 12), hence, multipath effects are considered negligible for the RCS measurements of L3.

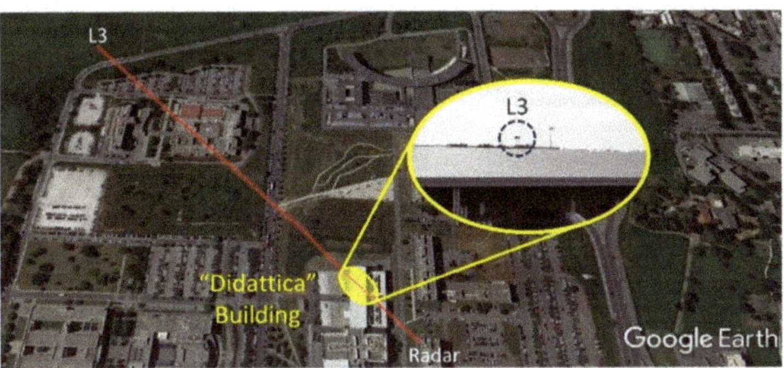

Figure 12. Reflection point on the ground for L3.

We collected the measurements obtained over the following four different days: 16 May 2019, 4 June 2019, 7 June 2019, and 26 June 2019. Figure 13 shows the measured RCS (in dB square meter) of these four lampposts, with a varying antenna elevation (five steps, 0.25 m each).

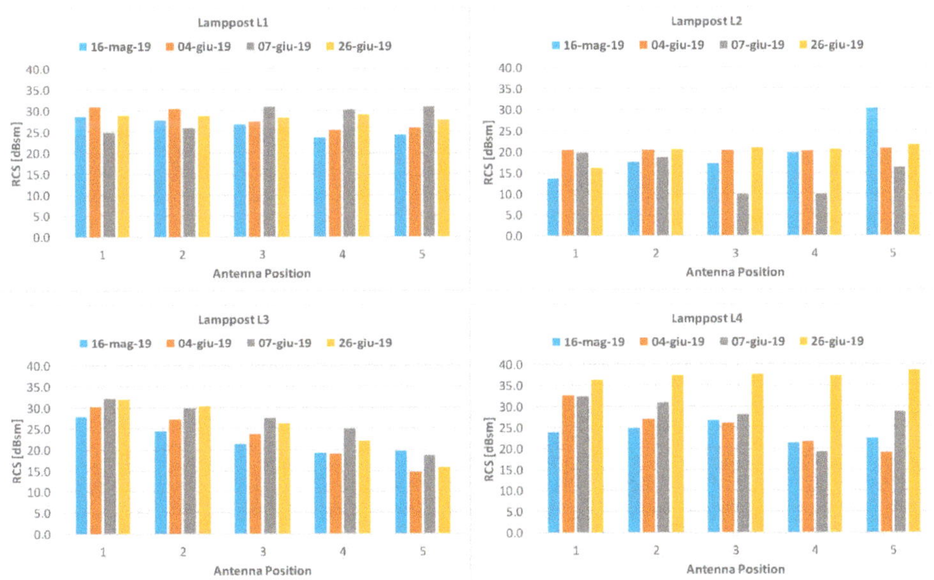

Figure 13. Measured RCS (dBsm) of the lampposts L1, L2, L3, and L4 with the antenna position varying (one step 0.25 m).

Lamppost L3, which is the reflection-masked target practically unaffected by the multipath effects, shows the reflection behavior different from the remaining three lampposts. During the different days, the RCS of the lamppost L3 always decreases from a maximum (step 1) to a minimum (step 5), as shown in Figure 14, due to an increased occlusion by the "Didattica" building. The temporal variation of the multipath effect is described by the parameter ρ, which has been evaluated as follows:

$$\rho = \frac{\sum_{i=1}^{n}\left(X_i - \overline{X}\right)\left(Y_i - \overline{Y}\right)}{\sqrt{\sum_{i=1}^{n}\left(X_i - \overline{X}\right)^2 \sum_{j=1}^{n}\left(Y_j - \overline{Y}\right)^2}} \tag{6}$$

where $n = 5$ denotes the number of positions for each day, while $\overline{X} = \frac{1}{n}\sum_{i=1}^{n} X_i$ and $\overline{Y} = \frac{1}{n}\sum_{i=1}^{n} Y_i$ are the sample means.

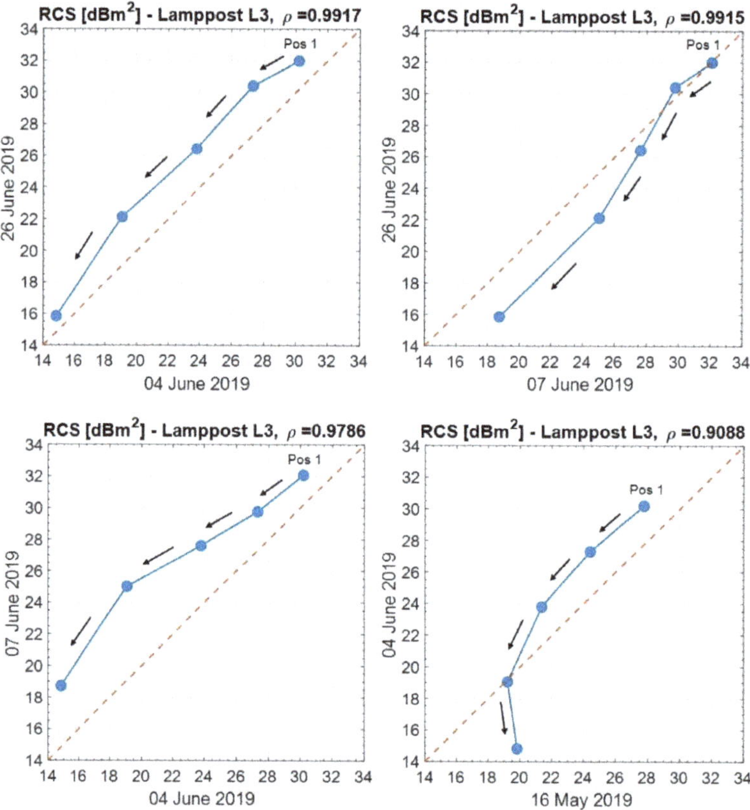

Figure 14. Comparison of the measured RCS of lamppost L3 on different days, varying the antenna position. Starting from the maximum antenna elevation (Position 1), the arrows denote the successive positions of the antenna up to the minimum elevation.

The measurement of the RCS of lamppost L3 does vary a little from day to day, making the correlation coefficient ρ of the measured RCS on two different days (X, Y) close to the unit (Figure 14).

A different situation arises for the lampposts L1, L2, and L4, where the measured RCSs (at the same position on different days) show differences that can reach 10 dB or more. These significant variations must be attributed to the propagation effects, that is, to a time-varying multipath.

For lamppost L1 and L2, Figure 15 shows a comparison of the measured RCS, on different days, varying the antenna position.

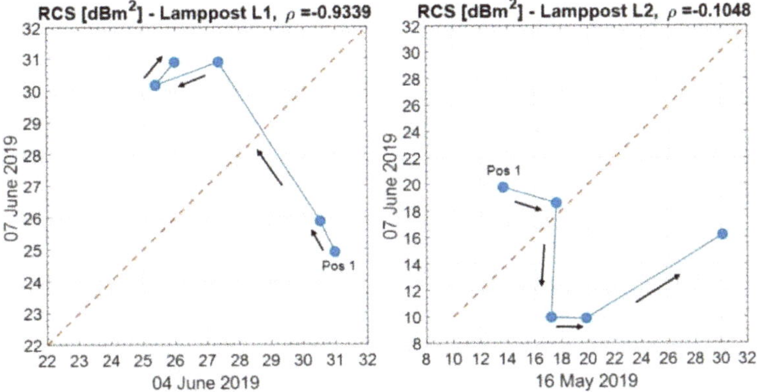

Figure 15. Comparison of the measured RCS of lampposts L1 and L2 on different days, varying the antenna position. Starting from the maximum antenna elevation (Position 1), the arrows denote the successive positions of the antenna up to the minimum antenna elevation.

For L1, the values show a negative correlation ($\rho = -0.9339$), while for L2, they are quite de-correlated ($\rho = -0.1048$). These results are strongly affected by the time variation of the reflector coefficient (moving vehicles on the road, day-to-day variation, etc.)

7. Comments and Conclusions

Field experiments have been performed in suburban and country environments to better understand radar propagation and scattering effects, with particular attention to multipath. To this aim, it has been necessary to calibrate the radar. In general, radar calibration is a cumbersome task, requiring standard reflectors (such as corner reflectors and metal-coated spheres) and the compensation of the backscatter from the underlying surface (i.e., land) and of the multipath in the vertical plane (assuming the horizontal plane free from multipath). In Tor Vergata trials we used a non-coherent radar (with Doppler measurements not available) operating in the marine X-band with horizontal polarization. Some preliminary indications from the experimental results are synthesized as follows.

First, the unwanted reflection and scattering may be mitigated using clutter fences, with the calibration reflectors positioned in order to be seen from the radar "just above the fence".

Second, the multipath (when not avoided by a fence) may strongly affect the calibration measurements, even when due to an apparently a more "benign" surface (slightly wet land, limited vegetation) that the well-known, strongly reflecting sea surface. The multipath compensation calls for estimating the reflection coefficient by a set of measurements (order of five to ten). In a not uniform environment (e.g., with different vegetation, roads, trees, etc.), it is necessary to vary both the height of the reflector and of the radar antenna, in order to keep the reflection point fixed on ground.

Furthermore, different experiments have proven that the multipath effect is independent of the kind of radar emission, that is, pulsed or continuous emission. Instead, it is the range resolution of the signals that is important. Future investigations might determine the influence of the radar bandwidth on the RCS variations created by the multipath effect.

As a side effect of the calibration procedure after multipath compensation, a significant "bonus", which is obtained, is the "black box" evaluation of the radar equipment total losses.

Author Contributions: The authors have contributed to this work in the following parts. G.G. proposed the idea and coordinated the overall writing. G.P. designed the experiments Sections 4 and 6, and collected and analyzed the related data. C.W. implemented the experimental set of Section 5, performed the expeiments, collected the data and analyzed them. All authors have read and agreed to the published version of the manuscript.

Funding: This research, internally funded by CNIT, received no external funding.

Acknowledgments: The authors are warmly grateful to Sergio Pandiscia who took care of all of the logistic operations for the trials. They are also grateful to the RBU of the Navico Company (Alberto Baroncelli), which provided the radars on loan. Finally, a sincere thanks goes to the whole staff of the Botanic Garden, Tor Vergata University, in particular, to Antonella Canini and to Roberto Braglia, for the on-site activities on the vegetation.

Conflicts of Interest: The authors declare no conflict of interest.

Appendix A. The Phenomenon of Multipath

Let M be the point of reflection on the ground (Figure A1). The ground distance $y = OM$ of the point from the radar is a fraction of the distance $D = OO'$ of the target, as follows:

$$y = \frac{h_a}{h_a + h_t} D \tag{A1}$$

For smooth flat Earth, h_a and h_t are the heights of the radar antenna and of the target above the land surface, respectively. The antenna pattern is considered to be uniform in the elevation angle. R denotes the distance (direct path AB) between the radar and the target (Figure A1). The change of the field strength $|\vec{E}|$ (V/m) at the target due to the multipath is expressed by the following ratio [4]:

$$\eta = \frac{|\vec{E}|}{|\vec{E}_0|} \tag{A2}$$

where $|\vec{E}|$ is the amplitude of the field at the target in the presence of the Earth's surface, and $|\vec{E}_0|$ is the amplitude of the target in free space. In the literature, the coefficient η is also known as the "propagation factor" F when it includes the effects of the antenna pattern on the elevation coverage, [4].

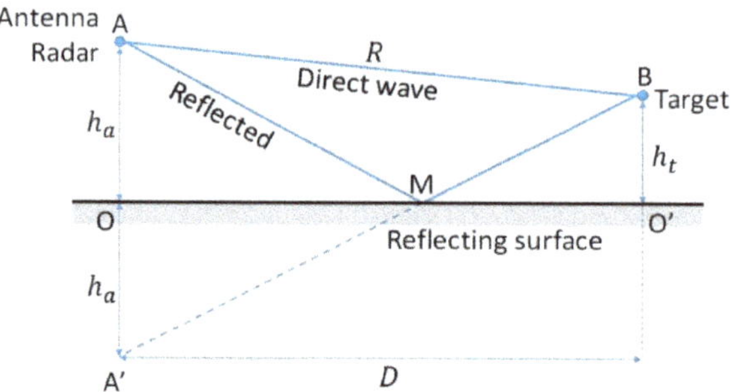

Figure A1. Geometry of the radar propagation on a plane-reflecting surface.

The length difference between the direct (AB) ray and the reflected one (AMB = A′MB) causes a phase difference between both signals, hence affecting their coherent sum. The differences in the amplitude and in the phase of the direct and reflected signals depend on the complex reflection coefficient, namely: $\Gamma = \rho e^{\psi_r}$. From Figure A1, with the approximation of $\sqrt{1+x^2} \cong 1 + \frac{x^2}{2}$ for $x \ll 1$ that is, where $|h_a - h_t| \ll D$, as in our experiments, the path difference is approximated as follows:

$$\Delta = D\left[\sqrt{1 + \frac{(h_a + h_t)^2}{D^2}} - \sqrt{1 + \frac{(h_a - h_t)^2}{D^2}}\right] \cong \frac{2h_a h_t}{D}. \tag{A3}$$

The sum signal is $\overline{X}_{sum} = \overline{X}_{dir}\left[1 + \rho e^{j(\psi_r+\psi_\Delta)}\right]$, with $\psi_\Delta = \frac{2\pi}{\lambda}\Delta$. Hence,

$$\left|\frac{\overline{X}_{sum}}{\overline{X}_{dir}}\right| = |1 + \rho e^{j(\psi_r+\psi_\Delta)}| \overset{\text{def}}{=} \eta. \tag{A4}$$

The multipath factor is the fourth power of η, as follows:

$$\eta^4 = \left[1 + 2\rho \cdot cos(\psi_r + \psi_\Delta) + \rho^2\right]^2. \tag{A5}$$

For low grazing angles, that is, with $\psi_r \cong \pi$, where $\psi_\Delta \cong \frac{2\pi}{\lambda}\frac{2h_a h_t}{D}$, we have the following:

$$\eta^4 = \left[(1-\rho)^2 + 4\rho \cdot sin^2\left(\frac{2\pi}{\lambda}\frac{h_a h_t}{D}\right)\right]^2. \tag{A6}$$

The maximum of η^4 is $\eta^4_{max} = (1+\rho)^4$, while the minimum is $\eta^4_{min} = (1-\rho)^4$. Hence, the variation in the "apparent" radar cross section of the target is (in decibel) as follows:

$$\Delta RCS\ [dB] = 10 \cdot log_{10}\left(\frac{\eta^4_{max}}{\eta^4_{min}}\right) = 40 \cdot log_{10}\left(\frac{1+\rho}{1-\rho}\right). \tag{A7}$$

Table A1 shows some values of $\Delta RCS\ [dB]$ for different values of ρ.

Table A1. Maximum target RCS variation due to multipath.

ρ	1.0	0.8	0.7	0.6	0.5	0.3	0.2	0.1
ΔRCS_{dB}	$+\infty$	38.1	30.1	24.1	19.1	10.8	7.0	3.5

The multipath factor, Equation (A6), is a periodic function. A full cycle is achieved when $\frac{2\pi}{\lambda}\frac{h_a h_t}{D} \doteq \pi$, that is, $h_a h_t = \frac{\lambda D}{2}$. Hence, setting h_a while varying h_t (typical experimental set-up with standard calibrated targets, such as corner reflectors and metallized spheres), the period is $\frac{\lambda D}{2h_a}$. On the other hand, setting h_t (typical experiments using large and heavy opportunity targets such as lampposts, buildings, towers, and tanks) and varying h_a, the period is $\frac{\lambda D}{2h_t}$. For the latter case, Table A2 shows the period of η^4 for $\lambda = 0.03$ m, $D = 1000$ m, and various heights of the target.

Table A2. Period of η^4 for $\lambda = 0.03$ m and $D = 1000$ m.

h_t [m]	2.0	5.0	10.0	15.0	20.0	30.0
$\frac{\lambda D}{2h_t}$ [m]	7.5	3.0	1.5	1.0	0.75	0.5

For $\rho = 1$ ($\psi_r = \pi$), Equation (A6) becomes the following:

$$\eta^4 = 16 \cdot sin^4\left(\frac{\pi\Delta}{\lambda}\right) \cong 16 \cdot sin^4\left(\frac{2\pi}{\lambda} \cdot \frac{h_a h_t}{D}\right) \tag{A8}$$

The theoretical range for η^4 in a linear scale is [0, 16], and in decibels is $(-\infty, 12]$. From Equation (A1), posing $r = \frac{h_a}{h_t}$, the sensitivity w.r.t. h_a (resp. h_t), is easily computed as follows:

$$\begin{aligned} S^y_{h_a} &= \frac{h_a}{y} \cdot \frac{dy}{dh_a} = \frac{h_t}{h_a+h_t} = \frac{1}{1+r} \\ S^y_{h_t} &= \frac{h_t}{y} \cdot \frac{dy}{dh_t} = -\frac{1}{1+r} \end{aligned}. \tag{A9}$$

When both heights are allowed to vary, the overall percentage variations of the reflection point distance *OM* (Figure A1) are as follows:

$$\Delta y(\%) \cong \Delta h_a(\%)\frac{1}{1+r} - \Delta h_t(\%)\frac{1}{1+r}. \quad \text{(A10)}$$

In order to maintain the reflection point fixed on the ground (i.e., ensure that $\Delta y = 0$), equal percentage variations shall be applied to both h_a and h_t, according to Equation (A10).

References

1. Nathanson, F.E.; Reilly, J.P.; Cohen, M.N. *Radar Design Principles: Signal Processing and the Environment*, 2nd ed.; SciTech Publishing: Mendham, NJ, USA, 1999.
2. Barton, D.K. *Radar Clutter (Radars Volume 5)*; Artech House: Norwood, MA, USA, 1975.
3. Billingsley, J.B. *Low-Angle Radar Land Clutter Measurements and Empirical Models*; William Andrew Publishing SciTec: Norwich, NY, USA, 2002.
4. Skolnik, M.I. *Introduction to Radar Systems*, 3rd ed.; McGraw Hill Education: New York, NY, USA, 2015.
5. Galati, G.; Pavan, G. High Resolution Measurements and Characterization of Urban, Suburban and Country Clutter at X-band and Related Radar Calibration. In Proceedings of the 9th International Conference on Ultrawideband and Ultrashort Impulse Signals (UWBUSIS), Odessa, Ukraine, 4–7 September 2018; pp. 20–27. [CrossRef]
6. Galati, G.; Pavan, G. Calibration of an X-Band Commercial Radar and Reflectivity Measurements in Suburban Areas. *IEEE Aerosp. Electron. Syst. Mag.* **2019**, *34*, 4–11. [CrossRef]
7. Nitzberg, R. Clutter Map CFAR Analysis. *IEEE Trans. Aerosp. Electron. Syst.* **1986**, *4*, 419–421. [CrossRef]
8. Musicki, D.; Evans, R. Clutter map information for data association and track initialization. *IEEE Trans. Aerosp. Electron. Syst.* **2004**, *40*, 387–398. [CrossRef]
9. Galati, G.; Pavan, G. Generation of Land-Clutter Maps for Cognitive Radar Technology. In Proceedings of the World Conference on Information Systems and Technologies, Naples, Italy, 27–29 March 2018; pp. 1463–1470. [CrossRef]
10. Galati, G.; Pavan, G. Radar Environment Experimental Analysis for Optimal Siting. In Proceedings of the IRS 2018, Bonn, Germany, 20–22 June 2018. [CrossRef]
11. Galati, G.; Pavan, G. Radar environment characterization by signal processing techniques. In Proceedings of the 2017 IEEE International Symposium on Signal Processing and Information Technology (ISSPIT), Bilbao, Spain, 18–20 December 2017. [CrossRef]
12. Unmanned Aircraft Systems. Available online: https://www.faa.gov/uas (accessed on 13 January 2020).
13. Schroder, A.; Renker, M.; Aulenbacher, U.; Murk, A.; Boniger, U.; Oechslin, R.; Wellig, P. Numerical and Experimental Radar Cross Section Analysis of the Quadrocopter DJI Phantom 2. In Proceedings of the IEEE Radar Conference, Johannesburg, South Africa, 27–30 October 2015; pp. 463–468. [CrossRef]
14. Duque de Quevedo, Á.; Ibañez Urzaiz, F.; Gismero Menoyo, J.; Asensio López, A. Drone Detection and RCS Measurements with Ubiquitous Radar. In Proceedings of the Conference on Radar 2018, Brisbane, Australia, 27–31 August 2018. [CrossRef]
15. Ritchie, M.; Fioranelli, F.; Griffiths, H.; Torvik, B. Monostatic and Bistatic Radar Measurements of Birds and Micro-Drone. In Proceedings of the IEEE Radar Conference, Philadelphia, PA, USA, 2–4 May 2016. [CrossRef]
16. Torvik, B.; Olsen, K.E.; Griffiths, H.D. X-band measurements of radar signatures of large sea birds. In Proceedings of the Radar Conference, Lille, France, 13–17 October 2014; pp. 1–6. [CrossRef]
17. Galati, G.; Pavan, G.; De Palo, F.; Latini, D.; Carbone, F.; Del Frate, F.; Pietrobono, F. Visibility Trials of Unmanned Aerial Vehicles (Drones) by commercial X-Band Radar in Sub-urban Environment. In Proceedings of the AEIT International Annual Conference, Cagliari, Italy, 20–22 September 2017. [CrossRef]
18. Galati, G.; Pavan, G.; Wasserzier, C. Multipath Effect on Radar Cross Section Measurements in Natural Environment and Related Correction. In Proceedings of the IRS 2019, Ulm, Germany, 26–28 June 2019.

19. Galati, G.; Pavan, G.; Wasserzier, C. Environmental Effects on Ground-based Radar Measurements. In Proceedings of the IEEE 6th International Workshop on Metrology for AeroSpace, Turin, Italy, 19–21 June 2019; pp. 335–340.
20. Ulaby, F.; Moore, R.; Fung, A. *Microwave Remote Sensing—Active and Passive—Vol. II, Radar Remote Sensing and Emission Theory*; Artech House: Norwood, MA, USA, 1982.

© 2020 by the authors. Licensee MDPI, Basel, Switzerland. This article is an open access article distributed under the terms and conditions of the Creative Commons Attribution (CC BY) license (http://creativecommons.org/licenses/by/4.0/).

Article

Secondary Radar Beacons for Local Ad-Hoc Autonomous Robot Localization Systems †

Martin Schuetz *, Tatiana Pavlenko and Martin Vossiek

Institute of Microwaves and Photonics, University of Erlangen-Nuremberg, 91058 Erlangen, Germany; tatiana.pavlenko@fau.de (T.P.); martin.vossiek@fau.de (M.V.)

* Correspondence: martin.schuetz@fau.de

† This paper is an extended version of our paper published in 2019 IEEE International Workshop on Metrology for AeroSpace (MetroAeroSpace).

Received: 19 October 2019; Accepted: 9 December 2019; Published: 12 December 2019

Abstract: In this paper, we present a detailed analysis and implementation of secondary radar beacons designed for a local ad-hoc localization and landing system (LAOLa) to support the navigation of autonomous ground and aerial vehicles. We discuss a switched linear feedback network as a virtually coherent oscillator and show how to use it as a secondary radar transponder. Further, we present a signal model for the beat signal of the transponder response in an FMCW radar system, which is more detailed than in previously published papers. An actual transponder realization in the 24 GHz ISM band is presented. Its RF performance was evaluated both in the laboratory and in the field. Finally, we put forward some ideas on how to overcome the range measurement inaccuracy inherent in this transponder concept.

Keywords: wireless local positioning; secondary radar; radar beacons; autonomous robot navigation; autonomous landing; space exploration

---

1. Introduction

Secondary radar systems have proven to be a reliable and performant solution for target localization and identification, especially in environments with weak or denied GNSS performance. They can be used for both indoor and outdoor localization, going down to centimeter precision and accuracy with update rates of more than 10 Hz. Fusing time-of-arrival (TOA), time-difference-of-arrival (TDOA), and angle-of-arrival (AOA) localization methods improves measurement precision and reliability in secondary radar setups with multiple stations [1–5].

A drawback of this system concept is that every participating radar station needs to provide a complex base band processing system for receive signal evaluation and transmit signal generation, even if a station acts only as a static reference. For the latter, a simple backscatter transponder with modulated response is acceptable and can be realized with much less effort. Several architectures for radar backscatter transponders, such as simple passive systems with modulated antenna base impedance [6], active transponders with heterodyne or homodyne down conversion [7], or active backscatters based on direct modulation of the RF signal, have been published in the past [8].

An alternative circuit for realizing active backscatter transponders is the super regenerative oscillator [9], whose internal output signal is radiated back over the receive antenna. By switching the oscillator on and off periodically, the signal appears as a virtually coherent response to the radar system. This concept was previously published as the switched-injection locked oscillator (SILO) in [10,11]. It is a simple but high-performance circuit for creating miniaturized, active backscatter transponders for radar localization systems [12]. The SILO was used for both localization and communication and a series of realizations for a great number of frequencies are available in the literature. Figure 1 summarizes a series of related publications [13–21].

Figure 2 shows the general secondary radar setup with a SILO-based backscatter transponder as published in the past [13,14]. The fact that transient oscillator behavior severely degrades the FMCW range measurement accuracy is ignored in prior publications. Further, exactly how the transponder response beat signal emerges in an FMCW radar system was not satisfactorily established.

We discuss the SILO as a linear feedback oscillator and derive the dependence of its oscillation amplitude and phase on the incidental radar signal in the following section. In Section 3, we show how to use this phase dependence to create virtually coherent, regenerative backscatter transponders for FMCW radar systems, and discuss some of the disadvantages of this system. A formally published SILO transponder system is presented in detail in Section 4, along with a lab characterization and field measurements. The transponder was used in our paper published at the 2019 IEEE International Workshop on Metrology for AeroSpace (MetroAeroSpace) [15]. In the last section, we outline the future trajectory of robot navigation and SILO-based backscatter transponders.

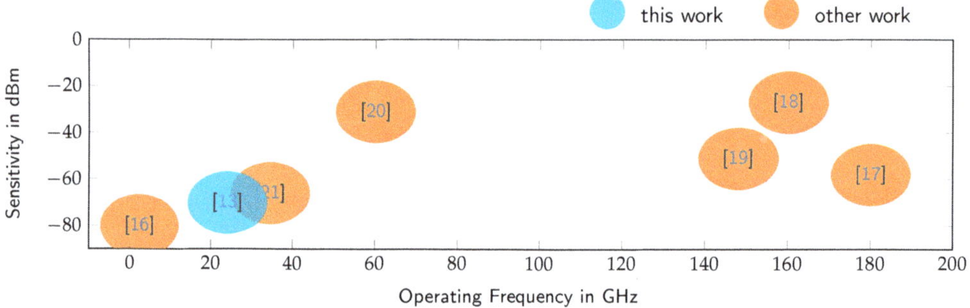

Figure 1. Previously published implementations of the switched-injection locked oscillator.

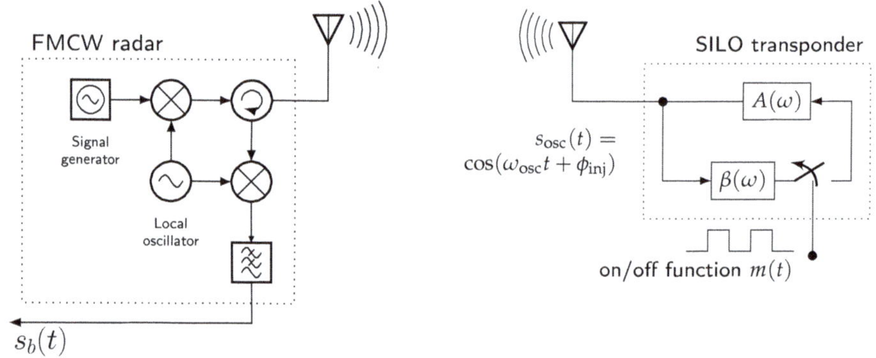

Figure 2. The general setup of a switched injection-locked transponder and an FMCW radar system.

2. Switched Oscillators as Virtually Coherent Transponders

In this section, we review some fundamental equations relating to the super regenerative oscillator and consider, in particular, its sensitivity to the receiving signal phase. A good grounding in these fundamentals is needed to understand how this concept can be used to build coherent transponders for secondary radar applications in a straightforward manner.

2.1. The Linear Feedback Oscillator

As a basis for super regenerative receiving, we consider the linear feedback oscillator, which consists of an amplifier and a linear system $F(s)$ in the feedback path as shown in Figure 3. The linear system is typically a bandpass filter at the desired oscillation frequency ω_0 with bandwidth B. The oscillator output is given by

$$s_o(t) = g\left(s_o(t) * f(t) + s_i(t)\right). \tag{1}$$

This equation is transformed into the Laplace domain where it can be analyzed easily. After Laplace transformation, Equation (1) reads as follows:

$$S_o(s) = g\left(S_o(s) \cdot F(s) + S_i(s)\right). \tag{2}$$

By reordering according to $S_o(s)$, this leads to the well-known transfer function for feedback systems:

$$S_o(s) = \frac{S_i(s)}{1 - gF(s)}. \tag{3}$$

Considering $F(s)$ as a second-order bandpass filter with bandwidth B_f, center frequency ω_0 and passband attenuation α, given by

$$F(s) = \alpha \frac{2\pi B_f \frac{s}{\omega_0^2}}{\frac{s^2}{\omega_0^2} + 2\pi B_f \frac{s}{\omega_0^2} + 1} = \alpha \frac{Bs}{s^2 + Bs + \omega_0^2}, \quad B = 2\pi B_f, \tag{4}$$

Equation (3) is written as

$$S_o(s) = \frac{S_i(s)}{1 - \alpha g \frac{Bs}{s^2+Bs+\omega_0^2}} = S_i(s) \frac{s^2 + Bs + \omega_0^2}{s^2 + Bs(1-D) + \omega_0^2}, \quad D = \alpha g. \tag{5}$$

By exciting the feedback system with a Dirac impulse, i.e., $S_i(s) = 1$, the impulse response is obtained:

$$\begin{aligned} S_o(s) &= \frac{s^2 + Bs + \omega_0^2}{s^2 + Bs(1-D) + \omega_0^2} \\ &= \frac{s^2}{s^2 + Bs(1-D) + \omega_0^2} + \frac{Bs}{s^2 + Bs(1-D) + \omega_0^2} + \frac{\omega_0^2}{s^2 + Bs(1-D) + \omega_0^2}. \end{aligned} \tag{6}$$

The latter is transformed to the time domain by well-known transformation pairs. Considering $D = 1$, Equation (6) is written as

$$S_o(s)\Big|_{D=1} = \frac{s^2}{s^2 + \omega_0^2} + \frac{Bs}{s^2 + \omega_0^2} + \frac{\omega_0^2}{s^2 + \omega_0^2} \quad , \tag{7}$$

which, after inverse Laplace transformation, is written in the time domain as

$$\begin{aligned} s_o(t)\Big|_{D=1} &= -\sin(\omega_0 t)\varepsilon(t) + B\cos(\omega_0 t)\varepsilon(t) + \sin(\omega_0 t)\varepsilon(t) \\ &= B\cos(\omega_0 t)\varepsilon(t). \end{aligned} \tag{8}$$

Equation (8) describes a stable oscillation at the center frequency of the bandpass filter. More interestingly, the oscillation amplitude is a function of the filter bandwidth. This is because more energy from the broad band Dirac impulse is injected into the system, which continues oscillating without damping.

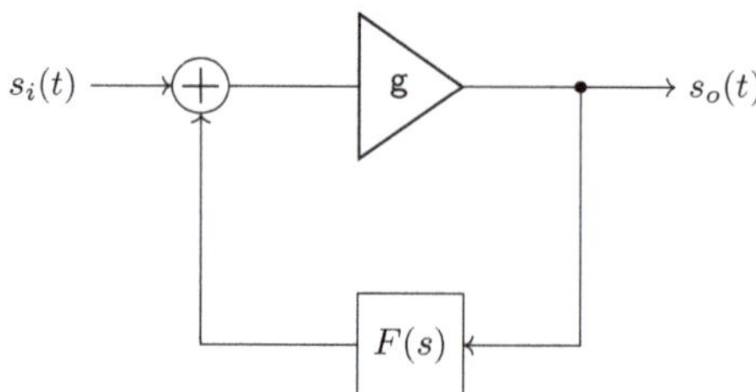

Figure 3. The input–output relationship of the linear feedback oscillator.

For the general case $D \in \mathbb{R}$, the amplitude of the oscillator output signal is either stable, rising, or damping, depending on the ratio of α and g:

$$S_o(s) = \frac{s^2}{s^2 + Bs(1-D) + \omega_0^2} + \frac{Bs}{s^2 + Bs(1-D) + \omega_0^2} + \frac{\omega_0^2}{s^2 + Bs(1-D) + \omega_0^2} \tag{9}$$

$$\bullet\!\!-\!\!\circ$$

$$\begin{aligned} s_o(t) &= \frac{e^{-B(1-D)t/2}}{\omega_M}\left(\left(\frac{1}{4}B^2(1-D)^2 - \omega_M^2\right)\sin(\omega_M t) - B(1-D)\omega_M\cos(\omega_M t)\right) \\ &+ \frac{B}{\omega_M}e^{-B(1-D)t/2}\left(\left(-\frac{B(1-D)}{2}\right)\sin(\omega_M t) + \omega_M\cos(\omega_M t)\right) \\ &+ \frac{\omega_0^2}{\omega_M}e^{-B(1-D)t/2}\sin(\omega_M t) \\ &= \frac{e^{-B(1-D)t/2}}{\omega_M}\left(\left(\omega_0^2 - \omega_M^2 + \frac{1}{4}B^2(1-D)^2 - \frac{1}{2}B(1-D)\right)\sin(\omega_M t)\right. \\ &\quad \left. + \omega_M\left(1 - B(1-D)\right)\cos(\omega_M t)\right) \\ &= e^{-B(1-D)t/2}\left(A_{\sin}\sin(\omega_M t) + A_{\cos}\cos(\omega_M t)\right). \end{aligned} \tag{10}$$

With the modified resonant frequency,

$$\omega_M = \sqrt{\omega_0^2 - \left(\frac{B(1-D)}{2}\right)^2}. \tag{11}$$

It is clear that, for $D \to 1$, Equation (10) converges to the undamped oscillation with fixed frequency described by Equation (8). This happens in practice when the amplifier is saturated and the oscillator reaches its maximum output power.

2.2. Signal-Injection Into the Linear Feedback Oscillator

In the above subsection, we examine the output signal of the linear feedback oscillator only for an exciting pulse, i.e., $S_i(s) = 1$. Another interesting case is the excitation by a harmonic signal with a specific frequency ω_i and phase φ_i, which causes the oscillation to start. Here, the harmonic signal may force the feedback oscillator to its injecting frequency ω_i in the case of high input amplitude. In the case of weak input signals, they are just settling the starting condition for the oscillation (amplitude and phase). The latter, used as a receiver, is known as the super regenerative receiver, as its output signal depends directly on the phase and amplitude of the injected signal as show in Figure 4.

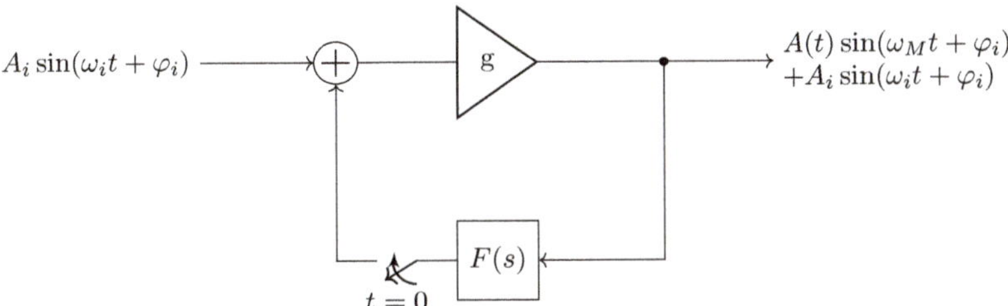

Figure 4. By exciting the linear feedback oscillator with a harmonic signal, the phase of the output signal is coherent to the input at $t = 0$.

The Laplacian of a harmonic input signal with arbitrary frequency ω_i, phase φ_i, and amplitude A_i is given by

$$\begin{aligned} s_i(t) &= A_i \sin(\omega_i t + \varphi_i) \\ &= A_i \sin(\omega_i t)\cos(\varphi_i) + A_i \cos(\omega_i t)\sin(\varphi_i) \end{aligned} \tag{12}$$

$$S_i(s) = A_i \frac{\omega_i}{s^2 + \omega_i^2} \cos(\varphi_i) + A_i \frac{s}{s^2 + \omega_i^2} \sin(\varphi_i). \tag{13}$$

For this case, we reevaluate Equation (5), which yields

$$S_{o,inj}(s) = \frac{s^2 + Bs + \omega_0^2}{s^2 + Bs(1-D) + \omega_0^2} \cdot A_i \left(\frac{\omega_i}{s^2 + \omega_i^2} \cos(\varphi_i) + \frac{s}{s^2 + \omega_i^2} \sin(\varphi_i) \right). \tag{14}$$

The inverse transformation reads as follows:

$$\begin{aligned} s_{o,inj}(t) = {} & A_{\text{osc}} A_i e^{-B(1-D)t/2} \left(A_{\text{osc,sin}}(\varphi_i) \sin(\omega_M t) + A_{\text{osc,cos}}(\varphi_i) \cos(\omega_M t) \right) \\ & + A_i \left(A_{i,\cos}(\varphi_i) \cos(\omega_i t) + A_{i,\sin}(\varphi_i) \sin(\omega_i t) \right), \end{aligned} \tag{15}$$

again with the modified resonant frequency

$$\omega_M = \sqrt{\omega_0^2 - \left(\frac{B(1-D)}{2} \right)^2} \tag{16}$$

and with the injection-phase dependant amplitudes $A_{\text{osc,sin}}(\varphi_i), A_{\text{osc,cos}}(\varphi_i), A_{\text{i,sin}}(\varphi_i), A_{\text{i,cos}}(\varphi_i)$, which can be approximated, as $\omega_M \approx \omega_0$ and $\omega_0 \approx \omega_i$, with very good accuracy by

$$\begin{aligned} A_{\text{osc}} &= \frac{1}{\omega_i{}^4 + (-2\omega_0{}^2 + B^2 D^2 - 2B^2 D + B^2)\,\omega_i{}^2 + \omega_0{}^4} \\ &= \frac{1}{(B^2 D^2 - 2B^2 D + B^2)\omega_i^2} \end{aligned} \tag{17}$$

$$\begin{aligned} A_{\text{osc,sin}}(\varphi_i) &= \Big(\Big(\Big(-2BD\,\omega_0{}^2 + B^3\,D^3 - 2B^3\,D^2 + B^3D\Big)\,\omega_i{}^2 + 2BD\,\omega_0{}^4\Big)\sin(\varphi_i) \\ &+ \Big(\Big(B^2\,D^2 - B^2D\Big)\,\omega_i{}^3 + \Big(B^2\,D^2 - B^2D\Big)\,\omega_0{}^2\omega_i\Big)\cos(\varphi_i)\Big) \\ &/ \Big(\sqrt{4\omega_0{}^2 - B^2\,D^2 + 2B^2D - B^2}\Big) \\ &\approx \Big(B^2\,D^2 - B^2D\Big)\,\omega_i{}^2\cos(\varphi_i) = \widetilde{A}_{\text{osc}}\cos(\varphi_i) \end{aligned} \tag{18}$$

$$\begin{aligned} A_{\text{osc,cos}}(\varphi_i) &= \Big(\Big(B^2\,D^2 - B^2D\Big)\,\omega_i{}^2\sin(\varphi_i) + \Big(BD\,\omega_i{}^3 - BD\,\omega_0{}^2\omega_i\Big)\cos(\varphi_i)\Big) \\ &\approx \Big(B^2\,D^2 - B^2D\Big)\,\omega_i{}^2\sin(\varphi_i) = \widetilde{A}_{\text{osc}}\sin(\varphi_i) \end{aligned} \tag{19}$$

$$\begin{aligned} A_{\text{i,sin}}(\varphi_i) &= \frac{(BD\,\omega_i{}^3 - BD\,\omega_0{}^2\omega_i)\sin(\varphi_i) + (\omega_i{}^4 + (-2\omega_0{}^2 - B^2D + B^2)\,\omega_i{}^2 + \omega_0{}^4)\cos(\varphi_i)}{\omega_i{}^4 + (-2\omega_0{}^2 + B^2\,D^2 - 2B^2D + B^2)\,\omega_i{}^2 + \omega_0{}^4} \\ &\approx \cos(\varphi_i) \end{aligned} \tag{20}$$

$$\begin{aligned} A_{\text{i,cos}}(\varphi_i) &= \frac{((\omega_i{}^4 + (-2\omega_0{}^2 - B^2D + B^2)\,\omega_i{}^2 + \omega_0{}^4)\sin(\varphi_i) + (BD\,\omega_0{}^2\omega_i - BD\,\omega_i{}^3)\cos(\varphi_i))}{\omega_i{}^4 + (-2\omega_0{}^2 + B^2\,D^2 - 2B^2D + B^2)\,\omega_i{}^2 + \omega_0{}^4} \\ &\approx \sin(\varphi_i) \end{aligned} \tag{21}$$

Based on the previous approximations, the solution of the time domain signal reads as follows:

$$\begin{aligned} s_{o,inj}(t) &\approx A_{\text{osc}}\widetilde{A}_{\text{osc}}A_i e^{-B(1-D)t/2}\left(\cos(\varphi_i)\sin(\omega_M t) + \sin(\varphi_i)\cos(\omega_M t)\right) \\ &+ A_i\left(\sin(\varphi_i)\cos(\omega_i t) + \cos(\varphi_i)\sin(\omega_i t)\right) \\ &= \overline{A}_{\text{osc}}A_i e^{-B(1-D)t/2}\sin(\omega_M t + \varphi_i) + A_i\sin(\omega_i t + \varphi_i). \end{aligned} \tag{22}$$

Equation (22) shows a very important result for super regenerative receivers, namely, that the oscillation of the feedback linear oscillator depends on the injected signal's amplitude and phase. In particular, the oscillation is coherent to the injected signal at the time of excitation $t = 0$.

2.3. Simulation Results

To confirm the previous derivation, the time domain behavior of the switched linear feedback oscillator was simulated. We used a realistic parameter set, i.e.

$$\begin{aligned} f_{\text{osc}} &= 24.125\,\text{GHz}, \\ f_{\text{inj}} &= 24\,\text{GHz}, \\ B_f &= 250\,\text{MHz}, \\ D &= 10. \end{aligned}$$

Figure 5 shows the simulation results and, as expected, the oscillator is coherent to the injection signal at $t = 0$. The oscillation amplitude quickly rises and stays virtually coherent to the radar signal. In real applications, the oscillator is switched off after a short period and switched on again after full decay.

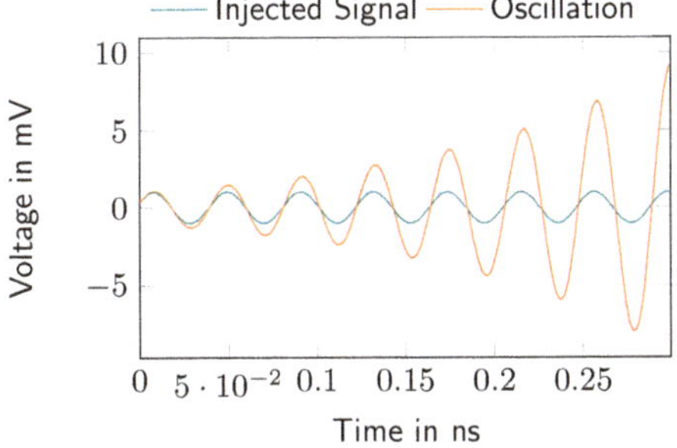

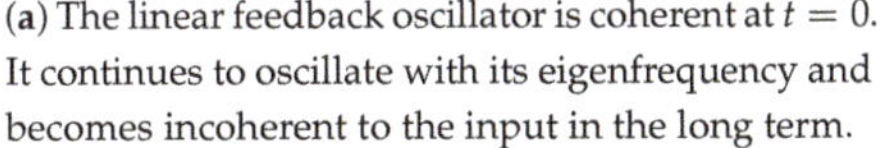
(a) The linear feedback oscillator is coherent at $t = 0$. It continues to oscillate with its eigenfrequency and becomes incoherent to the input in the long term.

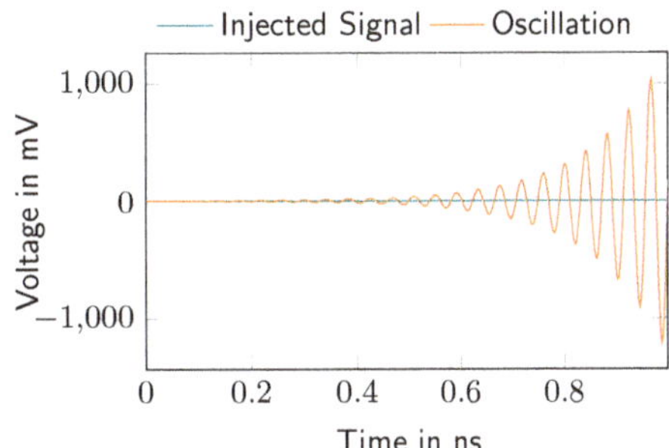

(b) After a short period, the amplitude of the oscillation rises several orders of magnitude above the input signal.

Figure 5. Simulation results for the linear feedback oscillator with harmonic signal injection.

3. Secondary Radar Beacons Based on Switched Oscillators

Based on the results from the previous section, we now show how the switched oscillator is used as a modulated backscatter transponder for secondary radar applications, also known as the switched-injection locked oscillator (SILO). For this, we review the SILO from [10] and derive a more sophisticated signal model for the case of an FMCW radar localization system. Here, we use several simplifications, which hold in real applications with very high accuracy.

3.1. Assumptions

We assume that the amplitude A_i of the injected signal is negligible, i.e.,

$$\begin{aligned} A_i &\ll \overline{A}_{\text{osc}} A_i e^{-B(1-D)t/2}, \quad t > 0, \\ \Leftrightarrow 1 &\ll \overline{A}_{\text{osc}} e^{-B(1-D)t/2}, \quad t > 0, \end{aligned} \tag{23}$$

which is reasonable in radar applications with mid to large ranges. Thus, we write the oscillator signal of Equation (22) as

$$s_{\text{o,inj}}(t) = \overline{A}_{osc} A_i e^{-B(1-D)t/2} \cdot \sin(\omega_M t + \varphi_i). \tag{24}$$

For the initial signal model, we even omit the exponential envelope, thus the oscillator output signal is modeled by

$$s_{\text{osc}}(t) = \sin(\omega_M t + \varphi_i). \tag{25}$$

Further, we set

$$\omega_M \approx \omega_0 = \omega_{\text{osc}} \tag{26}$$

as for real applications, the center frequency of the oscillator varies only about a few MHz in the 24 GHz or 77 GHz operation bands.

3.2. The Beat Signal of a Single Oscillator Cycle

In the following section, we want to derive the beat signal of a SILO transponder in an FMCW radar during a single oscillator on-cycle, as shown in Figure 6. For simplicity, we use IQ signals without loss of generality.

An FMCW radar system transmits a linear chirp with rate $\mu = B_f/T$, where B_f is the chirp bandwidth and T the sweep interval. The transmit signal is given by

$$s_{\text{tx}}(t) = \exp\{j(\omega_{\text{tx}} + \pi\mu t)t\}. \tag{27}$$

The transponder receives the radar signal transmitted at time t_0 after the delay time τ_{tx} and is switched on at time t_{on}. Without loss of generality, we set $t_0 + \tau_{\text{tx}} = t_{\text{on}}$. The oscillator turns on with the phase of the received radar signal, the injected phase

$$\begin{aligned} \varphi_{\text{inj}}(t_{\text{on}}) &= \arg\{s_{\text{tx}}(t_{\text{on}} - \tau_{\text{tx}})\} \\ &= \arg\{s_{\text{tx}}(t_0)\} \\ &= (\omega_{\text{tx}} + \pi\mu t_0)t_0. \end{aligned} \tag{28}$$

Thus, the SILO transmit signal during a single on-cycle with duration T_s is given by Equation (25)

$$\begin{aligned} s_{\text{osc}}(t) &= \exp\{j\omega_{\text{osc}}(t - t_{\text{on}}) + j\varphi_{\text{inj}}(t_{\text{on}})\} \\ &= \exp\{j\omega_{\text{osc}}(t - t_{\text{on}}) + j(\omega_{\text{tx}} + \pi\mu t_0)t_0\} \\ t &\in [t_{\text{on}}; t_{\text{off}}]. \end{aligned} \tag{29}$$

The transponder signal $s_{\text{osc}}(t)$ is delayed by τ_{rx} and received by the radar system

$$\begin{aligned} s_{\text{rx}}(t) = &\quad s_{\text{osc}}(t - \tau_{\text{rx}}) \\ = &\quad \exp\{j\omega_{\text{osc}}(t - t_{\text{on}} - \tau_{\text{rx}}) + j(\omega_{\text{tx}} + \pi\mu t_0)t_0\} \\ t \in &\quad [t_{\text{rx}}; t_{\text{rx}} + T_s]. \end{aligned} \tag{30}$$

In the FMCW receiver, $s_{\text{rx}}(t)$ is down-converted with the local oscillator signal $s_{\text{tx}}(t)$

$$\begin{aligned} s_{\text{b}}(t) = &\quad s_{\text{tx}}(t) \cdot s_{\text{rx}}^*(t) \\ = &\quad \exp\{j(\omega_{\text{tx}} + \pi\mu t)t - j\omega_{\text{osc}}(t - t_{\text{on}} - \tau_{\text{rx}}) - (\omega_{\text{tx}} + \pi\mu t_0)t_0\} \\ t \in &\quad [t_{\text{rx}}; t_{\text{rx}} + T_s]. \end{aligned} \tag{31}$$

Equation (31) is the beat signal model of a single oscillator on-cycle within the FMCW sweep. At the time of reception $t_{\text{rx}} = t_0 + \tau_{\text{tx}} + \tau_{\text{rx}} = t_0 + \tau_{\text{rt}}$, this evaluates to

$$\begin{aligned} s_{\text{b}}(t_{\text{rx}}) &= s_{\text{b}}(t_0 + \tau_{\text{rt}}) \\ &= \exp\{j(\omega_{\text{tx}} + \pi\mu(t_0 + \tau_{\text{rt}}))(t_0 + \tau_{\text{rt}}) - j\omega_{\text{osc}}(t_0 + \tau_{\text{rt}} - t_{\text{on}} - \tau_{\text{rx}}) - j(\omega_{\text{tx}} + \pi\mu t_0)t_0\} \\ &= \exp\{j(\omega_{\text{tx}}t_0 + \omega_{\text{tx}}\tau_{\text{rt}} + \mu\pi(t_0^2 + 2t_0\tau_{\text{rt}} + \tau_{\text{rt}}^2) - \omega_{\text{tx}}t_0 - \pi\mu t_0^2)\} \\ &= \exp\{j(2\pi\mu\tau_{\text{rt}}t_0 + \omega_{\text{tx}}\tau_{\text{rt}} + \pi\mu\tau_{\text{rt}}^2)\} \\ &\approx \exp\{j(2\pi\mu\tau_{\text{rt}}t_0 + \omega_{\text{tx}}\tau_{\text{rt}})\}. \end{aligned} \tag{32}$$

Equation (32) shows that the beat signal of the SILO transponder at $t_{\text{rx}} = t_0 + \tau_{\text{rt}}$ has the same phase as a signal of a passive scatterer with the same round-trip delay time τ_{rt}. Within the oscillator on-time T_s, a specific bandwidth μT_s is traversed:

$$\begin{aligned} s_{\text{b}}(t_{\text{rx}} + T_s) &= s_{\text{b}}(t_0 + (\tau_{\text{rt}} + T_s)) \\ &= \exp\{j(2\pi\mu(\tau_{\text{rt}} + T_s)t_0 + \omega_{\text{tx}}(\tau_{\text{rt}} + T_s) + \pi\mu(\tau_{\text{rt}} + T_s)^2 - \omega_{\text{osc}}T_s)\} \\ &\approx s_{\text{b}}(t_0 + \tau_{\text{rt}}) \cdot \exp\{j(2\pi\mu T_s t_0 + \omega_{\text{tx}}T_s - \omega_{\text{osc}}T_s) \\ &= s_{\text{b}}(t_0 + \tau_{\text{rt}}) \cdot \exp\{j(2\pi\mu T_s t_0 + \Delta\omega T_s). \end{aligned} \tag{33}$$

It also shows up that the beat signal's phase is a function of the oscillator on-time T_s and of the offset $\Delta\omega$ of the transmit frequency ω_{tx} and oscillator frequency ω_{osc}.

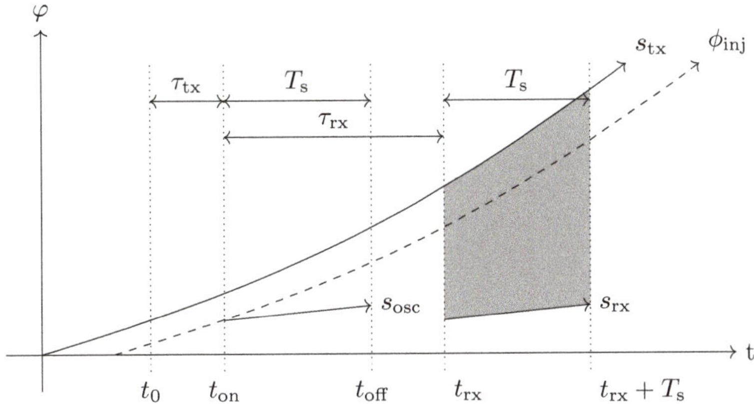

Figure 6. The instantaneous phases and induced bandwidth in the FMCW base band within one oscillator on-cycle.

3.3. The Beat Signal of the Entire Frequency Sweep

Equations (32) and (33) describe the beat signal during a single oscillator on-cycle. Within the entire frequency sweep, the beat signal consists of multiple on-cycles at discrete times t_i, where in each cycle the signal is equal to a passive scatterer for the singular moment t_{rx} and induces an additional frequency bandwidth $2\pi\mu T_s$ until the oscillator is switched off. This is interpreted as a discrete time sampling process of a frequency band with bandwidth $2\pi\mu T_s$, starting at the actual beat frequency $2\pi\mu\tau_{rt}$ with phase $\omega_{tx}\tau_{rt}$ and stopping with an additional phase $\Delta\omega T_s$. Spectral replicas, which need to be filtered in the baseband, appear at frequencies $1/T_s$ due to the discrete time sampling. During the oscillator on-time, the beat signal is a chirp starting from the beat-frequency $2\pi\mu\tau_{rt}$ with bandwidth $2\pi\mu T_s$. This chirp is approximated in the frequency domain with good accuracy by a rect-function with the same bandwidth. The reconstruction filter for the sampled signal is a lowpass filter with a cutoff frequency of at least the on/off frequency $1/T_s$. This is normally realized in real FMCW systems by the anti-aliasing filter before the analog-to-digital conversion stage.

As a result, the frequency spectrum of the beat signal of a SILO transponder within a linear FMCW ramp is written (with the window function $W(\omega)$) as:

$$\begin{aligned} S_b(\omega) &= W_T(\omega) * \text{rect}\left(\frac{\omega - 2\pi\mu\tau_{rt} - 2\pi\mu T_s/2}{2\pi\mu T_s}\right) \cdot \exp\left\{j\left(\omega_{tx}\tau_{rt} + \Delta\omega T_s \frac{\omega - 2\pi\mu\tau_{rt}}{2\pi\mu T_s}\right)\right\} \\ &= W_T(\omega) * \text{rect}\left(\frac{\omega}{2\pi\mu T_s}\right) e^{j\omega\frac{\Delta\omega}{2\pi\mu}} * \delta\left(\omega - 2\pi\mu\tau_{rt} - 2\pi\mu T_s/2\right) e^{j\omega_{tx}\tau_{rt} - j\Delta\omega\tau_{rt}}. \end{aligned} \tag{34}$$

The following time-domain model is obtained from Equation (34) (considering $\omega_{tx}\tau_{rt} - \Delta\omega\tau_{rt} = \omega_{osc}\tau_{rt}$):

$$s_b(t) = w_T(t) \cdot \text{sinc}\left(\frac{2\pi\mu T_s(t - \frac{\Delta\omega}{2\pi\mu})}{2}\right) \cdot \exp\left\{j2\pi\left(\mu\tau_{rt} + \frac{\mu T_s}{2}\right)t + j\omega_{osc}\tau_{rt}\right\} \tag{35}$$

For simplicity, all amplitude values are normalized to 1 as they carry no relevant information.

If we assume that ω_{osc} is in the middle of the operating band (what is preferable for real applications) and if $t \in [0; T]$, the delta $\Delta\omega$ from the oscillator frequency to the FMCW start frequency is given by:

$$\Delta\omega = 2\pi\mu\frac{T}{2}. \tag{36}$$

By modulating the amplitude of the transponder output signal, either by on/off keying or by controlling the output power of the oscillator with a periodic function with frequency ω_m, the beat spectrum is shifted to an arbitrary center frequency. This allows the reader to distinguish the transponder signal clearly from primary radar targets and clutter. Thus, the beat signal is written as

$$s_b(t) = w_T(t) \cdot \text{sinc}\left(\frac{2\pi\mu T_s(t - T/2)}{2}\right) \cdot \exp\left\{j2\pi\left(\mu\tau_{\text{rt}} + \frac{\mu T_s}{2}\right)t + j\omega_{\text{osc}}\tau_{\text{rt}}\right\} \cdot \cos\left(\omega_m t + \varphi_m\right). \tag{37}$$

Figure 7 shows the measured beat frequency spectrum of a real transponder with modulated reflection.

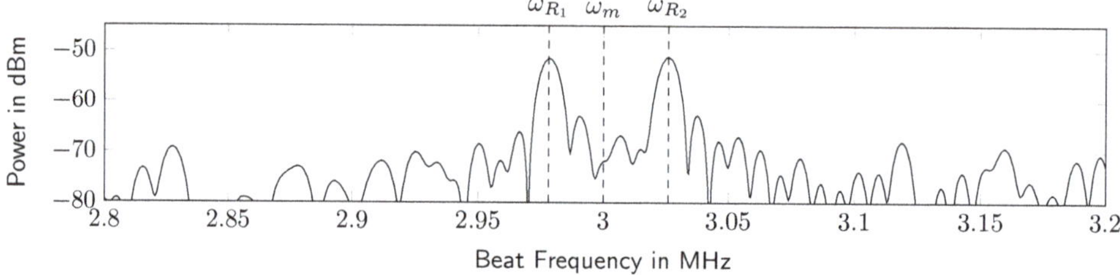

Figure 7. The transponder signal spectrum in the radar baseband. The center frequency ω_m is determined by the amplitude modulation of the transponder signal.

3.4. Measuring the Absolute Distance to a SILO Transponder

By transforming a measured beat signal into the frequency domain, the signal round trip time τ_{rt} and thus the transponder distance d is measured by the frequency difference between the two AM sidebands, as shown in Figure 7:

$$d_\omega = c\frac{\Delta\omega_{\text{R}}}{4\mu} - c\frac{T_s}{4}, \quad \Delta\omega_{\text{R}} = \omega_2 - \omega_1. \tag{38}$$

In practical systems, the correction term $c\frac{T_s}{4}$ depends on the transient behavior of the oscillator and cannot be calibrated accurately, because the latter is a function of the temperature and especially of the strength of the injected power, which depends on the other hand on the transponder distance and on the orientation because of the antenna pattern. It is good practice to calibrate the distance error at the middle of the operational range and accept the range inaccuracy. Below, in the Outlook Section 6, we present some ideas on how to include the range error correction term in a multi-static radar setup.

4. Realization of a Miniaturized Secondary Radar Beacon

In this section, we give a detailed description of a previously presented, miniaturized secondary radar transponder based on the switched injection locked oscillator in the 24 GHz ISM-Band. The transponder comprises of a four-layer PCB, whose dimensions are 35 mm × 15 mmand which uses a standard FR4 combined with Rogers RO4003 as a layer stack. The Rogers substrate exhibits high RF performance because of its homogeneous electrical permittivity and low dielectric losses; it is used to realize the microwave oscillator in microstrip technology.

4.1. Overview

Figure 8 gives an overview of the essential devices on the top and bottom layers of the transponder hardware. The actual switched-injection locked oscillator, consisting of a feedback amplifier in the form of a microwave field-effect transistor of type CEL CE3520K3, is located on the top layer. The SILO is described in detail in Section 4.2. The RF signals are fed over an SMP connection from and into the system. SMP provides a low-cost but performant RF link at 24 GHz without the need for bolting. This is a small footprint feature. A chip of type MAX8663 provides power management for the transponder. It converts the primary voltages to the target voltages. It also maintains a balance between the USB power supply and the battery, whereby the latter is charged automatically when there is sufficient USB power. The RF oscillator voltage is produced apart from the peripheral supply because it needs a high degree of linearity along with low noise. For this, a linear regulator of type LT3042 is used, which provides a high PSRR of about 80 dB between 0 and 2 MHz. A good PSRR number is crucial, as any amplitude modulation on the power supply will create unexpected sidebands in the RF oscillation and thus ghost targets in the radar base band. Furthermore, the power supply noise affects the oscillator's spectral purity and its sensitivity to the receive signal's phase, which is directly related to the SNR of the transponder response in the base band. Finally, there are minor devices on the top layer such as a balun to create a single-ended modulation signal out of the symmetric DDS output.

All transponder digital periphery is located on the bottom layer. An STM32L100 microcontroller and an AD9102 DDS contain the transponder response modulation IP. Further, the STM32L100 provides a USB interface for online modulation parameter configuration. The transponder can be powered either by USB or by a LiPo battery, connected through a PHR2 socket.

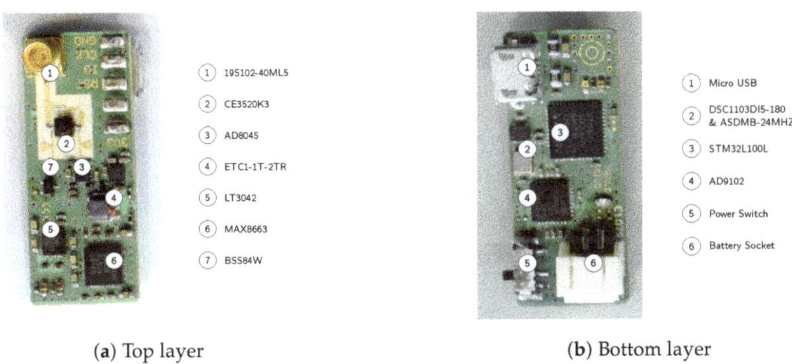

(**a**) Top layer (**b**) Bottom layer

Figure 8. The most important parts of the transponder on the top and bottom layers.

4.2. The SILO Core

The switched-injection locked oscillator is built, as shown in Figure 9, of an amplifier with positive feedback at the target frequency, which is the center of the 24 GHz ISM band in this case. The amplifier is realized by a GaAs RF FET of type CE3520K3 as a grounded-emitter circuit. The latter provides the highest power amplification among transistor circuits and is convenient for realizing the oscillator.

The periodic on/off switching action is realized by push/pulling the oscillator's supply voltage. The push/pull operation is implemented by a second p-channel FET. By pulling the supply voltage, the oscillator starves and becomes ready for the next on-cycle very quickly. Thus, a fast switching frequency can be realized by maintaining the phase sensitivity at the same time. The transistor gate voltage is defined at the switch-on time solely by the injected signal and thermal noise. By pushing the supply voltage, the loop gain becomes positive and the oscillation starts with the receive signal phase, as explained in Section 2. The oscillation is radiated over the antenna and the oscillator is switched off again after a short period.

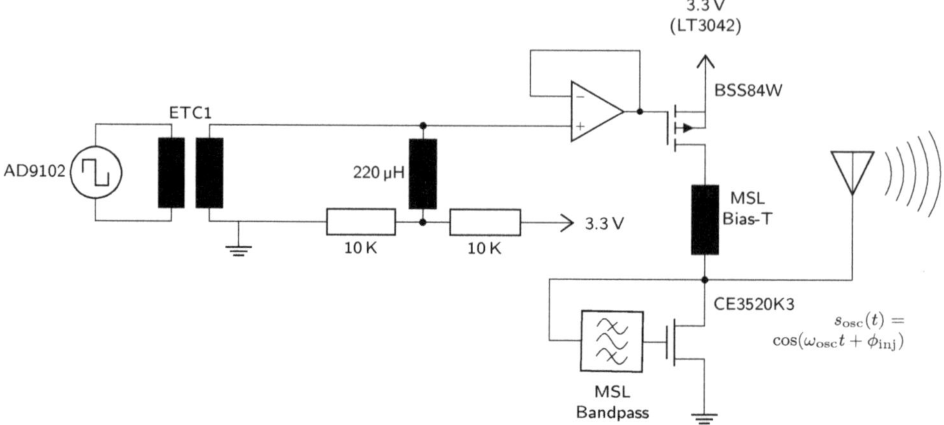

(**a**) The circuit diagram of the SILO core.

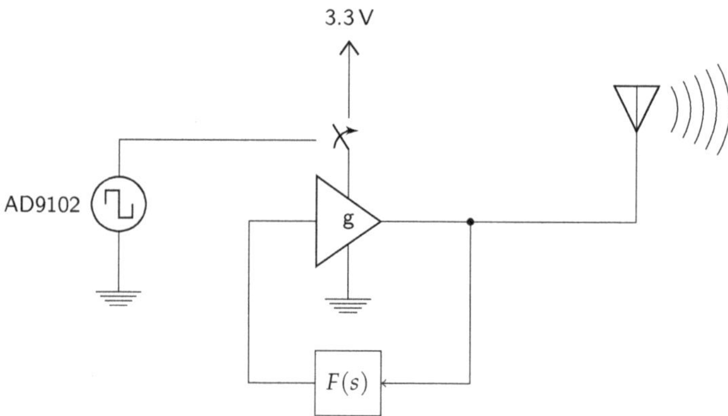

(**b**) The equivalent circuit of the SILO core. It is a feedback amplifier, whose power supply is switched on and off periodically.

Figure 9. The switched injection locked oscillator core of the transponder. The modulation signal is generated by direct digital synthesis.

5. Evaluation of the Transponder

The transponder was tested and assessed both in the lab and in the field, to characterize its performance and test its operation as a secondary radar beacon. First, the transponder was tested *in the loop*, i.e., without antennas, but with a separate signal source to produce the injecting signal, and a high-speed oscilloscope to verify the switching and locking performance. Further, the circuit sensitivity, i.e., the minimum power needed for injection locking, was measured. The transponders were then tested with antennas, but without signal injection and without switching. The steady-state output power and center frequency can be measured easily in this setup. The transponder was tested finally with a real FMCW radar unit in order to verify its function as a radar beacon. The response spectrum was examined in this context and a real distance measurement was made and compared with total station reference data.

5.1. Hardware in the Loop

The transponder was evaluated in the lab with the setup shown in Figure 10. The injection signal was generated by a Rohde & Schwarz SMF100A signal generator, and a high-speed Agilent DSO-X 925504A oscilloscope was connected to the transponder to measure both the injection signal and the transponder response. All three devices were put together through an Agilent 11667C power splitter. The minimum output power of the signal generator is −20 dBm. Series of attenuators were connected before the power splitter in order to reduce the input power.

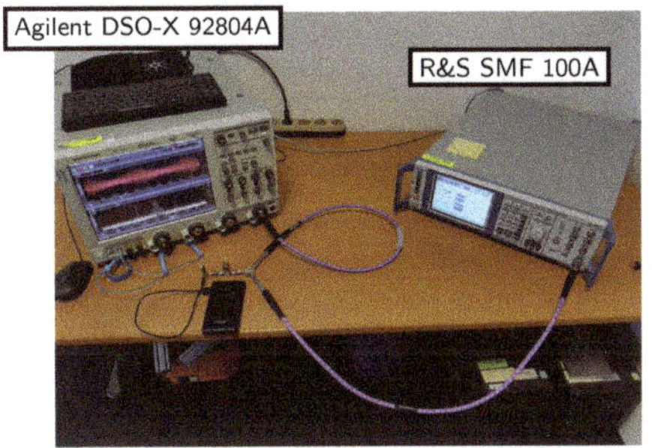

(**a**) A signal source and a high-speed oscilloscope were used to test the transponder.

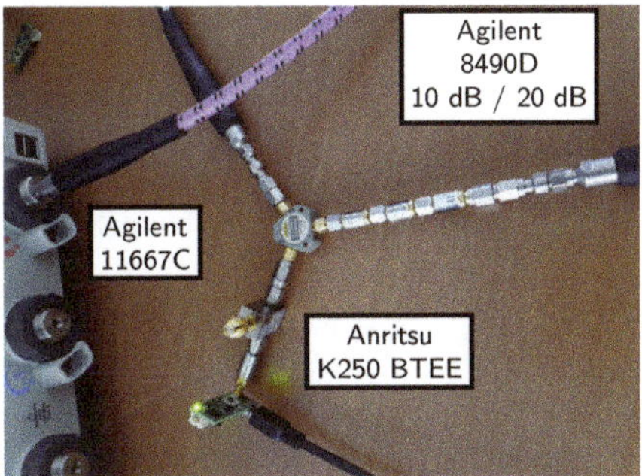

(**b**) Additional attenuators were used to decrease the power of the injection signal.

Figure 10. The setup for measuring the transponder hardware in the loop.

5.1.1. Evaluation of the Free-Running, Non-Switched Oscillator

The transponder was connected to the measurement setup and operated without modulation, only with the oscillator fixed power supply. Thus, the output power and the center frequency can be measured easily. Further, possible spurious signals can be identified and characterized. Figure 11a shows the amplitude spectrum of the measured output signal. There is a stable oscillation at approximately 24.165 GHz with 5 dBm power and within the total spectrum there is no interference higher than −70 dBc. Figure 11b shows a detailed view of the transponder spectrum, which gives the oscillator output power and center frequency. The spectrum was calculated from only a few samples to mitigate the oscillator drift.

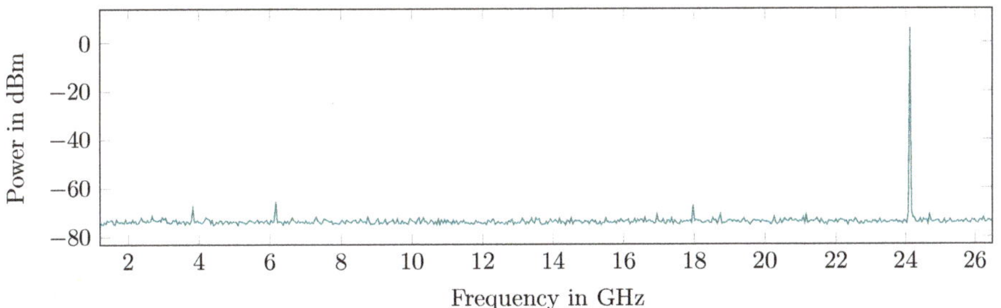

(**a**) The overall spectrum from DC through 26 GHz. The transponder signal is located at the center of the 24 GHz ISM band with a power of about 5 dBm.

Figure 11. *Cont.*

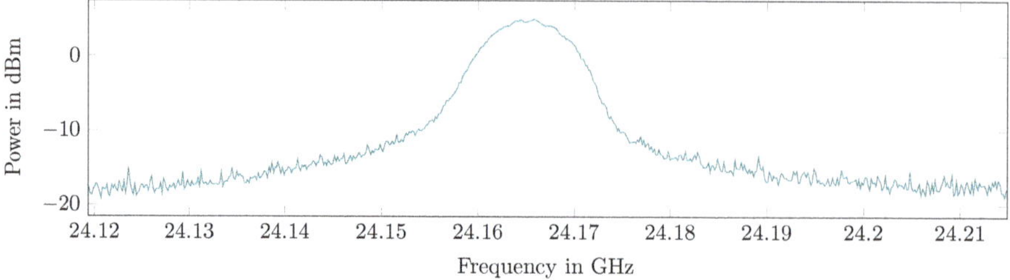

(**b**) The spectrum of the output signal in detail. It was calculated with a very short measurement to mitigate the drift of the free-running oscillator.

Figure 11. The spectrum of the free-running, non-switched oscillator.

5.1.2. Evaluation of the Free-Running, Switched Oscillator

Figure 12a shows the amplitude spectrum of the output signal for a switched oscillator, but without signal injection. It appears that there is no oscillation at a fixed frequency, but a raised noise floor around the oscillator center frequency. This is caused by the fact that, for every oscillator on-cycle, the oscillation is coherent to a random phase and, thus the overall output signal is a band-limited noise signal. It shows the frequency transfer function of the feedback loop and the frequency dependence of the loop gain. An important result is that no self-locking appears, which means that the oscillation of one on-cycle does not disturb the phase of the successive cycle, which gives the SILO its maximum sensitivity.

5.1.3. Evaluation of the Free-Running, Switched Oscillation with Signal Injection

The last evaluation in the lab was the signal injection into the switched oscillator in order to test the coherent oscillation. Since the signal source has a minimum power level of −20 dBm, the signal was damped with a series of attenuators to set the injection power down to −80 dBm. The output spectrum for injection locking is shown in Figure 12c. The oscillator locks virtually to the frequency of the injected signal, because the phase in each on-cycle is coherent at $t = 0$. Sidebands appear at multiples of the switching frequency (90 MHz) because of the periodic on-off-switching. The SILO sensitivity, where injection locking is barely seen, was evaluated by decreasing the injected power. It is evident that the injection locking vanishes for signals below approximately −70 dBm and the noise output, as shown in Figure 12a, remains.

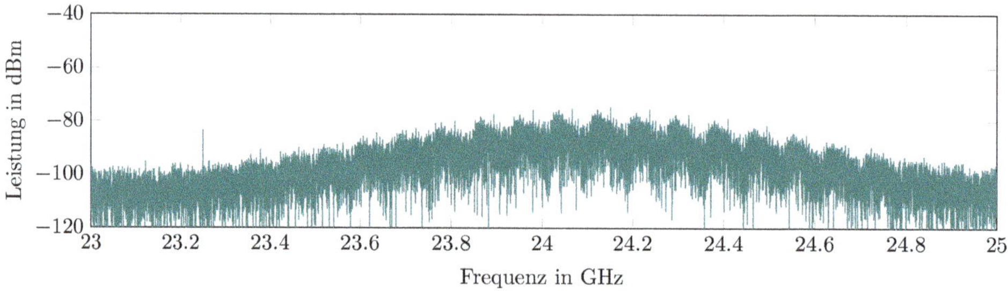

(**a**) The output spectrum of the switched oscillator without signal injection. The output power is spread over the operational bandwidth because of the random phase in every on-cycle. The signal at 23.25 GHz is a spur caused by the oscilloscope.

Figure 12. *Cont.*

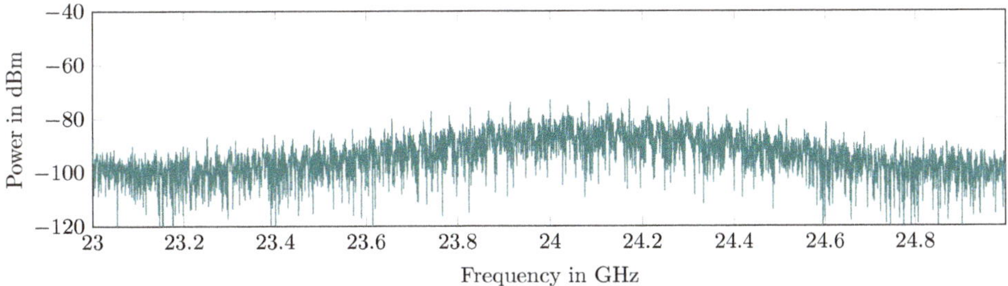

(**b**) The injection locking is barely seen at −70 dBm, as the noise component of the phase is very high here.

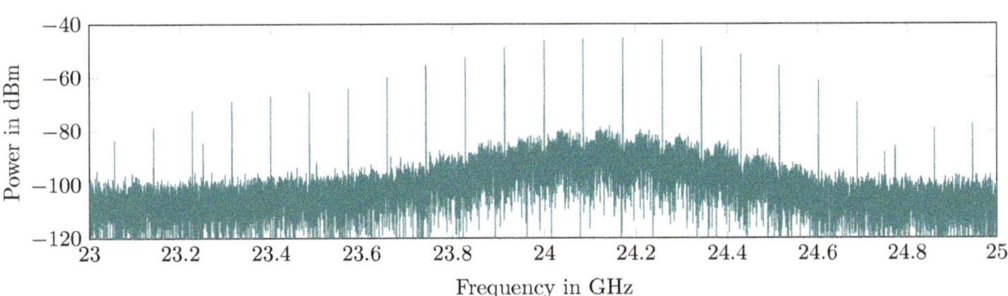

(**c**) The oscillator locks to the injection signal at injection levels above −70 dBm. Further, sidebands at multiples of the switching frequency appear.

Figure 12. The evaluation of the oscillator output signal for different injection power levels.

The time domain output of a single oscillator on-cycle is shown for different injection power levels in Figure 13. In the latter, the coherent start of the oscillation is clearly seen. It also proves that the SILO pulse form depends on the input power, which becomes important for localization applications.

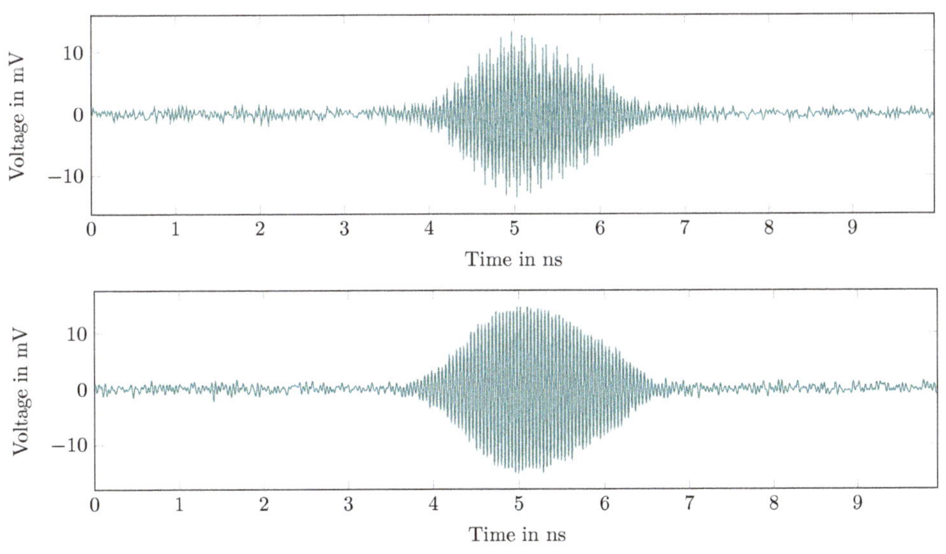

Figure 13. *Cont.*

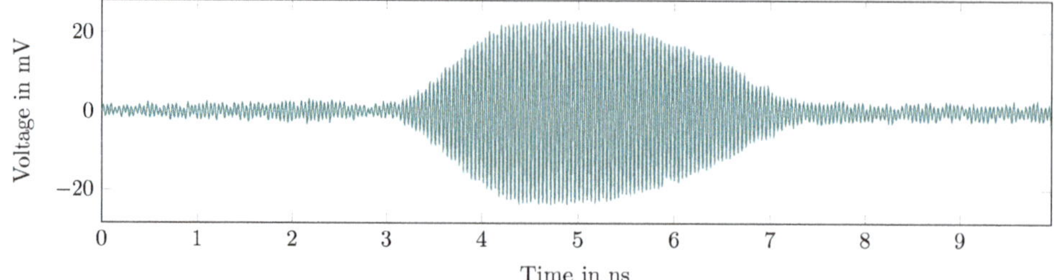

Figure 13. The evaluation of a single oscillator pulse for different injection power levels.

5.2. The Free-Running, Switched and Modulated Transponder Without Injection

In the next step, the transponder hardware was equipped with an antenna, as it is used in the real localization scenario. Additionally, a second antenna was connected to the oscilloscope to measure the transponder radiated signal. Mismatches in the RF and antenna hook-ups, which may cause center frequency deviations or even total oscillator malfunction, can thus be detected. Further, in this setup, the low frequency amplitude modulation of the output signal was configured and measured. This is crucial for setting the spectral position of the transponder signal in the radar base band. Figure 14 shows the measured output signal in the time domain. Another option is to use on–off-keying as low frequency modulation and maximum signal power in the fundamental frequency, but it would create harmonics in the radar base band.

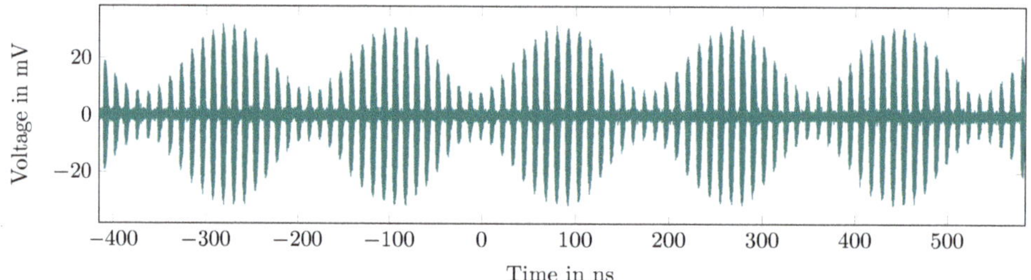

Figure 14. The modulated RF output of the transponder. It shows the high-frequency on–off switching and the low frequency amplitude modulation as explained in Equation (37).

5.3. Test with a Real Radar System

The transponder was tested in a real distance measurement scenario using the radar system that was built for the LAOLa project. Reference data were provided by a Leica TS30 total station in tracking mode. Figure 15a shows the range spectrum of the modulated response and Figure 15b the distance measured with the radar system compared to the total station. There is very good accordance between both plots, with the measurement update rate of the radar system being superior to that of the total station. The outlier at sample 120 might have been caused by a multipath distortion, e.g., by a ground reflection. The distance error at the center of the operational range was used as the distance error correction term, and it was assumed to be constant over the entire range. The distance error is shown in Figure 15c with an RMSE of 5.64 cm.

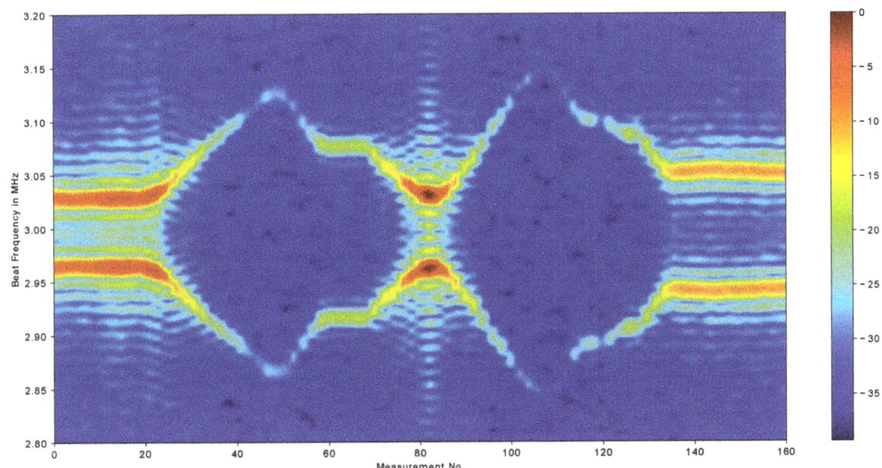

(**a**) The range spectrum at the modulation frequency of the measurements.

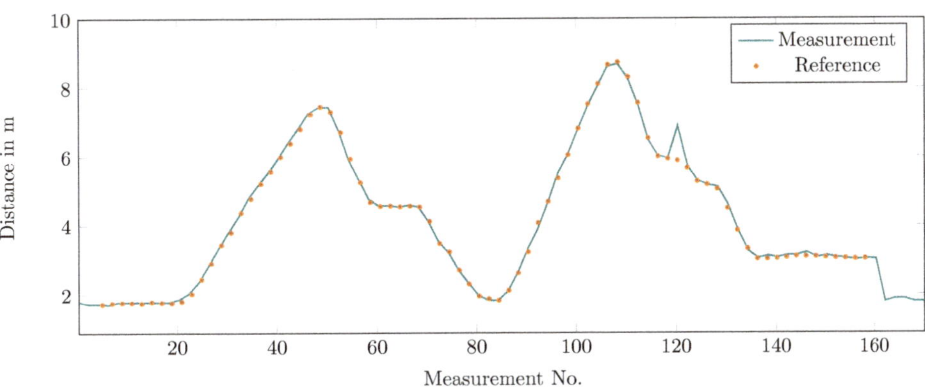

(**b**) The measured transponder distance compared to the total station reference.

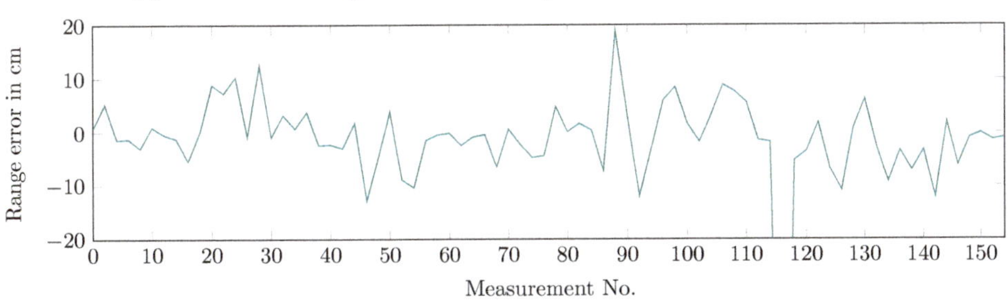

(**c**) The range error with an RMSE of 5.64 cm.

Figure 15. The transponder test with a real radar system.

6. Outlook

6.1. The SILO Transponder in a Cooperative Radar Setup

As mentioned, the transponder concept presented in this paper introduces a range uncertainty that is a function of the oscillator's transient behavior. In general, this error cannot be measured with a single radar channel, because it is independent of all radar parameters such as the ramp duration or bandwidth.

However, by introducing a second receive channel or an additional, cooperative radar system, respectively, the difference of the time of arrival/distance can be calculated:

$$\begin{aligned}\Delta d = d_1 - d_2 &= c\frac{\Delta\omega_{\mathrm{R},1}}{4\mu} - c\frac{T_\mathrm{s}}{4} - c\frac{\Delta\omega_{\mathrm{R},2}}{4\mu} + c\frac{T_\mathrm{s}}{4} \\ &= c\frac{\Delta\omega_{\mathrm{R},1}}{4\mu} - d_\mathrm{err} - c\frac{\Delta\omega_{\mathrm{R},2}}{4\mu} + d_\mathrm{err} \\ &= c\frac{\Delta\omega_{\mathrm{R},12}}{4\mu}\end{aligned} \tag{39}$$

which results again in an accurate measurement without the influence of the transient behavior of the oscillator. With Δd and the positions of the radar sensors, a solution for the absolute distance can be found by solving the general TDOA equation [22].

By comparing the latter solution to the direct measurements, an estimate for the error is obtained

$$\hat{d}_\mathrm{err} = d_1 - d_\mathrm{TDOA} \tag{40}$$

This estimate can be computed, e.g., recursively in a Kalman filter during the measurements, whereby the filter output will have the accuracy of the TDOA evaluation and the precision of the TOA evaluation. A Kalman filter that relies only on TDOA measurements can be deployed as well for localization [23].

6.2. Active Backscatter Transponders for Automotive Radar Systems

There is still a gap in the frequency range for which SILO transponders have been realized. There are no valuable transponders available at this time—especially for the 77 GHz automotive band. It would provide a great opportunity for building low-cost radar target simulators with high RCS for radar calibration and verification [6].

7. Conclusions

Switched-injection locked oscillators (SILOs) provide a simple but powerful circuit for creating active radar transponders with modulated backscatter. They are used as secondary radar beacons in local ad-hoc localization systems for autonomous robot navigation. The phase coherence of the SILO to the radar signal and a corresponding FMCW beat signal of the response are derived based on a liner system model. A working system and its performance for the 24 GHz ISM band are shown. Because of the transient behavior of the switched oscillator, a range uncertainty is introduced, which needs to be calibrated or considered, and estimated online in the overall localization algorithm.

Author Contributions: Conceptualization, M.S., T.P., and M.V.; Data curation, M.S.; Formal analysis, M.S.; Funding acquisition, M.V.; Investigation, M.S.; Methodology, M.S.; Project administration, M.V.; Resources, M.S.; Software, M.S.; Supervision, M.V.; Validation, M.S., T.P., and M.V.; Visualization, M.S.; Writing—original draft, M.S.; and Writing—review and editing, M.S., T.P., and M.V.

Funding: This study was financially supported by the German Space Agency (DLR) in the project LAOLa, registration no. 50NA1528.

Conflicts of Interest: The authors declare no conflict of interest.

Abbreviations

The following abbreviations are used in this manuscript:

AOA	Angle-of-arrival
DDS	Direct digital synthesis
FET	Field effect transistor
FMCW	Frequency modulated continuous wave radar system
IC	Integrated circuit
ISM	Industrial, scientific and medical frequency band
LAOLa	Local ad-hoc localization and landing system
RF	Radio frequency
SILO	Switched-injection locked oscillator
TOA	Time-of-arrival
TDOA	Time-difference-of-arrival

References

1. Roehr, S.; Vossiek, M.; Gulden, P. Method for High Precision Radar Distance Measurement and Synchronization of Wireless Units. In Proceedings of the 2007 IEEE/MTT-S International Microwave Symposium, Honolulu, HI, USA, 3–8 June 2007; pp. 1315–1318. [CrossRef]
2. Staudinger, E.; Zhang, S.; Dammann, A.; Zhu, C. Towards a radio-based swarm navigation system on mars—Key technologies and performance assessment. In Proceedings of the 2014 IEEE International Conference on Wireless for Space and Extreme Environments (WiSEE), Noordwijk, The Netherlands, 30–31 October 2014; pp. 1–7. [CrossRef]
3. Dobrev, Y.; Vossiek, M.; Shmakov, D. A bilateral 24 GHz wireless positioning system for 3D real-time localization of people and mobile robots. In Proceedings of the 2015 IEEE MTT-S International Conference on Microwaves for Intelligent Mobility (ICMIM), Heidelberg, Germany, 27–29 April 2015; pp. 1–4. [CrossRef]
4. Dobrev, Y.; Vossiek, M.; Christmann, M.; Bilous, I.; Gulden, P. Steady Delivery: Wireless Local Positioning Systems for Tracking and Autonomous Navigation of Transport Vehicles and Mobile Robots. *IEEE Microw. Mag.* **2017**, *18*, 26–37. [CrossRef]
5. Dobrev, Y.; Dobrev, Y.; Gulden, P.; Lipka, M.; Pavlenko, T.; Moormann, D.; Vossiek, M. Radar-Based High-Accuracy 3D Localization of UAVs for Landing in GNSS-Denied Environments. In Proceedings of the 2018 IEEE MTT-S International Conference on Microwaves for Intelligent Mobility (ICMIM), Munich, Germany, 15–17 April 2018; pp. 1–4. [CrossRef]
6. Scheiblhofer, W.; Feger, R.; Haderer, A.; Stelzer, A. A low-cost multi-target simulator for FMCW radar system calibration and testing. In Proceedings of the 2017 47th European Microwave Conference (EuMC), Nuremberg, Germany, 11–13 October 2017; pp. 1191–1194. [CrossRef]
7. Lorenzo, J.; Lazaro, A.; Villarino, R.; Girbau, D. Active Backscatter Transponder for FMCW Radar Applications. *IEEE Antennas Wirel. Propag. Lett.* **2015**, *14*, 1610–1613. [CrossRef]
8. Dadash, M.S.; Hasch, J.; Voinigescu, S.P. A 77-GHz active millimeter-wave reflector for FMCW radar. In Proceedings of the 2017 IEEE Radio Frequency Integrated Circuits Symposium (RFIC), Honolulu, HI, USA, 4–6 June 2017; pp. 312–315. [CrossRef]
9. MacFarlane, G.G.; Whitehead, J.R. The Theory of the Super-Regenerative Receiver Operated in the Linear Mode. *J. Inst. Electr. Eng. Part III Radio Commun. Eng.* **1948**, *95*, 143–157. [CrossRef]
10. Vossiek, M.; Gulden, P. The Switched Injection-Locked Oscillator: A Novel Versatile Concept for Wireless Transponder and Localization Systems. *IEEE Trans. Microw. Theory Tech.* **2008**, *56*, 859–866. [CrossRef]
11. Vossiek, M.; Schafer, T.; Becker, D. Regenerative backscatter transponder using the switched injection-locked oscillator concept. In Proceedings of the 2008 IEEE MTT-S International Microwave Symposium Digest, Atlanta, GA, USA, 15–20 June 2008; pp. 571–574. [CrossRef]
12. Carlowitz, C.; Vossiek, M.; Strobel, A.; Ellinger, F. Precise ranging and simultaneous high speed data transfer using mm-wave regenerative active backscatter tags. In Proceedings of the 2013 IEEE International Conference on RFID (RFID), Penang, Malaysia, 30 April–2 May 2013; pp. 253–260. [CrossRef]

13. Schuetz, M.; Dobrev, Y.; Carlowitz, C.; Vossiek, M. Wireless Local Positioning with SILO-Based Backscatter Transponders for Autonomous Robot Navigation. In Proceedings of the 2018 IEEE MTT-S International Conference on Microwaves for Intelligent Mobility (ICMIM), Munich, Germany, 15–17 April 2018; pp. 1–4. [CrossRef]
14. Schütz, M.; Dobrev, Y.; Pavlenko, T.; Vossiek, M. A Secondary Surveillance Radar with Miniaturized Transponders for Localization of Small-Sized UAVs in Controlled Air Space. In Proceedings of the 2018 5th IEEE International Workshop on Metrology for AeroSpace (MetroAeroSpace), Rome, Italy, 20–22 June 2018; pp. 107–111. [CrossRef]
15. Pavlenko, T.; Schütz, M.; Vossiek, M.; Walter, T.; Montenegro, S. Wireless Local Positioning System for Controlled UAV Landing in GNSS-Denied Environment. In Proceedings of the 2019 IEEE 5th International Workshop on Metrology for AeroSpace (MetroAeroSpace), Torino, Italy, 19–21 June 2019; pp. 171–175. [CrossRef]
16. Schulz, M.; Strobel, A.; Joram, N.; Ellinger, F. Design of a switched injection-locked oscillator at 2.45 GHz for a local positioning system. In Proceedings of the 2014 9th European Microwave Integrated Circuit Conference, Rome, Italy, 6–7 October 2014; pp. 116–119. [CrossRef]
17. Ghaleb, H.; Carlowitz, C.; Fritsche, D.; Carta, C.; Ellinger, F. A 180-GHz Super-Regenerative Oscillator with up to 58 dB Gain for Efficient Phase Recovery. In Proceedings of the 2019 IEEE Radio Frequency Integrated Circuits Symposium (RFIC), Boston, MA, USA, 2–4 June 2019; pp. 131–134. [CrossRef]
18. Ghaleb, H.; Testa, P.V.; Schumann, S.; Carta, C.; Ellinger, F. A 160-GHz Switched Injection-Locked Oscillator for Phase and Amplitude Regenerative Sampling. *IEEE Microw. Wirel. Components Lett.* **2017**, *27*, 821–823. [CrossRef]
19. Ghaleb, H.; El-Shennawy, M.; Carta, C.; Ellinger, F. A 148-GHz regenerative sampling oscillator. In Proceedings of the 2017 12th European Microwave Integrated Circuits Conference (EuMIC), Nuremberg, Germany, 8–10 October 2017; pp. 65–68. [CrossRef]
20. Ferchichi, A.; Ghaleb, H.; Carta, C.; Ellinger, F. Analysis and Design of 60-GHz Switched Injection-Locked Oscillator with up to 38 dB Regenerative Gain and 3.1 GHz Switching Rate. In Proceedings of the 2018 IEEE 61st International Midwest Symposium on Circuits and Systems (MWSCAS), Windsor, ON, Canada, 5–8 August 2018; pp. 340–343. [CrossRef]
21. Strobel, A.; Schulz, M.; Ellinger, F.; Carlowitz, C.; Vossiek, M. Analysis of phase sampling noise of switched injection-locked oscillators. In Proceedings of the 2014 IEEE Topical Conference on Wireless Sensors and Sensor Networks (WiSNet), Newport Beach, CA, USA, 19–23 January 2014; pp. 43–45. [CrossRef]
22. Bard, J.D.; Ham, F.M. Time difference of arrival dilution of precision and applications. *IEEE Trans. Signal Process.* **1999**, *47*, 521–523. [CrossRef]
23. Lipka, M.; Sippel, E.; Vossiek, M. An Extended Kalman Filter for Direct, Real-Time, Phase-Based High Precision Indoor Localization. *IEEE Access* **2019**, *7*, 25288–25297. [CrossRef]

© 2019 by the authors. Licensee MDPI, Basel, Switzerland. This article is an open access article distributed under the terms and conditions of the Creative Commons Attribution (CC BY) license (http://creativecommons.org/licenses/by/4.0/).

Article

Radiolocation Devices for Detection and Tracking Small High-Speed Ballistic Objects—Features, Applications, and Methods of Tests

Marek Brzozowski *, Mariusz Pakowski *, Mirosław Nowakowski, Mirosław Myszka and Mirosław Michalczewski

Laboratory Testing of Radar and Aviation Technology, Air Force Institute of Technology (AFIT), 01-494 Warsaw, Poland; miroslaw.nowakowski@itwl.pl (M.N.); miroslaw.myszka@itwl.pl (M.M.); miroslaw.michalczewski@itwl.pl (M.M.)

* Correspondence: marek.brzozowski@itwl.pl (M.B.); mariusz.pakowski@itwl.pl (M.P.)

Received: 24 October 2019; Accepted: 22 November 2019; Published: 5 December 2019

Abstract: This article describes radiolocation devices dedicated to the detection and tracking of small high-speed ballistic objects and multifunctional radars. This functionality is implemented by applying space search technology and adaptive algorithms for detection and tracking of air objects in parallel with classic search and tracking of objects in controlled airspace. This article presents examples of the construction of both types of devices produced by foreign companies and Polish industry. The following sections present methods for testing radars with the function of tracking small high-speed ballistic objects along with examples of results of observations of combat ammunition.

Keywords: radar; research radar; high-speed ballistic objects tracking; radar field tests; GPS applications

---

1. Introduction

We presented the issue of radars capable of detecting and tracking high-speed ballistic objects, as well as issues related to the specifics of research and testing of such devices at the conference Metro Aerospace 2019, in Turin. The presented issues were met with great interest from the conference participants and became a source of information for many people about radar devices manufactured by the Polish industry.

This article expands the subject of the publication "Radars with the function of detecting and tracking artillery shells—selected methods of field testing", published in Metro AeroSpace 2018 proceedings [1].

Despite continuous technological development, the role of classic artillery in a contemporary battlefield is not diminishing. If we observe this in relation to the conflicts conducted in recent years, it is easy to notice that most activities of infantry units take place under covering fire from lighter or heavier artillery. Both cases, which aim to destroy a detected enemy and protect our positions, still require using sufficient fire power to quickly and effectively incapacitate the enemy. The key to effective use of artillery is to quickly and precisely determine the coordinates of the position of targets and to verify the accuracy and effectiveness of the conducted firing [2].

In this area, modern technology offers the following two most popular solutions: Unmanned aerial vehicles (UAV) equipped with optoelectronic observation sensors and specialized radiolocation devices intended for detecting, tracking, and calculating the trajectory parameters of high-speed ballistic objects [3]. Both methods have their advantages and disadvantages. This paper only features the characteristics of radars. The discussion features methods of detecting high-speed ballistic objects using radiolocation sensors and examples of existing technological solutions used globally and in the Republic of Poland. This paper also contains a detailed discussion of the methods of testing specialized ballistic radars or multifunctional radars with the ability to detect and track high-speed ballistic objects viewed as one of their operating modes. It also features examples of the results of field testing conducted using real high-speed ballistic objects [1,4].

We present research methods and example results related to Polish radars developed for the needs of the Polish Armed Forces. From a scientific point of view, it would be desirable to present many technical parameters that characterize individual sensors and compare the effects of different sensors depending on the technical solutions used in them, but this is impossible because the detailed technical parameters and test results of these devices are secret.

2. Materials and Methods

The latter part of this paper, aside from a short characterization of radiolocation devices intended for detecting and tracking high-speed ballistic objects, includes a presentation of testing methods used at the Air Force Institute of Technology (AFIT) for the purpose of field testing to check the most important parameters of such devices. The most frequently used verifications are as follows: minimum and maximum distances, minimum and maximum height, maximum elevation angle of detecting for a given type of detected object, and the accuracies of determining the points of origin (POO) and points of impact (POI) [2].

2.1. Detection and Tracking of High-Speed Ballistic Objects Using Radiolocation Sensors

The issue of detecting and tracking high-speed ballistic objects has been a significant challenge for radiolocation. Because these objects achieve high velocities and move at relatively small heights above the horizon, the device that ensures their effective detection and tracking must quickly scan the horizon (several times per second) in order to immediately detect the flight object and track its flight path, i.e., observe space in a broad range of elevation angles with the radar beam [3].

These capabilities have appeared along with the development of antennas with an electronically controlled pencil beam. Such antennas allow for very fast beam movement both in the azimuth and elevation planes. This searches the entire observation sector with an information refreshing time of less than 1 s. The algorithms for searching ballistic objects are usually optimized to search a narrow elevation sector right above the horizon. This allows for the optimal use of the radar's time budget. Application of this method is possible because after detecting the echo, the radar automatically switches to the detection mode using additional lightings with a pencil beam, which obtains information about the high-speed ballistic object's flight trajectory, optimized in terms of the ballistic calculations [5,6]. On the basis of subsequent object detection, the ballistic calculator determines the appropriate ballistic model to calculate its full flight trajectory. The calculated trajectory is then applied to the terrain's digital model, which determines the POO and POI's coordinates. Antennas with an electronically controlled beam are made using various technologies, starting with passive antennas with phase-frequency control and ending with active antennas made in the form of transmit and receive module array with full digital control of the transmission and reception characteristics [7]. The only disadvantage of both types of older and modern antennas is that the electronic characteristic control is realized only in a limited sector of azimuth and elevation angles. The limitation of the azimuth sector, usually to approximately ±45° in relation to the antenna's normal aperture, is especially troublesome. This type of antenna is, therefore, usually rotated mechanically in the azimuth plane and such radar's operational use imposes the need to determine the sector of responsibility earlier. The POO and POI coordinates designated by the radar are usually transmitted to an automated command and control system.

One of the most popular examples of a radar intended for tracking high-speed ballistic objects and detecting launchers positions is ARTHUR (ARTillery HUnting Radar, Ericsson Microwave Sytsem AB, Mölndal, Sweden) (Figure 1) developed and manufactured by Ericsson Microwave Systems (currently SAAB Microwave Systems) [2].

Figure 1. ARTillery HUNting Radar (www.radartutorial.eu) [8].

The radar is a device characterized by high mobility, able to quickly and effectively move in a combat operations area. It operates in the C band (4 to 8 GHz) and possesses a flat passive antenna with electronic beam scanning in both planes. In elevation, the antenna beam is controlled by changing the probing signal's frequency, whereas in the azimuth the control takes place using phase shifters [8].

The width of the azimuth sector in which the electronic beam scanning is conducted amounts to 90°. The radar's antenna setting towards the selected direction is executed by its mechanical rotation. The radar detections are analyzed by an advanced software that determines the POO and POI coordinates with consideration of the terrain's digital model and computation of commands for the co-operating military units. In order to execute the aforementioned functions, the radar must precisely determine its position in the terrain. This is done with the use of an advanced navigation system consisting of an inertial module and a GPS receiver [2].

Other examples of radars intended for detecting high-speed ballistic objects and launcher positions include: AN/TPQ-37, AN/TPQ-47, and AN/TPQ-50 developed by Thales Raytheon Systems, COBRA (COunter Battery Radar) [2] developed and manufactured in co-operation with French, German, and Turkish companies, as well as the Chinese SLC-2 Fire Finding Radar structure [2].

2.2. Polish Radars with the Function of Detecting and Tracking Small High-Speed Ballistic Objects

Radars produced by Polish companies also detect and track small high-speed ballistic objects and determine the POO and POI. An example of a device especially dedicated for such tasks is the Radiolocation Artillery Reconnaissance Unit LIWIEC [9] (Figure 2). It is embedded on a wheeled chassis and is equipped with a passive flat antenna with a two-plane electronically controlled beam. The maximum width of the observed azimuth sector amounts to 90°. The frequency of scanning and refreshing information about each element in the sector amounts to 0.5 s. The antenna's mechanical rotation in the azimuth allows for sector control in a 270° width. The maximum observation angle of objects in elevation amounts to 20° [9].

The LIWIEC radar can be used to protect important objects and support artillery operation in the following scope:

- automatic detection and tracking of high-speed ballistic objects;
- determination of the coordinates of launcher positions (single and grouped);
- automatic classification of flight object type and launcher position and type;
- determination of the POI coordinates;
- transmission of information to automated command and control systems.

The main advantage of this radar is that its operator can adapt space scanning algorithms to current needs, and, in addition, the automatic tracking system of detected objects can force more frequent scanning of new objects, which greatly accelerates the calculation of the ballistic trajectory of the tracked object. On the basis of the calculated trajectory, the radar identifies the type of detected object (including its launcher type) and accurately determines the POO and POI before the detected object reaches its target.

Aside from small high-speed ballistic objects, the LIWIEC radar detects and tracks aircrafts, helicopters, unmanned aerial vehicles, and land-based mechanical vehicles [9].

Figure 2. Radiolocation Artillery Reconnaissance Unit LIWIEC (www.pitradwar.com).

A different perspective on the problem of detecting small high-speed ballistic objects and especially mortars is applied in the SOŁA and BYSTRA radars. The radars are intended for air defense forces to control airspace in the area of land troops operations. Both devices search the airspace by mechanically rotating the antenna in the azimuth plane. The application of relatively high antenna rotation speeds (30 and 60 rpm) guarantee the ability to detect and track small high-speed ballistic objects with sufficient accuracy for the ballistic calculation algorithm to estimate the complete object trajectory based on the sample of over a dozen detections, and therefore determines the POO and POI with satisfactory accuracies. Especially good results of such a method for detection and tracking are achieved in relation to mortar [4].

The SOŁA redeployment-capable radiolocation station (Figure 3) is a short-range radar operating in the S band, intended for the SHORAD (short range air defense) systems. It ensures the detection of aerial objects, including unmanned air vehicles, helicopters, and small high-speed ballistic objects. It is a three-dimensional radar with an antenna rotated mechanically along the azimuth and a beam controlled electronically in the elevation. Depending on the operating mode, the radar's antenna can be rotated with a speed of 30 or 60 rpm. The radar is embedded in an armored vehicle with very good off-road performance. The data on the detected objects is transmitted from the radar to the automated command system via digital radio link [9].

Figure 3. Redeployment-capable radiolocation station SOŁA (www.pitradwar.com).

The BYSTRA (Figure 4) is a multifunctional and multitask radar with versatile capabilities and applications, possessing the ability to detect and track typical aerial threats, such as aircrafts and helicopters (also hovering), unmanned air vehicles, and small high-speed ballistic objects, particularly mortars. The radar uses state-of-the-art technological solutions, including: active antenna with semiconducting transmitting modules and electronically controlled transmission beam position, digital receiving beams formation, digital synthesis, signal encoding and matched filtering, coordinate estimation supported by the algorithm limiting the multipath effects, and finally tracking system using the multiple hypothesis algorithm. The applied solutions achieving quick radiolocation information refreshing of no more than 2 s [9].

Electronic control of the position of the antenna beam in the azimuth plane allows the system to automatically track air objects and generate re-scanning requests for newly detected objects in the same antenna rotation. This is a very important property of radar, which very quickly eliminates false ballistic trajectories based on the detection of passive or active interference.

Figure 4. Redeployment-capable radiolocation station BYSTRA (www.pitradwar.com) [9].

The SOŁA and BYSTRA radars are perfectly matched for covering and protecting important objects and areas because, during normal operation of controlling the airspace around the object, they additionally ensure the execution of the helicopter detection function, including hovering helicopters, as well as the detection and tracking of small high-speed ballistic objects including the POO and POI coordinates' estimation. This functionality provides the personnel of the protected object early warning about the threat detected.

2.3. Small High-Speed Ballistic Objects Detection Zone Verifications

The design and production of radars capable of detecting and tracking small high-speed ballistic objects and calculating their ballistic trajectory parameters requires developing specific research methods to objectively evaluate their technical parameters. Generally, such radars require conducting actual observations of shelling from many types of launchers. The planning of such tests requires simultaneous consideration of many various factors, including the following:

- expected values of the tested radar parameter;
- firing capabilities of the given launcher;
- available military trying area;
- safety zones military trying area;
- possibility of finding a proper place of radar operation in relation to launcher position and firing targets.

In the case of conducting testing to verify a small high-speed ballistic object detection zone, mortar shelling is most often used. This means of launcher fires ballistic objects to both small and large heights, in a broad range of distances and does not require too large of a safety zone. The most difficult element in preparing the tests for the conditions of a specific trying area is most often finding a suitable place of operation for the tested radar. Usually, the area of firing means that location of launchers and field of fire on a trying area are imposed by the adopted safety zones, whereas the terrain outside of the tactical strips is covered by forest. For this reason, the location of equipment in the field can only approximately correspond to the theoretical assumptions. Prior to the small high-speed ballistic object detection, each radar undergoes testing with the use of unmanned or classical aircrafts equipped with GPS receivers that record their flight trajectory. The comparison of radar detections with the recorded flight trajectory determines the accuracy of estimation of the detected object's coordinates (azimuth, distance, elevation angle, and height of flight). The sequence of testing analyzes the small high-speed ballistic object detection with consideration of the earlier designated estimation errors.

2.3.1. Verification of the Minimum Distance and Minimum Height of Small High-Speed Ballistic Objects Detection and Tracking

In the case of verification of the minimum distance and minimum height of small high-speed ballistic objects detection and tracking, the mortar's position must be located in the radar's dead zone. The ballistic objects must be fired into the radar's characteristic, at a small angle, so that the radar is capable of observing them continuously along the entire flight trajectory. Such a test requires realization of several dozen firings in order to evaluate the minimum ballistic objects detection distance and height in a statistical manner. The test sketch is presented below (Figure 5).

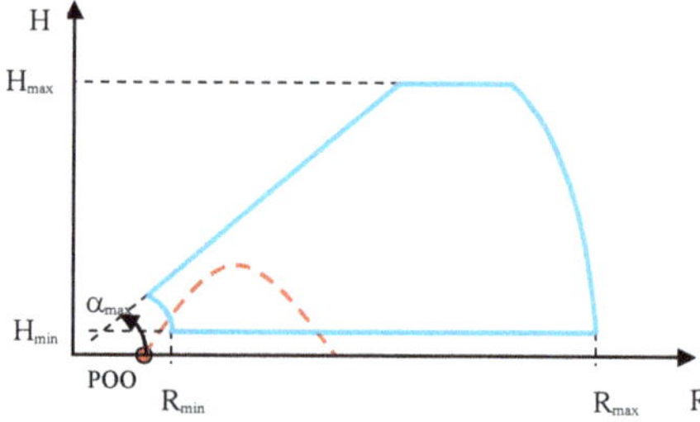

Figure 5. Sketch of the method of observing the mortar firings for the purpose of verifying the minimum detection distance and height. Start of the coordinates system in the radar position. Points of origin (POO), mortar position; radar detection characteristic, blue color; and desired ballistic objects flight trajectory, red color.

Figure 6 presents a set of data recorded during the testing of the minimum ballistic object detection distance. The graph presents the object detection distances from subsequent firings. Then, the minimum distance values were selected from the entire observation and this set of data was used to estimate the average value and standard deviation, as well as the bottom and top confidence interval limit. The calculated parameters allowed for an objective evaluation of the actual value of the minimum detection distance of the tested device.

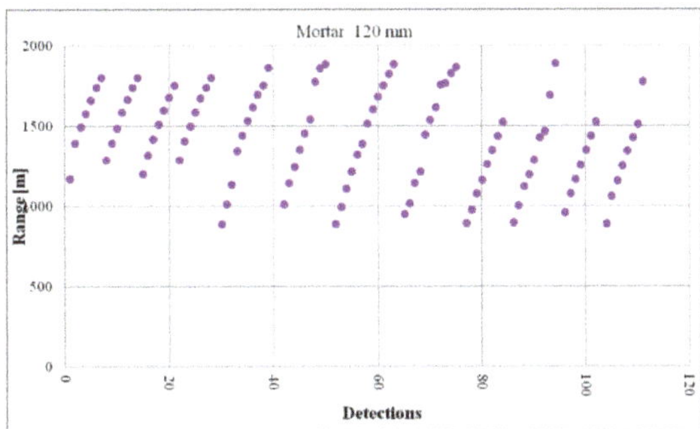

Figure 6. Example of mortars detections recorded during the verification of the minimum detection distances.

Figure 7 presents a set of data recorded during the testing of the minimum ballistic object detection height. The graph presents detections in the following coordinates: distance and height. Due to the verification's specificity, the radar's position in relation to the mortar's position and target of fire was especially important. The radar must be able to observe the ballistics' entire flight trajectory without any terrain obstacles. The detections of both ascending and descending flight objects were used for the evaluation of the parameter minimum detection height. From the complete set of detections, detections with minimum height values were selected. Such a dataset was used to conduct statistical calculations analogously to the minimum detection distance parameter.

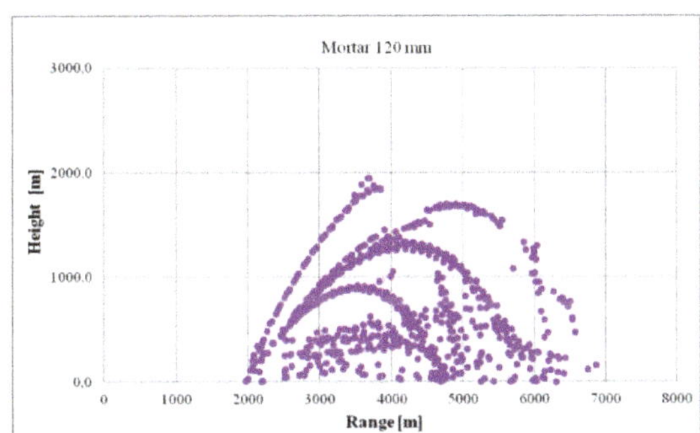

Figure 7. Example of mortar detections recorded during the verification of the parameter minimum detection heights.

2.3.2. Verification of the Maximum Elevation Angle for the Detection and Tracking of Small High-Speed Ballistic Objects

The sketch of firings conducted for the purpose of this test is presented in Figure 8. In order to obtain correct results from the observation of firings, it is necessary to carefully select the distance between the radar and the mortar's firing position, with consideration of the expected maximum radar detection angle in elevation and distance, as well as the maximum flight height of the fired ballistic objects.

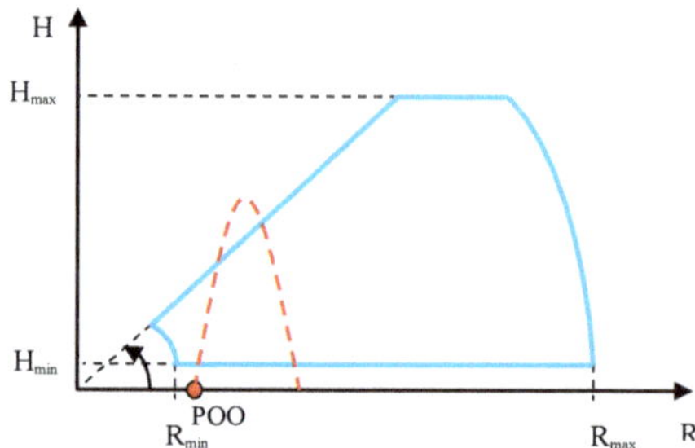

Figure 8. Sketch of the method of observing the mortars for the purpose of verifying the maximum angle of detection in elevation. Start of the coordinates system in the radar position. POO, mortar position; radar detection characteristic, blue color; desired flight trajectory, red color.

Figure 9 presents the results of firings conducted for the purpose of verifying the maximum elevation angle. Then, detections with maximum elevation angle values were selected from the entire observation and statistical calculations were conducted with reference to the obtained dataset by designating the estimation of the sought parameter.

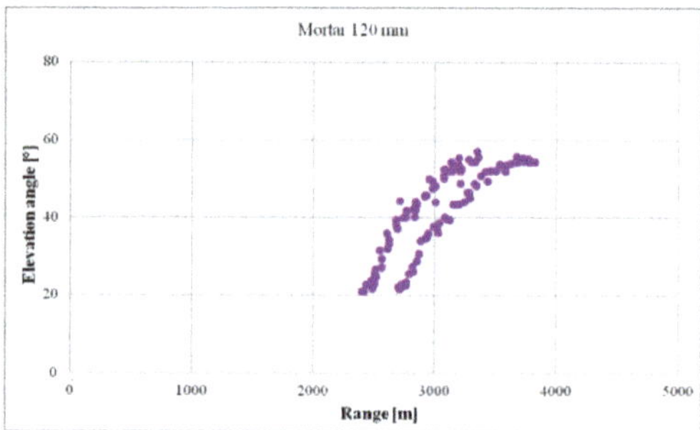

Figure 9. Example of mortar detections recorded during the verification of the maximum angle of detection in elevation.

2.3.3. Verification of the Maximum Height and Distance of Detection and Tracking of Small High-Speed Ballistic Objects

The sketch of verification of the maximum height and distance of detection and tracking of mortars is presented in Figure 10.

The firings conducted for the purpose of verifying the maximum detection height must feature a large angle of mortar barrel elevation, with simultaneous use of the maximum propelling charge. This type of firing does not ensure sufficient range. In practical testing, it is, therefore, necessary to conduct separate firings at the maximum flight altitude and at the maximum detection range. The last type of test requires setting the mortar's barrel at an angle near 45°.

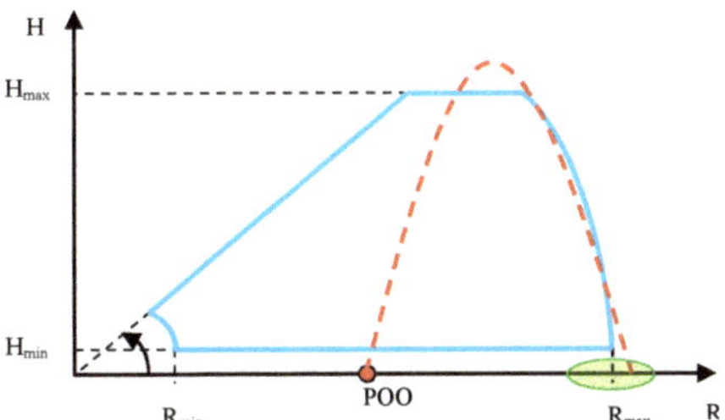

Figure 10. Sketch of the method of observing the mortars for the purpose of verifying the maximum detection height and distance. Start of the coordinates system in the radar position. POO, mortar position; radar detection characteristic, blue color; desired ballistics objects flight trajectory, red color; impact zone, green color.

Figure 11 presents detections of ballistics objects fired in a manner allowing them to reach the highest flight altitude. The lack of descending trajectory was caused by the applied operating mode of the tested radar in which the search and initialization of flight objects tracking were only conducted in a small value of elevation angles.

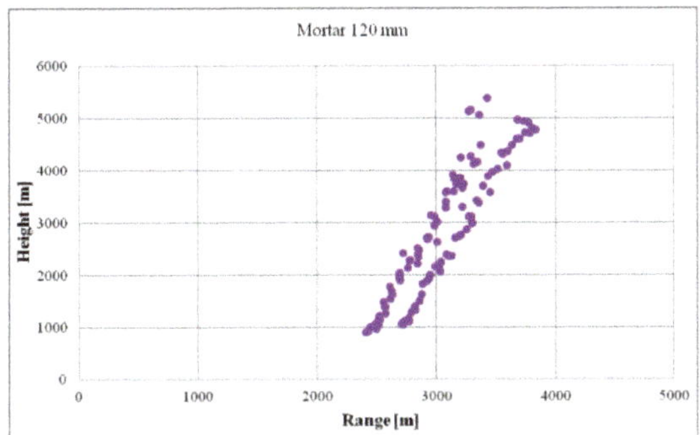

Figure 11. Example of mortar detections recorded during the verification of the parameter maximum detection height.

Figure 12 presents detections of ballistics objects fired with launcher settings that ensured the maximum flight range. As shown in the figure, firing in such a manner did not cause the ballistic objects to exit the radar's observation zone.

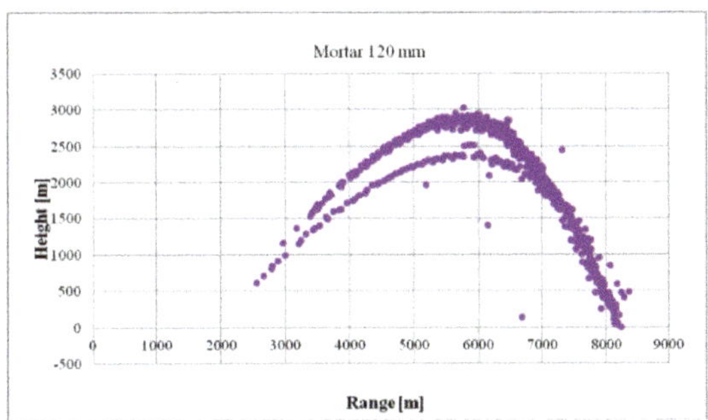

Figure 12. Example of results of verifying the maximum detection distance of mortars, i.e., height in the distance function.

2.4. Verifications of the Accuracy of Determining the Coordinates of POO and POI

Verification of the accuracy of determining the coordinates of launchers positions (POO) and points of impact (POI) of ballistic objects when the radar observes objects that are moving away. This method controls the execution of the firing task at much greater distances than by optical observation instrumentation. A radar deployed on an observation position far from the enemy is also more difficult to destroy than an observation UAV (unmanned air vehicle), which in order to evaluate the firing effectiveness must fly near the area of the shelled target, and thus can be easily destroyed by the enemy. The sketch of the method of executing the test is presented in Figure 13.

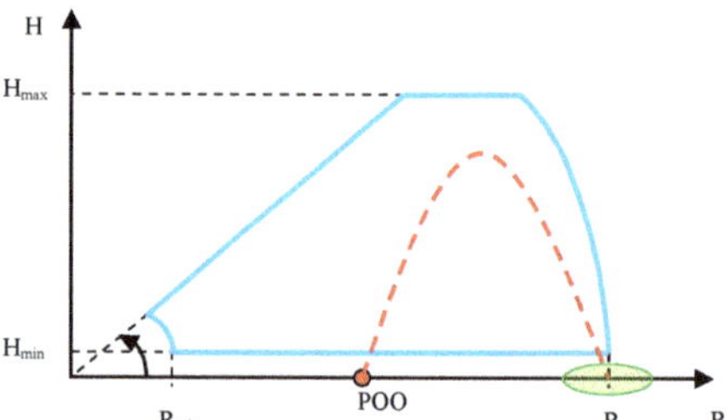

Figure 13. Sketch of the method of observing ballistic objects moving away from the radar for the purpose of verifying the accuracy of the POO and points of impact (POI). Start of the coordinates system in the radar position. POO, launcher position; radar detection zone, blue color; desired flight trajectory, red color; impact zone (POI), green color.

In order to evaluate the radar's POO estimation accuracy, the launcher position's coordinates are measured with the use of a surveying GPS receiver from Spectra Precision, type Epoch 50 (Figure 21). Measurements are conducted in the differential mode with the use of data derived from the ASG-EUPOS network's reference station. Such a measurement allows for achieving positioning accuracy of no less than 20 to 30 cm, and thereby assumes the measured coordinates as true coordinates. The coordinates obtained this way are then compared with the results of the POO estimation calculated by the tested radar. The estimation accuracy analysis is conducted in the UTM (Universal Transverse Mercator) rectangular coordinates system.

Figure 14 presents the results of POO estimation conducted based on the observation of a 120 mm mortar firing. The red color is used to mark the mortar positions, whereas the blue markers are the radar's POO coordinates estimation conducted based on the observation of subsequently fired ballistic objects.

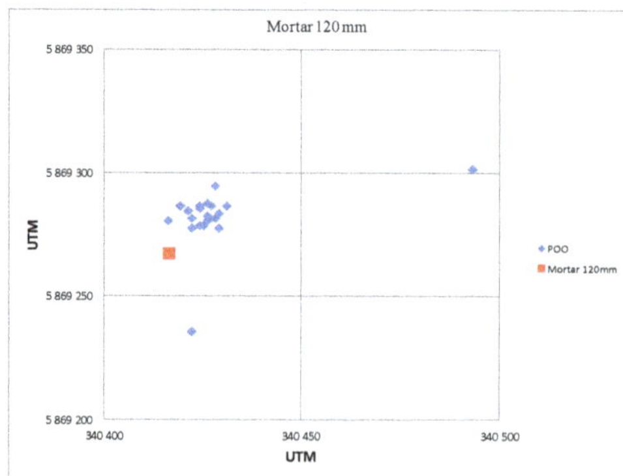

Figure 14. Example of estimations of the POO of 120 mm mortars. Firing in the direction from the radar.

Figure 15 presents results of the radar's observations of firings of a 122 mm caliber cannon subdivision. In the figure, red markers represent the positions of particular cannons, measured by the GPS receiver. Other markers illustrate subsequent POO estimations and are assigned with colors to particular cannons. In principle, the flight trajectories of ballistic objects fired from the cannon are flatter than mortar flight trajectories. This impairs the conditions of radar observation of the flying objects and the conditions of estimating the ballistic trajectory, which due to its flattening, makes it more difficult to precisely designate the intersections with the digital terrain map. Consequently, the estimated POO demonstrates a relatively large spread in the axis compliant with the firing direction.

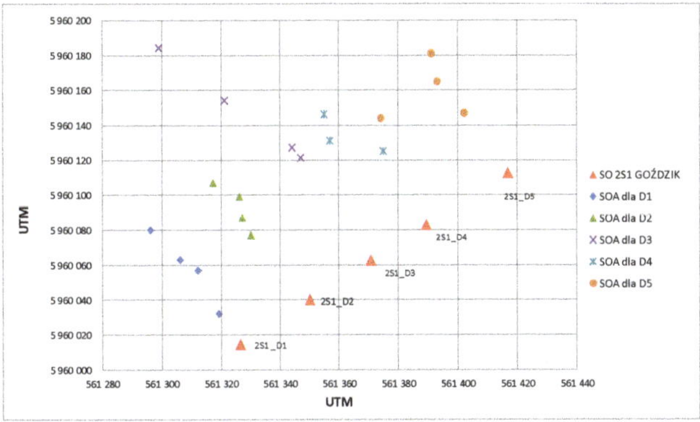

Figure 15. Example of estimations of the POO of 122 mm caliber cannon. Firing in the direction from the radar.

The analysis of the POI coordinates estimation accuracy of the tested radar is conducted similarly to the POO accuracy analysis. However, it features an essential difficulty in the form of the necessity to find the particular impact points after the firing and to measure their positional coordinates using the GPS receiver. The measurement itself is conducted similarly as in the case of measuring the firing mean position with the use of the surveying differential Epoch 50 receiver.

In order to find and identify the point of impact of each of the fired object, the firings must be executed in a special manner, i.e., subsequent objects must be fired not at the same target, but with a slight displacement, for example, 20 to 30 m left or right. Additionally, it is necessary to note the results of localizations conducted by military observers with the use of laser rangefinders. Such tests are always conducted by the polygon services in order to control compliance with safety conditions. The practice of field testing also shows that if the tested radar allows for stable tracking of the ballistic objects, it is necessary to note the estimated coordinates of particular POIs and use them to find particular points of impact. Such data can be entered into the manual GPS receiver and used in combination with the GO TO function to find the firing area and particular impact points in a relatively short time. During the searching of impact points, it is necessary to take special care, comply with the safety principles in force on the training area, and not take any actions without making arrangements with persons responsible for safety during the executed firings.

Figure 16 presents the illustration of the position of POIs estimated by the radar, impact points determined by observers, actual points of impact found in the area and measured with the GPS receiver, as well as the matching markers of particular POI estimations for GPS measurement purposes. The matching of particular POIs with actual points of impact requires a detailed analysis of all datasets, because the actual points of impact can differ substantially from the adopted targeting scheme and the occurring POI estimation errors can in some cases suggest a better matching of the impact points of another object than the object actually observed.

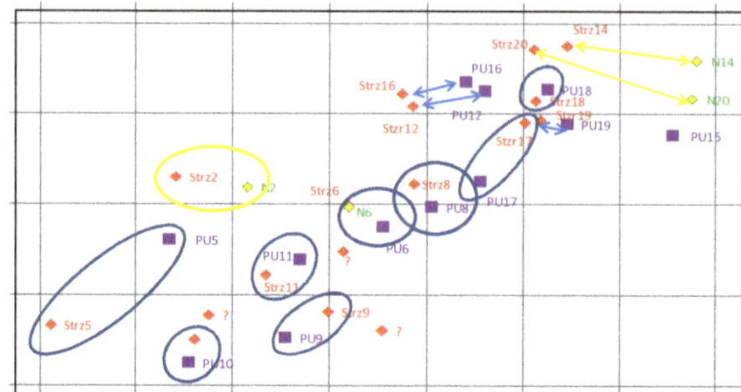

Figure 16. Illustration of the POIs of 120 mm mortars (purple color) estimated by the radar, GPS coordinates of the found points of impact (red color), and explosion localization by the military observers (yellow color).

Verification of the accuracy of determining the coordinates of the launcher position (POO) and the points of impact (POI) of ballistic objects when the radar observes objects that are moving towards it. The scheme of such verifications is presented in Figure 17.

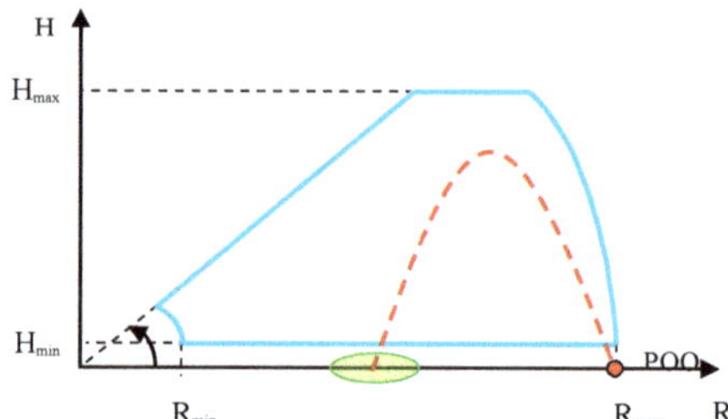

Figure 17. Sketch of the method of observing ballistic objects moving towards the radar for the purpose of verifying the accuracy of the POO and POI localization. Start of the coordinates system in the radar position. POO, mortar position; radar detection characteristic, blue color; desired flight trajectory, red color; impact zone (POI), green color.

The observation of firings executed in the direction of the radar requires particularly careful planning. On one hand, the radar must be placed in a suitable position, free of any terrain obstacles in the observation sector, at a distance ensuring the ability to detect flying objects. On the other hand, the test safety conditions must be respected, which means that the radar must be located at a suitable distance from the impact zone and the point of maximum theoretical range of the tested launcher. The reconciliation of these conditions is sometimes difficult and requires strict cooperation of the testing team with the trying area services.

Figure 18 presents the results of POO estimation of a 98 mm mortar, obtained during the observation of objects flying in the direction of the radar.

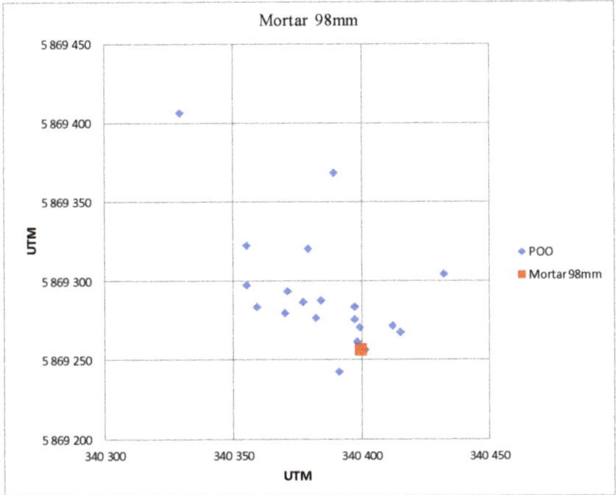

Figure 18. Example of estimations of the POO of 98 mm caliber mortar. Firing in the direction of the radar.

Figure 19 presents the results of POI estimation of a 98 mm mortar, obtained during the observation of ballistic object fired to the direction of the radar.

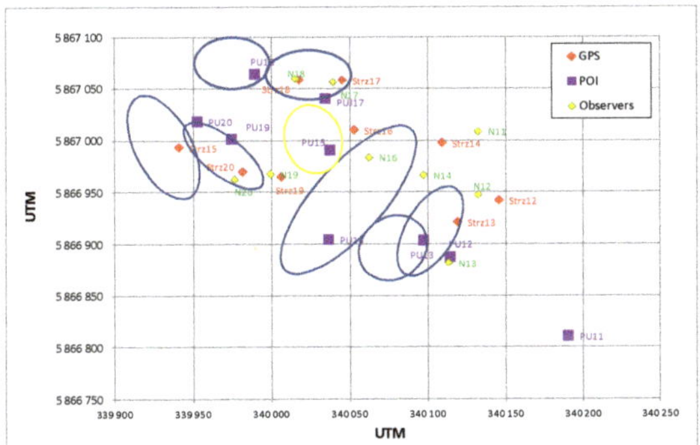

Figure 19. Illustration of the POIs of 98 mm mortars (purple color) estimated by the radar, GPS coordinates of the found points of impact (red color), and explosion localization by the observers (yellow color).

Figure 20 presents the comparison of the results of estimation of POI coordinates obtained during the observations of firings of a 122 mm caliber cannon. The firing was executed at four different targets that are also marked in the figure.

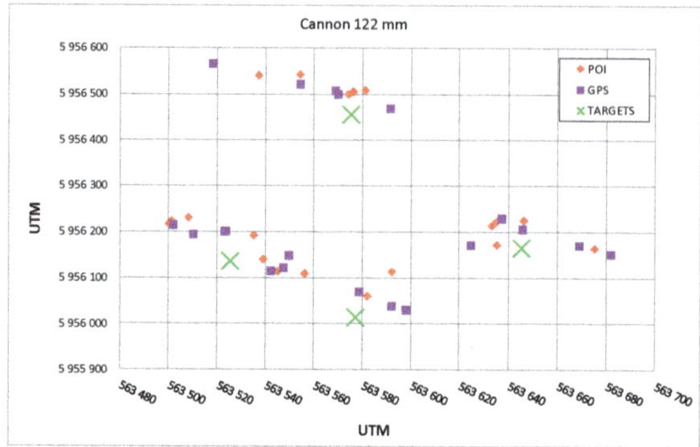

Figure 20. Illustration of the POIs of 122 mm caliber cannon (purple color) estimated by the radar, GPS coordinates of the found points of impact (red color), and coordinates of the targets (green color).

The presented results of field testing are illustrative, and their aim is only to illustrate the specifics of particular tests. The detailed results of verifications of the tactical and technical parameters of particular devices are confidential and will not be published.

2.5. Special Equipment

The basic measurement tool used in the polygon testing of radars is the surveying Epoch 50 GNSS receiver from Spectra Precision. The receiver interoperates with the Nomad type field controller, including the installed Survey Pro field measurement software. The receiver features 220 receiver channels and can operate with the use of the following signals:

- GPS L1/L2/L2C/L5;
- GLONASS L1/L2.

The Epoch 50 can operate in the following modes: autonomous, code differential, and phase differential. The differential modes can be realized in real time with the use of differential corrections from the local reference station (radio modem) or with the use of corrections sent via the Internet, for example, from the networks of the ASG-EUPOS reference stations. During field testing practice, the most common solution is the ability to record the measurements with a single receiver and their latter specification in the Spectra Precision Survey Office software. Such a specification only requires downloading from the ASG-EUPOS network of the files including the data recorded by the reference station located nearest to the testing location. The scheme of distribution of the reference stations available in the ASG-EUPOS network is presented in Figure 21. The receiver is used to obtain precise coordinates of the field of position of the tested radar, launcher positions, and the points of impact of the ballistic objects (Figure 22) [10].

Figure 21. Distribution of the GNSS reference stations of the ASG-EUPOS network. (www.asgeupos.pl [11]).

Figure 22. Application of the Epoch 50 GNSS receiver during field measurements.

Figure 23 presents an illustrative box in the Spectra Precision Survey Office software with visible measurement points specified in the post processing mode, in relation to the TORU permanent reference station.

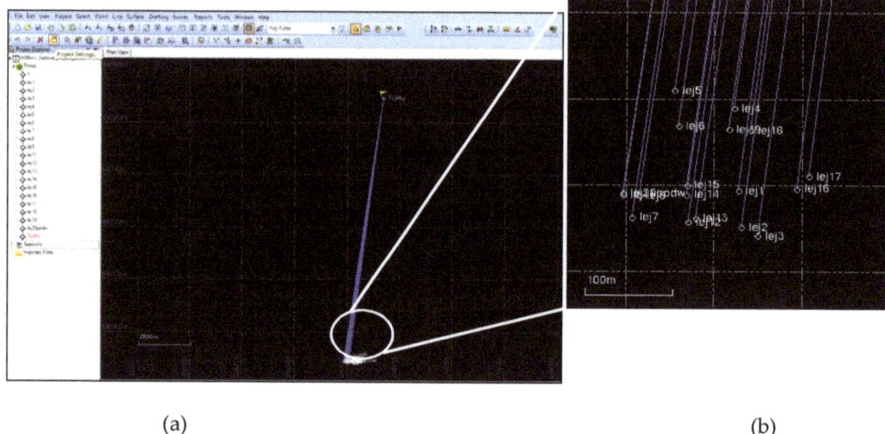

(a) (b)

Figure 23. Presentation of the GPS measurements in the Spectra Precision Survey Office software: (**a**) the main program window; (**b**) oom on measuring points.

3. Results

Research and testing of modern radars are complicated processes, both in technical and organizational terms. One of the dominant trends in radiolocation is the development of multifunctional radars. Such radars can detect and identify the following: aircrafts, helicopters, unmanned flying objects, and small high-speed ballistic objects. They require comprehensive knowledge of various fields of technology and a constant quest for new research methods from the research teams [12]. In particular, the specificity of field testing, conducted with the use of various types of launchers requires, apart from specialist knowledge in the domain of radiolocation, knowledge of artillery operations, the specificity of military training areas, and experience in using satellite geodesy. Our presented methods made it possible to check radar parameters with small high-speed ballistic object detection function accurately and objectively.

Technological progress in the field of radiolocation contributes not only to the improvement of the basic technical parameters of radars, forcing an increase in the accuracy of existing research and testing methods, but also leads to the creation of completely new functionalities [13].

In this presentation of parameters, we are aware that this article leaves unsatisfied the specific properties and test results of individual devices, but this is impossible due to their final military application.

Author Contributions: Conceptualization, M.B.; methodology, M.B., M.P. and M.M. (Mirosław Myszka); software, M.P.; validation, M.N. and M.M. (Mirosław Myszka); formal analysis, M.N.; investigation, M.P. and M.M. (Mirosław Michalczewski); resources, M.M. (Mirosław Michalczewski); data curation, M.M. (Mirosław Myszka); writing—original draft preparation, M.B.; writing—review and editing, M.B. and M.P.; visualization, M.P.; supervision, M.B.; project administration, M.N.

Funding: This research received no external funding.

Conflicts of Interest: The authors declare no conflict of interest.

References

1. Brzozowski, M.; Pakowski, M.; Nowakowski, M.; Myszka, M.; Michalczewski, M. Radars with the function of detecting and tracking artillery shells—Selected methods of field testing. In Proceedings of the 2019 IEEE International Workshop on Metrology for AeroSpace (MetroAeroSpace 2019), Torino, Italy, 19–21 June 2019.
2. Maini, A.K. *Handbook of Defence Electronics and Optronics: Fundamentals, Technologies and Systems*; John Wiley & Sons Ltd.: Hoboken, NJ, USA, 2018.

3. Barton, D.K. *Modern Radar System Analysis*; Artech House INC: Norwood, MA, USA, 1988.
4. Czekała, Z. *Parada Radarów [Radar Parade]*; Dom Wydawniczy Bellona: Warszawa, Poland, 1999.
5. Kokot, K. *Podstawy Radiolokacji [Basics of Radiolocation], cz. 1*; WAT: Warszawa, Poland, 1967.
6. Kokot, K. *Podstawy Radiolokacji [Basics of Radiolocation], cz. 2*; WAT: Warszawa, Poland, 1968.
7. Farina, A.; De Maio, A.; Haykin, S. *The Impact of Cognition on Radar Technology*; Scitech Publishing: Stevenage, UK, 2017.
8. Card Index of Radar Sets - Battlefield Radars. Available online: www.radartutorial.eu (accessed on 26 November 2019).
9. LIWIEC Radiolocation Artillery Recognition Kit. Available online: www.pitradwar.com (accessed on 26 November 2019).
10. Misra, P. *Global Positioning System, Signals, Measurements, and Performance*; Ganga-Jamuna Press: Lincoln, MA, USA, 2001.
11. Witamy na stronie systemu ASG-EUPOS. Available online: www.asgeupos.pl (acessed on 26 November 2019).
12. Brzozowski, M.; Pakowski, M.; Myszka, M.; Michalczewski, M.; Winiarska, U. *Testing of Contemporary Radiolocation Devices with the Use of Military Aircrafts Conducted at the AFIT*; Prace Naukowe ITWL: Warsaw, Poland, 2017; Volume 40.
13. Pakowski, M.; Brzozowski, M.; Myszka, M.; Michalczewski, M. Methods for testing military radars produced in Poland. In Proceedings of the 2018 IEEE International Workshop on Metrology for AeroSpace (MetroAeroSpace 2018), Rome, Italy, 20–22 June 2018.

© 2019 by the authors. Licensee MDPI, Basel, Switzerland. This article is an open access article distributed under the terms and conditions of the Creative Commons Attribution (CC BY) license (http://creativecommons.org/licenses/by/4.0/).

Article

LiDAR-Based System and Optical VHR Data for Building Detection and Mapping †

Silvia Liberata Ullo [1,*,‡], Chiara Zarro [1,*,‡], Konrad Wojtowicz [2,*,‡], Giuseppe Meoli [3,*,‡] and Mariano Focareta [3,*,‡]

1 Engineering Department, University of Sannio, 82100 Benevento, Italy
2 Faculty of Mechatronics and Aerospace, Military University of Technology, 00-908 Warsaw, Poland
3 Mapsat, 82100 Benevento, Italy
* Correspondence: ullo@unisannio.it (S.L.U.); chiara.zarro@unisannio.it (C.Z.); konrad.wojtowicz@wat.edu.pl (K.W.); g.meoli@mapsat.it (G.M.); m.focareta@mapsat.it (M.F.)

† This paper is an extended version of High-resolution topographic surveys and earth features extraction through LiDARs. Discussion of some Case Studies, 6th IEEE International Workshop on Metrology for Aerospace, MetroAeroSpace 2019.
‡ All the authors contributed equally to this work.

Received: 12 January 2020; Accepted: 22 February 2020; Published: 27 February 2020

Abstract: The aim of this paper is to highlight how the employment of Light Detection and Ranging (LiDAR) technique can enhance greatly the performance and reliability of many monitoring systems applied to the Earth Observation (EO) and Environmental Monitoring. A short presentation of LiDAR systems, underlying their peculiarities, is first given. References to some review papers are highlighted, as they can be regarded as useful guidelines for researchers interested in using LiDARs. Two case studies are then presented and discussed, based on the use of 2D and 3D LiDAR data. Some considerations are done on the performance achieved through the use of LiDAR data combined with data from other sources. The case studies show how the LiDAR-based systems, combined with optical Very High Resolution (VHR) data, succeed in improving the analysis and monitoring of specific areas of interest, specifically how LiDAR data help in exploring external environment and extracting building features from urban areas. Moreover the discussed Case Studies demonstrate that the use of the LiDAR data, even with a low density of points, allows the development of an automatic procedure for accurate building features extraction, through object-oriented classification techniques, therefore by underlying the importance that even simple LiDAR-based systems play in EO and Environmental Monitoring.

Keywords: LiDAR technology; 2D LiDARs; 3D LiDARs; optical VHR data; WorldView-2; environmental monitoring; analysis and classification; building feature extraction; Object-Based Image Analysis (OBIA)

1. Introduction

An efficient exploration of the surrounding environment is becoming the key point for the future world where human functionalities are going to be substituted by automatic or semiautomatic systems. Self-driving cars and charging robots for electrical cars are just some examples where the knowledge of the space in which to move becomes essential for the success of complex operations. In this context, Light Detection and Ranging (LiDAR)-based systems are going to play an increasingly more important role because of their higher capability to discriminate smaller objects with respect to other active systems, such as RAdio Detection And Ranging (RADAR) systems. In fact, LiDAR is an active object detection method similar to RADAR, but it makes use of other parts of the electromagnetic spectrum, predominantly infrared light rather than radio waves. LiDAR measures reflected light emitted by a laser, which has a much smaller wavelength than radio signals, and therefore its capability to discriminate smaller objects is much higher, and consequently also its capability to solve the navigation problems.

Since the 1990s LiDAR technology spread and continued to search for ts commercial use, until its rapid growth in 2010s. In the last decade, it has found an increasing use in combination with other data sources for land surface cover analysis and classification [1–8].

An extensive additional list of references, highlighting the main LiDAR characteristics, its advantages and disadvantages, and the different fields of applications with several case studies, can be found in [9–13], showing how LiDAR has contributed significantly in the EO and Remote Sensing fields. The above references can be regarded as useful guidelines for researchers interested in using LiDARs.

In this paper, we aim to further extend the analysis of LiDARs systems, by enriching the list of references, and proposing a classification algorithm , showing the performance that can be achieved through the use of LiDAR data, when combined with data from other sources, in exploring external environments and extracting building features from urban areas.

Relevant literature about the building detection with the employment of LiDAR can be found in [14], where the most important articles for building extraction are listed, by highlighting the approaches and the auxiliary sources used in combination with LiDAR data. Further important literature about the building detection, based on the fusion of LiDAR data with optical remote sensing imagery, can be found in [15–18] .

Although there are studies that have concerned the building detection, by combining LiDAR data and optical imagery, the use of OBIA, in combination with optical satellite remote sensing data and low density LiDAR point cloud on extended areas, is not common in a simple LiDAR-based system version. For instance, the review in [14], summarizing the LiDAR point cloud density and covered areas for the most relevant papers, highlights that the dataset size is always less than one km^2, and in the few cases in which it is greater than one, also the point cloud density increases, far beyond the unity. In this paper, we applied the proposed algorithm to an area of 5.3 km^2, after testing it on a smaller area of only 200 square meters, with a LiDAR point cloud density equals approximately to 1.5 points per square meter (ppsm). Note that, although the work in [14] was published in 2015, if we look for other applications of the OBIA technique, in recent years, in similar conditions, no other results are comparable with those reached with the proposed algorithm, where such a low LiDAR point cloud density has been used in combination with optical imagery on wide urban areas. Moreover, the transferability from the tested area to the wider area has been carried out without any modification of the proposed algorithm.

Two Case Studies, based on the use of 2D and 3D LiDARs (2-dimensional and 3-dimensional LiDARs) are presented and analyzed in an extensive way.

The Case Studies show how the LiDAR-based systems succeed in improving the analysis and monitoring of specific areas of interest, specifically how LiDAR data help in exploring external environment and extracting building features from urban areas.

In particular, the case study of Lioni demonstrates that the use of the LiDAR data, even with a low density of points, allows the development of an automatic procedure for accurate building feature extraction, through object-oriented classification techniques, therefore by underlying the importance that even simple LiDAR-based systems play in EO and Environmental Monitoring. In the second case study, a measurement campaign has been carried out to test among others the 3D LiDAR MRS 6000 by the SICK company, a new device very useful for research purposes, resulting in an efficient tool for monitoring external environments, made by buildings, trees, roads, and other objects.

2. The LiDAR System

LiDAR is an active remote sensing technique used for performing high-resolution topographic surveys that operates in the visible or near-infrared regions of the electromagnetic spectrum. The survey could be carried out in flight (data collected using drones or airplanes), in space (data collected using satellites), or land-based (data collected from the ground). The very high-speed data acquisition and high ground resolution are the main characteristics of a LiDAR system [19]. The light pulse emitted by the transmitting device is reflected by a target to the sensor, and the system calculates the relative distance between them. By scanning a surface of interest, a cloud of points is created, that discriminates the points relative to the ground (represented through a Digital Terrain Model (DTM)) and those relative to the "objects" on the ground (represented through a Digital Surface Model (DSM)). By measuring the vegetative cover, and penetrating up to the ground, information is obtained on the altitudes with a centimetric accuracy. Many examples of LiDAR applications can be examined, for instance, from the punctual morphology representation of hydrogeological hazard areas, to the urban and infrastructure modeling and planning, to the design of power and energy distribution lines, and inventory and management of forests [9–13]. Moreover, the combination of LiDAR information with data from other sources can greatly help in improving the features' extraction and classification for Earth Monitoring and Remote Sensing as highlighted in [20] and discussed later in this paper.

In the most common cases, the LiDAR sensor synchronously scans the distance to the obstacles in one plane. The result of a single scan is therefore a cloud of points in one plane only, and the LiDAR in this case will be referred to as a 2D LiDAR scanner. It is the most often used device in obstacle detection systems [21] and simultaneous localization and mapping (SLAM) [22]. It is then possible to obtain the third dimension by using a 2D LiDAR scanner, and, for example, by making the sensor's head move in the direction perpendicular to the scanned plane.

LiDARs can be also mounted on the top of a quadrotor, above the rotors line. The results are merged with the current height values and in this way the number of dimensions can be extended, if necessary, for the measurements acquisition. LiDAR-based systems can be therefore used also in vertical take-off and landing (VTOL) of Unmanned Aerial Vehicles (UAVs), by making these operations completely autonomous. An example of an autonomously flying system is that developed at the University of Warwick [23] for outdoor applications, where a LiDAR on the top of an Unmanned Aerial Inspection Vehicle (UAIV) creates a 3D map of surroundings and it allows the UAIV to return to the starting point by following the same path. Another interesting work for vertical take-off and landing (VTOL) of UAVs is presented in [24] for indoor applications. Another example similar to this latter is given by the Hokuyo UGR-4LX LiDAR, which has been integrated with a switching mechanism and provides an ordinary 2D LiDAR with the 3D LiDAR features. The function of sweeping the scanning planes can be embedded inside the module of a 2D LiDAR by simulating what a 3D LiDAR does: scanning a few planes simultaneously. In this way, a 3D point cloud is provided by a 2D LiDAR, without external mechanical elements [25].

Other examples of laser-based systems, mounted on UAVs, are those where aerial photogrammetry solutions need to be employed [26]. In this case, the use of LiDARs allows the realization of simple visual navigation systems able to monitor large areas of interest with a high-speed data acquisition and a high ground resolution. One of the main concerns with UAVs is that fast movements in air require good performances in avoiding obstacles. A very interesting work has been done and presented by the Massachusetts Institute of Technology in [27] to this end. A collision avoidance system is realized, based on a visual alert that is delivered to an operator when the UAV is approaching an obstacle and the warning is computed according to data from a laser range scanner. Therefore, LiDARs demonstrate to give UAVs a large and important contribution for navigation.

As above introduced the so-called 3D LiDARs are laser sensors integrated with synchronously moving platforms for efficient performing of a scan. Realistically, it is the next step of the single-layer sensors mounted on moving stands of a 2D LiDAR module or embedded inside. In a "pure" 3D LiDAR, instead, the sensor can acquire data points simultaneously in many layers. Data acquisition is faster and more comfortable and results into a broader number of possible applications. Robotics seems to be the main beneficent of the rapid development of 3D LiDARs [28–30] and this is in accordance with initial observations on the significant role that LiDAR-based systems are going to play for an efficient exploration of the surrounding environment, to allow human functionalities to be substituted by automatic or semiautomatic systems. Other examples of important applications for 3D LiDAR-based systems can be found in [13,31–33], where 3D LiDAR data are used in combinations with cameras and radars for the indoor exploration of inside tunnels, underground mines, and caves.

3. Environmental Monitoring and Extraction of Features in Urban Areas

Exploration of the environment and its knowledge must be timely and precise to be efficient. Understanding urban dynamics, growth, and changes, brought by the urbanization, is fundamental for managing the land resources and providing services responding to the requests, in these rapidly changing environments [34]. Moreover, accurate and timely information on the coverage of the urban soil is essential for the government policies. However an urban environment is extremely complex and heterogeneous, typically composed of built structures (buildings and transport areas), various vegetation coverings (parks, gardens, and agricultural areas), areas of bare soil, and water bodies.

Remote sensing resources come to help in this regard, especially VHR images and LiDAR data, along with new analysis techniques, that allow to greatly improve mapping and classification of urban areas.

Among the different classification techniques, OBIA has attracted significant attention in recent years. Specifically, OBIA is a technique in which the semantic information is not enclosed in the single pixel but is found in an image-object, or groupings of pixels, that have similar characteristics, such as color, texture, and brightness. The aim of this technique is, therefore, to determine the specific characteristics of an object, its geometric, structural and relational features, with the corresponding thresholds for classifying different objects in the image. Numerous studies have shown that environmental monitoring and extraction of features in urban areas highly benefit when OBIA is jointly used with LiDAR [35–39]. In all these papers, an object-oriented approach has been applied for urban cover classification, using LiDAR data. The difference is in the parameter selection or in the number of classes, that can refer for instance to building, pavement, bare soil, fine textured vegetation, and coarse textured vegetation, resulting in a knowledge base of rules, potentially applicable to other urban areas. In the last paper, [39], some considerations on shadows affecting the areas under analysis, are also given, by showing how this may limit their correct classification.

In this paper, as already underlined in the introduction, we aimed to show how the LiDAR-based systems succeed in improving the analysis and monitoring of specific areas of interest, specifically, how LiDAR data help in exploring external environment and extracting building features from urban areas.

Two Case Studies are presented, and, in particular, the Case Study of Lioni (Avellino, Italy) demonstrates that LiDAR data, even with a low density of points, used in combination with VHR satellite imagery, allow the development of an automatic procedure for accurate building feature extraction, through the OBIA technique, therefore by underlying the importance that even simple LiDAR-based systems play in EO and Environmental Monitoring, whereas in the second case study, the use of a new 3D LiDAR device is discussed and tested through a measurement campaign carried out at the premises of Military University of Technology (MUT) in Warsaw, Poland.

The work done to this end will be presented in the next sessions.

4. Extraction of Building Features from LiDAR Data and WorldView-2 Images Through OBIA

This session is going to highlight how the combined use of LiDAR data and the VHR optical satellite WorldView-2 images, processed through OBIA, can produce a detailed map of the buildings in an urban area.

Moreover, this specific case study has been chosen with the aim of presenting a simple and precise workflow for building feature extraction, to provide decision makers with a tool capable of producing information related to the territory and useful to different needs. The next sub-sessions will describe the proposed algorithm.

4.1. Methods and Data

4.1.1. Study Areas

This Case Study focuses on the urban areas of Lioni, a municipality in the province of Avellino (AV), in the Campania region, located in the south of Italy. Lioni is located 550 meters above sea level (expressed at the point where it is located the Municipal House). The geographical coordinates of Lioni are Latitude 40°52′ North and 15°11′ East.

The proposed algorithm was first applied to an area of Lioni, chosen randomly as tested area and extended approximately 200 m^2.

Later, the same algorithm was applied to the whole area of Lioni, covering 5.3 km^2, to verify its reliability and transferability. Results demonstrate the transferability of the proposed algorithm, without any parameter modification, as shown ahead in this paper.

The town of Lioni and its surrounding area are shown in the Figure 1, where the reference Google Map has been retrieved in October 2019.

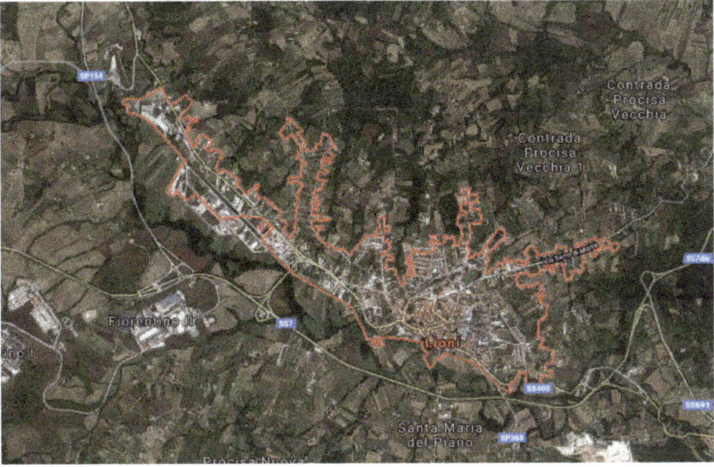

Figure 1. Lioni: Immages@2019 Google, Immages@2019 CNES/Airbus, Landsat/Copernicus, Maxar Technologies, Cartographic data (October 2019).

4.1.2. Datasets

The proposed classification algorithm is shown in the Figure 2, where LiDAR data, WorldView-2 images, the Regional Technical Chart (CTR), and orthophotos are the inputs, as discussed briefly in the next paragraphs.

- **LiDAR data**
  The Digital surface model (DSM) and the Digital Terrain Model (DTM) of Lioni were acquired from LiDAR dataset, property of the Italian Ministry for the Environment and the Protection of the Territory and the Sea, in conjunction with the Extraordinary Plan of Environmental Remote Sensing (EPRS-E) [40]. The acquisition year dates back to 2011 and the point cloud density is approximately 1.5 ppsm.

  Some considerations must be made about the ppsm of LiDAR data. Some case studies, presented in the literature, highlight the difficult to provide accurate models for buildings when low-density LiDAR data are used [41,42]. For example, in [41] the authors showed that is necessary a density point cloud higher than 5 ppsm in order to have better classification results. However, the data provided free of charge in many databases often cover vast areas but they have a point cloud density below 5 ppsm (such as the database of the Italian Ministry in the context of EPRS-E). Therefore, it is necessary to develop a specific strategy for features' extraction over large areas with this type of data. Furthermore, the transferability of the strategy must also be carefully analyzed and discussed when these data are used.

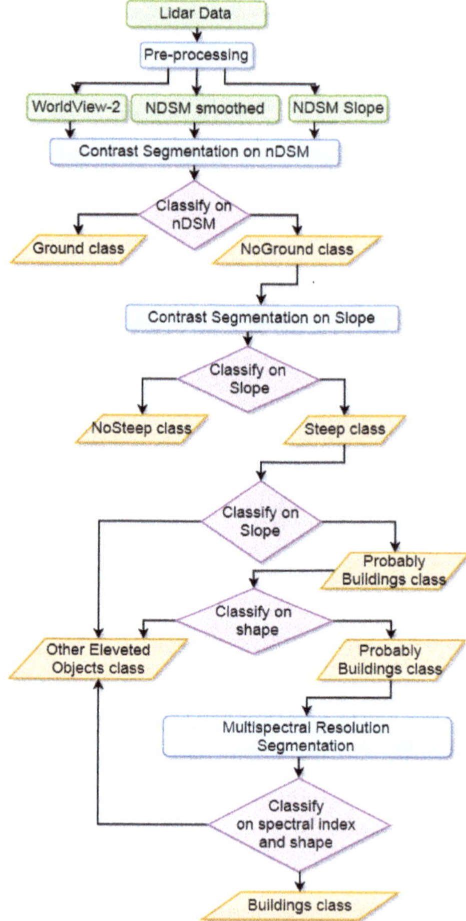

Figure 2. Workflow of the developed rule-based classification algorithm.

- **WorldView-2 image**
The WorldView-2 image on the area of Lioni was acquired by the satellite on 5 June 2017 at an angle of 16°, with cloud cover equal to 1% . The WorldView-2 image includes a panchromatic image with eight multispectral bands (Coastal Blue, Blue, Green, Yellow, Red, Red Edge, Near-Infrared-1, and Near-Infrared-2).
The data were purchased by Mapsat (the company involved as a partner in this research) within the project "Asbesto 2.0", and downloaded from the DigitalGlobe image archive, as a standard Ortho-Ready product projected on a plane with a UTM projection (Universal Transverse of Mercator) and a WGS84 datum. WGS84 is the abbreviation for World Geodetic System 1984,

a geodesic, worldwide geographic coordinate system.
A pan-sharpening and orthorectification procedure was performed by Mapsat. The image was pan-sharpened using the algorithm developed by Yun Zhang at the University of New Brunswick, in New Brunswick, Canada [43]. The pan-sharpening was performad to "transfer" the geometric resolution of the panchromatic datum, 0.5 meters (m), to the multispectral datum (with a 2 m resolution), to generate a final image with a resolution of 0.5 m, and improve the spatial information associated with the different bands. The orthorectification was applied using the using the CE90 standards [44] and the rigorous Toutin's model [45], in accordance with the UTM 33 projection system, to have an error less than 4 m.
The flowchart in the Figure 3 presents the main processing steps, as above described.

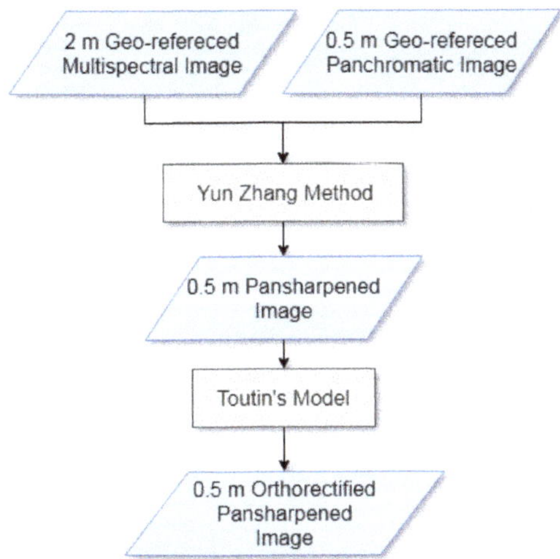

Figure 3. Flowchart for WorldView-2 image preprocessing.

- **Regional Technical Chart (CTR) and orthophotos**
  To validate the results from the proposed algorithm, both the CTR of the Campania region and orthophotos, obtained under the license granted by the Agricultural Dispensing Agency of the Ministry of Agricultural and Forestry Policies, have been used. This is usually done, because the available public data are often not recent. In this case, the CTR dated back to 2004, at a scale of 1:5000, and the orthophotos were instead acquired in 2011, at a scale of 1:10,000 [46]. Therefore some buildings present in the orthophotos were not present in the CTR, because of new construction. On the other hand, some buildings in the CTR were demolished before the orthophotos survey, and therefore were not present in the 2011 data. The validation process will be discussed later in the paper.

4.1.3. Processing LiDAR Data

The LiDAR DSM and DTM were processed in a QGIS enviroment [47], to generate three separate raster datasets: a normalized DSM (nDSM), a nDSM smoothed (Figure 4), and a slope map (Figure 5). A nDSM is generated by subtracting the DTM from the DSM, and it is necessary to identify high-rise objects in a DSM. High-rise objects in fact have a local nature due to the topography of the terrain surface [16]. Therefore, to eliminate the effects of topography from the DSM, the latter must be normalized [48]. Moreover, to get a better image quality for the nDSM, a smoothed nDSM is obtained by using a complex interpolation method, such as the *cubic B spline*. The objective of this method is to make the interpolation curve more smoothly and the image edges more defined. The smoothing operation is also necessary to adapt the geometric resolution of the LiDAR data (1 m) to the geometric resolution of the WordView-2 image (0.5 m), for matching the information from different data sources. Before the smoothing, groups of neighboring pixels in the WordView-2 image correspond to the same elevation (Figure 6a); after the smoothing, a different elevation is associated to each pixel (Figure 6b). In addition, as also shown in the figure, the magnified image, obtained by using the *cubic B spline* interpolation method, has no longer the so-called *saw tooth phenomenon* [49].

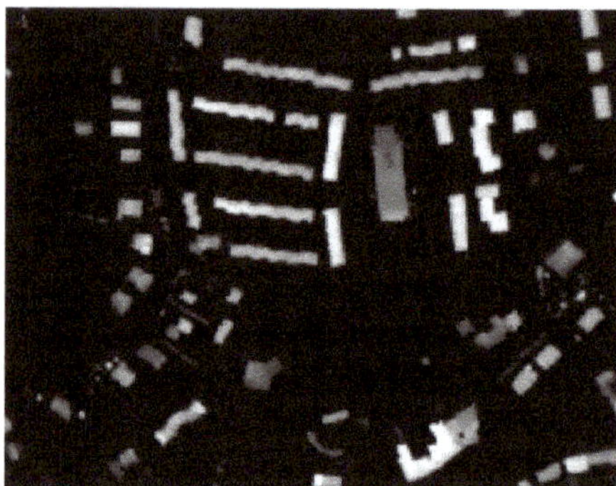

Figure 4. The normalized Digital Surface Model (nDSM) smoothed.

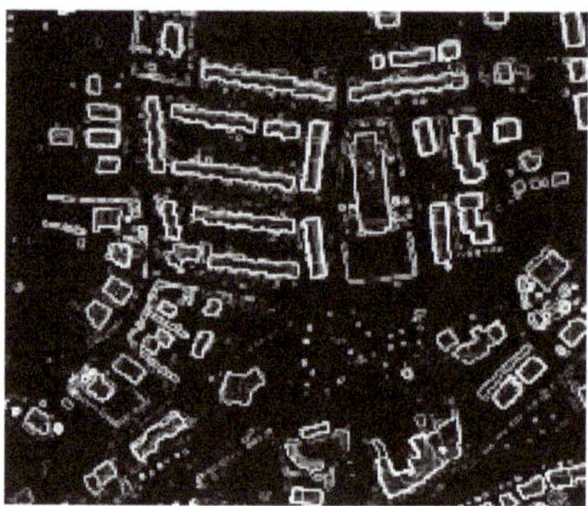

Figure 5. The slope map.

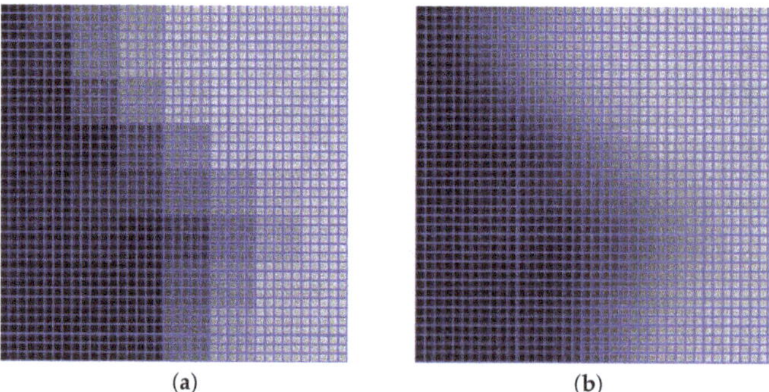

(a) (b)

Figure 6. Pixels of the WordView-2 image before (a) and after the smoothing (b). Courtesy of the authors of [50].

Processing the smoothed nDSM is also necessary in order to obtain information about the slopes of the objects present in the image under analysis. The slope map is calculated from the smoothed nDSM using the Zevenbergen Thorne method [51], implemented in QGIS. This method allows to have greater detail in the calculation of the curvature and is preferable with respect to other algorithms, like, for instance, the "Horn method", in those applications where the processing of numerical data requires an exact definition of the parameters [52].

The slope represents a very important information, since it allows all buildings to be discriminated in the area of interest. In fact (with some exceptions in specific cultures which, for instance, tents are used for dwellings) all above ground artificial objects provide a strong and sudden change in height, forming a crisp edge around them [53]. Figures 4 and 5 show, respectively, the nDSM and the slop map. The processing of LiDAR data, carried out and described above, has represented the starting point for the development of the proposed classification algorithm, used for the extraction of buildings' features in the urban areas.

The flowchart in the Figure 7 summarizes the main processing steps above described.

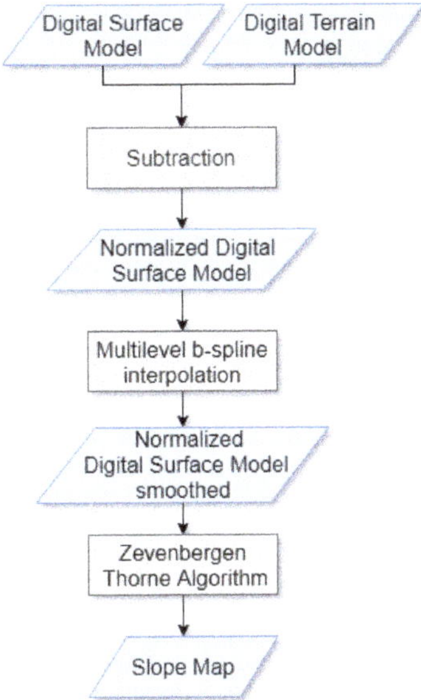

Figure 7. Flowchart for preprocessing LiDAR data.

4.1.4. Object-Based Classification Method

In recent years, OBIA has proved to be a very efficient method for classifying high-resolution images, because, as already highlighted in the paper, this technique relies on semantic information not enclosed in the single-pixel, but in an object of the image, that is in groups of pixels that have similar characteristics, such as color, texture, brightness.

The main steps in the development of the rule set for achieving the object-based classification are (1) the segmentation and (2) the feature extraction and classification.

The segmentation divides the image into separate and homogeneous regions (objects). The particularity of this operation is to use not only the spectral characteristics, but also the geometrical, structural and relational properties of an object. After the segmentation, the pixels of the image are grouped and, consequently, much information is available for each object (for example the spectral signature, the shape, and the size or the context), and in addition, features concerning the interrelations between the objects will also be available. All these attributes can be applied and combined for the development of a rule set to classify the objects in the scene of interest.

The feature extraction is the next step necessary to detect the objects of interest. As each object is characterized by particular features that distinguish it from other objects in the image, the feature extraction works in identifying these characteristics.

The rule-set creation is based on the analyst's knowledge of the spatial, spectral, textural, and elevation characteristics of each feature. For instance, buildings have a different elevation pattern with respect to roads, roads are elongated compared with other features, trees have a coarser texture than grass, and various roof materials vary in texture and spectral characteristics. This type of human knowledge and reasoning is applied to the OBIA processing to separate the targeted objects from unwanted features [54]. In the rule-based classification phase, once the attributes have been chosen, thresholds must therefore be defined, on the basis of which objects are assigned to each class. In this case, the characteristics, chosen for the building extraction, belongs to the following three main categories: Average band values, Geometric features, and Spectral Indexes, summarized in the following Table 1.

Table 1. Categories and features chosen for the building extraction rule-set.

Average Band Values	Geometric Features	Spectral Indexes
Mean (NIR1)	Area (pxl)	
Mean (Red)	Area (m^2)	$NDVI = \frac{Mean(NIR1) - Mean(Red)}{Mean(NIR1) + Mean(Red)}$
Mean (nDSM)	Lenght/Width	
Mean (nDSM smoothed)		
Mean (Slope)		

The software used for the classification has been eCognition Developer, a product developed by Trimble for the analysis and interpretation of images, which allows semi-automatic information to be extracted, through an object-oriented classification [55].

To avoid burdening the manuscript, the rules adopted for the building extraction have been collected in an Appendix A, and numerical values are presented in the Appendix A, as well. They refer mainly to the workflow shown in the Figure 2. Note that the choice of the specific features for the proposed algorithm belongs to the authors' expertise, based on the state-of-the-art knowledge and good sense and sound judgment, developed in practical matters related to the image analysis. Moreover, it must be underlined that, among the features shown in Table 1, some are more important than others. In fact, nDSM and slope, obtained from LiDAR data, represent the starting point for the development of the proposed classification algorithm.

The development of the set of carried out rules is, in fact based on the concept that the buildings are characterized by a sudden change of elevation at their edges and show a very different height, when compared to the surrounding environment. Therefore, for the extraction of buildings, the use of the information from the nDSM smoothed raster and the nDSM slope raster, previously described, is particularly important. Regarding other features, like, for instance, the Normalized Difference Vegetation Index (NDVI), the latter has been identified as a needful index for distinguishing vegetation, but this does not mean that other indexes cannot be added to improve the classification.
Extending the number of possible indexes can certainly represent a future development of this work.

4.2. Results Analysis and Accuracy Assessment

Figures 8 and 9 show the segmentation and classification of the first Lioni area used for testing the proposed algorithm.

To evaluate the accuracy of the buildings' detection, the classification results have been compare with CTR and orthophotos, in a GIS environment, but before analyzing the results, it is important to make some considerations.

The orthophotos and the CTR used for the validation refer to two different years, 2011 and 2004, respectively. As already explained, this is usually done, because depending on the specific Italian region, the available public data are often not recent. Therefore, in the specific case, some buildings present in the orthophotos were not present in the CTR, because of new construction. On the other hand, some buildings in the CTR were demolished before the orthophotos survey, and therefore were not present in the 2011 data. For these reasons, both data have been considered as reference data for the validation. Moreover, the total number of buildings, assumed as ground truth, has been considered equal to the sum of the buildings in the CTR (excluding those demolished) and the new buildings from the orthophotos. The total number of classified buildings has been instead calculated as the sum of the correctly classified buildings and the objects incorrectly classified as buildings (false positives).

Figure 8. Result of the contrast segmentation for a first area of Lioni.

Figure 9. Result of buildings extraction over the selected first area of Lioni.

The results of the classification and the references data have been reported in a confusion matrix from which the Producer Accuracy (PA) and the User Accuracy (UP) have been calculated.

PA has been calculated by the ratio of the correctly classified buildings and the total number of buildings. UP has been calculated by the ratio of the correctly classified buildings and the total number of classified buildings.

The developed strategy, performed first on the tested area, led to detection of 200 correctly classified buildings and 14 false positives, while the amount of buildings in the reference data was 220, by resulting into a PA and PU of 91% and 93.5%, respectively, as easily retrieved from the confusion matrix represented in the Table 2.

Table 2. Confusion matrix.

		Result of Building Extraction		
		Buildings	**Other**	**Total**
Reference data	**Buildings**	200	20	220
	Other	14	0	14
	Total	214	20	234

Figure 10 shows the results of the developed rule-set building extraction algorithm on the entire municipality of Lioni.

Figure 10. The WorldView-2 image and results of the proposed buildings extraction algorithm over Lioni town.

The amount of buildings in the reference data is 1538 and the objects classified as buildings were 1479 (1372 correctly classified and 107 false positives). Therefore, from the confusion matrix, as represented in the Table 3, PA and PU for the second area are 89% and 93%, respectively.

Table 3. Confusion matrix.

		Result of Building Extraction		
		Buildings	**Other**	**Total**
Reference data	**Buildings**	1372	166	1538
	Other	107	0	107
	Total	1479	166	1645

The low density of LiDAR points may have affected the refining of the edges of the buildings. Yet, the developed strategy remains valid since it provides new information at urban scale. In fact, the results can be used to produce or update a municipal database as shown in the Figure 11 and it can be used also for updating the CTR as shown in the Figure 12.

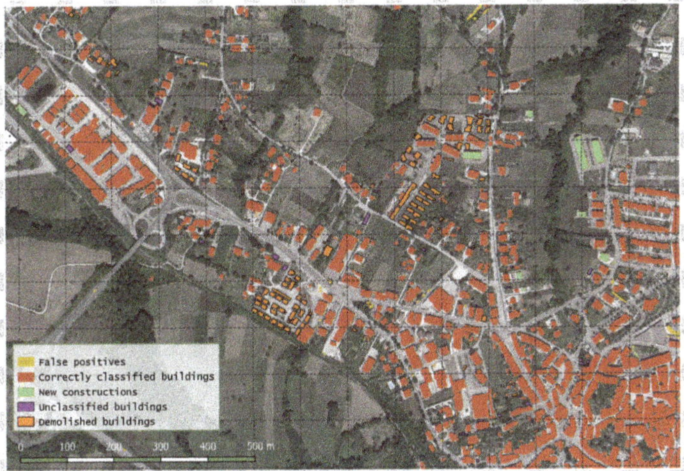

Figure 11. Example of updated municipal map and database.

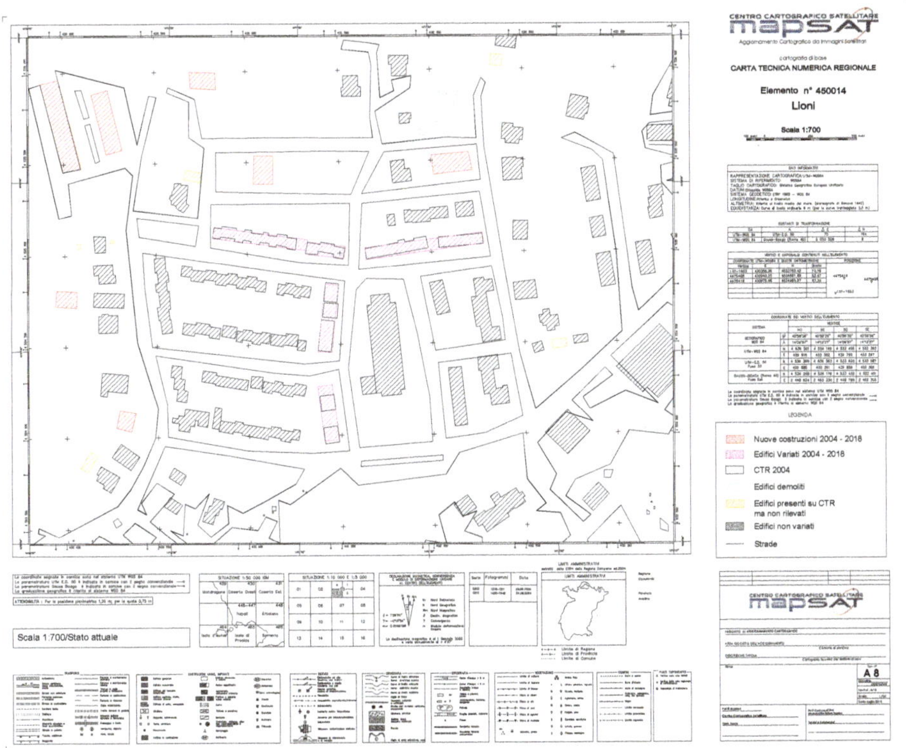

Figure 12. Updating of the CTR based on the results of proposed algorithm.

Also, to give an idea of the amount of resources required for running this algorithm for building extraction, computational cost (Table 4) and computer characteristics (Table 5) have been registered. However, the time complexity depends on the computer that is used and on the considered area.

Table 4. Computational cost.

Main Steps of Building Extraction	Sw	Mean Computation Time
Pre-processing	QGIS	5:15.376 (5 min and 15.376 s)
Processing	eCognition	21:59.784 (21 min and 59.784 s)

Table 5. Computer characteristics.

Processor	Ram	System Type
Intel(R) Core(TM) i3 CPU 530 @2.93 GHz	4.00 GB (2.99 GB available)	32-bit operating system

The obtained results have demonstrated how the use of the LiDAR data, through the OBIA technique, can allow to obtain useful information concerning the urban land cover. Moreover, the low density of LiDAR points did not significantly affect accuracy in the algorithm developed. Furthermore, the development of a rule-set mainly based on objects information (such as height and slope), obtained from the LiDAR survey, has made possible to transfer the proposed approach to study areas of different sizes. Note that, in applying the proposed algorithm to the second area of the whole Lioni city, the rule-set has not been modified and parameters have not been adjusted. In addition, the LiDAR data used in this case are free-of-charge, because they were downloaded from public databases or they were collected through direct measurement acquisitions, as in the second Case Study presented in the next section. Unfortunately, this is the reason why one has to settle for the low density of the LiDAR point cloud in the case of public databases. Yet, it has been demonstrated that good performances have been achieved in any case, if results from the extended literature presented are used for comparison.

Future work will use more information from LiDAR data, for example, taking into account also the intensity, in order to further improve the building extraction on large areas, only using free-of-charge LiDAR data.

5. External 3D Scanning with a LiDAR Located at a Low Altitude

In this section, the objective is to show how 3D LiDARs can further help in monitoring the surrounding environment, and in identifying objects, such as buildings, trees, roads, etc. In this second case study, 3D LiDARs and other devices are presented and tested. Among them, the 3D LiDAR MRS 6000 by the SICK company, shown in Figure 13, is a very useful tool for research purposes. It will be referred to as MRS 6000 or simply LiDAR later in this section.

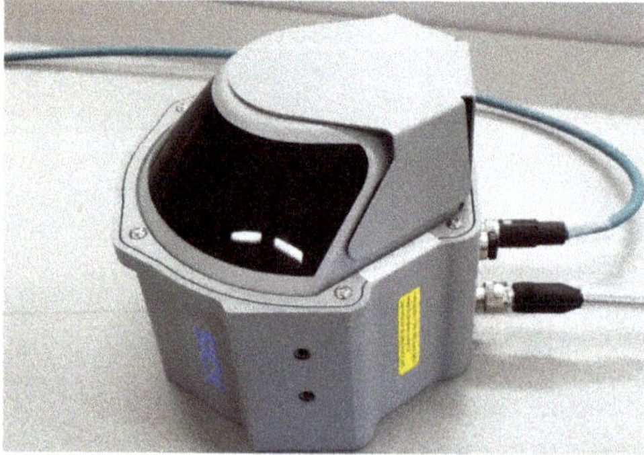

Figure 13. 3D LiDAR MRS 6000 by the SICK company.

MRS 6000 has a multilayer sensor able to scanning simultaneously in all the layers. This device is an optoelectronic LiDAR sensor that uses a 24-layer non-contact laser beam, as shown in Figure 14, with a frequency of 10 Hz, which makes a 3D point cloud and creates an accurate representation of its surroundings in real-time. A unique mirror technology makes MRS 6000 a sensor that has a high scanning field stability. Using a Polygon Mirror, the LiDAR generates a straight scan field line without distortion. The LiDAR provides nearly gap-free detection through the ability to generate up to 880,000 measurement points per second (mpps). In this way, a scanning field, created on the distance of 25 meters from a sensor, contains 72 ppsm, throughout an entire aperture angle of 120 degrees horizontally and 15 degrees vertically. With a scan point density at 0.13 degrees horizontally and 0.625 degrees vertically, the distance between two individual measuring points is thereby 4.3 millimeters. LiDAR has a minimum range of 0.5 meters and a maximum range of 200 meters, but only 30 meters range at 10 percent of emission (Figure 15). In the sensor, a multi-echo evaluation is applied for increasing reliability. The LiDAR uses a calculation of time-in-flight of the emitted pulse to set up distance between the sensor and an object. Multi-echo characteristic allows to evaluate up to four echo signals for each measuring beam, and this helps in working under unfavorable weather or adverse environmental conditions, like fog, rain, or dust in the air. The data interface is 1 Gb/s Ethernet with a web app embedded.

Besides the use of the MRS 6000 LiDAR, additional scans of the earth's surface may be supplemented through the use of small UAVs moving at the height of several meters above the ground. They may allow scans by using high-resolution multilayer LiDAR that provides detailed information about objects in the area of interest. During the test, scans of the surface were made using a multilayer LiDAR and a photo of the area covered by the scan. Another Sick device, the MRS 6124R LiDAR, was used and the SOPAS ET software for visualization of the measurements was employed (Figure 16).

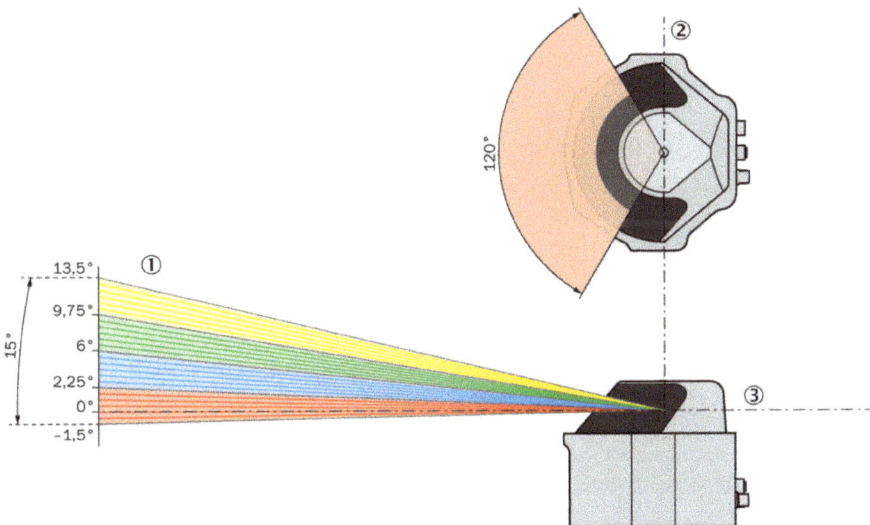

Figure 14. MRS 6000 range of scanning. Source: https://cdn.sick.com.

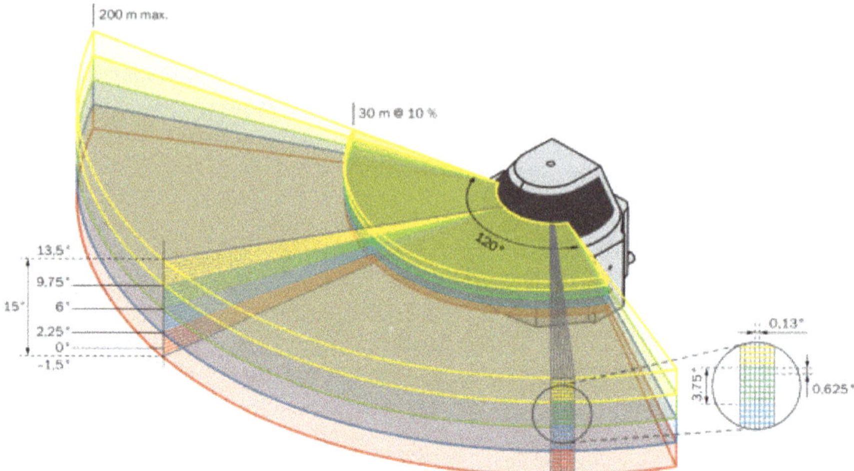

Figure 15. MRS 6000 range of scanning. Source: https://cdn.sick.com.

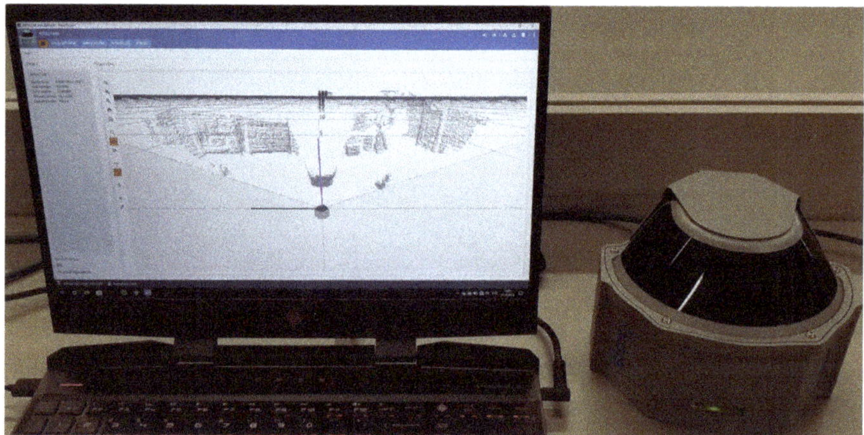

Figure 16. MRS 6124R testing with SOPAS ET software.

Scans and photos were taken from a height of 9 meters, which may correspond to the measurement from the sensors of a low-flying drone. The main idea of the tests was to determine the quality of the scan by assessing the measurement of range in low-altitude scans almost parallel to the earth's surface. Such experiments were to imitate the measurement from a drone flying at a low altitude. Another reason was to learn about the possible form of object visualization in the acquired scan. Particular attention was paid to building identification, which is the essential task in this paper. Some Test Areas at the premises of MUT (Military University of Technology) in Warsaw, Poland, have been selected.

In the Test Area 1, the reference point was a corner of the building located 57 meters from the scanner (Figure 17). The distance was determined using a laser rangefinder and shown on the image retrieved from Google maps (Figure 18). The building's wall 8 meters away from the scanner and clearly visible in the scan foreground was also marked. It was found that the building elements are well identifiable, and the distances designated to the characteristic points of the building are consistent with the actual values (Figure 19).

In Test Area 2, the construction site buildings are located between 50 and 70 meters from the scanner. Cars are parked at the construction site. The gate and fence of the construction site are made of metal painted with matte paint. One of the vehicles (blue) has a matte paint. Other cars have metallic paint (Figure 20). The scans show the construction site fence and the open gate. The side of the blue van is also visible on the scan. The other two cars with metallic paint are not visible on the scan (Figure 21). On both scans, you can also see how the trees are visible, which, in this case, can be considered noise.

Figure 17. Panoramic picture of the Test Area 1.

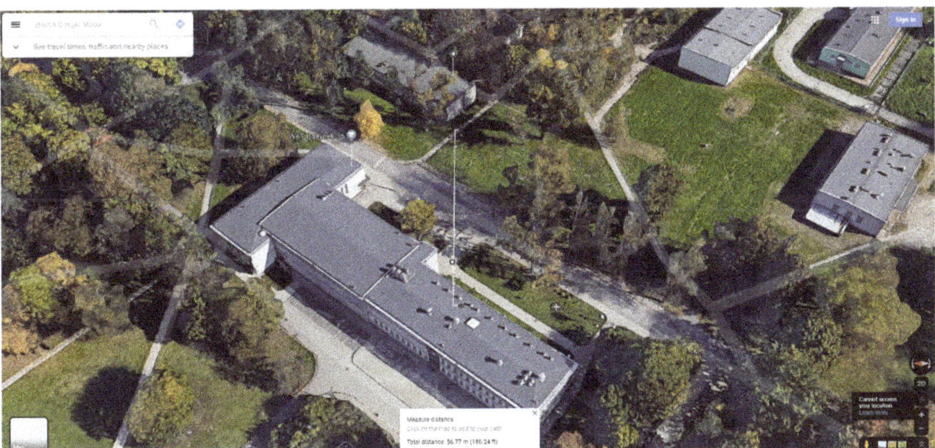

Figure 18. Google Maps view of the Test Area 1 with the marked distance.

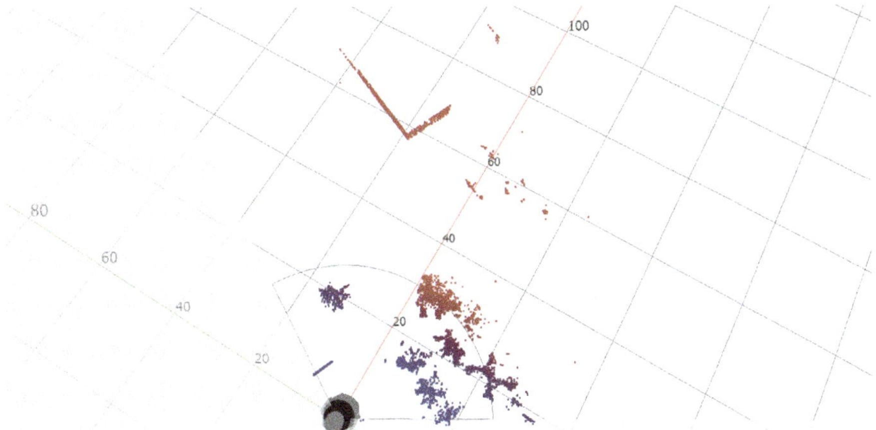

Figure 19. *Cont.*

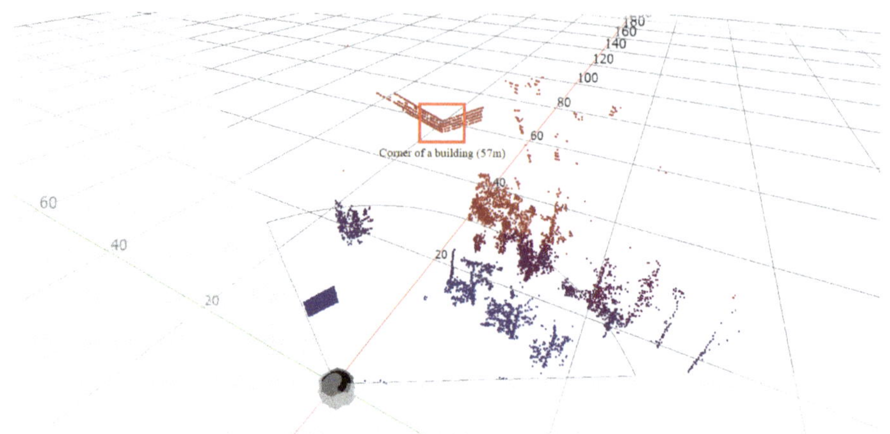

Figure 19. Scan of the Test Area 1.

Figure 20. Panoramic picture of the Test Area 2.

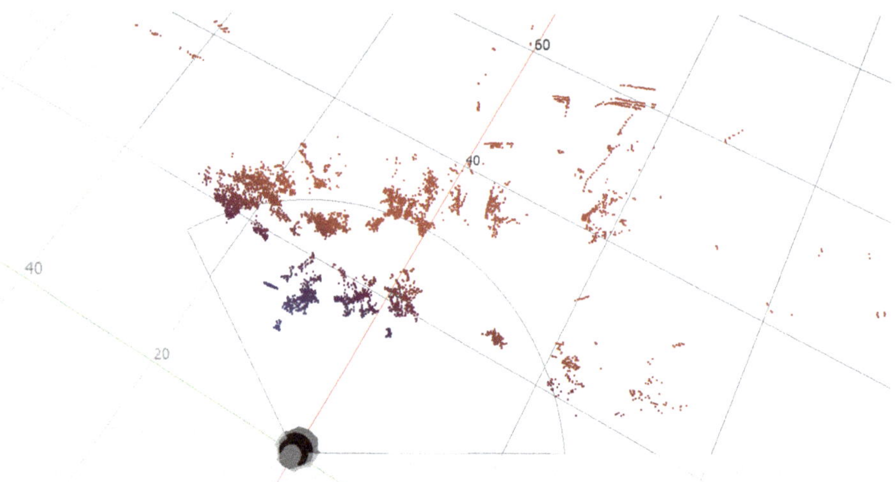

Figure 21. *Cont.*

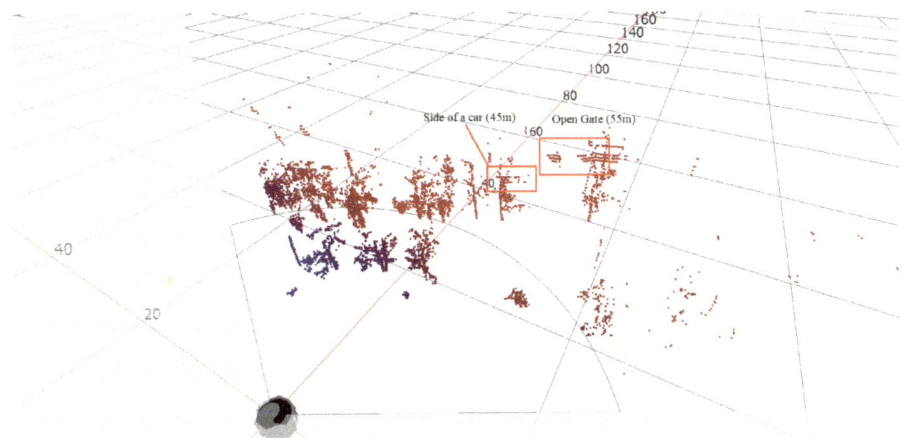

Figure 21. Scan of the Test Area 2.

Analysis of scans and the corresponding photos showed that the laser beam is almost wholly deflected on reflective surfaces. As a result, reflective objects are partially mapped or utterly invisible on the scan. Objects visible through the glass are correctly mapped on the scan unless other objects are reflected in the glass. In this case, the mapped objects are superimposed on the mapped objects behind the glass. On rough surfaces, part of the energy is lost due to shadowing. Curved surfaces produce instead a higher diffusion, and finally dark surfaces reflect the laser beam worse than clear ones. Note that the surface characteristics reduce the scanning range of the device, in particular for surfaces with low emission values. This second Case Study has allowed demonstrating how 3D LiDARs can be applied to analyze the external environment by further helping in discriminating the several objects present there, with respect to 2D LiDARs. Specific devices by Stick have been used for the tests by showing that they perform correctly, despite the presence of objects, such as cars, trees, and others, that can be regarded as "noise".

6. Conclusions

In this paper, we aimed to present a detailed analysis of LiDARs systems by pointing out the performance that can be achieved through the use of LiDAR data when combined with data from other sources. Two Case Studies, based on the use of 2D and 3D LiDARs (2- and 3-Dimensional LiDARs), have been presented and analyzed in an extensive way.

The Case Studies have allowed to show how the LiDAR-based systems succeed in improving the analysis and monitoring of specific areas of interest, specifically how LiDAR data can help in exploring external environment and extracting building features from urban areas. Moreover, the Case Study of Lioni has demonstrated that the use of the free-of-charge LiDAR data, even with a low density of points, allows the development of an automatic procedure for accurate building features extraction, through object-oriented classification techniques, therefore once again underlining the importance that even simple LiDAR-based systems play in EO and Environmental Monitoring.

Author Contributions: All the authors contributed equally to this work. All authors have read and agreed to the published version of the manuscript.

Funding: This research received no external funding.

Conflicts of Interest: The authors declare no conflict of interest.

Appendix A

As anticipated in Section 4.1.4, in this appendix, the detailed steps of the proposed algorithm are presented and discussed in detail.

After the creation of the smoothed nDSM, the first necessary step is to divide luminous and dark objects through a threshold that maximizes the contrast of luminous objects (consisting of pixel values above the fixed threshold) and dark objects (consisting of pixel values below the threshold).

This choice is made because the nDSM raster has high brightness in correspondence of the regions that have a certain elevation. Therefore, through the "Contrast Split Segmentation" on nDSM, it is possible to separate high objects and low objects.

After the first segmentation on the smoothed nDSM, all the objects with a value of Mean (nDSM smoothed) layer lower than 2 m are assigned to the class *Ground*, to exclude in the following steps all the objects that certainly do not belong to the building class. The remaining objects are assigned to the class *NoGround*.

Another Contrast Split Segmentation has been carried out on nDSM slope, as the generated slope layer from LiDAR data has strong contrasts of brightness on the edge of the buildings and the steep and flat areas are easily recognizable: the flat areas are associated with a very dark color, whereas the steep areas are tending to white. As reported also in other studies [56,57], the segmentation is based on the information of slopes related to the real footprint of the buildings .

All above-ground artificial objects provide a strong and sudden change in height, forming a crisp edge around them. Therefore, to fully encompass the outer building outline, the slope map is been reclassified into two classes, representing weak and strong edge in accordance to [57].

In particular, two passages of segmentation are necessary on the nDSM slope: one to obtain the objects with a strong slope, and a second one to separate those with an average slope from the very steep ones.

Therefore, after this second segmentation, based on the slope layer of objects classified as *NoGround*, all those with an average slope lower than 15° are assigned to the class *NoSteep areas*, because all those objects are characterized by a weak edge. The remaining objects are assigned to the class *Steep areas*. Next, all objects with an average slope grater than 30° are assigned to the temporary class *Probable buildings*. From the *Probable buildings* class, all the objects with an area with a number of pixels less than 4 (too few to be considered buildings), are excluded.

The next segmentation algorithm is the "Multiresolution segmentation", that groups areas of similar pixels (pixels with equal values) into objects. Consequently, homogeneous areas result in larger objects, heterogeneous areas in smaller ones. The "Multiresolution segmentation" algorithm joins together existing pixels or image objects. This type of segmentation in this study was realized by exploiting the information related to the WorldView-2 image bands to create objects characterized by a strong spectral similarity, to be used in the further classification phases.

Over the objects classified as *Probable buildings*, a multi-resolution segmentation is carried out in order to create a final class *Buildings*, and to improve the geometry of the buildings. To assign the objects to the final class of *Buildings*, the height has been considered again, and a further consideration has been undertaken, based on the NDVI. The high objects, probable buildings, with a high value of NDVI are considered as *Other Elevated objects*, as they are very likely to be trees.

Moreover in the final class, buildings with an area less than 14 square meters are not considered in the analysis. Therefore garages or smaller sheds are excluded from the classification, in order to avoid misclassification errors.

References

1. Available online: http://www.asprs.org/a/society/committees/lidar/Downloads/Vertical_Accuracy_Reporting_for_Lidar_Data.pdf (accessed on 24 May 2004).
2. Available online: https://www.marketsandmarkets.com/Market-Reports/airborne-lidar-market-171546900.html (accessed on 10 March 2020).

3. Lemmens, A. Airborne LiDAR sensors. *GIM Int.* **2007a**, *21*, 24–27.
4. Lemmens, A. Airborne LiDAR processing software. *GIM Int.* **2007b**, *21*, 52–55.
5. Lemmens, A. Airborne LiDAR sensors. *GIM Int.* **2009**, *23*, 16–19.
6. Lemmens, A. Airborne LiDAR processing software. *GIM Int.* **2010**, *24*, 14–15.
7. Renslow, M.S. *Manual of Airborne Topographic LiDAR*; American Society for Photogrammetry and Remote Sensing (ASPRS): Bethesda, MD, USA, 2012.
8. Rottensteiner, F.; Sohn, G.; Gerke, M.; Wegner, J.D.; Breitkopf, U.; Jung, J. Results of the ISPRS benchmark on urban object detection and 3D building reconstruction. *ISPRS* **2013**, *93*, 256–271. [CrossRef]
9. Zarro, C.; Ullo, S.L. High-resolution topographic surveys and earth features extraction through LiDARs. Discussion of some Case Studies. In Proceedings of the IEEE Metrology for Aerospace, Turin, Italy, 19–21 June 2019.
10. Yang, W.Y.; Shaker, A.; El-Ashmawy, N. Urban land cover classification using airborne LiDAR data: A review. *Remote Sens. Environ.* **2015**, *158*, 295–310. [CrossRef]
11. Fiscante, N.; Focareta, M.; Galdi, C.; Ullo, S.L. Analysis and validation of high-resolution satellite DEMs generated from EROS-B data for Montaguto landslide. In Proceedings of the IEEE International Geoscience and Remote Sensing Symposium, 2012, IGARSS 2012, Munich, Germany, 22–27 July 2012.
12. Ullo, S.L.; Addabbo, P.; Di Bisceglie, M.; Galdi, C.; Focareta, M.; Maffei, C. Combination of LANDSAT and EROS-B satellite images with GPS and LiDAR data for land monitoring. A case study: The Sant'Arcangelo Trimonte dump. In Proceedings of the IEEE International Geoscience and Remote Sensing Symposium, 2015, IGARSS 2015, Milan, Italy, 26–31 July 2015.
13. Rochala, Z.; Wojtowicz, K.; Kordowski, P.; Brzozowski, B. Experimental Tests of the Obstacles Detection Technique in the Hemispherical Area for an Underground Explorer UAV. *IEEE Aerosp. Electron. Syst. Mag.* **2019**, *34*, 18–26. [CrossRef]
14. Tomljenovic, I.; Höfle, B.; Tiede, D.; Blaschke, T. Building Extraction from Airborne Laser Scanning Data: An Analysis of the State of the Art. *Remote Sens.* **2015**, *7*, 3826–3862. [CrossRef]
15. Jamali, A.; Kumar, P.; Rahman, A.A. Automated extraction of buildings from aerial LiDAR point clouds and digital imaging datasets. In Proceedings of the International Archives of the Photogrammetry, Remote Sensing and Spatial Information Sciences, Volume XLII-4/W16, 6th International Conference on Geomatics and Geospatial Technology (GGT 2019), Kuala Lumpur, Malaysia, 1–3 October 2019.
16. Ekhtari, N.; Zoej, M.J.V.; Sahebi, M.R.; Mohammadzadeh, A. Automatic building extraction from LIDAR digital elevation models and WorldView imagery. *J. App. Remote Sens.* **2009**, *14*. [CrossRef]
17. Hasani, H.; Samadzadegan, F.; Reinartz P. A metaheuristic feature-level fusion strategy in classification of urban area using hyperspectral imagery and LiDAR data. *Eur. J. Remote Sens.* **2017**, *50*, 222–236. [CrossRef]
18. Zhang, J.; Lin, X. Advances in fusion of optical imagery and LiDAR point cloud applied to photogrammetry and remote sensing. *Int. J. Image Data Fusion* **2017**, *8*, 1–31. [CrossRef]
19. Available online: http://ambiente.regione.emilia-romagna.it/it/geologia/temi/costa/il-rilievo-lidar#autotoc-item-autotoc-7 (accessed on 22 December 2011).
20. Hadas, E.; Estornell, J. Accuracy of tree geometric parameters depending on the LiDAR data density. *Eur. J. Remote Sens.* **2016**, *49*, 73–92. [CrossRef]
21. Jackson, K.F. Development and Evaluation of a Collision Avoidance System for Supervisory Control of a Micro Aerial Vehicle. Master's Thesis, Massachusetts Institute of Thechnology, Cambridge, MA, USA, May 2012.
22. Lange, S.; Sunderhauf, N.; Neudert, P.; Drews, S.; Protzel, O. Autonomous Corridor Flight of a UAV Using a Low-Cost and Light-Weight RGB-D Camera. In *Advances in Autonomous Mini Roobots*; Springer: Berlin/Heidelberg, Germany, 2012; pp. 183–192.
23. Gageik, G.; Muller, T.; Montenegro, S. Obstacle detection and collision aviodance using ultrasonic distance sensors for an autonomous quadrocopter. In Proceedings of the 1st microdrones International Research Workshop UAVWeek 2012, Siegen, Germany, 20–21 November 2012.
24. Rochala, Z.; Wojtowicz, K.; Kordowski, P.; Brzozowski, B. Distance measurement technique in hemispherical area for indoor vertical take-off and landing unmanned aerial vehicle. In Proceedings of the IEEE Metrology for Aerospace (MetroAeroSpace), Benevento, Italy, 29–30 May 2014; pp. 62–67.
25. Gageik, N.; Strokmeier, M.; Montenegro, S. An Autonomous UAV with an Optical Flow Sensor for Positioning and Navigation. *Int. J. Adv. Robot. Syst.* **2013**, *10*, 341. [CrossRef]

26. Krajnik, T.; Nitsche, M.; Pedre, S.; Preucil, L.; Mejail; M.E. A simple visual navigation system for an UAV. In Proceedings of the 9th International Multi-Conference on Systems, Signals and Devices, Chemnitz, Germany, 20–23 March 2012; pp. 1–6.
27. Zsedrovits, T.; Zarandy, A.; Vanek, B.; Peni, T.; Bokor, J.; Roska, T. Collision avoidance for UAV using visual detection. In Proceedings of the IEEE International Symposium on Circuits and Systems 2011, Rio de Janeiro, Brazil, 15–18 May 2011; pp. 2173–2176.
28. Hsu, C.-M.; Shiu, C.-W. 3D LiDAR-Based Precision Vehicle Localization with Movable Region Constraints. *Sensors* **2019**, *19*, 942. [CrossRef]
29. Blanco-Claraco, J.L.; Mañas-Alvarez, F.; Torres-Moreno, J.L.; Rodriguez, F.; Gimenez-Fernandez, A. Benchmarking Particle Filter Algorithms for Efficient Velodyne-Based Vehicle Localization. *Sensors* **2019**, *19*, 3155. [CrossRef]
30. Im, J.-H.; Im, S.-H.; Jee, G.-I. Extended Line Map-Based Precise Vehicle Localization Using 3D LIDAR. *Sensors* **2018**, *18*, 3179. [CrossRef] [PubMed]
31. Huang, L.; Chen, S.; Zhang, J.; Cheng, B.; Liu, M. Real-Time Motion Tracking for Indoor Moving Sphere Objects with a LiDAR Sensor. *Sensors* **2017**, *17*, 1932. [CrossRef]
32. Zhang, W.; Qiu, W.; Song, D.; Xie, B. Automatic Tunnel Steel Arches Extraction Algorithm Based on 3D LiDAR Point Cloud. *Sensors* **2019**, *19*, 3972. [CrossRef]
33. Ren, Z.; Wang, L.; Bi, L. Robust GICP-Based 3D LiDAR SLAM for Underground Mining Environment. *Sensors* **2019**, *19*, 2915. [CrossRef]
34. Yang, X.; Lo, C.P. Using a time series of satellite imagery to detect land use and land cover changes in the Atlanta, Georgia metropolitan area. *Int. J. Remote Sens.* **2002**, *23*, 1775–1798. [CrossRef]
35. Zhou, W.; Troy, A. An object-oriented approach for analysing and characterizing urban landscape at the parcel level. *Int. J. Remote Sens.* **2008**, *29*, 3119–3135. . [CrossRef]
36. Zhou, W. An Object-Based Approach for Urban Land Cover Classification: Integrating LiDAR Height and Intensity Data. *IEEE Geosci. Remote Sens. Lett.* **2013**, *10*, 928–931. [CrossRef]
37. Samal, D.R.; Gedam, S.S. Monitoring land use changes associated with urbanization: An object based image analysis approach. *Eur. J. Remote Sens.* **2015**, *48*, 85–99. [CrossRef]
38. Blaschke, T. Object based image analysis for remote sensing. *ISPRS J. Photogramm. Remote Sens.* **2010**, *65*, 2–16. [CrossRef]
39. Zhou, W.; Huang, G.; Troy, A.; Cadenasso, M.L. Object-based land cover classification of shaded areas in high spatial resolution imagery of urban areas: A comparison study. *Remote Sens. Environ.* **2009**, *113*, 1769–1777. [CrossRef]
40. Available online: http://www.pcn.minambiente.it/mattm/progetto-piano-straordinario-di-telerilevamento (accessed on 10 March 2020).
41. Tomljenovic, I.; Rousell, A. Influence of point cloud density on the results of automated object-based building extraction from ALS data. In Proceedings of the AGILE 2014 International Conference on Geographic Information Science, Castellon, Spain, 3–16 January 2014.
42. Africani, P.; Bitelli, G.; Lambertini, A.; Minghetti, A.; Paselli, E. Integration of LiDAR data into a municipal GIS to study solar radiation. In Proceedings of the International Archives of the Photogrammetry, Remote Sensing and Spatial Information Sciences, Volume XL-1/W1, ISPRS Hannover Workshop 2013, Hannover, Germany, 21–24 May 2013.
43. Zhang, Y.; Mishra, R.K. From UNB PanSharp to Fuze Go—The success behind the pan-sharpening algorithm. *Int. J. Image Data Fusion* **2014** , *5*, 39–53. [CrossRef]
44. Available online: https://dg-cms-uploads-production.s3.amazonaws.com/uploads/document/file/38/DG_ACCURACY_WP_V3.pdf (accessed on 10 March 2020).
45. Toutin, T. Review article: Geometric processing of remote sensing images: Models, algorithms and methods. *Int. J. Remote Sens.* **2004**, *25*, 1893–1924. [CrossRef]
46. Available online: https://sit2.regione.campania.it/geoportal/catalog/search/resource/details.page?uuid=r_campan%3A%7BA33AF02A-B85B-46FA-A1EE-A70A29AB1F5D%7D (accessed on 10 March 2020).
47. Available online: https://www.qgis.org/en/site/ (accessed on 10 March 2020).
48. Ekhtari, N.; Sahebi, M.R.; Valadan Zoej M.J.; Mohammadzadeh, A. Automatic building detection from LIDAR point cloud data. In 21st ISPRS Congress, Commission, WG IV/3, Beijing, China, 3–11 July 2008.

49. Han, D. Comparison of Commonly Used Image Interpolation Methods. In Proceedings of the 2nd International Conference on Computer Science and Electronics Engineering (ICCSEE 2013), Hangzhou, China, 22–23 March 2013.
50. Available online: https://docs.ecognition.com/v9.5.0/eCognition_documentation/Modules/7%20Tutorials/Tutorial%20Overview.htm (accessed on 10 March 2020).
51. Zevenbergen, L.W.; Thorne, C.R. Quantitative analysis of land surface topography. *Earth Surf. Process. Landf.* **1987**, *12*, 47–56. [CrossRef]
52. Ciampalini, R.; Carnicelli, S. *Analisi Comparata di Algoritmi Morfometrici in GRASS*; Dipartimento di Scienza del Suolo e Nutrizione della Pianta, Università degli Studi di Firenze: Florence, Italy, 2003.
53. Lang, S.; Blaschke, T. Hierarchical object representation—Comparative multi-scale mapping of anthropogenic and natural features. In Proceedings of the ISPRS Archives XXXIV (Part 3/W8), Munich, Germany, 17–19 September 2003.
54. Hamedianfara, A.; Zulhaidi, H.; Shafria, M.; Mansora, S.; Ahmadc N. Improving detailed rule-based feature extraction of urban areas from WorldView-2 image and lidar data. *Int. J. Remote Sens.* **2014**, *35*, 1876–1899. [CrossRef]
55. Available online: http://www.ecognition.com/ (accessed on 10 March 2020).
56. Franci, F.; Lambertinia, A.; Bitella G. Integration of different geospatic data in urban areas: A casa study. In Proceedings of the Second International Conference on Remote Sensing and Geoinformation of the Environment (RSCy2014), Pafos, Cyprus, 7–10 April 2014.
57. Tomljenovic, I.; Tiede, D.; Blaschke, T. A building extraction approach for Airborne Laser Scanner data utilizing the Object Based Image Analysis paradigm. *Int. J. Appl. Earth Obs. Geoinf.* **2016**, *52*, 137–148. [CrossRef]

© 2020 by the authors. Licensee MDPI, Basel, Switzerland. This article is an open access article distributed under the terms and conditions of the Creative Commons Attribution (CC BY) license (http://creativecommons.org/licenses/by/4.0/).

Article

Attitude Sensor from Ellipsoid Observations: A Numerical and Experimental Validation †

Dario Modenini *, Alfredo Locarini and Marco Zannoni

Department of Industrial Engineering, University of Bologna, 47121 Forlì, Italy; alfredo.locarini@unibo.it (A.L.); m.zannoni@unibo.it (M.Z.)

* Correspondence: dario.modenini@unibo.it

† This paper is an extended version of our paper published in: D. Modenini, M. Zannoni "A High Accuracy Horizon Sensor for Small Satellites". In Proceedings of the 2019 IEEE 5th International Workshop on Metrology for AeroSpace (MetroAeroSpace), Turin, Italy, 19–21 June 2019; pp. 451–456. doi:10.1109/MetroAeroSpace.2019.8869676.

Received: 10 December 2019; Accepted: 8 January 2020; Published: 13 January 2020

Abstract: The preliminary design and validation of a novel, high accuracy horizon-sensor for small satellites is presented, which is based on the theory of attitude determination from ellipsoid observations. The concept consists of a multi-head infrared sensor capturing images of the Earth limb. By fitting an ellipse to the imaged limb arcs, and exploiting some analytical results available from projective geometry, a closed form solution for computing the attitude matrix is provided. The algorithm is developed in a dimensionless framework, requiring the knowledge of the shape of the imaged target, but not of its size. As a result, the solution is less sensitive to the limb shift caused by the atmospheric own radiance. To evaluate the performance of the proposed method, a numerical simulator is developed, which generates images captured in low Earth orbit, including also the presence of the atmosphere. In addition, experimental validation is provided due to a dedicated testbed, making use of a miniature infrared camera. Results show that our sensor concept returns rms errors of few hundredths of a degree or less in determining the local nadir direction.

Keywords: attitude determination; horizon sensor; ellipsoid

---

1. Introduction

The determination of spacecraft attitude from a set of vector observations is a recurrent problem, which has been extensively studied for many decades. One of the vector directions that can be profitably exploited for the attitude determination of an Earth orbiter is the local nadir. Indeed, horizon or limb sensors have been traditionally used for this purpose in many missions [1–4], as they offer reliability and relatively low cost. Standard algorithms for limb sensors rely on the knowledge of the horizon height to compute the nadir direction [3]; as a result, their accuracy, especially in low Earth orbit (LEO), is limited by the variability of the atmospheric layer being detected as the surface of the infrared Earth's spheroid [2].

In recent years, horizon sensors received renewed interest in the aerospace community, as they are becoming a convenient choice for attitude sensing in small satellite platforms, due to the availability of MEMS infrared sensors. In particular, thermopiles have been the most popular choice for miniaturized horizon sensors onboard of CubeSats and micro-satellites [5–8]. The reported implementations often require accurate modeling of the thermopile response, and the knowledge of the size and range to the imaged target, the Earth. Rated errors are quite variable, spanning from few degrees down to 0.1 deg.

In this paper, we present a novel horizon sensor prototype, which leverages on the availability of low-cost, highly miniaturized infrared cameras having increasingly higher image resolution, for enhancing its accuracy. A short version of this paper with preliminary results validated only through simulations was presented at IEEE MetroAero 2019 [9]: experimental tests are presented in this paper, along with new analysis.

The idea underlying our horizon sensor consists of using an infrared camera to capture one (or more) images of the Earth, and to solve the mathematical problem of attitude determination from the imaged ellipse generated by a target ellipsoid. In particular, it will be shown that exploiting the knowledge of the target shape (i.e., the oblateness), not only the local nadir direction (pitch and roll angles) can be retrieved, but in principle also the rotation about nadir (yaw angle), thus constraining the full attitude. If an image of the planet is made available, the first step is to detect the limb points on the image and fit those points to an ellipse. To this end, it is desirable to capture the entire planetary limb within the camera field of view (FoV), however, the feasibility of such an assumption depends on the orbit altitude. Indeed, for a LEO satellite, the Earth disk may cover more than 130°, which cannot fit within a wide camera FoV. We take, nevertheless, the LEO scenario as a reference to develop a horizon sensor concept, showing that very good attitude determination accuracies can be achieved by combining multiple partial views of the same planetary limb.

Note that the task of computing the attitude of a camera with respect to a celestial body is closely related to the optical navigation problem: this last aims at providing an estimate of the relative position between the spacecraft and an imaged planetary target, assuming the attitude known. Here we deal with the opposite situation, i.e., estimating the attitude when the relative position is known. Despite the optical navigation problem has received lots of interest [10–13], the attitude determination from images of celestial bodies seems to have received fewer attention [14,15]. In particular, the authors of [14,15] provide solutions to the three-axes attitude determination from the joint observation of the center of the (partially) illuminated planet and of the terminator limb. Such observations fix two directions in the camera frame, namely the observer-to-target direction and the target-to-sun direction. Once these are known, the attitude can be reconstructed using any method available for the classic attitude determination problem from vector observations, i.e., the so-called Wahba's problem [16]. In this work, instead, we aim at computing the attitude from limb fitting only, without knowledge of the terminator and of the Sun position. Furthermore, our method benefits from the detection of the full limb, thus it is best suited for images in the infrared spectrum, while the cited methods benefit from images in the visible spectrum, as they rely also on the detection of the terminator limb.

The theory upon which our horizon sensor measurement principle relies leads to a quite simple solution for estimating the attitude matrix, which closely resembles an orthogonal Procrustes problem [17]. With respect to some earlier results by the authors [18], the method is refined here in that it now provides the attitude without knowing the size of the target, rather only the shape is required, e.g., the polar and equatorial flattening coefficients. Such a feature is especially useful when dealing with planets having atmospheres, whose impact is to produce a limb shift, i.e., an increase in the effective size of the planet as seen by the sensor.

To the knowledge of the authors, the existing horizon sensor implementation closest to the one proposed herein is found in [19]. In that recent work, a monolithic thermopile array is used to capture a single, partial view of Earth limb, which is assumed as a perfect sphere. Then, two different approaches are considered for estimating the attitude from the extracted limb. The first one relates the attitude to the location of the limb within the image plane. This way, the attitude information depends on the knowledge of the nominal location of the limb in the image at the reference attitude, which in turns depends on the size of the Earth and range to the target. Alternatively, the limb can be fitted to a circle, and the orientation inferred by the location of the circle's center. This second option is in principle insensitive to the range and target size; however, it relies on the assumption of a spherical target and neglects the effects of the perspective transformation. Indeed, the center of the imaged circle/ellipse is not, in general, the projection of the center of the target (this is true only for a perfectly nadir-pointing camera). Therefore, any method that seeks to measure the attitude straight from the location of the imaged circle/ellipse center would suffer from such an error. As a final drawback, a single, partial view of the limb hardly provides a reliable fit of the entire "Earth disk".

The method proposed herein attempts to overcome all the above limitations, providing an algorithm for attitude determination, which: (a) makes no restrictive assumptions on the shape of the target, which can be a tri-axial ellipsoid; (b) properly accounts for the perspective transformation; and (c) does not depend on the target size and range to the target, thus being less sensitive to the apparent limb shift due to the atmosphere. As a by-product, the algorithm provides estimate also for the angle about nadir, hence constraining the full attitude, by making use of the non-sphericity of the target. However, this estimate is very coarse for an Earth orbiter, due to the very low flattening of our planet.

The manuscript is organized as follows: first, we recall the mathematical background of pinhole projective transformation and its action onto quadric surfaces. Then, we formulate the problem of attitude determination from an imaged ellipsoid by studying the transformation between an ellipsoid and the resulting imaged ellipse (Section 2). A covariance analysis of the estimate is also provided in Section 2. Sensor architecture is discussed in Section 3, taking as a reference micro- and nano-satellites as hosting platforms. Section 4 presents the simulation environment developed for testing the attitude determination algorithm through synthetically generated images, while the experimental testbed implemented for the validation of a sensor prototype is described in Section 5. Results of the numerical and experimental test campaigns are then provided in Section 6. Finally, a discussion on the main outcomes of the study is found in Section 7.

2. Theory of Attitude Determination from Imaged Ellipsoids

The theory of attitude determination from imaged ellipsoids relies on the algebraic representation of a known result from perspective geometry: the apparent contour of a quadric surface imaged by a projective camera is the intersection of the tangent cone to the ellipsoid, and whose vertex is lying on the camera center, with the image plane; this intersection is a conic.

In particular, under the roto-translation $T = R\begin{bmatrix} I & \boldsymbol{t}_w \end{bmatrix}$, the quadric Q transforms to the conic C on the image plane, according to [20]:

$$K^{-1}C^{*}K^{-T} \propto TQ^{*}T^{T}. \tag{1}$$

In Equation (1) Q is the matrix:

$$Q = \begin{bmatrix} 1/a^2 & 0 & 0 & 0 \\ 0 & 1/b^2 & 0 & 0 \\ 0 & 0 & 1/c^2 & 0 \\ 0 & 0 & 0 & -1 \end{bmatrix}, \tag{2}$$

a, b, and c, being the ellipsoid semi-axes lengths (for Earth c = polar radius, and $a = b$ = equatorial radius; even though the numerical and experimental validations presented in this manuscript take as a reference scenario the Earth spheroid, all the following theory holds for an arbitrary triaxial ellipsoid); R is the orthogonal attitude matrix mapping from the world frame, defined by the ellipsoid axes, to the camera frame (the unknown to be estimated); $\boldsymbol{t}_w$ is the translation vector from the camera center to the origin of the target ellipsoid expressed in world frame (assumed known); C^* is the inverse (or the adjugate) of the conic matrix C computed from the coefficients of the ellipse quadratic equation:

$$\begin{gathered} Ax^2 + Bxy + Dy^2 + Ex + Gy + H = 0 \leftrightarrow [x\ y\ 1]C\begin{bmatrix} x \\ y \\ 1 \end{bmatrix} = 0, \\ C = \begin{bmatrix} A & \frac{B}{2} & \frac{E}{2} \\ \frac{B}{2} & D & \frac{G}{2} \\ \frac{E}{2} & \frac{G}{2} & H \end{bmatrix}, \end{gathered} \tag{3}$$

and K is the intrinsic camera matrix, representing the projective transformation. For a pinhole camera with optical axis coincident with body axis z and the pixel array plane coincident with body $x - y$ plane:

$$K = \begin{bmatrix} f_x & \alpha & p_x \\ 0 & f_y & p_y \\ 0 & 0 & 1 \end{bmatrix}, \tag{4}$$

where f_x and f_y is the focal vector, p_x and p_y are the coordinates of the principal point, and α is the skew-angle (equal zero for orthogonal x-y axes).

Note that C matrix represents an ellipse in the image plane in homogeneous coordinates, as such it is invariant to a common scaling of all its elements.

By defining matrix $C = K^T C K$, Equation (1) can be rewritten as:

$$C^* \propto R\left(Q_3^* - \boldsymbol{t}_w \boldsymbol{t}_w^T\right)R^T = R\left(Q_3^* - \rho^2 \boldsymbol{v}_w \boldsymbol{v}_w^T\right)R^T, \tag{5}$$

where $C^* = K^{-T} C^* K^{-1}$, ρ is the range from the camera to the target, and $\boldsymbol{v}_w = \boldsymbol{t}_w/\rho$ is the line-of-sight unit vector. An expression for the projection of the point quadric Q into the conic C may also be derived, but it is more complicated than Equation (5). It retains, however, a clearer physical interpretation, as it relates the target ellipsoid to the tangent cone centered at the camera [21].

Defining $B^* = Q_3^* - \rho^2 \boldsymbol{v}_w \boldsymbol{v}_w^T$, Equation (5) becomes:

$$B^* \propto R^T C^* R. \tag{6}$$

Equation (6) is a modified form of the well-known orthogonal Procrustes problem, and is also known as the hand-eye calibration problem [22]. Now, assume that we do not wish the overall target size to be known, rather only the semi-axes ratios. We define a dimensionless matrix $\widetilde{B}^*$:

$$\widetilde{B}^* = \widetilde{Q}^* - \widetilde{\rho}^2 \boldsymbol{v}_w \boldsymbol{v}_w^T, \tag{7}$$

where $\widetilde{Q}^* = \operatorname{diag}\left(\left[1, \frac{b^2}{a^2}, \frac{c^2}{a^2}\right]\right)$ and $\widetilde{\rho} = \rho/a$ is an unknown dimensionless range. Equation (6) is rewritten as:

$$\widetilde{B}^* = \widetilde{Q}^* - \widetilde{\rho}^2 \boldsymbol{v}_w \boldsymbol{v}_w^T \propto R^T C^* R. \tag{8}$$

This matrix equation has now four unknowns, namely $\widetilde{\rho}$ and the three independent components of the attitude matrix. The idea is to split the solution in two steps:

1. Find $\widetilde{\rho}$ such that $\widetilde{B}^*$ and C^* are orthogonally similar matrices apart from an unknown scaling.
2. Once $\widetilde{\rho}$ and thus $\widetilde{B}^*$ are known, compute the attitude matrix as the solution of a modified orthogonal Procrustes problem.

To solve step 1, we recognize that a necessary condition for Equation (6) to hold is that any dimensionless rational function of the eigenvalues for the two matrices $\widetilde{B}^*$ and C^* shall be the same. One of such functions is, for example:

$$0 < \frac{\operatorname{tr}^2\left(\widetilde{B}^*\right)}{\operatorname{tr}\left(\widetilde{B}^{*2}\right)} = \frac{\operatorname{tr}^2(C^*)}{\operatorname{tr}(C^{*2})} = \frac{\left(\lambda_{c,1} + \lambda_{c,2} + \lambda_{c,3}\right)^2}{\lambda_{c,1}^2 + \lambda_{c,2}^2 + \lambda_{c,3}^2} = k_c, \tag{9}$$

where:

$$\operatorname{tr}\left(\widetilde{B}^*\right) = \operatorname{tr}\left(\widetilde{Q}^*\right) - \widetilde{\rho}^2, \operatorname{tr}\left(\widetilde{B}^{*2}\right) = \operatorname{tr}\left(\widetilde{Q}^{*2}\right) + \widetilde{\rho}^4 - 2\widetilde{\rho}^2 k_q, \tag{10}$$

and $k_q = \boldsymbol{v}_w^T \widetilde{Q}^* \boldsymbol{v}_w \leq 1$. Substituting Equation (10) into Equation (9) leads to the following second order equation in $t = \widetilde{\rho}^2$:

$$(1 - k_c)t^2 - 2\left[\operatorname{tr}\left(\widetilde{Q}^*\right) - k_q k_c\right]t + \left[\operatorname{tr}^2\left(\widetilde{Q}^*\right) - k_c tr\left(\widetilde{Q}^{*2}\right)\right], \tag{11}$$

which admits analytical solutions. Equation (11) will have two real roots, whether of different signs or both positive. In the former case, the correct root is the positive one. If both roots are positive, then one should check the one which better matches another dimensionless rational function of the eigenvalues of $\widetilde{B}^*$ and C^*, such as $\frac{\mathrm{tr}^3(\widetilde{B}^*)}{\det(\widetilde{B}^*)} = \frac{\mathrm{tr}^3(C^*)}{\det(C^*)}$. Knowing $\widetilde{\rho}^2$, matrix $\widetilde{B}^*$ is fully determined. Now, following the method devised in [18], we consider the spectral decompositions of C^* and $\widetilde{B}^*$:

$$C^* = VD_CV^T; \widetilde{B}^* = WD_BW^T, \tag{12}$$

with V, W orthogonal matrices. Then, it is easy to verify that by setting

$$R = VW^T. \tag{13}$$

Equation (16) is satisfied, provided that the eigenvalues are arranged in the same order relative to each other. Actually, any matrix of the form:

$$R = VSW^T, \tag{14}$$

with $S = \mathrm{diag}\{\pm 1 \pm 1 \pm 1\}$ being signature matrices, will be a solution, too. In [18], it was proved that Equation (13) provides an optimal estimate of R in a least squares sense. Of the eight possible solutions given by Equation (14) only four will have determinant equal to +1, thus being proper rotation matrices. Of these four, only two corresponds to the camera pointing towards the ellipsoid. As a result, there will be a two-fold ambiguity left in the solution that cannot be resolved, unless some additional independent information is available, e.g., past attitude history or angular information obtained from other sensors.

Note that the rotation matrix computed according to Equation (14) allows, in principle, to constrain the full attitude of the spacecraft when imaging a tri-axial ellipsoid. For a spherical target, the yaw angle (about nadir) is clearly unobservable due to spherical symmetry. When imaging a spheroid (i.e., an ellipsoid of revolution, such as the Earth), the capability of detecting the yaw angle depends on the vantage point: whenever the local vertical is aligned to the axis of revolution, then the symmetry of the target prevents yaw observability [18]. Furthermore, the detectability of the orientation about nadir is also degraded when the difference between the polar and equatorial semi-axes gets smaller: unfortunately, that is the case for a spacecraft horizon sensor targeting the Earth, which features a very low flattening (1/298). Nevertheless, it is interesting to assess whether at least a coarse estimate of the yaw angle can be achieved in such a scenario.

Covariance Analysis

The value of an estimate is as equally as important as the knowledge of its uncertainty. To provide an estimate of the estimated R matrix accuracy, we first aimed at a first-order perturbation analysis of Equation (14) subject to perturbations $\delta C^*, \delta\widetilde{B}^*$. The former accounts for error in the best-fitted ellipse to the detected limb points, the latter accounts for errors in the ellipsoid model and in the observation direction.

The derivatives of eigenvalues and eigenvectors of real symmetric matrices admit quite simple analytical formulations. In particular, for any real symmetric matrix A, having eigenvalue–eigenvector couples $(v_i, \mathbf{u}_i)$, subject to a perturbation δA, it holds [23,24]:

$$\begin{aligned} \delta\mathbf{u}_i &= (v_i I - A)^+ \delta A \mathbf{u}_i, \\ \delta v_i &= \mathbf{u}_i{}^T \delta A \mathbf{u}_i, \end{aligned} \tag{15}$$

where the superscript + denotes the Moore-Penrose pseudoinverse. Equation (15) allows us to compute the variation of the spectral decomposition of C^*; $\widetilde{B}^*$, as a function of some applied perturbations $\delta C^*, \delta\widetilde{B}^*$. Note that these equations indicate that the eigenvector perturbation lies in the plane orthogonal to the unperturbed eigenvector. Furthermore, they hold only for eigenvalues with multiplicity one, as multiple eigenvalues bifurcate, thus are non-differentiable [23]. The variation in the estimated R follows directly by differentiation of Equation (14):

$$\delta R = \delta VSW^T + VS\delta W^T. \quad (16)$$

When dealing with perturbed matrices for covariance computation, it is convenient to switch to the vectorized matrix representation using the "vec" operator, which stacks the columns of a matrix one underneath the other. To this end, we reshaped the first of Equation (15) in a vectorized form:

$$\begin{aligned} \text{vec}(\delta V) &= \begin{bmatrix} (\lambda_1 I - C^*)^+ \otimes \mathbf{v}_1{}^T \\ (\lambda_2 I - C^*)^+ \otimes \mathbf{v}_2{}^T \\ (\lambda_3 I - C^*)^+ \otimes \mathbf{v}_3{}^T \end{bmatrix} \text{vec}(\delta C^*) = M\text{vec}(\delta C^*), \\ \text{vec}(\delta W) &= \begin{bmatrix} (\mu_1 I - B^*)^+ \otimes \mathbf{w}_1{}^T \\ (\mu_2 I - B^*)^+ \otimes \mathbf{w}_2{}^T \\ (\mu_3 I - B^*)^+ \otimes \mathbf{w}_3{}^T \end{bmatrix} \text{vec}\left(\delta \widetilde{B}^*\right) = N\text{vec}\left(\delta \widetilde{B}^*\right), \end{aligned} \quad (17)$$

where $\otimes$ denotes the Kronecker product and $(\lambda_i, \mathbf{v}_i)$, $(\mu_i, \mathbf{w}_i)$ are the eigenvalue–eigenvector couples of C^*, $\widetilde{B}^*$, respectively. It is worth to be noted that, as we deal with 3×3 symmetric matrices, their eigen-decomposition admits a simple analytical solution [25].

Vectorization of Equation (16) leads to:

$$\text{vec}(\delta R) = (WS \otimes I)\text{vec}(\delta V) + (I \otimes VS)\text{vec}\left(\delta W^T\right), \quad (18)$$

where $\text{vec}\left(\delta W^T\right) = \text{vec}(\mathcal{P}\delta W)$, $\mathcal{P}$ being a permutation matrix. Equation (18) allows computing the desired attitude error covariance matrix defined as $P_{RR} = E\left[\text{vec}(\delta R)\text{vec}(\delta R)^T\right]$, through:

$$\begin{aligned} P_{RR} &= (WS \otimes I)P_{VV}(WS \otimes I)^T + (I \otimes VS)P_{WW}(I \otimes VS)^T \\ &= (WS \otimes I)MP_{CC}M^T(WS \otimes I)^T + (I \otimes VS)\mathcal{P}NP_{BB}N^T\mathcal{P}^T(I \otimes VS)^T, \end{aligned} \quad (19)$$

starting from the covariance of the errors affecting matrices C^* and B^*:

$$P_{CC} = E\left[\text{vec}(\delta C^*)\text{vec}(\delta C^*)^T\right];\ P_{BB} = E\left[\text{vec}\left(\delta \widetilde{B}^*\right)\text{vec}\left(\delta \widetilde{B}^*\right)^T\right]. \quad (20)$$

In deriving Equation (19) use has been made of the reasonable assumption of uncorrelated noise sources: $E\left[\text{vec}(\delta C^*)\text{vec}\left(\delta \widetilde{B}^*\right)^T\right] = 0$.

To check the consistency of the analytic covariance formula, we performed a set of Monte Carlo simulations, as follows. We generated 1000 synthetic images of an Earth-like spheroid as captured by a nadir-pointing camera, using Matlab® 3D scene control. All images are created assuming the same relative position and attitude between the camera and the target, and differ one from each other only by some additive random noise. For each image, the limb is detected and fitted to an ellipse. Then, the attitude determination algorithm is run, and the error δC^* between the true and estimated C^* matrices is recorded, along with the estimated attitude error, δR. Once all images are processed, we numerically estimated the covariance matrix P_{CC} from the recorded 1000 samples of δC^*. Having P_{CC} available, we computed for each test case the analytical covariance P_{RR} according to Equation (19) to be compared to the actual estimation errors.

The outcome of this process is summarized in Figure 1, where histogram distributions are shown for three off-diagonal elements of the δR matrix, $r_{i,j}$ $(i \neq j)$. Since the assumed nominal R matrix is the identity (null attitude) $r_{i,j}$ is equal, to first order, to the error angle about k-axis, so that $r_{1,2}$ $r_{1,3}$ and $r_{2,3}$ correspond approximately to the error angles about the third (yaw), second (pitch), and first (roll) axis, respectively. Superimposed to each histogram, the curve of a Gaussian distribution probability function is shown, having variance equal to the corresponding element in the main diagonal of the analytic P_{RR}.

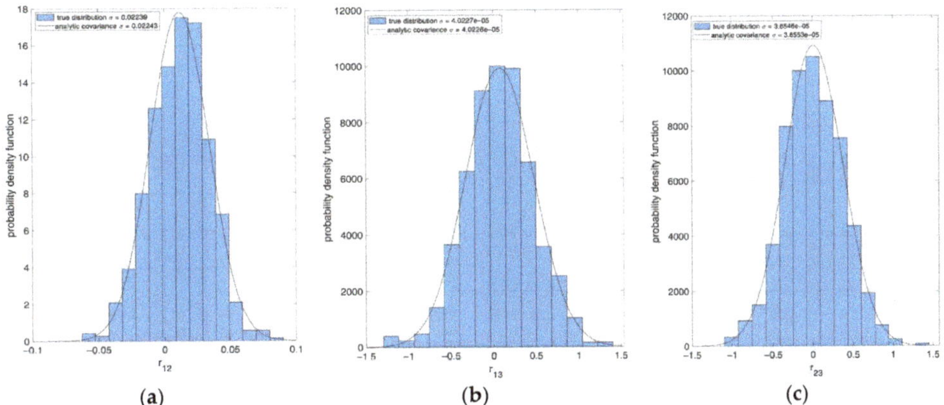

Figure 1. Comparison between the numerically retrieved error distributions and analytic variances for three off-diagonal elements of the attitude matrix. (**a**) $r_{1,2}$; (**b**) $r_{1,3}$; and (**c**) $r_{2,3}$.

Comparison of the numerical distributions with the theoretical variances shows very good matching. The simulations confirm also that the yaw angle (leftmost panel) was poorly observable, as expected for an imaged spheroid having very low flattening.

3. Sensor Architecture

Since computing the attitude from ellipsoid images relies on fitting an ellipse to a set of edge points, it is desirable to have the entire planetary limb lying within the camera field of view. The feasibility of such an assumption depends on the orbit altitude with respect to the planet size and, for LEO, it is hardly to be fulfilled, unless a very wide FoV camera is available, which would induce severe lens distortion. We take, however, this scenario as a target, since a small satellite platform in LEO orbit is deemed to be the most common application for such a kind of attitude sensor.

To cope with a target that exceeds the camera FoV, a multi-head sensor concept was envisaged: here, we required that three simultaneous views of limited limb portions were made available to properly constrain the full "Earth circle" [9]. We therefore considered a sensor assembly formed by a set of three miniaturized infrared camera cores. Two alternative implementations were envisaged, according to the hosting spacecraft platform: configuration (a) for the microsatellite class, and configuration (b) for the nanosatellite class, both being based on commercial, uncooled microbolometer. In particular, a FLIR Boson 320® and a FLIR Lepton® 3 were chosen as the reference for configuration (a) and (b), respectively. Their main characteristics are collected in Tables 1 and 2 [26,27].

Table 1. Physical characteristics of the IR camera core, FLIR Boson 320, taken as a baseline for a microsatellite targeted implementation of our horizon sensor prototype.

Image Size (px)	HFoV (deg)	Power (mW)	Weight (g)	Envelope (mm^3)
320 × 256	50	500 (operating) + 330 (shutter)	20 g	21 × 21 × 30

Table 2. Physical characteristics of the IR camera core, FLIR Lepton® 3, taken as a baseline for a nanosatellite targeted implementation of our horizon sensor prototype.

Image Size (px)	HFoV (deg)	Power (mW)	Weight (g)	Envelope (mm^3)
120×160	57	150 (operating) + 500 (shutter)	0.9 g	$10.5 \times 12.7 \times 7.14$

Measurement Principle: Limb Detection and Fitting

In our sensor conceptual design, the measurement is an ellipse extracted from the image, which consists of five independent scalar parameters. Estimating the ellipse parameters from an image requires two processing steps, namely:

(a) Edge detection;
(b) Edge fitting to an ellipse.

Both these topics are extensively studied in the image processing community, however, since they are not the core of this work, we will not pursue them in detail.

The multi-head limb sensor demands for an extension of the theory described in Section 2 (originally developed to handle a single ellipsoid view), to fuse the observations from different images into one best-fit ellipse, as follows. Let us first define a body reference frame, with respect to which the orientation of the three sensor heads is specified through matrices $R_{1/b}, R_{2/b}, R_{3/b}$, respectively. Then, it is easy to verify that, for each image i, the following transformed homogeneous limb pixel coordinates $\widetilde{x_i}$:

$$\widetilde{x_i} = R^T_{i/b} K^{-1} x_i, \tag{21}$$

are referenced to the common body frame. We then stack the transformed homogeneous coordinates of the limb points detected in the three images into a single vector, $\widetilde{x_{st}}$:

$$\widetilde{x_{st}} = \begin{bmatrix} \widetilde{x_1} \\ \widetilde{x_2} \\ \widetilde{x_3} \end{bmatrix}. \tag{22}$$

Fitting an ellipse to the stacked vector, the corresponding adjoint matrix will satisfy Equation (6), which provides the desired generalization of the attitude estimation method from multiple partial views of the same ellipsoid.

It is worth to note that, for the multi-head sensor concept applicability, the three sensor cores do not need to be clustered together in a single mechanical assembly: one just need to know the relative orientation of each camera with respect to the others, and to the common spacecraft reference frame.

To test the consistency of the developed theoretical framework, we set up both a numerical simulation scenario and an experimental testbed, as described in the following sections.

4. Numerical Simulator

As a first step towards the validation of the horizon sensor concept, we focused on the testing of the algorithm for attitude determination from multiple views of the limb of an Earth-like target. To this end, we developed a Matlab®-based simulation environment, which generates synthetic images of an ellipsoid target with the same flattening of the Earth. The simulator accounts for the main error source affecting the attitude determination from infrared Earth images, namely, the presence of a diffuse, inhomogeneous limb shift, due to the atmosphere. The consequence of a diffuse limb is that of resembling a slightly larger Earth, or a slightly smaller camera distance to the target. A non-homogeneous limb shift, instead, will result to a detected target shape, which differs from the solid Earth ellipsoid. While the former effect is rejected by the proposed algorithm, the latter is expected to induce some errors. The simplified atmospheric model implemented in our simulator aims at assessing, at least approximately, the magnitude of such errors, rather than providing a high-fidelity imaging simulation tool.

Several studies are found in the literature, assessing the issue of the stability of our planetary atmosphere over the infrared spectrum, both from a theoretical and experimental standpoint. A widespread descriptor used to characterize the atmospheric stability are the curves of the atmospheric radiance variation as a function of the tangent height, defined as the minimum altitude of the radiometric line of sight, see Figure 2. The main features can be summarized as [2,4]:

- The normalized atmospheric radiance profile decreases with the tangent height roughly following an inverse S-shaped curve.
- The spatial variation of the apparent limb shift has a systematic component, depending mainly on the latitude, plus a stochastic component. The two are almost equally important in magnitude.
- Overall, the variability of the atmospheric radiance induces changes in the detected infrared limb height of about ±10 km.

These features were taken as a guideline to implement a simple model for simulating the effects of a planetary limb on an infrared gathered image, as follows. First, we assumed the atmospheric infrared radiance profile width to be the sum of a nominal, constant value, $\overline{w}$, plus a variable term, dw. The nominal width $\overline{w}$ depends on the band of interest within the IR spectrum; for our study, we assumed $\overline{w} = 76$ km. The variation dw is modeled as a discrete first-order Markov process function of the latitude λ of each subpixel point:

$$dw(\lambda_{k+1}) = dw(\lambda_k)e^{-\frac{\Delta\lambda}{\tau}} + u_k, \tag{23}$$

where the correlation length τ and the standard deviation of the normally distributed random process u_k, were tuned for leading to dw variations of up to about ±10 km.

Then, we generated synthetic images through the following steps. Given the assumed relative attitude and position from the camera to the target ellipsoid:

(a) For each pixel p of the sensor, compute the nearest point on the target ellipsoid to the line-of-sight vector stemming from the camera centre to the pixel (i.e., the sub-pixel point). This task was accomplished through JPL's SPICE toolkit routine cspice_npedln.

(b) Compute the distance from that ellipsoid point to the pixel line, i.e., the tangent height, and the latitude of the sub-pixel point.

(c) To each sub-pixel point latitude, an inverse S-shaped atmospheric radiance profile is assigned, having width equal to $\overline{w} + dw$, as predicted by Equation (23);

(d) The image intensity level at the given pixel, *I(p)*, is assumed to be proportional to the normalized atmospheric radiance profile computed at step (c).

The resulting synthetic images were convoluted using a Gaussian kernel having standard deviation of 1.5 pixels, to resemble the blur effect expected in the infrared spectrum. Then, Gaussian noise was added, and images were finally converted to 16-bit grayscale.

The process above was repeated for generating a cluster of three images (Figure 3, left panel) captured by the three sensor heads. A zoom of the limb region of one of the images is given in Figure 3, right panel, which highlights the variable limb height across the image.

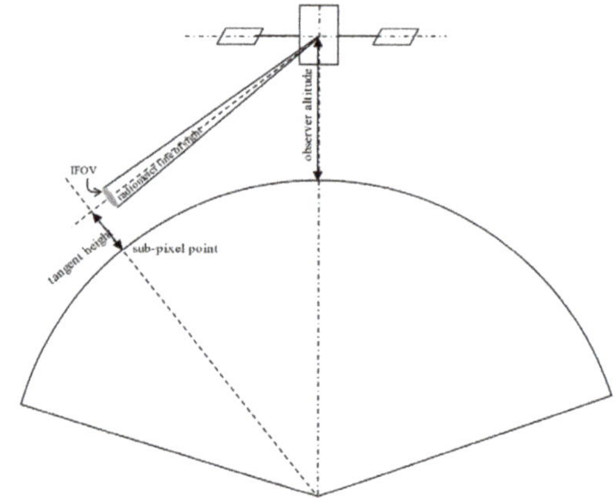

Figure 2. Definition of tangent height for atmosphere modeling.

Figure 3. (**a**) Cluster of three simulated limb images and (**b**) detail of the edge region highlighting the blur induced by the atmosphere model [9].

5. Experimental Testbed

The conceptual design of the horizon sensor was verified through a five degrees-of-freedom experimental testbed designed and assembled in our laboratory, making use of opto-mechanics components from Thorlabs Inc. (see Figure 4). The testbed was equipped with one camera and implements a virtual multi-head sensor, which is described in the following. Indeed, using a single camera with variable orientation allowed for a simpler and more cost-effective implementation of the testbed than by using three separated physical cameras, without undermining the algorithm validation. The sensor employed consists of a Flir Lepton® 3 camera (see Table 2), thus representative of a lower resolution sensor implementation targeted to nano-satellites. Three rotational degrees of freedom are provided by micrometric stages, having 10 arcmin of resolution. The target consists of a spheroid made of PVC plastic having nominal radii equal to 68.75 and 69.00 mm, respectively and mounted with the axis of symmetry orthogonal to the test-bed plane. The remaining two translational degrees of freedom allows for fine alignment between the camera and the target center. Rotations are implemented as a 3-1-2 (yaw-roll-pitch) rotation sequence, with angular excursions of 360° about yaw (ψ), ±5° and ±10° for the roll (φ, inner) and pitch (ϑ, outer) axes. Yaw axis is directed towards the spheroid center, pitch axis is orthogonal to the test-bed plane (positive downward), and roll axis directed to create a right-handed triad. The camera itself is then mounted on a PVC support tilted by 45° about pitch axis, for having a portion of the spheroid limb in view at nominal orientation.

The implemented rotation kinematics allows for the camera to perform a conical scan of the target limb when rotating around the yaw axis. Capturing three successive images at three different camera yaw orientations, allows us to virtually simulate the multi-camera scenario. More precisely, since the conical scan is the first angle of the rotation sequence (performed before the pitch-roll rotations), such configuration is equivalent to prescribe variable camera orientations $R_{b/i}$ with respect to a fixed body-to-world R attitude, rather than assuming a fixed camera-to-body orientation $R_{b/i}$ w.r.t. a rotating body-to-world attitude, as in an actual operative scenario. However, implementing this last option requires the conical scan to be performed after the pitch-roll rotations, as in a 1-2-3 sequence, which, in turn, is not achievable with the mounting options offered by the three rotational stages in our testbed.

The theory developed in Section 2 assumes an undistorted pinhole camera model, with known intrinsic matrix K; therefore, a calibrated camera is required for the experimental tests. In this work, camera calibration has been performed using Matlab Camera Calibration Toolbox, with a second order radial distortion model. The calibration device consists of a checkerboard pattern printed on an aluminum board (checkerboard size = 15 mm, see Figure 5). To allow for sufficient thermal contrast, the checkerboard is heated up using a heat gun. An index commonly used for evaluating the quality of a camera calibration is the mean reprojection error (MRE) of the corner points': in this work, an MRE = 0.13 pixels has been achieved after processing 12 checkerboard images. Such MRE value is quite in line with published data on thermal cameras calibration employing standard checkerboard patterns [28], even though lower MRE values can be achieved with dedicated calibration devices [29]. The estimated intrinsic camera parameters are reported in Table 3.

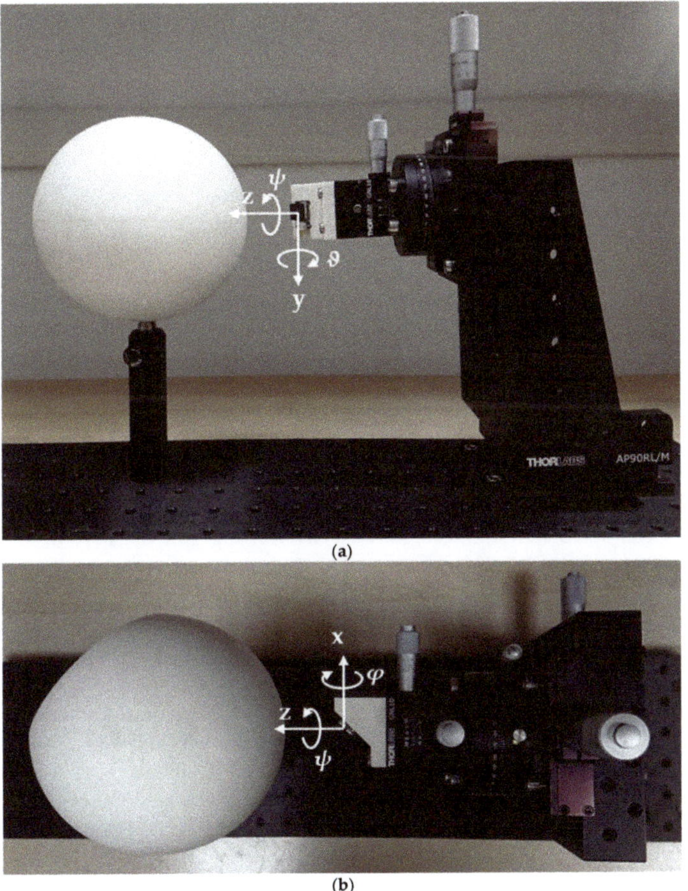

Figure 4. (**a**) Experimental testbed (side-view). (**b**) Experimental testbed (top-view).

(a)

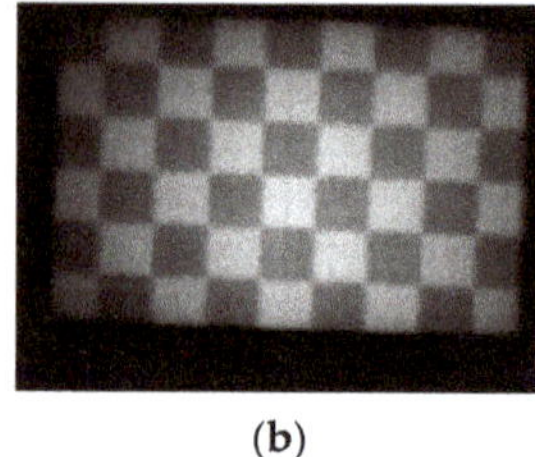

(b)

Figure 5. Sample image checkerboard image used for infrared camera calibration. (**a**) Raw and (**b**) undistorted using the estimated radial distortion coefficients.

Table 3. Estimated intrinsic camera parameters after calibration procedure (length unit in pixel).

Focal Lengths f_x, f_y	Principal Point Coordinates	Radial Distortion Coefficients
160.42 ± 1.22	80.70 ± 0.57	−0.333 ± 0.012
160.35 ± 1.22	63.29 ± 0.41	0.212 ± 0.051

6. Results

6.1. Results of Numerical Validation

The Matlab tool described in Section 4 was used for numerical validation through extensive simulations, for assessing the expected accuracy of the attitude determination algorithm when capturing images of an Earth-like spheroid with atmosphere, under varying points of observation. In particular, three orbit altitudes were considered, corresponding to different ratios of altitude to Earth radius, namely $h/R_E = 0.1$, 0.2, and 0.3 (note that in the short conference version of this paper [9], the dimensionless orbit altitudes were incorrectly reported as 0.01, 0.02, and 0.03, due to typos). For each altitude, vantage points were generated at latitudes ranging from 0 to 85°, with 5° of angular step. For each vantage point, a set of three images was generated (as if they were captured by the three sensor heads), prescribing a nominally nadir-pointing spacecraft attitude, with some randomly generated, normally distributed off-nadir angles (φ, ϑ having mean and standard deviation $\mu = 0°$, $\sigma = 1°$, respectively). Images resolution corresponds to the one of a micro-satellite targeted camera, as found in Table 1.

For each triple of images, raw limb points were detected using the method in [30], transformed according to Equation (21), and then fitted to an ellipse using the robust iterative ellipse fitting algorithm of [31]. Then, the attitude determination algorithm was run applying the steps from Equation (11) to (14), and the estimated attitude matrix was compared to the one used to generate the image triple, to compute the attitude error angles. The above procedure was repeated for 30 times, thus providing 30 image triples, and the corresponding attitude errors were root-mean-squared (rms) across the batch. The resulting rms error angles are displayed in Figures 6–8, for $h/R_E = 0.1$, 0.2, and 0.3, respectively.

Inspection of the figures suggest that very high accuracy was reached in the estimation of the nadir direction, with rms errors spanning from 10^{-3} to 10^{-2} degrees for pitch and roll angles. On the other hand, the angle about nadir (yaw) could be retrieved only very coarsely, with increasing error as the vantage point approaches the symmetry axis of the Earth spheroid (i.e., for latitudes close to 90°). This is an expected outcome, according to the discussion at the end of Section 2. Overall, the rms yaw error remained below 10° only up to a latitude of 60°. On the other hand, pitch and roll angles accuracies were only marginally degraded when approaching Earth symmetry axis.

Note that the worse accuracy in estimating the yaw angle has a clear physical interpretation, which is well-known also in star tracking sensors [32]: while the off-nadir angles depends mostly on the location of the center of the ellipse in the image, which can be determined with sub-pixel precision, yaw accuracy is related to how well the orientation of the ellipse axes within the image can be determined.

If the target resembles a sphere (i.e., it has very low flattening ratios), the imaged ellipse will resemble a circle, and the yaw angle becomes barely detectable: indeed, this is the situation encountered in an Earth sensor scenario.

Finally, performance was also sensitive to orbit altitude: the accuracy tended to improve when increasing the altitude, as a result of the wider portions of the Earth limb being captured by the cameras and, at the same time, of the apparently thinner atmosphere.

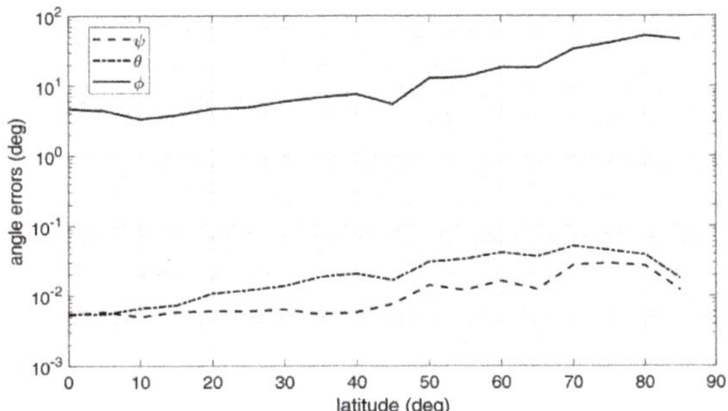

Figure 6. Attitude errors as a function of latitude for an Earth-like spheroid with atmosphere, $h/R_E = 0.1$ [9].

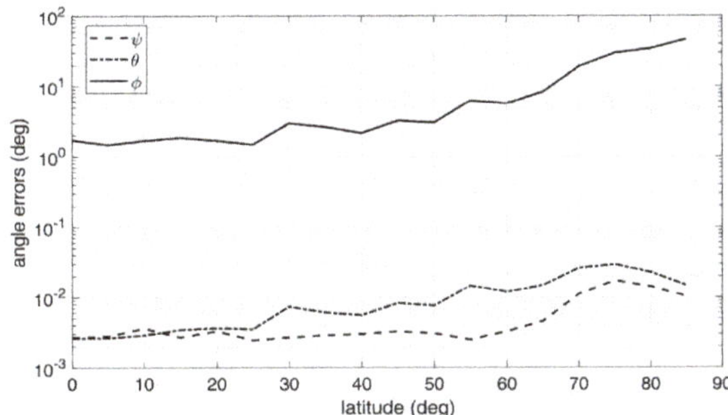

Figure 7. Attitude errors as a function of latitude for an Earth-like spheroid with atmosphere, $h/R_E = 0.2$ [9].

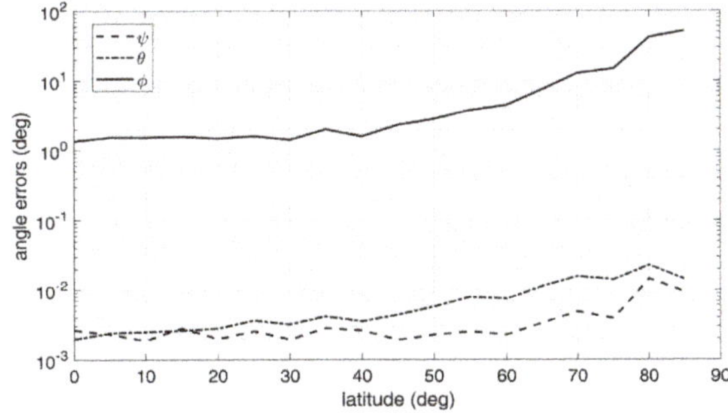

Figure 8. Attitude errors as a function of latitude for an Earth-like spheroid with atmosphere, $h/R_E = 0.3$ [9].

6.2. Results of Experimental Validation

Four test cases were considered for the preliminary validation of the sensor prototype, corresponding to different pitch-roll configurations. For all the tests, the camera height was kept aligned to the spheroid center, which corresponds to setting $\boldsymbol{v}_w = \begin{bmatrix} 1 & 0 & 0 \end{bmatrix}^T$ in Equation (7) when computing $\widetilde{B}^*$ matrix. For each pitch-roll setting, three images were gathered at yaw orientations separated by 120° one from each other. For each image, the limb was extracted using a standard Sobel operator. Limb pixels' coordinates from raw pictures were undistorted according to the estimated distortion coefficients, using a built-in Matlab function. The undistorted limb coordinates are then transformed according to Equation (21), using the calibrated intrinsic camera parameters in Table 3, and setting matrices $R_{b/i}$ according to:

$$\begin{aligned} R_{b/1} &= E_2[45° - \vartheta]E_1[\varphi]E_3[0°], \\ R_{b/2} &= E_2[45° - \vartheta]E_1[\varphi]E_3[-120°] \;, \\ R_{b/3} &= E_2[45° - \vartheta]E_1[\varphi]E_3[120°], \end{aligned} \tag{24}$$

where E_j $j = 1, 2, 3$ denotes the elementary rotation matrix around the j-th coordinate axis. Finally, an ellipse was fitted to the stacked vector of transformed points from the three views, using the method in [31]. Having the ellipse matrix available, the attitude determination algorithm was run applying the steps from Equations (11)–(14). The attitude matrix R output of this process was compared to the zero-reference attitude matrix R_0 computed running the algorithm for $\varphi = \vartheta = 0°$. We defined the estimation errors as the angles of the 3-1-2 rotation sequence computed from the error attitude matrix RR_0^T. The process was repeated for four test-cases, corresponding to different pitch-roll combinations: the outcome is summarized in Table 4. A schematic representation of the overall measurement workflow, from image capturing to attitude matrix estimation, is given in Figure 9.

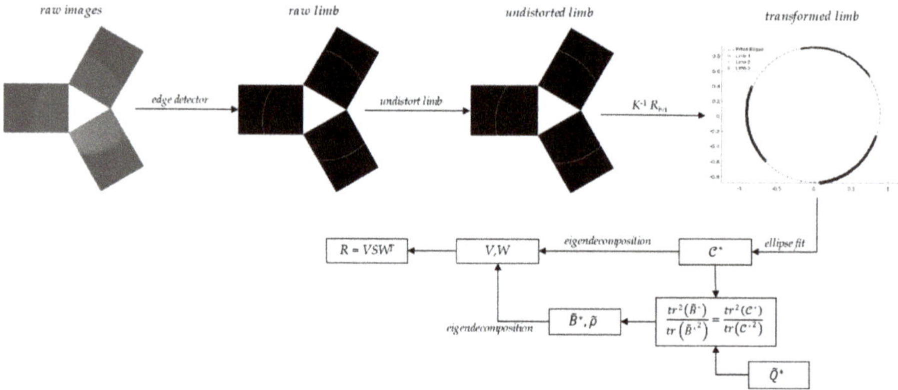

Figure 9. Schematic workflow for the attitude estimation from multiple limb images of an ellipsoid.

Table 4. Error angles retrieved during experimental tests.

Test Case	ψ Error (°)	φ Error (°)	ϑ Error (°)
1: $\varphi = 5°$, $\vartheta = 0°$	−3.29	0.03	0.04
2: $\varphi = 0°$, $\vartheta = 5°$	14.20	−0.01	−0.006
3: $\varphi = 5°$, $\vartheta = 5°$	8.05	−0.02	0.02
4: $\varphi = 10°$, $\vartheta = 5°$	−6.88	0.07	−0.001

By inspection of Table 4, it emerged that the errors in the determination of the off-nadir angles lay in the range 10^{-2} deg, i.e., well below the accuracy limit of the experimental platform. This is evidence that the foreseen measurement procedure leads to results that are consistent with the previous theoretical and numerical analyses, providing high accuracy nadir direction determination. This is a relevant outcome, especially when considering the additional error sources affecting the experimental set-up, such as the imperfect camera distortion compensation, and imperfect knowledge of the relative orientation of the camera when capturing the three images. On the other hand, the estimation of the yaw angle was very poor, and worse than what was predicted by the numerical simulator, which is, nevertheless, not surprising, given the lower image resolution of the FLIR Lepton® 3 camera.

7. Discussion

Some new theoretical results for the attitude determination from multiple limb views of an ellipsoid, together with the increased availability of highly miniaturized infrared cameras, allowed us to design a high accuracy horizon sensor for spacecraft attitude determination in Low Earth Orbit. The sensor makes use of images of the Earth gathered in the infrared spectrum, and on fitting the detected limb to an ellipse. For satellites in LEO, where the target exceeds significantly the camera FoV, a single, partial, limb view is not enough to provide accurate ellipse fitting; therefore, the proposed sensor concept exploits a multi-head implementation, where limb views from three different images are combined for providing a single best-fitted ellipse. The full camera attitude is then estimated starting from the analytic projection of an ellipsoid onto the image plane of a pinhole camera. Once the Earth shape and the direction of observation are known, the camera attitude can be retrieved as the solution to a modified orthogonal Procrustes problem. The proposed solution, being invariant to an unknown scaling of the target size, is less affected by errors due to the atmosphere own radiance.

The attitude determination algorithm was first checked against synthetically generated images of an Earth-like target, embedding a simple model for a diffuse, inhomogeneous limb to resemble the presence of the atmosphere. Sensitivity of the algorithm accuracy under varying vantage points was assessed through numerical simulations. Obtained results lead to rms errors in the order of 10^{-2} deg (or less) for the off-nadir angles in most operating conditions. The capability of measuring the orientation about nadir was instead very coarse, and strongly dependent on the observation latitude and camera resolution.

The sensor concept was then validated through an experimental testbed, designed to virtually recreate a multi-head sensor, making use of only one low-resolution COTS infrared camera equipped with three rotational DoF. The experiments allowed us to verify the nadir direction estimation down to the resolution of the testbed rotational stages (<10 arcmin). On the other hand, the orientation about nadir can be estimated only very coarsely: its practical usefulness in an operative scenario might be achieved using higher resolution cameras, which remains a topic for further investigations.

Our results look promising when compared to existing static horizon sensors for LEO satellites, and motivated us to further develop the conceptual design presented herein. For example, Selex-Galileo IRES-C and Sodern ST-12 are reported to having 3-sigma errors of 1° and 0.1° respectively [33,34]; Maryland Aerospace MAI-SES has a resolution of 0.25 deg [35]; the thermopile array sensor flown on-board of PSSCT-2 picosatellite from the Aerospace Corporation has a rated error within 0.5° [36].

Future efforts will be mainly devoted to the manufacturing of a complete sensor prototype with higher camera resolution, and to its thorough experimental validation. In parallel, a more comprehensive error budget shall be developed, which accounts for additional error sources, such as the misalignment between the different sensor heads.

Author Contributions: Conceptualization, D.M. and M.Z.; methodology, D.M.; software, M.Z. and D.M.; experimental validation, A.L. and D.M.; writing, D.M. and A.L. All authors have read and agreed to the published version of the manuscript.

Funding: This research received no external funding.

Acknowledgments: D.M. and A.L. wishes to acknowledge Paolo Proli (University of Bologna) for having manufactured the pvc spheroid. D.M. is highly grateful to Claudio Bianchi (Sitael S.p.A.) for the fruitful discussions, and for having provided an accurate measurement of the spheroid dimensions through 3D scanning.

Conflicts of Interest: The authors declare no conflict of interest.

References

1. Astheimer, R. High precision attitude determination by sensing the earth and lunar horizon in the infrared. *Automatica* **1971**, *7*, 83–88. [CrossRef]
2. Jalink, A., Jr.; Davis, R.E.; Dodgen, J.A. *Conceptual Design and Analysis of an Infrared Horizon Sensor with Compensation for Atmospheric Variability*; Technical Report; NASA Langlely Research Center: Washington, DC, USA, 1972.
3. Hotovy, S.G. Horizon sensor models. In *Spacecraft Attitude Determination and Control*, 1st ed.; Wertz, J.R., Ed.; Kluwer Academic Publishers: Dordrecht, The Netherlands, 1978; pp. 230–237.
4. Hashmall, J.A.; Sedlak, J.; Andrews, D.; Luquette, R. Empirical correction for earth sensor horizon radiance variation. *Adv. Astronaut. Sci.* **1998**, *100*, 453–468.
5. Nguyen, T.; Cahoy, K.; Marinan, A. Attitude determination for small satellites with infrared earth horizon sensors. *J. Spacecr. Rockets* **2018**, *55*, 1466–1475. [CrossRef]
6. García Sáez, A.; Quero, J.M.; Angulo Jerez, M. Earth sensor based on thermopile detectors for satellite attitude determination. *IEEE Sens. J.* **2016**, *16*, 2260–2271. [CrossRef]
7. Vertat, I.; Linhart, R.; Masopust, J.; Vobornik, A.; Dudacek, L. Earth's thermal radiation sensors for attitude determination systems of small satellites. *Contrib. Astron. Obs. Skalnaté Pleso* **2017**, *47*, 157–164.
8. Kuwahara, T.; Fukuda, K.; Sugimura, N.; Sakamoto, Y.; Yoshida, K.; Dorsa, A.; Pagani, P.; Bernelli Zazzera, F. Design and implementation of a thermopile-based earth sensor. *Trans. JSASS Aerosp. Tech. Jpn.* **2016**, *14*, 77–81. [CrossRef]
9. Modenini, D.; Zannoni, M. A high accuracy horizon sensors for small satellites. In Proceedings of the 2019 Workshop on Metrology for Aerospace (MetroAero), Torino, Italy, 19–21 June 2019.
10. Lightsey, G.E.; Christian, J.A. Onboard image-processing algorithm for a spacecraft optical navigation sensor system. *J. Spacecr. Rockets* **2012**, *49*, 337–352. [CrossRef]
11. Mortari, D.; de Dilectis, D.; Zanetti, R. Position estimation using the image derivative. *Aerospace* **2015**, *2*, 435–460. [CrossRef]
12. Mortari, D.; D'Souza, C.N.; Zanetti, R. Image processing of illuminated ellipsoid. *J. Spacecr. Rocket.* **2016**, *53*, 448–456. [CrossRef]
13. Modenini, D.; Zannoni, M.; Manghi, R.L.; Tortora, P. An analytical approach to autonomous optical navigation for a CubeSat mission to a binary asteroid system. *Adv. Astronaut. Sci.* **2018**, *163*, 139–149.
14. Sekignchi, T.; Yamamoto, T.; Iwamaru, Y. 3-Axes attitude estimation experiments using CCD earth sensor. *IFAC Proc. Volumes* **2004**, *37*, 741–746. [CrossRef]
15. Park, K.-J.; Mortari, D. Planet or moon image processing for spacecraft attitude estimation. *J. Electron. Imaging* **2008**, *17*, 023020.
16. Wabha, G. A least squares estimate for spacecraft attitude. *SIAM Rev.* **1965**, *7*, 409.
17. Markley, F.L. Attitude determination using vector observations and the singular value decomposition. *J. Astronaut. Sci.* **1988**, *36*, 245–258.
18. Modenini, D. Attitude solution from ellipsoid observations: a modified orthogonal procrustes problem. *J. Guid. Control Dyn.* **2018**, *41*, 2320–2325. [CrossRef]
19. Wessels, J.H. Infrared Horizon Sensor for CubeSat Implementation. Master's Thesis, University of Stellenbosch, Stellenbosch, South Africa, 2018.
20. Hartley, R.; Zisserman, A. *Action of a Projective Camera on Quadrics*, 2nd ed.; Cambridge University Press: New York, NY, USA, 2004; pp. 201–202.
21. Christian, J.A.; Robinson, S.B. Non-Iterative horizon-based optical navigation by cholesky factorization. *J. Guid. Control Dyn.* **2016**, *39*, 2757–2765. [CrossRef]
22. Wu, J.; Sun, Y.; Wang, M.; Liu, M. Hand-eye calibration: 4D procrustes analysis approach. *IEEE Trans. Instrum. Meas.* **2019**. [CrossRef]
23. Magnus, R.M. On differentiating eigenvalues and eigenvectors. *Econ. Theory* **1985**, *1*, 179–191. [CrossRef]
24. Liounis, A.; Christian, J.; Robinson, S. Observations on the computation of eigenvalue and eigenvector jacobians. *Algorithms* **2019**, *12*, 245. [CrossRef]

25. Kopp, J. Efficient numerical diagonalization of Hermitian 3 × 3 matrices. *Int. J. Mod. Phys. C* **2008**, *19*, 523–548. [CrossRef]
26. Flir Compact Lwir Thermal Camera Core Boson. Available online: https://www.flir.com/products/boson/ (accessed on 20 November 2019).
27. Flir Lwir Micro Thermal Camera Module Lepton®. Available online: https://www.flir.com/products/lepton/ (accessed on 20 November 2019).
28. Ellmauthaler, A.; da Silva, E.A.; Pagliari, C.L.; Gois, J.N.; Neves, S.R. A novel iterative calibration approach for thermal infrared cameras. In Proceedings of the 2013 IEEE International Conference on Image Processing, Melbourne, Australia, 15–18 September 2013; pp. 2182–2186.
29. Usamentiaga, R.; Garcia, D.F.; Ibarra-Castanedo, C.; Maldague, X. Highly accurate geometric calibration for infrared cameras using inexpensive calibration targets. *Measurement* **2017**, *112*, 105–116. [CrossRef]
30. Christian, J.A. Accurate planetary limb localization for image-based spacecraft navigation. *J. Spacecr. Rockets* **2017**, *54*, 708–730. [CrossRef]
31. Szpak, Z.L.; Chojnacki, W.; van den Hengel, A. Guaranteed ellipse fitting with a confidence region and an uncertainty, measure for centre, axes, and orientation. *J. Math. Imaging Vis.* **2015**, *52*, 173–199. [CrossRef]
32. Lai, Y.; Gu, D.; Liu, J.; Li, W.; Yi, D. Low-frequency error extraction and compensation for attitude measurements from STECE Star Tracker. *Sensors* **2016**, *16*, 1669. [CrossRef] [PubMed]
33. Coarse InfraRed Earth Sensor IRES-C. Available online: https://artes.esa.int/projects/coarse-infrared-earth-sensor-ires-c (accessed on 5 January 2020).
34. Van Herwaarden, A.W. Low-cost satellite attitude control sensors based on integrated infrared detector arrays. *IEEE Trans. Instrum. Meas.* **2001**, *50*, 1524–1529. [CrossRef]
35. MAI-SES Static Earth Sensor. Available online: https://www.maiaero.com/ir-earth-horizon-sensor (accessed on 5 January 2020).
36. Janson, S.W.; Hardy, B.S.; Chin, A.Y.; Rumsey, D.L.; Ehrlich, D.A.; Hinkley, D.A. Attitude control on the pico satellite solar cell testbed-2. In Proceedings of the 26th Annual AIAA/USU Conference on Small Satellites, Logan, UT, USA, 13–16 August 2012.

© 2020 by the authors. Licensee MDPI, Basel, Switzerland. This article is an open access article distributed under the terms and conditions of the Creative Commons Attribution (CC BY) license (http://creativecommons.org/licenses/by/4.0/).

Article

Preliminary Design of a Model-Free Synthetic Sensor for Aerodynamic Angle Estimation for Commercial Aviation †

Angelo Lerro *, Alberto Brandl, Manuela Battipede and Piero Gili

Department of Mechanical and Aerospace Engineering, Politecnico di Torino, C.so Duca degli Abruzzi 24, 10129 Torino, Italy; alberto.brandl@polito.it (A.B.); manuela.battipede@polito.it (M.B.); piero.gili@polito.it (P.G.)

* Correspondence: angelo.lerro@polito.it

† This paper is an extended version of our paper published in Lerro, A.; Battipede, M.; Gili, P.; Ferlauto, M.; Brandl, A.; Merlone, A.; Musacchio, C.; Sangaletti, G.; Russo, G. The Clean Sky 2 MIDAS Project—an Innovative Modular, Digital and Integrated Air Data System for Fly-by-Wire Applications. In Proceedings of the 2019 IEEE 5th International Workshop on Metrology for AeroSpace (MetroAeroSpace), Torino, Italy, 19–21 June 2019.

Received: 19 October 2019; Accepted: 19 November 2019; Published: 23 November 2019

Abstract: Heterogeneity of the small aircraft category (e.g., small air transport (SAT), urban air mobility (UAM), unmanned aircraft system (UAS)), modern avionic solution (e.g., fly-by-wire (FBW)) and reduced aircraft (A/C) size require more compact, integrated, digital and modular air data system (ADS) able to measure data from the external environment. The MIDAS project, funded in the frame of the Clean Sky 2 program, aims to satisfy those recent requirements with an ADS certified for commercial applications. The main pillar lays on a smart fusion between COTS solutions and analytical sensors (patented technology) for the identification of the aerodynamic angles. The identification involves both flight dynamic relationships and data-driven state observer(s) based on neural techniques, which are deterministic once the training is completed. As this project will bring analytical sensors on board of civil aircraft as part of a redundant system for the very first time, design activities documented in this work have a particular focus on airworthiness certification aspects. At this maturity level, simulated data are used, real flight test data will be used in the next stages. Data collection is described both for the training and test aspects. Training maneuvers are defined aiming to excite all dynamic modes, whereas test maneuvers are collected aiming to validate results independently from the training set and all autopilot configurations. Results demonstrate that an alternate solution is possible enabling significant savings in terms of computational effort and lines of codes but they show, at the same time, that a better training strategy may be beneficial to cope with the new neural network architecture.

Keywords: air data system; flight dynamics; state observer; synthetic sensor; virtual sensor; analytical redundancy; avionics; neural network

---

1. Introduction

Air data systems (ADSs) are adopted on air vehicles to measure a set of data from the external environment. Generally speaking, a simplex ADS is made up of external (i.e., installed externally on the aircraft (A/C) fuselage) probes and vanes able to measure a full set of air data:

- local static pressure, P_s;
- local total pressure, P_0;
- local air temperature (static, outside air temperature (OAT), or total, total air temperature (TAT));
- local angle of attack (AoA), α;

- local angle of sideslip (AoS), β.

The ADS probes/vanes are connected (or integrated) with a corresponding measuring module (air data modules (ADMs)) encompassing suitable transducers able to convert the measure into analog (or digital) signals. If those ADM are all embedded in a single box, it usually refers to a central processor unit (air data computer, (ADC)). The ADC is able to provide pilots, or flight control computers, FCCs, with the more relevant air data information necessary for piloting, navigation and control purposes. In recent years, the air data algorithms are often implemented into the FCC and the ADC is removed. In both cases, the air data functionalities shall be able to calculate the following parameters:

- Pressure altitude
- Pressure altitude, baro-corrected (Kollsman)
- Vertical speed
- Calibrated airspeed (CAS)
- Equivalent airspeed (EAS)
- True airspeed (TAS) (only if OAT or TAT is available)
- Mach number
- Air temperature, T_∞, (only if OAT or TAT is available)
- AoA
- AoS

Generally speaking, for each element of the previous list except for AoA and AoS, the standard AS8002A [1] sets operative performance and environmental requirements. For aerodynamic angles, AoA and AoS, there are not clear and well-defined performance requirements. Even though the AoA measurements shall satisfy the standard AS403A [2], this standard sets prescriptions only for stall protection purposes and not for the entire range of flyable angle of attack. There is no standard applicable for AoS. In fact, when AoA and AoS are required to the ADS, the functional requirements are usually derived from other functionalities as described later in Section 3.2.

According to the function allocated to each air data measured, the A/C integrator with failure hazard analysis (FHA) will classify the criticality of the loss of each one of the air data measured or calculated. There are some parameters, e.g., the CAS, whose loss is always classified as catastrophic and therefore the corresponding air data become safety-critical. The ADS, therefore, is one of the safety-critical systems on board that should be redundant in order to meet the A/C category safety requirement: for example, triplex solution is the common standard in commercial aviation.

With modern technologies, recently the ADS moved towards digital solutions for a better integration with modern digital avionics. Fly-by-wire (FBW) paradigm is successfully applied to large aircraft and therefore a more electrical aircraft is a technological transition necessary to achieve the goals defined by the European Community, EC, within the FlightPath 2050 [3]. The aeronautical industry has launched several programs to cope with the fly-by-wire challenge, even for those systems that seemed less involved in this revolution, such as the ADS. In fact, in order to overcome drawbacks to connect probes and vanes to ADMs pneumatically and then each ADM to the flight control system (FCS), many recent large FBW A/C are equipped with integrated probes that embed transducers within the probe (or vane) itself [4].

The same approach is shared with those air vehicles of reduced size, e.g., small aircraft transport (SAT) category, unmanned aircraft system (UAS), and urban air mobility vehicles (UAM), where the FBW is necessary for a more-integrated system. SAT aircraft is a crucial segment for travelers in Europe because it is the only means of transportation that has those characteristics able to fill a gap that could not be done in another way. Short-range flights, local communities' connections, door-to-door within four hours [3] are only a few examples that highlight the importance of the SAT segment in the European infrastructures. Efforts spent worldwide by governments and aviation safety agencies to regulate the air work of UAS and UAM vehicles over populated areas is a clear evidence that what has been studied by many companies could take off in the next years. As far as ADS is concerned, SAT, UAS and UAM categories have common drawbacks, e.g., heterogeneity of ADS requirements due to high range of A/C mission (altitude, speed, etc.) and need to optimize ADS's line replaceable units (LRUs) installation on board in terms of space and weight due to reduced fuselage size (if compared with civil aircraft). Each aircraft will have its own dedicated probes/vanes to satisfy safety and performance requirements. This particular aspect would suggest to rely on a solid core in order to have a multi-platform ADS with interchangeable external probes.

ADS's safety is another crucial aspect to be taken into account. In order to increase the reliability of ADS, a physical redundancy is applied. Moreover, there are other requirements from airworthiness authorities that should be taken into account, e.g., those related to the bird strike, that set some constraints on the fuselage installations. Analytical redundancy [5–7] is a concept used more and more frequently in recent years because the avionic background is mature to welcome such innovations. This approach enables the replacement of physical sensors (used for redundancy) with analytical ones for aerodynamic angles [8–10] and airspeed [11] with several benefits in terms of weight, power consumption, reliability, maintainability, and emissions.

The advent of distributed avionics, e.g., ARINC 664 networks on Airbus A380, A400M, A350 and Boeing 787, has been seen as a significant booster for a better exploitation of onboard data to be used, for instance, by other subsystems for redundancy purposes.

Within this scenario, innovative ADS for FBW applications as part of a redundant ADS is introduced with the MIDAS project funded in the SAT category of Clean Sky 2 programme [12]. The MIDAS ADS will be driven by an integrative, modular and digital approach. Following the market trend and EC guidelines, the project outcome will be a fully-integrated probe (air data probes integrated with electronics) with digital outputs that can be interfaced with modern avionic bus onboard FBW A/C and several classes of external Pitot tubes and TAT probes.

The main innovation behind the MIDAS ADS lays on a patented technology [13], named smart-air data system, attitude and heading reference system (ADAHRS), firstly by Politecnico di Torino and later under AeroSmart S.r.l. [14] responsibility. This solution, basically a state observer obtained with a data-driven methodology, is able to estimate AoA and AoS with analytical sensors [15–17] exploiting A/C flight dynamic equations and onboard data fusion.

The MIDAS equipment will be qualified (both for hardware [18] and environmental [19] aspects) and, therefore, even the virtual sensors. This aspect makes the MIDAS project a fundamental milestone in the certification process of analytical estimators for civil applications. Therefore, the main outcome of this project is to provide a qualifiable electronics able to be integrated into modern avionic bus (e.g., ARINC 664 network) that can be interfaced with several COTS probes (Pitot/static and TAT probes already certified by their own manufacturers).

This work deals with preliminary design activities accomplished to define analytical sensors (or virtual ones) for AoA and AoS estimation exploiting simulated flight data. The final design will deliver a qualifiable software module to be integrated on a demo board with AoA and AoS estimators guaranteeing at least the same performance of state-of-the-art sensors. Mainly for certification reasons, moving from previous and consolidated practice, a more efficient (in terms of computational cost and number of lines of code) virtual sensor will be designed for civil operative scenarios. Side findings emerge from the present work and they will affect the future steps of the MIDAS projects. The paper aims to highlight the original elements emerged during the present work and, therefore, neural network architecture optimization will not be discussed here because similar to previous works [20,21].

Considering the current project's development stage, a complete and detailed reliability analysis cannot be conducted yet. For the same reason, the technology has been compared with simulated data. Even though the comparison with classical sensors is not described in this work, some brief considerations on expected performances will be given in Section 3.1.

This paper begins with an overview of the MIDAS ADS and comparison between the state-of-the-art in Section 2. Section 3 describes the preliminary design of the MIDAS ADS solution with details about the virtual sensor design (Section 3.1) and the certification aspects (Section 3.2) that can be condensed in well-established target performance. After the reference A/C is introduced in Section 4, the training strategy adopted in this work is presented in Section 5. Section 6 collects preliminary results obtained using only test maneuvers that are completely independent of the training pattern. The paper concludes with Section 7.

2. Approach

Possible certifiable architectural solutions are made up of certified probes, vanes, and air data units (ADUs) and/or ADC. The adjective *certifiable* refers to a system suitable for commercial flights, and not only for experimental purposes, according to applicable airworthiness regulations [22]. The Figure 1 shows schematically realistic solutions. A brief description of the three kinds of realistic architectures will be given in this section.

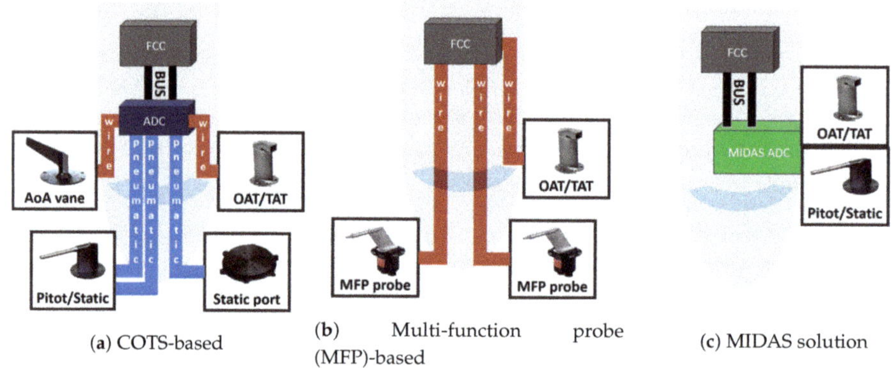

(**a**) COTS-based (**b**) Multi-function probe (MFP)-based (**c**) MIDAS solution

Figure 1. Schematic view of three realistic simplex air data system (ADS) architectures able to provide a complete set of air data.

2.1. ADS Based on COTS

Pressure probes and static ports are pneumatically connected to a central air data computer whereas the flow angle vanes and the TAT sensor are usually electronically connected to the ADC. Traditionally, the ADC contains several ADUs, essentially pressure transducers, which convert pressures into digital signals. Usually, the ADC has computational capabilities that are used to calibrate measured air data (e.g., local to freestream correction) and to calculate other air data parameters (e.g., mach number, pressure altitude, etc.). Today, the ADC has commonly been replaced by several ADM located near, or integrated into, the reference probe/vane with dedicated transducers. ADM's main function is to convert data measured into digital ones.

The main drawbacks of ADS based on COTS LRUs are related to weight, encumbrance limitations and power consumption requirements (mainly for de-icing purposes).

2.2. ADS Based on Multi-Function Probes

Another solution for ADS is based on multi-function probes (MFPs). Generally speaking, MFP are digital LRUs able to embed Pitot tube, static port and one flow angle with an ADM. By means of the combination of two MFPs and two TAT sensors, it is possible to define a duplex ADS architecture (but simplex for AoS), while using three MFPs and three TAT sensors it is possible to define a triplex ADS architecture (but duplex for AoS). The MFPs and the TAT can be digitally connected to cockpit display systems and, in the case of FBW aircraft, it can be sent to the FCC. As only a few ADC functionalities are carried out at ADM level, the main ADC's work is done at FCC level. The main drawbacks of the MFP-based solution are related to high costs and limited availability from very few companies all over the world.

2.3. ADS Based on MIDAS Solution

The focus of the MIDAS Project is shifted from probe/vane to flight mechanics using machine learning techniques and flight data already available onboard through a patented technology [13]. The MIDAS's approach will introduce synthetic aerodynamic angle sensors more suitable for a dynamic aeronautical segment as the SAT, UAS or UAM. The MIDAS ambition is to provide an ADS solution that joins all benefits of flyable ADS architectures (Section 2.1 and 2.2) with practical elimination of their drawbacks. In particular, the main benefits will be:

- low weight and space;
- reduced power consumption with consequent lower emissions;
- no more than two external probes (COTS Pitot/static and TAT (or OAT) probes);
- easy to be installed on the fuselage: no requirements about front or nose installation;
- flexibility to be interfaced with a wide range of probes in order to cope with heterogeneity of ADS requirements from A/C of the CS23 category;
- possible ITAR free ADS system, thanks to the availability of necessary probes available from several suppliers all over the world.

3. MIDAS Technological Solution

Today, two distinct probes (certified) to measure pressure and temperature for flyable ADS are needed. There are few attempts worldwide to integrate both measures into a single probe, but there are no certified products so far. Therefore, MIDAS approach is to integrate the COTS probes (Pitot-static and temperature probes) as close as possible, as in Figure 2, with dedicated electronics (MIDAS ADC) in order to achieve objectives defined later in Section 3.2.

Figure 2. Preliminary MIDAS configuration. Courtesy of SELT A&D [23].

Therefore, the MIDAS air data system will be made up of:

- ADC—electronic unit (embedding estimator(s) for AoA and AoS);
- pitot-static probe;
- TAT probe.

The MIDAS project expects a tandem solution of the external probes according to preliminary considerations, but a more detailed aerodynamic study will be performed to find the optimal configuration.

3.1. Virtual Sensors

As stated before, the MIDAS project's outcome will provide a single LRU embedding virtual sensor (or sensors), dedicated to AoA and AoS estimation, based on a patented technology at TRL6 [24,25].

These virtual sensors are essentially state observers for which the A/C flight dynamic model is replaced by a model based on neural networks.

Exploiting the paradigm of the FBW aircraft, the air data probe (ADP) will receive, as input, consolidated data from other A/C equipment to be fused with measured ones (P_s, P_t and T_∞) in order to estimate AoA and AoS with high reliability.

The input and the output signals will be transmitted through the avionic bus of the A/C. As stated in the introduction, being the ADS a safety-critical system, the proposed solution must be considered valid also for redundant architectures in a hybrid framework, merging classical sensors and the MIDAS technology. The A/C integrator will be in charge to design a redundant ADS able to meet the applicable safety requirements with the best compromise merging COTS and synthetic sensors. The detailed reliability analysis of the MIDAS technology is non-trivial because it involves reliability analysis of other proprietary A/C systems (e.g., the FCC). Although some information can be deduced from previous works [15], this topic will be extensively dealt with in a future step of this project.

The virtual sensors proposed in the MIDAS project rely on the use of A/C data from attitude and heading reference system, primary surface commands/deflections and Global Navigation Satellite System (GNSS) (Figure 3b).

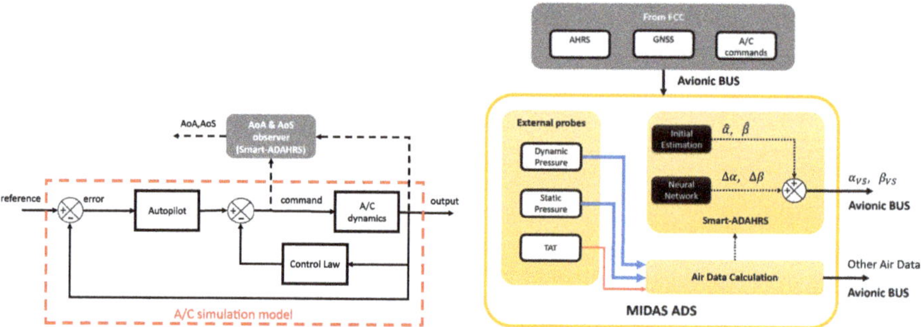

(**a**) Generic schematic of the aircraft (A/C) simulation involving autopilot modes and control laws

(**b**) High-level schematic of the smart-air data system, attitude and heading reference system (ADAHRS) (state observer)

Figure 3. Generic schematics of A/C simulation and angle-of-attack (AoA)/angle-of-sideslip (AoS) estimators.

Topologically speaking, it consists of a biased linear combination of non-linear activation functions, each activation function is driven by a biased linear combination of the output of the preceding nodes. Although the multilayer perceptron (MLP) can be described from several points of view, in this case, the best description is that it can represent a non-linear map between the input and the target. The point is to find those weights of the network such that the output fits the desired map. The validity of the approach is mathematically proven using the universal approximation theorem (UAT). In fact, it is proven that any continuous function of n real variables, with support in the unit hypercube, can be uniformly approximated by finite superposition of a fixed, univariate function that is discriminatory [26].

The Smart-ADAHRS project deals with a very straightforward model, suitable for real-time and cost effective innovative avionic systems. Consider valid the following assumption on AoA and AoS:

$$\alpha_{\mathrm{VS}} = \hat{\alpha} + \Delta\alpha \tag{1}$$

$$\beta_{\mathrm{VS}} = \hat{\beta} + \Delta\beta \tag{2}$$

where $\hat{\alpha}$ and $\hat{\beta}$ are initial estimation obtained with flight mechanics equations whereas $\Delta\alpha$ and $\Delta\beta$ are the differences between the linear estimations and the true nonlinear values. According to a patented procedure [13], $\hat{\alpha}$ and $\hat{\beta}$ is augmented with the evaluation of $\Delta\alpha$ and $\Delta\beta$ based on two MLPs, which process measurements obtained with non-protruding sensors (except for the Pitot tube and TAT).

$\hat{\alpha}$ and $\hat{\beta}$ can be evaluated as follows:

$$\hat{\alpha} = \theta - \gamma \tag{3}$$

$$\hat{\beta} = K\frac{n_y}{q_c} \tag{4}$$

where θ stands for the pitch angle, γ for the flight path angle, n_y is the proper acceleration as measured by the accelerometer along the Y_B axis and q_c is the impact pressure. K is an A/C constant derived from flight mechanic considerations (the order of magnitude for this category is $10^3\,\mathrm{kg\,m^{-2}}$).

The patented approach can bring to a neural network with limited output(s) that is crucial aspect when dealing with certification authorities. Generally speaking, using real or simulated flight data the value ranges of $\Delta\alpha$ and $\Delta\beta$ can be identified. Our approach is to apply the limited output only once the neural network is trained. This strategy allows to train the neural network without any limitations and to bound the estimated AoA and AoS to avoid any overshoot or spike values during the operative life.

Mathematical demonstrations exist [26–31] about the MLP performing as a universal approximator. During the training procedure, the weights of the linear combinations are estimated solving the non-convex problem of the error function optimization. Different heuristic rules exist and the Levenberg-Marquard, LM, algorithm is used in this work. The complete input vector needed by Smart-ADAHRS includes data from the GPS (providing V_{down}), the ADS (providing TAS) and the attitude and heading reference system (AHRS) (providing angular rates, Euler angles and linear accelerations), as can be seen in Figure 3b. Figure 3a shows a generic flight simulator model with autopilot and control laws in the loop. The Smart-ADAHRS is basically a data-based state observer exploiting neural functionalities and flight mechanic equations as described in Figure 3b.

Previous research [32] showed that analytic evaluation is indeed feasible for the evaluation of AoA and AoS, thanks to on board available data and with dedicated virtual sensors, one trained ad hoc for AoA and another for AoS. With the present application, where aircraft complexity is higher (for the auotpilot and control law presence itself), the scenario changes. Even though the presence of autopilot modes and control laws, as well known, affect the A/C dynamic behavior, they do not influence generic state observers if they are fed with current output and control surfaces, as showed in Figure 3a.

Another important aspect is related to real operating scenario: a common virtual sensor able to estimate at the same time both AoA and AoS (essentially a single neural network with double output) could be beneficial in terms of required computational time and for ceritification aspects as mentioned before. In order to provide a realistic comparison, three feed-forward predictors are designed, (i) virtual, analytical or synthetic sensor (VS)-AoA, (ii) VS-AoS and (iii) VS-A&S, with the same architecture, the same input vector and the same training path (as shown in Table 1).

Table 1. Required manoeuvres for training and test data collection and their scope.

Manoeuvre	Scope
Steady flight conditions subgroup 1	train
Steady flight conditions subgroup 2 (FT#1)	test
Sawtooth Glide subgroup 1	train
Sawtooth Glide subgroup 2 (FT#2)	test
Stall – Slow down	train
Pitch Hold	train
Pitch Sweep	train
Bank Hold subgroup 1	train
Bank Hold subgroup 2 (FT#3)	test
Bank Sweep	train
Flat Turn subgroup 1	train
Flat Turn subgroup 2 (FT#4)	test
Steady Heading and Sideslip	train
Dutch roll	train

The virtual sensors considered in this work have the following characteristics:

- feed-forward neural network;
- one hidden layer with 24 neurons;
- neurons with sigmoidal activation functions;
- one output layer with a single (or double for the VS-A&S) linear neuron;
- limited output during the operative life.

If compared with previous works, the current activity showed that both AoA and AoS need a complete set of input vector otherwise there is a lack of performances. This is mainly due to complexity of flight dynamics involved in the Piaggio flight simulator. The following input vectors are hence implemented:

$$\Delta\alpha = f_{\text{VS-AoA}}\left(TAS, \hat{\alpha}, n_x, n_y, n_z, \theta, \varphi, p, q, r, \delta_e, \delta_a, \delta_r, \delta_{th}, \Delta_{th}, \delta_{hs}\right) \quad (5)$$

$$\Delta\beta = f_{\text{VS-AoS}}\left(TAS, \hat{\alpha}, n_x, n_y, n_z, \theta, \varphi, p, q, r, \delta_e, \delta_a, \delta_r, \delta_{th}, \Delta_{th}, \delta_{hs}\right) \quad (6)$$

$$[\Delta\alpha, \Delta\beta]^T = \overline{f}_{\text{VS-A\&S}}\left(TAS, \hat{\alpha}, n_x, n_y, n_z, \theta, \varphi, p, q, r, \delta_e, \delta_a, \delta_r, \delta_{th}, \Delta_{th}, \delta_{hs}\right) \quad (7)$$

where TAS is the true airspeed, n_x, n_y, n_z are the accelerations measured by the accelerometers respectively in X_B, Y_B and Z_B axes, ψ, θ, ϕ are the Euler angles, p, q, r are the body angular rates, $\hat{\alpha}$ is the initial estimation for the AoA.

For the preliminary design, the synthetic sensors have been tested only with simulated data. The virtual sensors are compared in terms of measurement uncertainty required to COTS or MFP probes from Table 2. However, the Smart-ADAHRS technology has been already compared with vanes in [20] without providing any particular evidence of degradation of the ADS performance. However, this analysis will be conducted in future dedicated experiments. Previous research activities on simulated turbulent environment in [33] showed the possibility of considering previous time steps of the input vector in a time delay network. This practice however is not considered at this stage and will be considered as a further improvement step.

Table 2. High level performance requirements for the AoA and AoS in a limited area of the flight envelope, LFE, in the extended flight envelope, EFE, and in steady-state flight conditions, SSFC.

Data	2σ Error in LFE	2σ Error in EFE	Maximum Error in SSFC
α	0.75°	1.5°	0.5°
β	1.5°	2.5°	0.5°

3.2. Certification Consideration

As far as certification is concerned, the MIDAS system can be split in three topics: (i) external probes; (ii) electronic unit; (iii) virtual sensors for AoA and AoS.

The external probes, already certified by the supplier, satisfy the required regulations, i.e., TSO-C16A [34] for the Pitot-static probe and the AS793 [35] for the TAT probe.

Design, manufacturing and verification of the electronic unit is one of the main topic of the MIDAS project. In fact, aim of the MIDAS project is to achieve an equivalent DAL-B design assurance level that could be extended to DAL-A for future industrialization. Since all the analytical algorithms are integrated in the field programmable gate array (FPGA), the whole design and validation process will follow RTCA DO-254 [18] guidelines for product assurance and certification. The RTCA DO-178 is not applicable for the MIDAS project. The environmental features will be tested according to requirements established by the RTCA DO-160. In addition to DO-254 risk reduction, pre-qualified (by manufacturer) avionic components will reduce the DO-160 effort and risks as they already integrate lighting protections and other features.

As far as the MIDAS's virtual sensors for AoA and AoS is concerned, they will be treated as physical sensors and, therefore, they have to satisfy the applicable aeronautical standards. As before mentioned in the Section 1, performance requirements for AoA and AoS usually derive from other systems' specifications. For example, for autonomous navigation purposes AoA and AoS may be required by the control laws with defined uncertainty in order to achieve desired navigation performance. Therefore, flight mechanics will specify some requirements on the accuracy of AoA and AoS. For the MIDAS project, the AoA and AoS specification are defined by the project leader, Piaggio Aerospace, and published in a project deliverable [36].

Aiming to provide only necessary details for this work, the most significant target performance are summarised in Table 2 where, with a little abuse of notation, for 2σ is intended the value such that the probability $\Pr(-2\sigma \leq X \leq 2\sigma) = 95.4\%$ also in case the error is not normally distributed. Values reported in Table 2 come from project leader's system specifications for LFE and EFE.

During the normal and emergency flight conditions, the performance required for AoA and AoS are split into limited flight envelope, LFE, extended flight envelope, EFE, and steady-state flight conditions, SSFC. This latter was proposed by the authors and is more stringent because of the common lack of performance during steady operations [37].

4. Reference System

The present work is developed under regulations declared and published with the Grant Agreement number 821,140 of Clean Sky 2. Data used in this paper are provided by Piaggio Aerospace and they are based on the SAT aircraft model inspired to the Piaggio P180 Avanti aircraft. Therefore, all data used to produce this work are property of Piaggio Aerospace. In favour of the reader, some public information are reported here about the reference aircraft in order to understand better the content of the present work. The reference aircraft has a canard-wing-tail configuration with two pushing propellers. Figure 4 shows the body reference system $\{CG, X_B, Y_B, Z_B\}$, the true airspeed vector, V_∞, positive directions of attitude angles (roll, pitch, yaw), body angular rates (p,q,r), body linear velocities (u_B,v_B,w_B) and aerodynamic angles, AoA ($\alpha = \arctan \frac{w_B}{u_B}$) and AoS ($\beta = \arcsin \frac{v_B}{V_\infty}$).

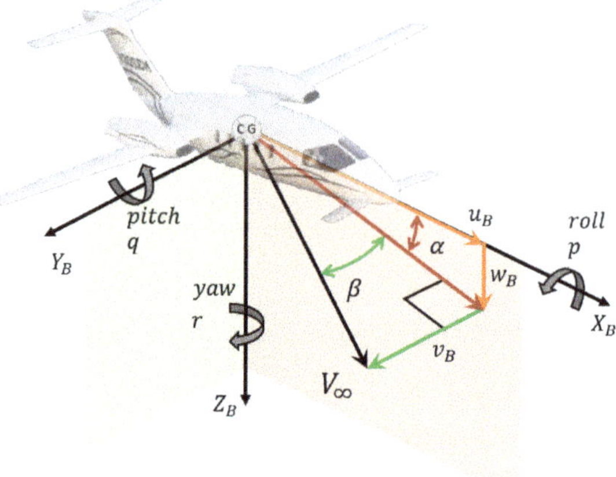

Figure 4. Body reference system.

In this present work, the aerodynamic angle estimators are fed with surface deflections and throttle. Because of the presence of autopilot modes and control laws, as described in Figure 3, the VS cannot be based on pilot reference commands. Control surfaces and throttle considered for the present work are:

- horizontal canard incidence, δ_{hs};
- elevator deflection, δ_e;
- aileron deflection, δ_a;
- rudder deflection, δ_r;
- differential throttle, $\Delta_{th} = throttle_{left} - throttle_{right}$;
- mean throttle, δ_{th}.

5. Training Strategy

The VS for AoA and AoS will be trained with consolidated strategy based on A/C flying characteristics. Firstly, a set of manoeuvres is defined in order to cover the most of the flight envelope. Once the operative speed range is defined, it is split in several flight regimes, e.g., according to the Mach number. Each flight regime is characterized in terms of its trim conditions: initial velocity, or Mach number, and attitude (usually uniform horizontal path, with null bank and sideslip angle). In each one of the flight regime, the following manoeuvres are simulated:

Each manoeuvre is performed with the aim to excite all A/C dynamic modes. Therefore, the single manoeuvre is repeated several times in order to populate the training pattern with adequate information. For example, the pitch hold is repeated from the minimum to the maximum values with steps of 10°.

Once all data are collected, all variables contained in the input vectors (Equation (7)) are normalized between ± 1 considering minimum and maximum values defined by the A/C manufacturer (and not the actual value flown). In this work these values are not shared because of intellectual property reasons. This aspect is crucial for a successful training stage because it will provide a feedback on the correctness of the flown manoeuvres. On the contrary, other flight data will be required to cover the entire region where the VS are defined. As an example, Figure 5 shows how the input variables and the outputs are distributed within the operative range provided by the A/C integrator.

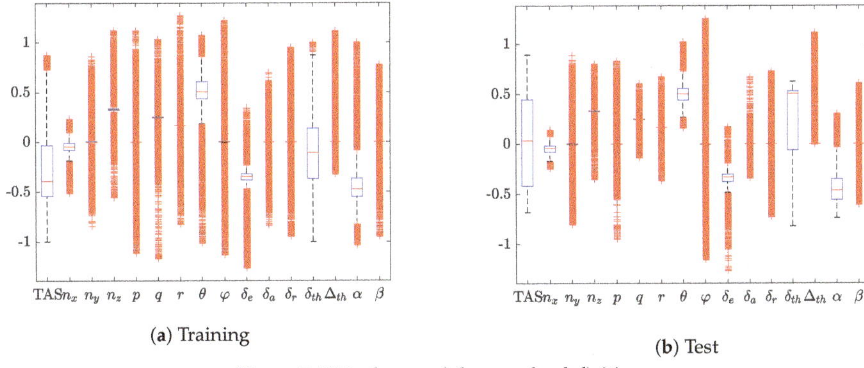

(**a**) Training (**b**) Test

Figure 5. Virtual sensor's hypercube definition.

For each single box, the central mark indicates the median and the bottom and top edges of the box indicate the 25th and 75th percentiles, respectively. The whiskers (dashed vertical lines) cover to the most extreme data points not considered outliers whereas the outliers are plotted using the + symbol. The distribution shown in Figure 5 demonstrates that test data collected during simulated flights are included in the training pattern.

6. Result

The AoA and AoS estimations rely on the input data provided by the A/C FCS (e.g., inertial data, primary and secondary control surface deflections). In this section results related to flight manoeuvres FT#1–4 are presented because they give a real feedback on the VS performances. The training results are only used for training purposes, e.g., to select the best training among several ones according to consolidated metrics, as showed in previous works [20] and not reported in this paper. Results will be presented as time histories of the errors whereas the current values are not shown for intellectual property reasons. Moreover, some important parameters are derived from result analysis that can be used as pass/fail criteria if compared with those in Table 2.

Figure 6 shows time histories of errors obtained with the VS-AoA on the four flight tests and corresponding error distribution. Results are within the required performance (Table 2).

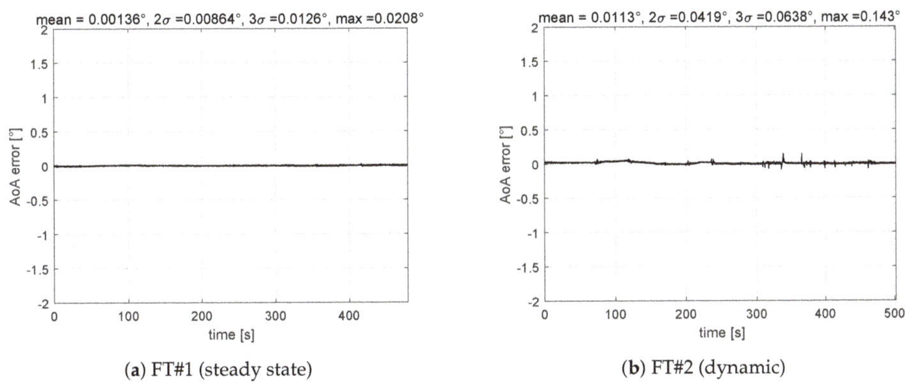

(**a**) FT#1 (steady state) (**b**) FT#2 (dynamic)

Figure 6. *Cont.*

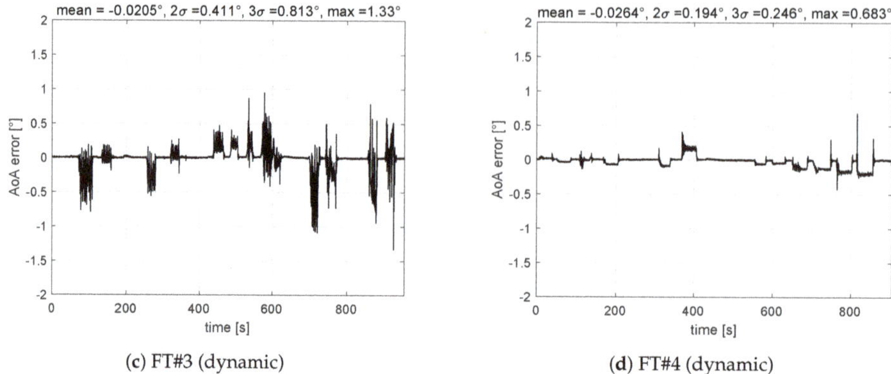

(**c**) FT#3 (dynamic) (**d**) FT#4 (dynamic)

Figure 6. Virtual, analytical or synthetic sensor (VS)-AoA validation results: maximum 2σ (dynamic) = 0.41°, maximum error (steady state) = 0.021°.

Figure 7 shows time histories of errors obtained with the VS-AoS on the four flight tests and corresponding error distribution. Even though errors of AoS estimation are larger than AoA, they are within the required performance (Table 2).

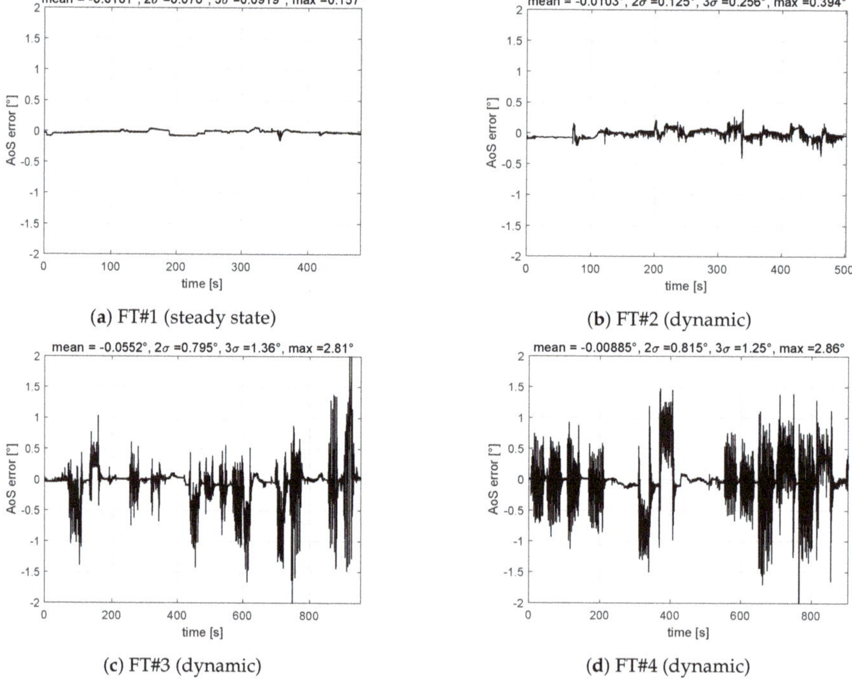

(**a**) FT#1 (steady state) (**b**) FT#2 (dynamic)

(**c**) FT#3 (dynamic) (**d**) FT#4 (dynamic)

Figure 7. VS-AoS validation results—maximum 2σ (dynamic) = 0.82°, maximum error (steady state) = 0.16°.

Figure 8 shows time histories of errors obtained with the VS-A&S on the four flight tests and corresponding error distribution. Performances of the double-output neural network are comparable with the two single neural networks (one dedicated to AoA and one to AoS) and, therefore, performance are still acceptable (Table 2).

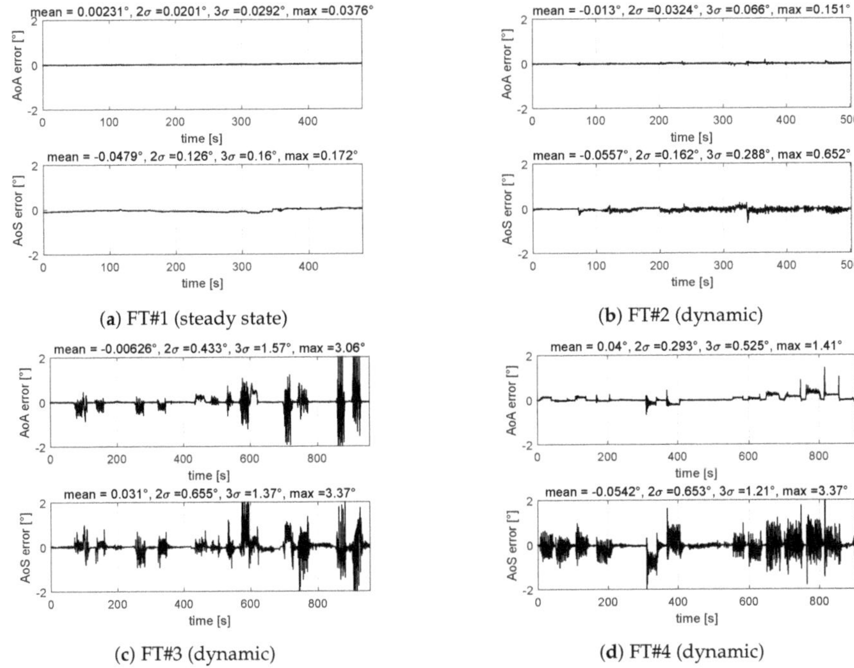

(**a**) FT#1 (steady state) (**b**) FT#2 (dynamic)

(**c**) FT#3 (dynamic) (**d**) FT#4 (dynamic)

Figure 8. VS-A&S validation results − AoA maximum 2σ (dynamic) = 0.43°, maximum error (steady state) = 0.038° − AoS maximum 2σ error (dynamic) = 0.66°, maximum error (steady state) = 0.17°.

Results shown in this section allow authors to select the VS-A&S as the candidate virtual sensor architecture because more suitable for real time operations and, more important, for certification aspects (a single software module to be qualified). Moreover, it is confirmed that the AoS estimation shows larger errors with respect to AoA estimation. To understand the reason behind, it was noted that training AoA and AoS at the same time requires a better balanced training pattern between longitudinal flight test points and lateral-directional ones. In fact, AoS was noted to be less than 1° for about 85% of the common training pattern. Therefore, as suggestion for the final design step, a review of the training strategy is required for the MIDAS objectives.

Table 3 collects all validation results obtained with the preliminary design about three aerodynamic angle estimators suitable for civil certification and compared with the most stringent requirements (LFE) provided by Piaggio Aerospace (project leader).

Table 3. Result comparison for the three virtual sensor architectures considered in this work for AoA and AoS estimation.

VS	Data	2σ (Dynamic)	Maximum (Steady State)
VS-AoA	AoA	0.41° < 0.75°	0.021° < 0.5°
VS-AoS	AoS	0.82° < 1.5°	0.16° < 0.5°
VS-A&S	AoA	0.43° < 0.75°	0.038° < 0.5°
	AoS	0.66° < 1.5°	0.17° < 0.5°

7. Conclusions

The present work introduces preliminary design activities for a reliable and ready-to-be-flown air data system for FBW applications within the MIDAS project funded in the frame of Clean Sky 2.

The MIDAS technology will enable to remove flow angles vanes or complex rotating/slotted probes that are hard to be procured on the market. The ambition is providing the SAT community with a digital and fully integrated ADS that joins all benefits of available flyable architectures and removes their drawbacks. It is introduced MIDAS projects' aims to improve the current state-of-the-art of ADS.

It is shown that the MIDAS ADS will be based on AoA and AoS analytical estimators based on neural network techniques, today at TRL6.

Training and test manoeuvres are introduced. Training manoeuvres are defined aiming to excite all dynamic modes of the A/C model whose complexity is increased by several autopilot modes and control laws. Test manoeuvres are collected with the scope to validate results independently from the training set and all possible autopilot configuration.

With respect to the previous works, it emerged that both AoA and AoS need a complete set of input pattern to show acceptable performances due to high complexity of A/C dynamics. According to previous works, AoA and AoS are estimated with dedicated virtual sensors using a single output neural network. This approach has two main drawbacks when applied to qualifiable avionics: (i) it requires a higher computational cost; (ii) two independent software modules to submit to a certification process (DO-178 or, as for the MIDAS project, DO-254). These two main drawbacks suggest to have a single neural network with double output, therefore the computational and certification effort can be drastically reduced.

The virtual sensor (VS-A&S) exhibits only slightly degraded performance for steady-state conditions whereas comparable errors for dynamic flight tests. This evidence makes the single VS (both AoA and AoS) the candidate solution in the next staged of development. Moreover, the A/C complexity conjugated with the new neural network architecture has introduced a new challenge: a common training pattern (both for AoA and AoS) will require, not only, a uniform distribution but even a balanced data between longitudinal and lateral-directional flight test points. This means that the hypercube definition of the neural network shall be uniformly populated as much as possible when collecting flight test data at the simulator. This topic will be discussed with the project leader in a next stage. The reliability analysis and the comparison of the MIDAS technology with classical solution will be studied with the other partners contributions in a next stage.

In conclusion, the selected VS (VS-A&S) exhibits good preliminary performances both for AoA and AoS and it is selected for the candidate VS architecture. Further investigation of larger errors on AoS estimation shall be investigated in the following works.

Author Contributions: Conceptualization, all; Methodology, all; Software, A.L., A.B.; Validation, A.L., A.B.; Formal Analysis, A.L., A.B.; Investigation, A.L., A.B.; Resources, all; Data Curation, A.L., A.B.; Writing-Original Draft Preparation, A.L., A.B.; Writing-Review & Editing, all; Visualization, A.L., A.B.; Supervision, P.G., M.B.; Project Administration, P.G., M.B.; Funding Acquisition, all.

Funding: This research is supported by H2020-Clean Sky 2 under SYS ITD area under Grant Agreement number 821140.

Conflicts of Interest: The authors declare no conflict of interest.

Abbreviations

The following abbreviations are used in this manuscript:

A/C	Aircraft
ADAHRS	Air data system, attitude and heading reference system
AHRS	Attitude and heading reference system
ADC	Air data computer
ADM	Air data module
ADP	Air data probe
ADS	Air data system

ADU	Air data unit
AoA	Angle-of-attack
AoS	Angle-of-sideslip
CAS	Calibrated airspeed
CG	Center of gravity
COTS	Commercial off the shelf
CS	Certification specifications
DAL	Design assurance level
FBW	Fly-by-wire
FCS	Flight control system
FHA	Failure hazard analysis
FPGA	Field programmable gate array
FT	Flight test
GNSS	Global Navigation Satellite System
LRU	Line replaceable unit
MFP	Multi-function probe
MLP	Multilayer perceptron
OAT	Outside air temperature
SAT	Small air transport
TAS	True airspeed
TAT	Total air temperature
UAM	Urban air mobility
UAS	Unmanned aircraft system
UAT	Universal approximation theorem
VS	Virtual, analytical or synthetic sensor

References

1. *Air Data Computer: Minimum Performance Standard, Aerospace Standard*; Standard SAE-AS8002A; SAE International: Troy, MI, USA, 1996.
2. *Stall Warning Instrument*; Standard SAE-AS403A; SAE International: Troy, MI, USA, 1993.
3. *Flightpath 2050—Europe's Vision for Aviation*; Publications Office of the European Union: Brussels, Belgium, 2011; ISBN 978-92-79-19724-6, doi:10.2777/50266. [CrossRef]
4. Collins Aerospace. SmartProbe Air Data Systems. 2018. Available online: https://utcaerospacesystems.com/wp-content/uploads/2018/04/SmartProbeR-Air-Data-Systems.pdf (accessed on 19 October 2019).
5. Sun, K.; Regan, C.D.; Gebre-Egziabher, D. Observability and Performance Analysis of a Model-Free Synthetic Air Data Estimator. *J. Aircr.* **2019**, *56*, 1471–1486, doi:10.2514/1.C035290. [CrossRef]
6. Oosterom, M.; Babuska, R. Virtual Sensor for the Angle-of-Attack Signal in Small Commercial Aircraft. In Proceedings of the 2006 IEEE International Conference on Fuzzy Systems, Vancouver, BC, Canada, 16–21 July 2006; pp. 1396–1403, doi:10.1109/FUZZY.2006.1681892. [CrossRef]
7. Samara, P.A.; Fouskitakis, G.N.; Sakellariou, J.S.; Fassois, S.D. Aircraft angle-of-attack virtual sensor design via a functional pooling narx methodology. In Proceedings of the 2003 European Control Conference (ECC), Cambridge, UK, 1–4 September 2003; pp. 1816–1821, doi:10.23919/ECC.2003.7085229. [CrossRef]
8. Rhudy, M.; Larrabee, T.; Chao, H.; Gu, Y.; Napolitano, M. UAV attitude, heading, and wind estimation using GPS/INS and an air data system. In Proceedings of the AIAA Guidance, Navigation, and Control (GNC) Conference, Boston, MA, USA, 19–22 August 2013.
9. Schettini, F.; Di Rito, G.; Galatolo, R.; Denti, E. Sensor fusion approach for aircraft state estimation using inertial and air-data systems. In Proceedings of the 2016 IEEE Metrology for Aerospace (MetroAeroSpace), Florence, Italy, 22–23 June 2016; pp. 624–629, doi:10.1109/MetroAeroSpace.2016.7573289. [CrossRef]
10. Lie, F.A.P.; Gebre-Egziabher, D. Sensitivity Analysis of Model-based Synthetic Air Data Estimators. In Proceedings of the AIAA Guidance, Navigation, and Control Conference, Kissimmee, FL, USA, 5–9 January 2015; American Institute of Aeronautics and Astronautics: Reston, VA, USA, 2015; pp. 1–18, doi:10.2514/6.2015-0081. [CrossRef]

11. Balzano, F.; Fravolini, M.; Napolitano, M.R.; d'Urso, S.; Crispoltoni, M.; Core, G. Air Data Sensor Fault Detection with an Augmented Floating Limiter. *Int. J. Aerosp. Eng.* **2018**, *2018*, 1072056, doi:10.1155/2018/1072056. [CrossRef]
12. Lerro, A.; Battipede, M.; Gili, P.; Ferlauto, M.; Brandl, A.; Merlone, A.; Musacchio, C.; Sangaletti, G.; Russo, G. The Clean Sky 2 MIDAS Project—An Innovative Modular, Digital and Integrated Air Data System for Fly-by-Wire Applications. In Proceedings of the 2019 IEEE 5th International Workshop on Metrology for AeroSpace (MetroAeroSpace), Torino, Italy, 19–21 June 2019; pp. 714–719, doi:10.1109/MetroAeroSpace.2019.8869602. [CrossRef]
13. Lerro, A.; Battipede, M.; Gili, P. System and Process for Measuring and Evaluating Air and Inertial Data. European Patent 3022565A2, 6 September 2017.
14. AeroSmart S.r.l. Available online: http://www.aerosmartsrl.it/ (accessed on 19 October 2019).
15. Brandl, A.; Battipede, M.; Gili, P.; Lerro, A. Sensitivity analysis of a neural network based avionic system by simulated fault and noise injection. In Proceedings of the AIAA Modeling and Simulation Technologies Conference, Kissimmee, FL, USA, 8–12 January 2018, doi:10.2514/6.2018-0122. [CrossRef]
16. Lerro, A.; Battipede, M.; Gili, P.; Brandl, A. Survey on a neural network for non linear estimation of aerodynamic angles. In Proceedings of the 2017 Intelligent Systems Conference (IntelliSys 2017), London, UK, 7–8 September 2017; pp. 929–935, doi:10.1109/IntelliSys.2017.8324240. [CrossRef]
17. Lerro, A.; Battipede, M.; Gili, P.; Brandl, A. Advantages of Neural Network Based Air Data Estimation for Unmanned Aerial Vehicles. *Int. J. Mech. Aerosp. Ind. Mechatron. Manuf. Eng.* **2017**, *19*, 1196–1205, doi:10.5281/zenodo.1130557. [CrossRef]
18. RTCA. *RTCA/DO-254, Design Assurance Guidance for Airborne Electronic Hardware*; RTCA, Inc.: Washington, DC, USA, 2005.
19. RTCA. *RTCA/DO-160, Environmental Conditions and Test Procedures for Airborne Equipment*; RTCA, Inc.: Washington, DC, USA, 2007.
20. Lerro, A.; Battipede, M.; Gili, P.; Brandl, A. Aerodynamic angle estimation: comparison between numerical results and operative environment data. *CEAS Aeronaut. J.* **2019**, doi:10.1007/s13272-019-00417-x. [CrossRef]
21. Battipede, M.; Gili, P.; Lerro, A.; Caselle, S.; Gianardi, P. Development of Neural Networks for Air Data Estimation: Training of Neural Network Using Noise-Corrupted Data. In Proceedings of the CEAS 2011—The International Conference of the European Aerospace Societies, Venice, Italy, 24–28 October 2011.
22. European Aviation Safety Agency (EASA). *Certification Specifications for Normal Utility, Aerobatic, and Commuter Category Aeroplanes—CS23*. Available online: http://www.easa.eu.int (accessed on 19 October 2019).
23. SELT A&D. 2019. Available online: http://www.selt-sistemi.com/ (accessed on 19 October 2019).
24. Lerro, A.; Battipede, M.; Gili, P.; Brandl, A. Comparison Between Numerical Results and Operative Environment Data on Neural Network for Air Data Estimation. In Proceedings of the 6th CEAS Air and Space Conference Aerospace Europe 2017, Bucharest, Romania, 16–20 October 2017.
25. Lerro, A.; Battipede, M.; Brandl, A.; Gili, P.; Rolando, A.L.M.; Trainelli, L. Test in Operative Environment of an Artificial Neural Network for Aerodynamic Angles Estimation. In Proceedings of the 28th Society of Flight Test Engineers European Chapter Symposium (SFTE-EC 2017), Milano, Italy, 13–15 September 2017.
26. Cybenko, G. Approximation by superpositions of a sigmoidal function. *Math. Control Signals Syst.* **1989**, *2*, 303–314, doi:10.1007/BF02551274. [CrossRef]
27. Bishop, C.M. *Neural Networks for Pattern Recognition*; Clarendon Press: Oxford, UK, 1995.
28. Castro, J.; Mantas, C.; Benítez, J. Neural networks with a continuous squashing function in the output are universal approximators. *Neural Netw.* **2000**, *13*, 561–563, doi:10.1016/S0893-6080(00)00031-9. [CrossRef]
29. Attali, J.G.; Pagès, G. Approximations of Functions by a Multilayer Perceptron: A New Approach. *Neural Netw.* **1997**, *10*, 1069–1081, doi:10.1016/S0893-6080(97)00010-5. [CrossRef]
30. Hornik, K. Approximation capabilities of multilayer feedforward networks. *Neural Netw.* **1991**, *4*, 251–257, doi:10.1016/0893-6080(91)90009-T. [CrossRef]
31. Haykin, S. *Neural Networks: A Comprehensive Foundation*; Prentice Hall PTR: Upper Saddle River, NJ, USA, 1994.

32. Battipede, M.; Cassaro, M.; Gili, P.; Lerro, A. Novel Neural Architecture for Air Data Angle Estimation. In *Engineering Applications of Neural Networks, Proceedings of the 14th International Conference on EANN 2013, Halkidiki, Greece, 13–16 September 2013*; Springer: Berlin/Heidelberg, Germany, 2013; pp. 313–322, doi:10.1007/978-3-642-41013-0_32. [CrossRef]
33. Battipede, M.; Gili, P.; Lerro, A. Neural Networks for Air Data Estimation: Test of Neural Network Simulating Real Flight Instruments. In *Engineering Applications of Neural Networks*; Jayne, C., Yue, S., Iliadis, L., Eds.; Springer: Berlin/Heidelberg, Germany, 2012; pp. 282–294,
34. Federal Aviation Administration, FAA. *Electrically Heated Pitot and Pitot-Static Tubes*; FAA: Washington, DC, USA, 2006.
35. *Total Temperature Measuring Instruments*; SAE International: Troy, MI, USA, 2008.
36. Piaggio Aerospace. *Air Data Probe Specification—Clean Sky 2 WP7.3—Deliverable D7,3,4-1*; Piaggio Aerospace: Genoa, Italy, 2018.
37. Brandl, A.; Lerro, A.; Battipede, M.; Gili, P. Air Data Virtual Sensor: A Data-Driven Approach to Identify Flight Test Data Suitable for the Learning Process. In Proceedings of the 5th CEAS Specialist Conference on Guidance, Navigation and Control, Milano, Italy, 3–5 April 2019.

© 2019 by the authors. Licensee MDPI, Basel, Switzerland. This article is an open access article distributed under the terms and conditions of the Creative Commons Attribution (CC BY) license (http://creativecommons.org/licenses/by/4.0/).

Article

Determining Wheel Forces and Moments on Aircraft Landing Gear with a Dynamometer Sensor †

Jaroslaw Pytka [1], Piotr Budzyński [1,*], Tomasz Łyszczyk [1], Jerzy Józwik [1], Joanna Michałowska [2], Arkadiusz Tofil [2], Dariusz Błażejczak [3] and Jan Laskowski [4]

1 Faculty of Mechanical Engineering, Lublin University of Technology, Nadbystrzycka 36, 20-618 Lublin, Poland; j.pytka@pollub.pl (J.P.); tomasz.lyszczyk@gmail.com (T.Ł.); j.jozwik@pollub.pl (J.J.)
2 The State School of Higher Education, The Institute of Technical Sciences and Aviation, 22-100 Chełm, Poland; jmichalowska@pwsz.chelm.pl (J.M.); atofil@pwsz.chelm.pl (A.T.)
3 Department of Construction and Usage of Technical Devices, West Pomeranian University of Technology in Szczecin, 70-310 Szczecin, Poland; Dariusz.Blazejczak@zut.edu.pl
4 Faculty of Management, Lublin University of Technology, Nadbystrzycka 36, 20-618 Lublin, Poland; jlasko@wp.pl
* Correspondence: p.budzynski@pollub.pl; Tel.: +48-602-459-881

† This paper is an extended version of the conference paper: Pytka, J.; Budzyński, P.; Józwik, J.; Łyszczyk, T.; Laskowski, J.; Gnapowski, E. Measurement of Forces and Moments Acting on Aircraft Landing Gear Wheel. Presented at the 6th IEEE International Workshop on Metrology for Aerospace Metro Aerospace 2019, Totino, Italy, 19–21 June 2019.

Received: 24 November 2019; Accepted: 25 December 2019; Published: 31 December 2019

Abstract: This paper describes airfield measurement of forces and moments that act on a landing gear wheel. For the measurement, a wheel force sensor was used. The sensor was designed and built based on strain gage technology and was embedded in the left landing gear wheel of a test aircraft. The sensor is capable of measuring simultaneously three perpendicular forces and three moments and sends data to a handheld device wirelessly. For the airfield tests, the sensor was installed on a PZL 104 Wilga 35A multipurpose aircraft. The aircraft was towed at a "marching man" speed and the measurements were performed at three driving modes: Free rolling, braking, and turning. The paper contains results obtained in the field measurements performed on a grassy runway of the Rzeszów Jasionka Aerodrome, Poland. Rolling resistance of aircraft tire, braking friction, as well as aligning moment were analyzed and discussed with respect to surface conditions.

Keywords: airfield performance; landing gear; wheel; force and moment measurement; strain gage; wheel force sensor; wireless data transfer; grassy airfield; GARFIELD System

1. Introduction

Airfield performance of an airplane plays an important role in the analysis of takeoff and landing and determines, among others, ground roll distances. Forces and moments that act on aircraft landing gear wheels are effects of gravitational acceleration, as well as surface reactions. These reactions are easily obtainable on paved runways, but difficulties arise when an airplane operates on unpaved, grassy, or gravel surfaces. One method is to monitor surface mechanical data, related to weather conditions. The *GARFIELD* information system [1] is based on a wheel–grass model that includes analysis of wheel–soil (wheel–grassy surface) interactions considering soil modeling with a special caution to loads by aircraft tires [2]. The model will respect non-linear, dynamic effects such as hyper-elastic tire deflection, rheological soil response to high rate deformation, and effects of grass and roots. The wheel–grass model also employs the analysis of weather impact upon mechanical characterization of runway surface, which can be identified and verified by means of wheel–force measurements in full-scale tests [3].

Researchers dealing with flight tests of aircraft, piloted or unmanned, pay significant attention to the importance of ground tests, including measurements for the validation and verification of landing gear systems [4–6].

Significant emphasis is placed on the development and research of new measuring technologies and, as an example, a new solution in the field of strain gauge sensors can be cited. Pecora et al. have presented an innovative strain gauge for monitoring of inflatable structures, suitable for an use in aerospace [7]. Another interesting example is given by Petritoli et al. (2019-1), who investigated uncertainty and measured errors of an inertial navigation systems for UAV (Unmanned Aerial Vehicle) [8]. Petritoli et al. (2018) and Petritoli et al. (2019-2) targeted the problem of collocation of the sensors to be installed in a UAV with a reduced space, together with the allocation of the payload [9,10]. This problem is vital for full size aircrafts, not only typically smaller UAVs, especially if a sensor has to be embedded in an existing part of an aircraft. Eling et al. (2015) presented an innovative georeferencing system for the guidance, navigation, and control of UAVs [11]. This is an example of an inertial measurement unit (IMU)-based system, the method of a wide spectrum of applications in aerospace.

Measurement of forces and moments that act on a road wheel is a well known practice in automotive research and development. It is usually done either on laboratory test stands or on a test vehicle. Either way, a device that provides wheel forces and moments is a wheel dynamometer. A wheel dynamometer is a measuring device that enables to measure forces and moments acting on a wheel in real time on a moving vehicle. The majority of wheel dynamometer solutions comes from the automotive industry or research, and a typical solution has a form of a 6-element sensor that measures three orthogonal forces, F_X, F_Y, and F_Z, together with corresponding moments, M_X, M_Y, and M_Z [12–15]. The measurement principle is based either on the strain gage or piezoelectric sensor technology. Some selected works that describe the use of such dynamometer systems for measurement of wheel forces and moments in various applications are included in the reference list [16–19].

The horizontal force, F_X, measured by a wheel dynamometer can be correlated with bending or shearing the grass plants, as well as soil deformation. The braking friction force can be correlated with sliding over the grass. The vertical force, Fz, depends the grass compression and can be used for coefficient calculations (braking friction, μ, and rolling resistance, k_{RR}, coefficients, as well as aligning moment, M_Z). This approach is alternative to commonly used laboratory drop tests of landing gear [20–22]. The aim of this study was to apply a wheel force and moment sensor for measurement of aircraft landing gear on grassy surface. The primary paper, entitled *Measurement of Forces and Moments Acting on Aircraft Landing Gear Wheel*, was presented during the 2019 IEEE International Workshop Metrology for Aerospace in Torino, Italy [23].

2. Materials and Methods

2.1. Wheel Dynamometer System

The wheel dynamometer system used in the experiment consisted of a sensor, modified wheel and tire, wireless data transfer unit, and software. The sensor core was made of steel, and strain gages were used as sensing elements. The instrumentation amplifiers, which prepare the outgoing signals for data acquisition, were placed outside the sensor in a small box, together with a board of the microprocessor data analysis and transfer system. Figure 1 shows a schematic of the sensor, but without the electronics. This schematic also includes the orientation of forces and moments.

As mentioned earlier, the sensor was designed based on the strain gage measurement technique. The sensing element was made of spring steel and linear type strain gages were glued onto it.

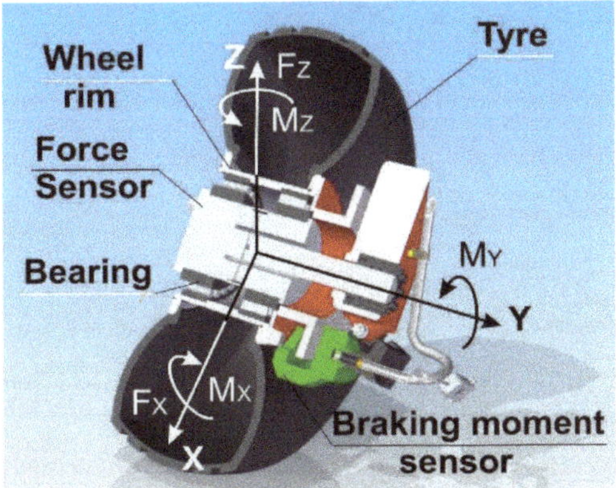

Figure 1. A schematic of the wheel sensor together with axes orientation and forces/moments system. The *X*, *Y*, and *Z* axes are mutually orthogonal.

The in-wheel transducer measures two forces, F_Z and F_X, and two moments, M_Z and M_X. Due to the restricted space mentioned above, a braking moment sensor was made as a separate, external device. The M_Y moment was measured with the strain gages sealed on the brake jig, which is bending proportionally to braking moment. Figure 2 shows the complete sensor system, embedded in an aircraft wheel.

Figure 2. The complete wheel force sensor.

The wireless data transfer unit was built as a microprocessor system (see Figure 3), with an 8-channel A/D converter and a Bluetooth radio device to transfer measured data wirelessly.

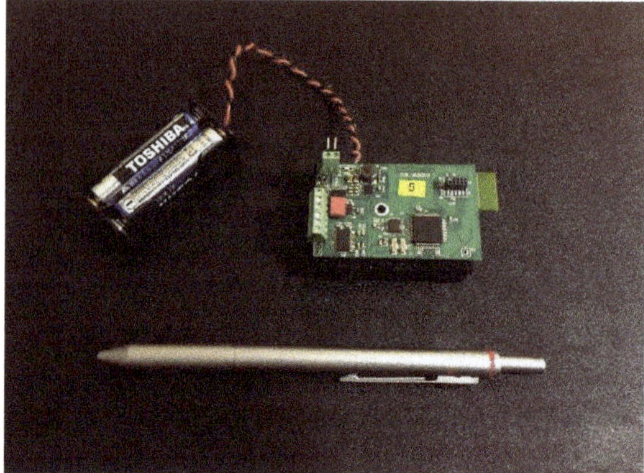

Figure 3. The heart of the microprocessor data transfer system, used in the wheel sensor system.

The complete and detailed description of the wheel force sensor system is given in the reference by Pytka et al. [3].

2.2. Airplane Used in the Experiment

The PZL 104 Wilga 35 multipurpose airplane was used as a test aircraft. This is a light, four-place, multipurpose aircraft, powered by a 192 kW radial piston engine. The Wilga has a good reputation as a STOL (Short Takeoff and Landing) or even the so-called "bush-plane", thanks to its landing gear. The aircraft has a tail-dragger-type landing gear and the main gears are rocker type with oleo-pneumatic shock absorbers. The landing gear wheels have low-pressure 500 × 200 mm tires with hydraulic brakes. The wheels are castored and have a positive rake angle of 18°; the axle offset is 400 mm. The airplane is 8.10 m in overall length, with a wingspan of 11.12 m, a wing area of 15.5 m^2, and an empty mass, equipped, of 900 kg. In the landing tests, there were four persons on board and the take-off weight was 1150 kg. The test airplane is depicted in Figure 4.

The wheel force sensor was installed on the test aircraft before measurements. Installation did not require any modification to the airframe or landing gear, since the sensor system was developed specially for the given aircraft. Simply, leveling up the aircraft, removing the wheel with its axle, and placing the sensor. Also, the brake hydraulic system had to be re-switched to the sensor, since the F_Y wheel moment was measured by means of a transducer embedded in the wheel brake.

Another element of the measuring system was the electronic unit, which consisted of five differential signal amplifiers, each for one of the five channels (two forces and three moments), as well as of a microcontroller-based, wireless data transfer unit. The electronic box was mounted on the landing gear with glue tape and the sensor was connected with the electronics with signal cables. The whole system, the wheel sensor, and the electronics power supply (7.2V 800 mAh NiMH accumulator) was placed in the cockpit. Figures 5 and 6 show the details of the test airplane with the installed measuring equipment.

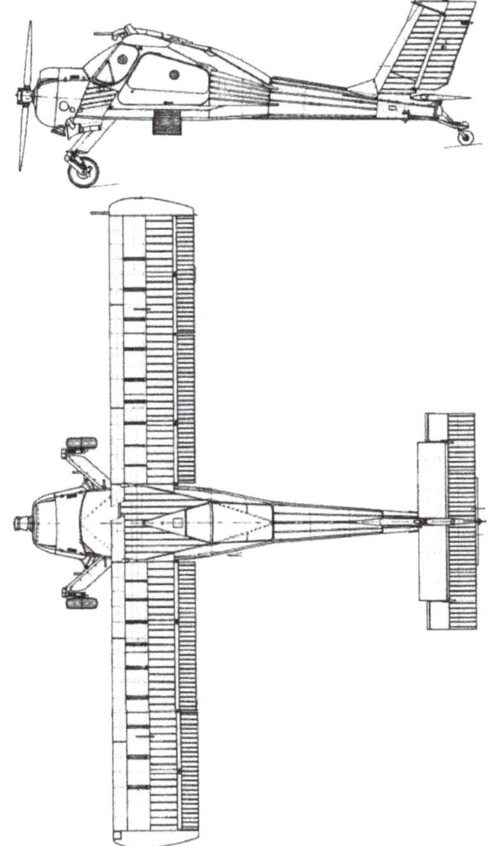

Figure 4. The PZL 104 Wilga 35A, used for the tests.

Figure 5. The wheel force sensor installed on the test airplane, ready for the tests.

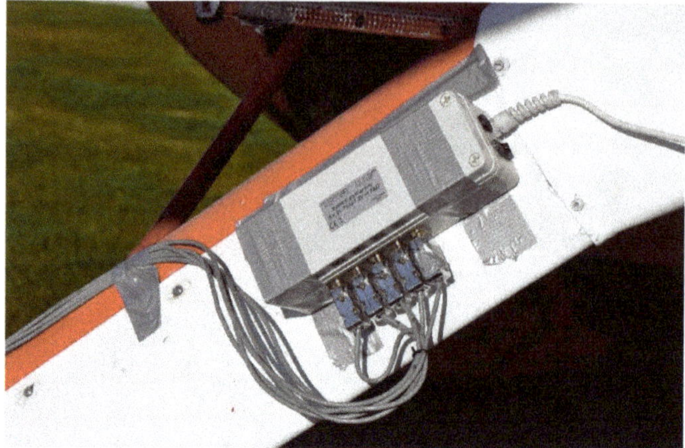

Figure 6. The electronics: Signal amplifiers and wireless data transfer system in the box, installed on the landing gear leg (a silver tape was used for quick installation).

2.3. Grassy Airfield Conditions

Measurements of wheel forces and moments were performed on a grassy surface of the Rzeszów Jasionka aerodrome, located in southeast Poland. The measurements were carried out in the area adjacent to the runway, with the surface and soil conditions identical to those on the grassy runway. The height of the grass was approximately 10 cm each time of the measurements.

The days in the autumn and spring months were selected, when there is a high probability of increasing humidity and lowering the traction properties of the grassland due to the influence of weather conditions (rainfall, low temperature, low sunshine). They were on 17 October, 5 November, 5 December, and 13 March. Table 1 gathers data describing atmospheric conditions prevailing on the days of measurement.

Table 1. Weather and soil conditions on the days of tests.

Weather	Air Temperature (°C)	Soil Moisture (%)	Soil Temperature (°C)
	17 October 2018		
Sunshine, wind 5–7m/s S, SE	15	8.80	n.a.
	5 November 2018		
Sunshine, wind 10-12 m/s W	12	18.80	18.3
	5 December 2018		
Cloudy, 7/8 Cu, wind 2–3m/s W, NW	2,5	23.67	4.6
	13 March 2019		
Cloudy, with break intervals	11	22.50	n.a.

Before the test runs, we also measured soil moisture using a handheld TDR moisture meter. TDR is an acronym for time-domain reflectometry, and this is a technique that utilizes the phenomenon of different electrical permeability of porous, three-phase media, such as soil. The TDR meter is very easy to use. Measurement requires simply inserting test electrodes into soil and the instrument quickly gives results, which are collected in Table 1. Figure 7 shows the test engineer performing soil moisture measurement on the grass field. See also the reference by Pytka et al. for more details on the use the TDR sensor in terramechanics studies [24].

Figure 7. Measurement of soil moisture with the use of a handheld time-domain reflectometry (TDR) sensor.

2.4. Field Procedures

Since the sensor system has not been certified for flight tests yet, we could only manage ground tests. In this study, it was aimed to determine wheel forces and moments acting on the aircraft's wheel at low speed of taxiing over the grassy runway. We used an all-terrain vehicle (ATV) as a tractor vehicle and the aircraft was connected to this vehicle by means of cables. As mentioned earlier, four adult persons were in the test aircraft during all tests. An approximate speed of taxiing was 5 km/h. The aircraft performed the following maneuvers:

- Accelerating from stop to the taxi speed;
- Taxiing with a constant speed (about 20–30 m);
- Turning left or right;
- Braking to a full stop or to block the wheels.

We performed a minimum of five repetitions of all the maneuvers for every measurement time. Figure 8 shows a sample test run on the grassy surface.

Figure 8. Starting a test run. The test airplane is pulled by an all-terrain vehicle (ATV). The test engineer is walking by the test airplane, collecting data wirelessly. Four persons onboard the airplane.

The wireless data transfer system enabled that the test engineer could be anywhere, within a 100 m range of the airplane. Since the measurements were performed with all four passengers on board, the test engineer was in the airplane during measurements. We tried to collect data on a laptop computer, but it was more comfortable to gather the results on a handheld smartphone.

3. Results

Results obtained in the field experiments include courses of wheel forces and moments acting on the left wheel of the Wilga airplane, and we collected data from four days of measurements: 17 October, 5 November, and 5 December 2018, as well as 13 March 2019. The following is a presentation and analysis of the results obtained in the autumn–winter part of the measurements campaign (a so-called "dead season" in sport aviation) that was performed as a full-season measurement for the verification of the wheel–turf interaction model of the *GARFIELD* information system [1,2,23,24].

3.1. A General Presentation of the Results

Figure 9 shows time courses of the vertical F_Z and horizontal F_X forces that act on the landing gear wheel. The F_Z shows almost constant values with low fluctuations, probably caused by vertical vibrations of the entire aircraft rolling over the grassy surface, and those values are close to the aircraft weight component on the wheel. The horizontal force F_X exhibits negative values since the sensor has a defined orientation and a force of the direction backwards is indicated as negative. Moreover, the course of the horizontal force F_X reveals effect of wheel function modes: Free rolling (the first second of the course), acceleration increasing of the force's values (between 1 and approximately 4 s of the course), braking to a full stop (5–9 s), and, again, acceleration (from approximately 9 s of the course). The peak maximum value of the horizontal force F_X reaches 2.7 kN, but the average value is much lower.

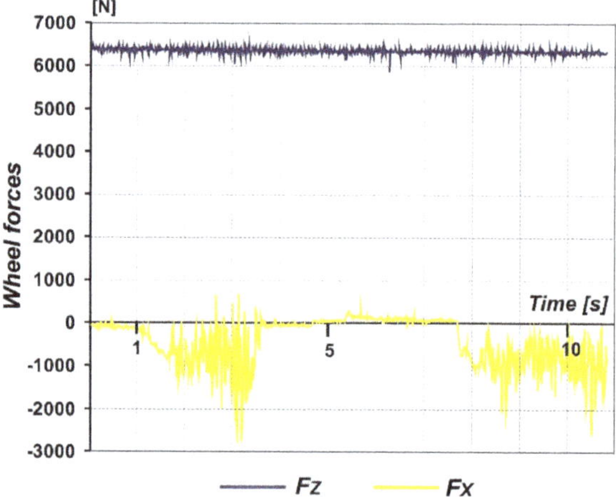

Figure 9. Sample result of the measurements: Time courses of wheel forces acting on the landing gear wheel.

Another sample result is shown in Figure 10, where we can see time courses of the three moments acting on the aircraft wheel. The character of these courses is very similar to those of Figure 9. These is a short period of free rolling, then acceleration, braking and stop, and finally acceleration again. The peak maximum values reach 150 Nm for the M_Z moment, 280 Nm for the M_X, and 250 for the M_Y moment. The values of the moments acting on the landing gear wheel changes their sign at a moment of a change between two function modes, for example, accelerating–braking.

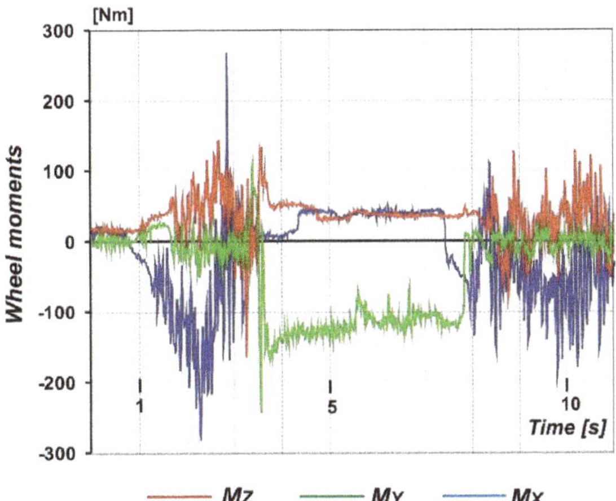

Figure 10. Sample result of the measurements: Time courses of three moments acting on the landing gear wheel.

In the next sections, an analysis of the measured forces and moments is presented in order to determine coefficients that describe wheel–surface interactions, which are of importance for the analysis and modeling of airfield performance of an aircraft.

3.2. Rolling Resistance Coefficient

Rolling resistance is a property of the tire–surface system and depends on many factors. In automotive technology, rolling resistance is determined for a tire rolling on a drum test stand and its coefficient takes essentially constant values, on the order of 0.010–0.012. In the case of a tire interacting with a deformable surface, such as grassy surface, rolling resistance is difficult to determine, mainly due to the changing conditions of the surface, which was mentioned in the introduction. In addition, the values of the rolling resistance coefficient of the tire–grass surface system can vary by as much as an order of magnitude, taking into account, for example, changes in soil moisture. The presence of vegetation (grass, roots), as well as the size and mass of green parts of vegetation, are also significant.

The rolling resistance coefficient, k_{RR}, is defined as the ratio of horizontal force during free turning to vertical force [25]:

$$k_{RR} = F_X/F_Z \quad (1)$$

Figure 11 shows an example of horizontal force, F_X, which is input data for calculations (upper graph) and the rolling resistance coefficient, k_{RR} (lower graph), value determined using Equation (1). Average values are marked on the graphs (F_X = 1719 N and k_{RR} = 0.275). When determining the k_{RR} coefficient, it is important to correctly select a portion of the F_X force waveform. Namely, it is a fragment where the waveform is determined (in Figure 11, up, this fragment begins after the first second of the wave).

The exemplary value of the k_{RR} coefficient was determined to be 0.275, which in comparison to the aforementioned value for a car tire on a paved road is almost 25 times [25]. This is the effect of a soft surface and a low-pressure aviation tire. The impact of weather conditions on the determined values of the k_{RR} coefficient will be presented in the next section.

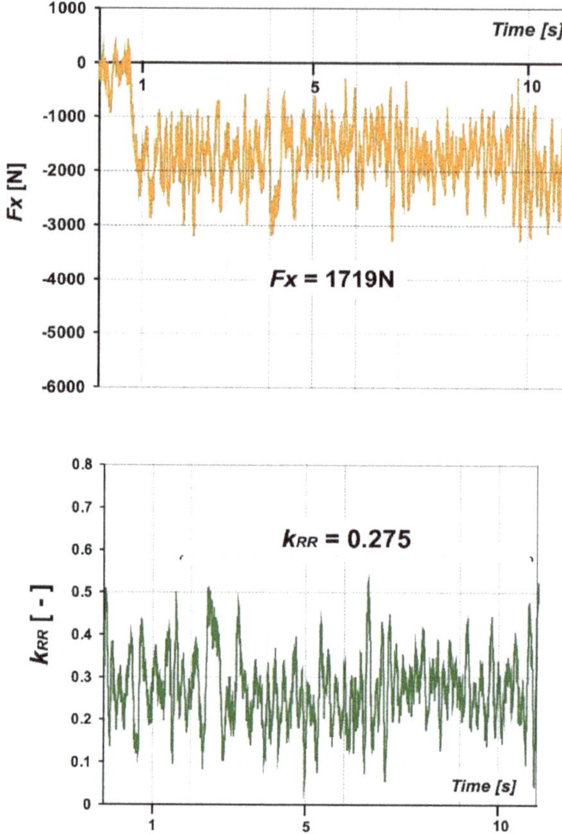

Figure 11. Sample courses of the measured horizontal force, F_X, and calculated rolling resistance coefficient, k_{RR}.

3.3. Braking Friction Coefficient

The braking friction coefficient μ is another important factor. It determines the course of the braking process and the landing distance of the aircraft. In the case of the tire–paved system (asphalt, concrete), the friction coefficient usually reaches values between 0.6–0.9. For unpaved surfaces such as grass turf, the friction coefficient is usually lower and may be 0.1–0.5. Factors affecting the μ value are grass length and humidity: The longer the grass and the more water it contains, the lower the friction coefficient. By definition, the friction coefficient is determined on the basis of the following relationship:

$$\mu = M_Y/F_Z \times r_d \tag{2}$$

where r_d is the dynamic radius of the tire.

Figure 12 shows an example of the braking moment, M_Y, course and the braking friction coefficient determined. As it can be seen, μ values oscillate around 0.22.

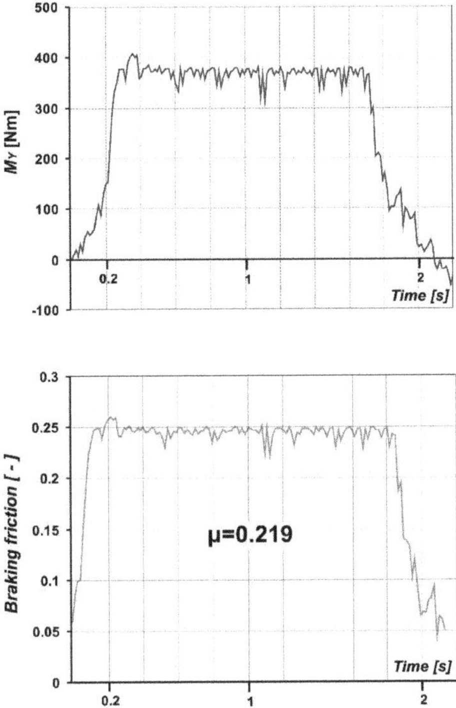

Figure 12. Sample courses of the measured braking moment, M_Y, and braking friction coefficient, μ.

3.4. Aligning Moment

When driving on a curved track, the road wheel moves in such a way that the longitudinal axis of the wheel is deflected from the direction of the instantaneous speed vector by an angle, called the side slip angle. The reaction of a pneumatic tire to side slip is to generate a moment that tends to reduce this angle. This moment is called the stabilizing or aligning moment.

The sensor used in the tests directly measures the value of the aligning moment, it is the moment marked as M_Z. Figure 13 shows the M_Z course recorded during one of the turning maneuvers. The importance of the M_Z moment is revealed during taxiing, especially at high speed. The higher the M_Z value, the easier it is to maintain a straight direction, while too high a stabilizing moment can cause significant steering forces when cornering, especially when taxiing at low speed.

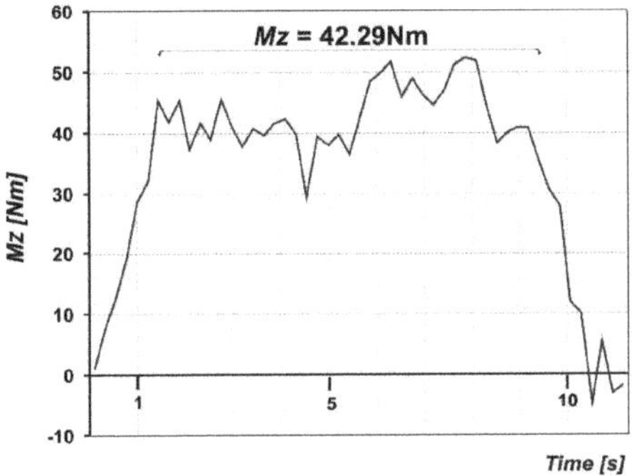

Figure 13. Sample course of the aligning moment, M_Z.

3.5. Effect of Weather Condidtions

As already mentioned, the study was carried out over four day—in October, November, December, and March. The purpose of such a research program was to check to what extent the worsening deterioration of weather conditions, in terms of lowering the external temperature and increasing humidity of air and soil, affects the measured values of forces and moments, as well as the resulting coefficients. Based on the measurements carried out, mean values of repetitions (minimum of five repetitions for each day) of individual measurements were determined. A collection of final data in the form of bar charts is presented in Figures 14–16, where average values of rolling resistance coefficient, k_{RR}, friction coefficient, μ, and stabilizing moment, M_Z, are shown for four days, representative for particular weather periods.

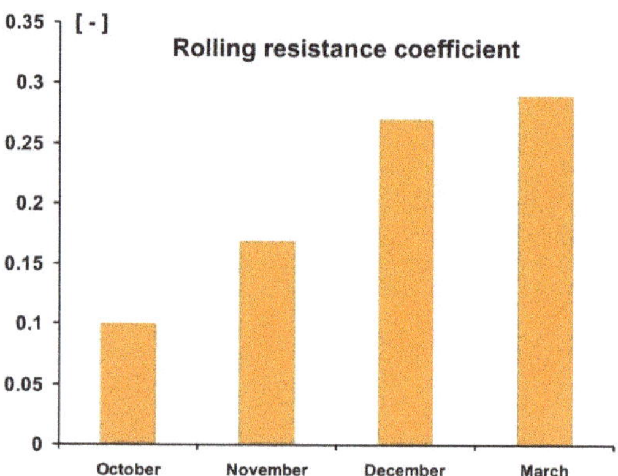

Figure 14. Average peak values of rolling resistance coefficient, k_{RR}, for the four measurements.

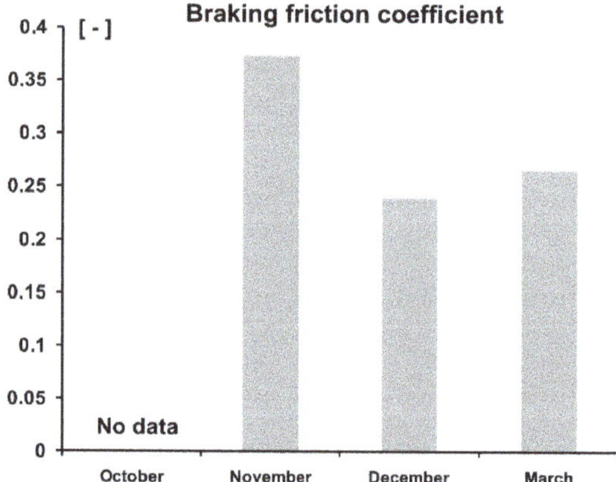

Figure 15. Average peak values of braking friction coefficient, μ, for the four measurements.

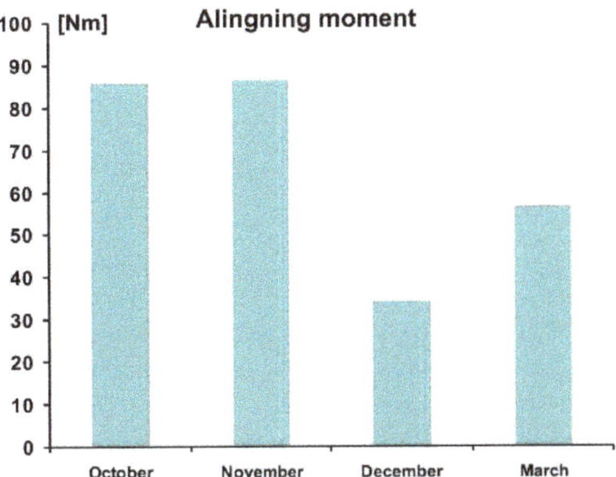

Figure 16. Average peak values of aligning moment, M_Z, for the four measurements.

The rolling resistance coefficient increases, and this is a significant increase. The explanation for that is quite obvious: In autumn, late autumn, winter, and early spring, there is a gradual increase in humidity of both air and soil, which causes increasing deformation of the surface and, as a result, an increase in rolling resistance. Although we did not construct any relationship between the rolling resistance coefficient and soil moisture (see Table 1), these parameters correspond well to each other.

The second term, braking friction coefficient shows an inverse relationship, i.e., it decreases during the period considered. In this case, the cause is similar, although the increasing humidity works slightly differently, namely, lower grip (=more slipperiness), especially in the case of grass-covered surfaces. A slight increase in the braking friction coefficient was observed for the measurement carried out in March. Perhaps the reason was a stronger solar operation, which dried the surface, especially grass.

The last of the parameters analyzed, the aligning moment, shows an interesting tendency (Figure 16). No significant impact of the weather period is observed during the months October–November, while a significant decrease in the M_Z moment value in December was recorded. In March, the M_Z value increased significantly compared to the December measurements.

4. Conclusions

An airfield experiment was conducted in order to measure landing gear wheel forces and moments. A five-element wheel force sensor was used to measure F_X and F_Z forces and M_X, M_Y, and M_Z moments acting on an aircraft wheel during a low-speed taxiing. The measurement was conducted using the PZL 104 Wilga 35A as a test aircraft. The wheel force sensor was installed on the left landing gear in place of the wheel. Forces and moments were measured on a grassy surface of the Rzeszów Jasionka aerodrome, four times on days of autumn and spring months. The results show changes in wheel force and moment values during various maneuvers: Acceleration, turning left–right, and deceleration (braking).

A significant influence of weather conditions on the values of determined coefficients k_{RR} and μ, as well as on the aligning moment M_Z, was noted. The rolling resistance coefficient increased during the considered weather period. However, the friction coefficient decreased, as did the stabilizing moment. The explanation for these trends is the increase in air and soil humidity, which translates into increased soil deformability and reduced adhesion.

Further research is planned in the area of applied machine learning to analyze the impact of unpaved airfield surface conditions on aircraft ground performance. First of all, we are planning to certify the wheel force system for flying in order to conduct flight tests. It is interesting how the forces and moments act during takeoff, as well as during landing, especially at touchdown. Also, we are planning to perform measurements on winter surfaces, different types of snow, ices, and at thawing conditions. Also, measuring wheel forces on higher grasses is also planned in the future.

Another planned research is dynamic modeling of the airplane motion during takeoff and landing ground roll.

Regarding the wheel force sensor design, we are planning to develop and build a smaller version, possible to install into a wheel of a very light aircraft.

Author Contributions: Conceptualization, J.P., P.B.; Methodology, J.P., P.B.; Software, P.B., T.Ł., Validation, J.M., Formal Analysis, A.T. and J.M.; Investigation, J.P., T.Ł. and J.L.; Resources, J.J.; Data Curation, J.P., T.Ł. and J.L.; Writing-Original Draft Preparation, J.P.; Writing-Review and Editing, P.B. and T.Ł., Visualization, J.L.; Supervision, A.T. and D.B.; Funding Acquisition, P.B. All authors have read and agreed to the published version of the manuscript.

Funding: The project/research was financed in the framework of the project *Lublin University of Technology—Regional Excellence Initiative*, funded by the Polish Ministry of Science and Higher Education (contract No. 030/RID/2018/19).

Acknowledgments: The authors are highly indebt to Grzegorz Piwowarski from the Aeroclub of Lublin and Zdzisław Nowak from the Aeroclub of Rzeszów, for their help in organizing the ground tests with the PZL 104 Wilga 35A airplane.

Conflicts of Interest: The authors declare no conflict of interest.

References

1. Pytka, J.; Łyszczyk, T.; Gnapowski, E. Monitoring Grass Airfield Conditions for the GARFIELD System. In Proceedings of the 5th International Workshop on Metrology for Aerospace (MetroAeroSpace), Rome, Italy, 20–22 June 2018.
2. Pytka, J.; Tarkowski, P.; Budzyński, P.; Józwik, J. Method for Testing and Evaluating Grassy Runway Surface. *J. Aircr.* **2017**, *54*, 229–234. [CrossRef]
3. Pytka, J.; Józwik, J.; Budzyński, P.; Łyszczyk, T.; Tofil, A.; Gnapowski, E.; Laskowski, J. Wheel dynamometer system for aircraft landing gear testing. *Measurement* **2019**, *148*, 106918. [CrossRef]
4. Hinüber, E.L.V.; Reimer, C.; Schneider, T.; Stock, M. INS/GNSS integration for aerobatic flight applications and aircraft motion surveying. *Sensors* **2017**, *17*, 941. [CrossRef] [PubMed]
5. Baiocchi, V.; Napoleoni, Q.; Tesei, M.; Costantino, D.; Andria, G.; Adamo, F. First tests of the altimetric and thermal accuracy of an UAV landfill survey. In Proceedings of the 5th International Workshop on Metrology for Aerospace (MetroAeroSpace), Rome, Italy, 20–22 June 2018; pp. 403–406.
6. Petritoli, E.R.; Leccese, F.; Cagnetti, M. Takagi-Sugeno Discrete Fuzzy Modeling: An IoT Controlled ABS for UAV. In Proceedings of the IEEE International Workshop on Metrology for Industry 4.0 and IoT, MetroInd 4.0 and IoT, Naples, Italy, 4–6 June 2019; pp. 191–195.
7. Pecora, A.; Maiolo, L.; Minotti, A.; de Francesco, R.; de Francesco, E.; Leccese, F.; Cagnetti, M.; Ferrone, A. Strain gauge sensors based on thermoplastic nanocomposite for monitoring inflatable structures. In Proceedings of the 4th International Workshop on Metrology for Aerospace (MetroAeroSpace), Benevento, Italy, 29–30 May 2014; pp. 84–88.
8. Petritoli, E.; Leccese, F.; Leccisi, M. Inertial navigation systems for UAV: Uncertainty and error measurements. In Proceedings of the 6th International Workshop on Metrology for Aerospace (MetroAeroSpace), Torino, Italy, 19–21 June 2019; pp. 218–222.
9. Petritoli, E.; Leccese, F. High accuracy attitude and navigation system for an autonomous underwater vehicle (AUV). *Acta IMEKO* **2018**, *7*, 3–9. [CrossRef]
10. Petritoli, E.; Leccese, F.; Cagnetti, M. A high accuracy buoyancy system control for an underwater glider. In Proceedings of the IEEE International Workshop on Metrology for the Sea; Learning to Measure Sea Health Parameters MetroSea, Bari, Italy, 8–10 October 2018; pp. 257–261.
11. Eling, C.; Klingbeil, L.; Kuhlmann, H. Real-time single-frequency GPS/MEMS-IMU attitude determination of lightweight UAVs. *Sensors* **2015**, *15*, 26212–26235. [CrossRef] [PubMed]
12. Burkard, H.; Calame, C. Rotating Wheel Dynamometer with High Frequency Response. *Tire Technol. Int.* **1998**, *1998*, 154–158.
13. Loh, R.; Nohl, F.W. Mehrkomponenten-Radmessnabe. Einsatzmoeglichkeiten und Ergebnisse. *ATZ* **1992**, *94*, 44–53, (In German with English Summary).
14. Kuchler, M.; Schrupp, R. Mehrkomponenten-Motorradmessnabe/Multiaxial Motorcycle Wheel Load Transducer. *VDI-FVT Jahrbuch VDI Verlag Dusseldorf* **2002**, *10*, 91–119, (In German with English Summary).

15. Pytka, J. *Wheel Dynamometer for Off-Road Vehicle Testing*; SAE Technical Paper Series; Paper No. 2008-01-0482; Society of Automotive Engineers: Warreendale, PA, USA, 2008.
16. Bigoš, M.P.P.; Kelemen, M.; Tonhajzer, R.; Šima, M. Measuring method for feedback provision during development of fuel map in hexadecimal format for high-speed racing engines. *Measurement* **2014**, *50*, 203–212.
17. Feng, L.; Chen, W.; Wu, T.; Wang, H.; Dai, D.; Wang, D.; Zhang, W. An improved sensor system for wheel force detection with motion-force decoupling technique. *Measurement* **2018**, *119*, 205–217. [CrossRef]
18. Meymand, S.Z.; Ahmadian, M. Design, development, and calibration of a force-moment measurement system for wheel–rail contact mechanics in roller rigs. *Measurement* **2016**, *81*, 113–122. [CrossRef]
19. Yuan, C.; Luo, L.P.; Yuan, Q.; Wu, J.; Yan, R.J.; Kim, H.; Shin, K.S.; Han, C.S. Development and evaluation of a compact 6-axis force/moment sensor with a serial structure for the humanoid robot foot. *Measurement* **2015**, *70*, 110–122. [CrossRef]
20. Lyle, K.H.; Jackson, K.E.; Fasanella, E.L. Simulation of Aircraft Landing Gears with a Nonlinear dynamic Finite Element Code. *J. Aircr.* **2002**, *38*, 142–147. [CrossRef]
21. Chester, D.H. Aircraft Landing Impact Parametric Study with Emphasis on Nose Gear Landing Conditions. *J. Aircr.* **2002**, *39*, 394–403. [CrossRef]
22. Wei, X.; Nie, H. Dynamic Analysis of Aircraft Landing Impact Using Landing-Region-Based Model. *J. Aircr.* **2005**, *42*, 1631–1637. [CrossRef]
23. Pytka, J.; Józwik, J.; Łyszczyk, T.; Budzyński, P.; laskowski, J.; Gnapowski, E. Measurement of Forces and Moments Acting on Aircraft Landing Gear Wheel. In Proceedings of the 6th International Workshop on Metrology for Aerospace (MetroAeroSpace), Torino, Italy, 19–21 June 2019.
24. Pytka, J.; Budzyński, P.; Kamiński, M.; Łyszczyk, T.; Józwik, J. Application of the TDR Moisture Sensor for Terramechanical Research. *Sensors* **2019**, *19*, 2116. [CrossRef] [PubMed]
25. Mitschke, M.; Wallentowitz, H. *Dynamik der Kraftfahrzeuge. 4. Auflage*; Springer: Berlin/Heidelberg, Germany, 2004.

© 2019 by the authors. Licensee MDPI, Basel, Switzerland. This article is an open access article distributed under the terms and conditions of the Creative Commons Attribution (CC BY) license (http://creativecommons.org/licenses/by/4.0/).

Article

Dynamic Modelling and Experimental Characterization of a Self-Powered Structural Health-Monitoring System with MFC Piezoelectric Patches †

Gianpietro Di Rito [1,*], Mario Rosario Chiarelli [1] and Benedetto Luciano [2]

1 Dipartimento di Ingegneria Civile ed Industriale-Università di Pisa, 56122 Pisa, Italy; mario.rosario.chiarelli@unipi.it
2 AESIS srl, 56017 San Giuliano Terme (PI), Italy; b.luciano@aesis-studio.it
* Correspondence: gianpietro.di.rito@unipi.it

† This manuscript is extension version of the conference paper: Di Rito, G., Luciano, B., Chiarelli, M.R., Galatolo, R. Condition monitoring of a morphing laminate with MFC piezoelectric patches via model-based approach. In proceedings of the 2019 IEEE 5th International Workshop on Metrology for AeroSpace, Torino, Italy, 19–21 June 2019.

Received: 23 November 2019; Accepted: 7 February 2020; Published: 11 February 2020

Abstract: The paper deals with theoretical and experimental studies for the development of a self-powered structural health monitoring (SHM) system using macro-fiber composite (MFC) patches. The basic idea is to integrate the actuation, sensing, and energy harvesting capabilities of the MFC patches in a SHM system operating in different regimes. As an example, during flight, under the effects of normal structural vibrations, the patches can work as energy harvesters by maintaining or restoring the battery charge of the stand-by SHM electronic board; on the other hand, if relevant/abnormal loadings are applied, or if local faults produce a noticeable stiffness variation of the monitored component, the patches can act as sensors for the power-up SHM board. During maintenance, the patches can then work as actuators, to stress the structure with pre-defined load profiles, as well as sensors, to monitor the structural response. In this paper, the investigation, based on the electromechanical impedance technique, is carried out on a system prototype made of a cantilevered composite laminate with six MFC patches. A high-fidelity nonlinear model of the system, including the piezoelectric hysteresis of the patches and three vibration modes of the laminate beam, is presented and validated with experiments. The results support the potential feasibility of the system, pointing out that the energy storage can be used for recharging a 3V-65mAh Li-ion battery, suitable for low-power electronic boards. The model is finally used to characterize a condition-monitoring algorithm in terms of false alarms rejection and vulnerability to dormant faults, by simulating built-in tests to be performed during maintenance.

Keywords: structural monitoring; health-management; piezoelectricity; actuators; sensors; energy harvesters; modeling; testing

---

1. Introduction

Piezoelectric materials are widely used in a variety of aerospace applications in both sensing and low-power actuation devices, and many research works discuss the use of such materials as actuators, sensors, or energy harvesters. An extensive literature documents the applications of piezoelectric actuators, with special focus on noise abatement and control [1–4], the design of smart mechanical systems [5,6], the development of self-shaping structures with vibration control capabilities [7–9], or the design of smart devices for aerodynamic flow control [10–16]. Important research works also focus on the study of the energy harvesting capability of piezoelectric materials [17–20], as well as on their application as sensors in structural health monitoring (SHM) systems [21–25]. A relevant example of piezoelectric devices is the macro-fiber composite (MFC), which consists of rectangular piezo-ceramic rods sandwiched between layers of adhesive films containing tiny electrodes [26,27]. The electrodes transfer a voltage directly to and from ribbon-shaped rods with thickness of a few tenths of a millimeter (Figure 1).

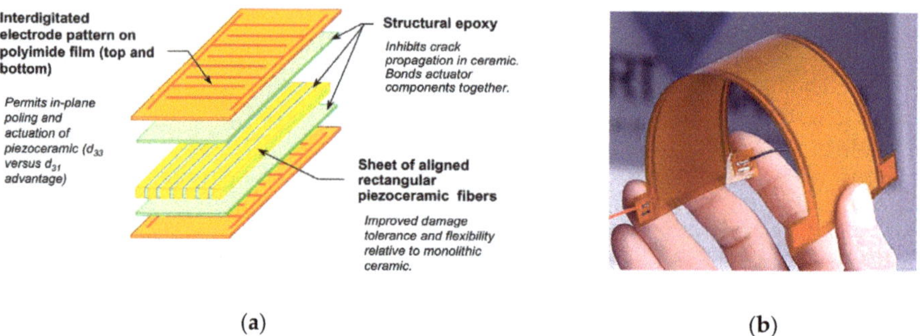

(a) (b)

Figure 1. Macro-fiber composite (MFC) piezoelectric devices [27]: (**a**) structural arrangement; (**b**) patch layout.

With reference to actuation capabilities, the MFC technology shows better performance and durability compared to other piezo-ceramic solutions, and can be used for both high and low frequency applications [28–30]. In the aerospace field, significant research efforts have been made on high frequency applications, with particular reference to vibration control [31,32] and wing ice protection systems [33]. In the low frequency domain, an interesting and challenging application is instead related to the so-called "morphing wings" (i.e., wing structures with the capability to dynamically modify their aerodynamic shape, to enhance efficiency, and/or to control the flight loads). Suitable and promising solutions were investigated by using MFC patches co-cured in or glued on substrates of composite material [34–37].

MFC piezoelectric devices also have relevant potentialities as sensors, and a strong interest is growing in the aerospace field for their application in SHM or health usage and monitoring systems (HUMS). Relevant research efforts have been made for developing HUMS for aerospace structures with embedded, distributed, and miniature PZT devices, taking advantage from the capability of integration in complex geometries as well as to detect hidden damages [21–25,38–40]. In addition, special attention has been dedicated to the potential self-powered characteristics of these systems [41–44], bringing to promising SHM solutions with airplane structures integrating MFC patches [45] (Figure 2). Nevertheless, literature information and research results are currently poor if stand-alone devices are addressed.

Figure 2. Example of integration of MFC patches in a wing structure [37].

This paper aims to provide a contribution within this research framework, with particular reference to the design of an electromechanical impedance-based SHM system [21–24] operating in different regimes. During flight, when normal loadings are applied, the patches work as energy harvesters by maintaining or restoring the battery charge of the stand-by SHM electronic board. If relevant or abnormal loadings occur, the patches instead act as sensors for the power-up SHM board, by acquiring in real-time relevant data on the component usage. Finally, during maintenance, the patches, operated by external high-voltage electronic equipment, can work as actuators, to stress the structure with pre-defined load profiles, as well as sensors, to monitor the structural response. It is worth noting that the vibration and/or loading levels strongly depend on the vehicle category (small or large airplane, turboprop or turbojet engines, helicopters, missiles, etc.) as well as on system location on the vehicle itself. Thus, the loading thresholds for the SHM board activation are expected to strongly depend on the reference application. Since this study is not focused on a particular aerospace case, the specifications/characteristics of loadings are not quantitatively given.

The work is articulated into three sections: in the first, the test rig used for the experimental characterization of the reference prototype is illustrated; successively, a model of the system dynamics is presented and validated against experimental data; finally, a model-based condition-monitoring algorithm is described and verified, by simulating different levels of system performance deviations.

2. Experimental Test System

As a reference structural element for the SHM studies, a carbon-epoxy composite laminate 400 × 45 × 2 mm × mm × mm is selected (Figure 3a and Table 1). Six MFC patches are installed and symmetrically arranged on the specimen (three per side), and the laminate is cantilevered to the rig frame. The patches are electrically connected and signaled to obtain a uniform distribution of bending torque along the laminate: the three patches on each side are connected in parallel and controlled by a dedicated channel, so that they are driven by opposite voltage levels.

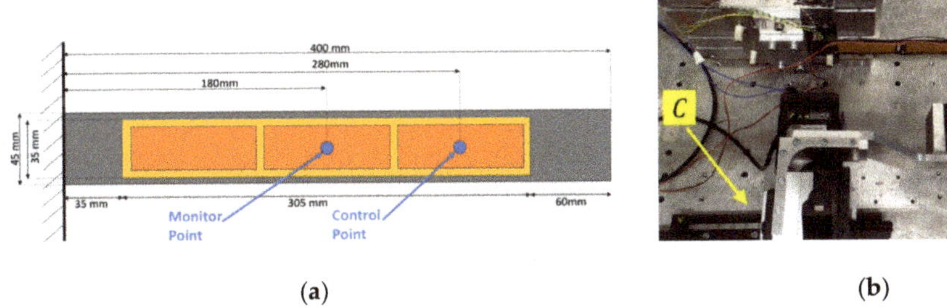

(a) (b)

Figure 3. Experimental set-up [40]: (**a**) morphing laminate with MFC patches; (**b**) rig top view: MFC laminate (A); external laser sensors (B); precision translation stage (C).

Table 1. MFC patch operational parameters.

Parameter	Unit	Value
Voltage demand range	V	−500; 1500
Temperature operating range	°C	−20; 80
Operational frequency range	kHz	DC; 10
Stall force (tensile stress)	N	454
Free strain	ppm	1800
Lifetime @ 1 kV peak-to-peak	cycles	10^9

The developed test system has two basic operating modes (Figure 4):

- Actuation mode (see black flows in Figure 4), in which the CRio embedded controller (B) receives from the user workstation (A) the demand voltages for the two channels of the morphing laminate, and a high-voltage amplifier (C) provides the MFC patches (D) with the electrical power;
- Sensing/energy harvesting mode (see red flows in Figure 4), in which the CRio embedded controller (B) receives from the user workstation (A) the demand vibration level for the modal shaker (G) connected to the laminate attachment, and a multimeter (F) measures the voltage generated by the MFC patches (D).

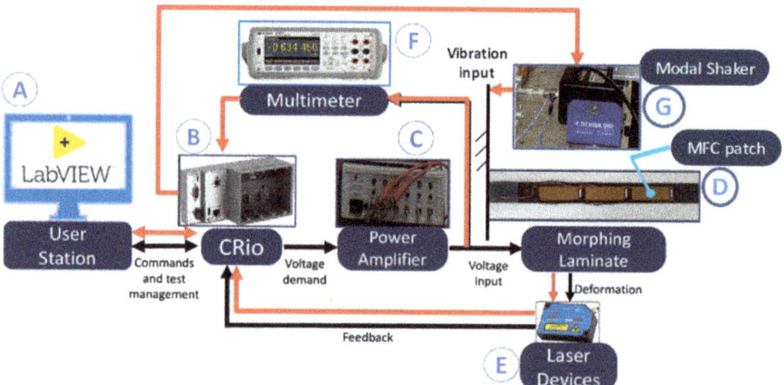

Figure 4. Experimental test system architecture.

The rig is also equipped with two laser sensors (E, in Figure 4), measuring the laminate deflections at two points (Figure 3b) to supervise the deformation as well as to reconstruct the morphed shape. The measurement points on the laminate can be adjusted or modified by means of a precision translation stage installed on the rig (Figure 3b).

3. Development and Validation of the Model of System Dynamics

3.1. Dynamic Modelling

The results of a preliminary study carried out by the authors on the prototype [40] demonstrated that the system dynamics is characterized by well-separated behaviors in the low-frequency (<0.1 Hz) and medium/high-frequency (from 1 to 100 Hz) ranges. In particular, the low-frequency range is dominated by hysteresis, while at high frequencies the system's response is characterized by structural resonances related to bending modes.

For these reasons, by taking into account the electrical and structural symmetry of the prototype, the dynamic modelling was developed with reference to the lumped-parameters scheme, shown in Figure 5, to include:

- three bending vibration modes of the laminate;
- the charge dynamics of the three couples of MFC patches, taking account of the piezoelectric hysteresis.

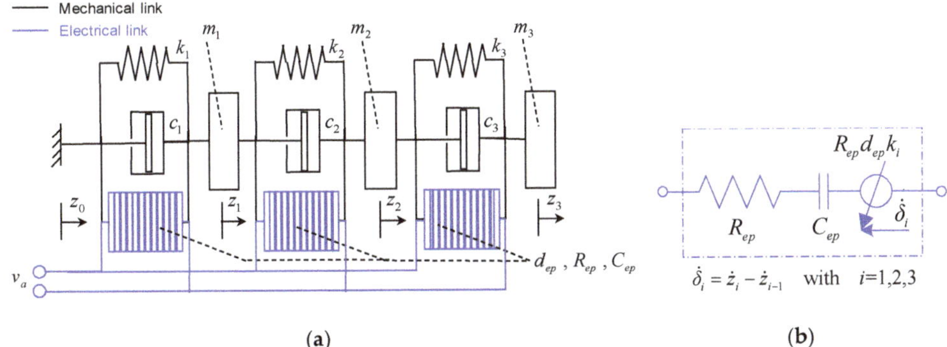

Figure 5. Lumped-parameters schematization: (**a**) complete prototype; (**b**) i-th patch.

Concerning the hysteresis modelling, several formulations are available in the literature to describe the phenomenon, and they are classified into two categories: static and dynamic models, depending on the use or not of ordinary differential equations (ODE) in the models themselves. Static models (e.g., Preisach [46], Prandtl-Ishlinskii [47], or polynomial models [48]) are typically very accurate, but they need a large number of experimental data for algebraic manipulations and interpolations, and their use for real-time condition monitoring is problematic [49]. On the other hand, dynamic models (e.g., Duhem [50], Maxwell [51], or Bouc-Wen [52,53] models) are often less accurate, but more adequate for control and real-time monitoring purposes. By selecting a Bouc-Wen hysteresis model, the system dynamics is described by the nonlinear twelfth-order ODE given by Equation (1),

$$\begin{cases} m_3\left(\ddot{\delta}_3+\ddot{\delta}_2+\ddot{\delta}_1\right) = -c_3\dot{\delta}_3 - k_3\delta_3 + \frac{k_3 d_{ep}}{C_{ep}} q_{h3} - m_3\ddot{z}_0 \\ m_2\left(\ddot{\delta}_2+\ddot{\delta}_1\right) = -c_2\dot{\delta}_2 - k_2\delta_2 + c_3\dot{\delta}_3 + k_3\delta_3 + \frac{k_2 d_{ep}}{C_{ep}} q_{h2} - \frac{k_3 d_{ep}}{C_{ep}} q_{h3} - m_3\ddot{z}_0 \\ m_2\ddot{\delta}_1 = -c_1\dot{\delta}_1 - k_1\delta_1 + c_2\dot{\delta}_2 + k_2\delta_2 + \frac{k_1 d_{ep}}{C_{ep}} q_{h1} - \frac{k_2 d_{ep}}{C_{ep}} q_{h2} - m_3\ddot{z}_0 \\ \left.\begin{array}{l} R_{ep}\dot{q}_i = -\alpha_h q_i / C_{ep} - R_{ep}k_i d_{ep}\dot{\delta}_i + v_a \\ \dot{h}_{qi} = A_h\dot{q}_i - \beta_h\left|\dot{q}_i\right|\left|h_{qi}\right|^{n-1}h_{qi} - \gamma_h\dot{q}_i\left|h_{qi}\right|^{n} \end{array}\right. \text{ with } i = 1,2,3 \end{cases} \quad (1)$$

where (with $i = 1, 2, 3$)

$$q_{hi} = \alpha_h q_i + (1-\alpha_h)h_{qi} \quad (2)$$

$$\delta_i = z_i - z_{i-1} \quad (3)$$

In Equations (1)–(3), $\ddot{z}_0$ is the base acceleration, z_i and δ_i are the displacement and the bending deformation of the i-th laminate section; m_i, k_i, and c_i are the mass, the stiffness, and the structural damping of the i-th section of the laminate; R_{ep}, C_{ep}, and d_{ep} are the equivalent resistance, capacitance, and piezoelectric coefficient of the patches; A_h, α_h, β_h, γ_h, and n are the Bouc-Wen model parameters; while q_i, h_{qi}, and q_{hi} are the linear, the hysteretic, and the total charge of the i-th patch. Finally, v_a depending on the operation regime of the system, represents the input voltage applied to the patches or the output voltage provided by them.

3.2. Experimental Validation of the Model

Aiming to collect the system identification database, the dynamics of the system prototype was experimentally characterized via an extensive test campaign, in terms of both time-domain and frequency-domain responses. The system response exhibited very good repeatability and the identification was developed in two main steps [54]:

- estimation of the model parameters (Table 2) by minimizing the model deviations from the hardware behavior with reference to sinusoidal frequency response tests, i.e.,
    - deformation responses to voltage inputs, with amplitudes from 150 to 600 V, ranging from 0.1 to 50 Hz, for a total of 120 tests and 200 simulations (Figure 6a);
    - stored voltage responses to acceleration inputs of amplitudes from 0.01 to 0.1 g, ranging from 1 to 50 Hz, for a total of 120 tests and 200 simulations (Figure 6b).
- model validation, by characterizing the model deviations from the hardware behavior with reference to time response tests, i.e.,
    - quasi-static deformation responses to high-amplitude/low-frequency sinewave voltage inputs (Figure 7a);
    - transient deformation responses to step voltage inputs (Figure 7b).

Table 2. Dynamic model parameters (identification method legend: TM, test measurement; MEM, model error minimization; LA: literature assumption from [54]).

Symbol	Definition	Unit	Value	Identification Method
R_{ep}	Patch resistance	Ohm	1500	TM
C_{ep}	Patch capacitance	F	$1.06\ 10^{-5}$	TM
d_{ep}	Patch piezoelectric coefficient	m/V	$1.37\ 10^{-5}$	MEM
m_1	Mass of the root section of the laminate	kg	$1.16\ 10^{-3}$	TM
m_2	Mass of the intermediate section of the laminate	kg	$2.32\ 10^{-3}$	TM
m_3	Mass of the tip section of the laminate	kg	$1.39\ 10^{-3}$	TM
k_1	Stiffness related to root section bending	N/m	332.98	MEM
k_2	Stiffness related to intermediate section bending	N/m	125.40	MEM
k_3	Stiffness related to tip section bending	N/m	21.28	MEM
c_1	Damping related to root section bending	N s/m	0.098	MEM
c_2	Damping related to intermediate section bending	N s/m	0.020	MEM
c_3	Damping related to tip section bending	N s/m	0.025	MEM
α_h	Bouc-Wen linearity factor	–	0.605	MEM
A_h	Bouc-Wen dissipation gain	–	−1	LA
n	Bouc-Wen shape factor	–	1	LA
β_h	Bouc-Wen shape factor	V^{-n}	8.307	MEM
γ_h	Bouc-Wen shape factor	V^{-n}	0	LA

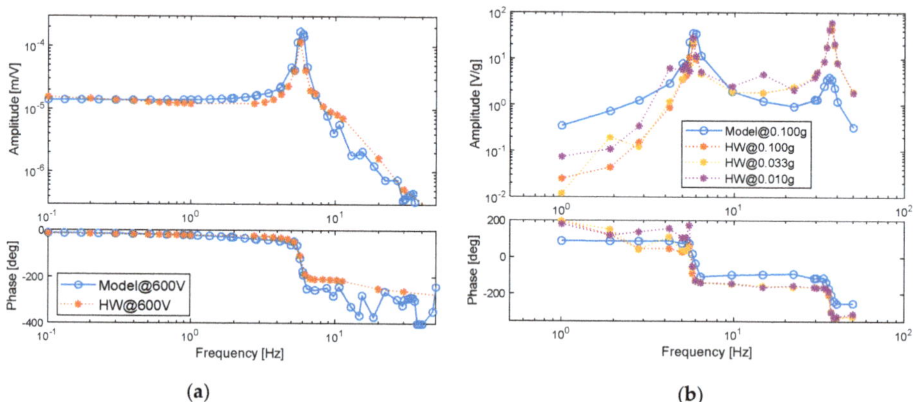

Figure 6. Model parameters' tuning via frequency-domain responses: (**a**) deformation response to voltage input; (**b**) stored voltage response to base acceleration input.

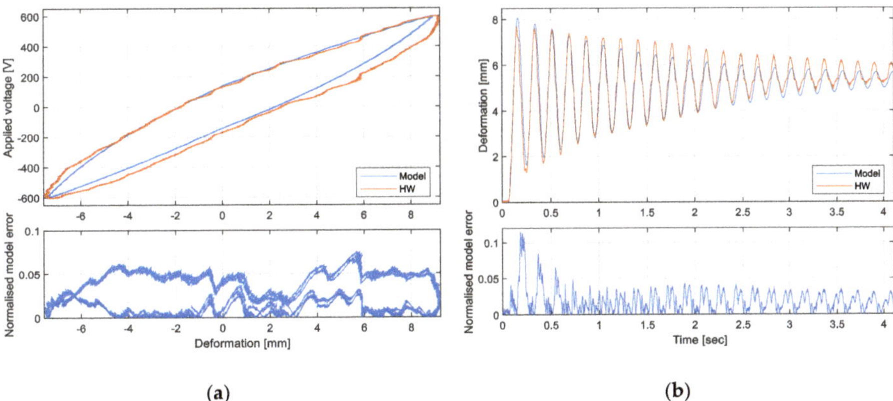

(a) **(b)**

Figure 7. Validation of the model via time-domain responses: (**a**) hysteresis deformation loop with sinewave voltage input at 0.01 Hz; (**b**) deformation response to step voltage input of 300 V.

It is worth noting that the experimental investigation pointed out that the deformation response weakly depends on applied voltage amplitude, and no significant variation of the system dynamic behavior is observed by changing the input voltage amplitude. On the other hand, the stored voltage response strongly depends on base acceleration amplitude. For these reasons, Figure 6a reports the deformation response at 600 V amplitude only, while Figure 6b shows the voltage storage results for different acceleration amplitudes (note that also simulation results are amplitude-dependent, but only one acceleration amplitude is given in Figure 6b for the clarity of the plot).

Both experiments and simulations clearly point out that the resonant peak related to the first bending mode occurs at 5.8 Hz, the resistance-capacitance electric dynamics induces a real pole at about 10 Hz, while the second bending mode implies a resonance at 37 Hz. Experiments also point out the relevant voltage storage capability of the prototype, which is characterized by an acceleration-to-voltage gain that depends on vibration amplitude and achieves 35 V/g at 5.8 Hz (first resonance) and 68 V/g at 37 Hz (second resonance). The behavior is well reproduced by simulations, even if the prediction accuracy in terms of stored voltage amplitude is lower at the second resonance. This is probably due to that fact that the lumped-parameters model concentrates the laminate deformations at only three points, and this implies a decrease of accuracy at higher modes, in which the deformation distribution along the laminate is more complex.

Concerning the model validation, the activity is here focused on the actuation behavior only, essentially because the model-based condition-monitoring algorithm is designed via maintenance built-in test (MBIT) simulations, when the patches work as actuators. Validation of the model addressing the sensing/energy harvesting behavior will be a future development of this work.

Preliminary comparisons between model predictions and experimental data highlighted that the prediction error of a generic system variable (y) is linearly correlated to the amplitude of the variable itself, Equation (4):

$$\left|y_{\mathrm{md}} - y_{\mathrm{hw}}\right| = e_{0y} + k_{erry}\left|y_{\mathrm{hw}}\right| \tag{4}$$

where y_{md} is the model prediction and y_{hw} is the hardware response, while e_{0y} and $k_{err\,y}$ are the bias and the gain characterizing the error on the variable y. For this reason, the evaluation of the prediction accuracy was made by normalizing the model error via Equation (5)

$$\varepsilon_{\mathrm{my}} = \frac{\left|y_{\mathrm{md}} - y_{\mathrm{hw}}\right|}{\left|y_{\mathrm{hw}}\right| + e_{0y}/k_{erry}} \tag{5}$$

Figure 7, related to deformation responses, actually demonstrates the effectiveness of the approach, since the normalized error does not significantly vary with the deformation amplitude for both quasi-static and fast transient tests. In the hysteresis test, the normalized deformation error is roughly constant during both discharge and charge phases, being 5% of the actual deformation in the charging phase and lower than 0.5% during discharge. Similar results are noted in the step response tests, where, after the initial transient, the mean value of the normalized error is 2% of the actual deformation, and its fluctuations are limited to 2%.

3.3. Assessment of the Energy Harvesting Capability

The experimentally-validated model of the prototype was used to assess its energy harvesting capability, aiming to support the feasibility of the self-powered system. The objective is to demonstrate that the MFC patches are capable of restoring the charge of a low-power battery, commonly used for supplying small electronic boards.

To perform the study, the open-circuit voltage characteristics (v_b) of the 3V-65mAh Maxell ML2032 Li-ion rechargeable battery (Figure 8a, [55]) are used to calculate the equivalent capacitance (C_b) of the Thevenin model of the battery (Figure 8b, [56]), Equation (6):

$$C_b = -\frac{\partial q_d}{\partial v_b} \Rightarrow C_b = C_b(q_d) = C_b(v_b) \tag{6}$$

where q_d is the discharged capacity of the battery in Amp·sec.

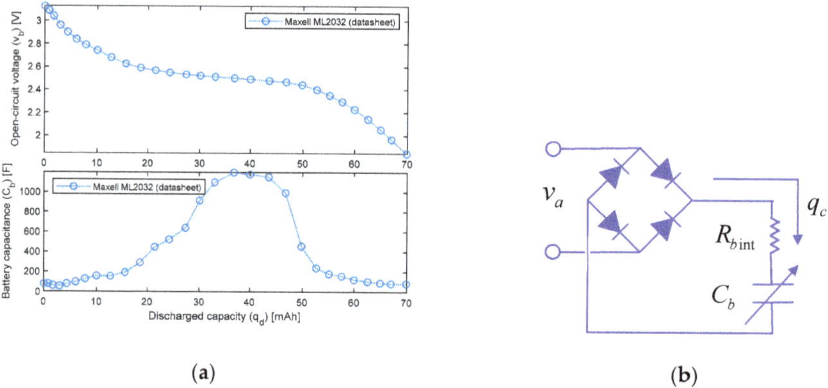

(a) (b)

Figure 8. Energy harvesting simulation: (**a**) Maxell ML2032 Li-ion rechargeable battery characteristics [55]; (**b**) Thevenin battery model connected to the patches' voltage output [56].

Once defined, the maximum capacity of the battery ($q_{\max}$) and its initial charge (q_0), the battery charge (q_b) is expressed by Equation (7):

$$q_b = q_{\max} - q_d = q_{\max} - q_0 + q_c \tag{7}$$

and the battery charge dynamics obtained through the rectified MFC system voltage output (v_a) is provided by Equation (8):

$$v_b + R_{bint}\dot{q}_b = |v_a| \tag{8}$$

By formulating the charge rate ($\dot{q}_b$) as a function of the battery voltage and voltage rate,

$$q_b = C_b(v_b)v_b \Rightarrow \dot{q}_b = \left(C_b + \frac{\partial C_b}{\partial v_b}v_b\right)\dot{v}_b = \left(C_b - \frac{\partial^2 q_d}{\partial {v_b}^2}v_b\right)\dot{v}_b \tag{9}$$

we finally obtain the nonlinear first-order model describing the charging behavior:

$$\dot{v}_b = \frac{|v_a| - v_b}{R_{bint}\left[C_b(v_b) - \frac{\partial^2 q_d}{\partial v_b{}^2} v_b\right]} \tag{10}$$

An example of the dynamics of the battery charge operated by the MFC patches is reported in Figure 9, where the simulation is carried out by imposing a 0.10 g sinusoidal base acceleration ($\ddot{z}_0$ in Equations (1)–(3)) at 5.8 Hz (first resonance), and by assuming an internal battery resistance ($R_{b\ \mathrm{int}}$) of 1 Ohm. It is worth noting that the prototype system is capable of harvesting the energy coming from the mechanical vibrations, and to restore the battery charge within 10 min (starting from a completely discharged battery, i.e., 2 V).

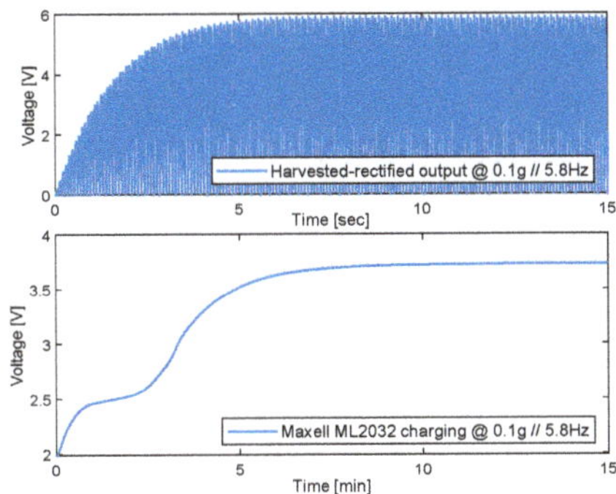

Figure 9. Energy harvesting simulation with sinusoidal base acceleration input ($R_{b\ \mathrm{int}}$ = 1 Ohm).

4. System Condition Monitoring

The experimentally-validated model of the prototype was finally used to test a model-based condition-monitoring algorithm developed by the authors [40,57–61], in order to assess the system applicability to SHM systems. In model-based condition-monitoring approaches, the detection of faults or anomalies is actually performed by evaluating the deviation of hardware responses from the predictions of a system model related to normal operating conditions.

4.1. Condition-Monitoring Algorithm

The proposed condition-monitoring logic, schematically reported in terms of flow chart in Figure 10, generates a fault flag ($\mathrm{Fault_{mon}}$, set to 1 in case of detection of detected anomalies) by elaborating the normalized model error ($\varepsilon_{\mathrm{mon}}$) sampled at the monitoring frequency (f_{mon}). If the normalized error exceeds a threshold (ε_{th}), a fault counter ($\mathrm{count_{mon}}$) is increased by 2; if the threshold is not exceeded, the fault counter is decreased by 1 when it is positive at the previous step, otherwise it is held to 0. The fault or anomaly is detected when the fault counter exceeds a pre-defined value ($\mathrm{count_{max}}$), a parameter which basically defines the fault latency of the algorithm.

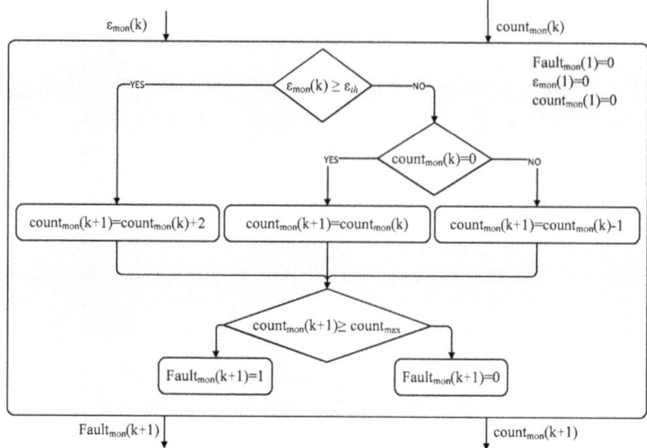

Figure 10. Flow chart of the condition-monitoring logic.

In the present study, the monitored variable is the displacement of the laminate tip due to the bending deformation (z_3 in Equations (1)–(3)), so the normalized model error is given by Equation (5), where $e_{0\,z} = 0.75$ mm and $k_{err\,z} = 0.075$.

$$\varepsilon_{\text{mon}} = \frac{|z_{3\text{md}} - z_{3\text{hw}}|}{|z_{3\text{hw}}| + e_{0z}/k_{errz}} \tag{11}$$

4.2. Maintenance Built-in-Test Simulations

The performances of the condition-monitoring algorithm are characterized by simulating the execution of two types of maintenance built-in tests (MBITs), i.e.,

- *MBIT.a*: response to sinewave voltage input at 0.1 Hz and 600 V (quasi-static test, Figure 7);
- *MBIT.b*: response to step voltage input with 300 V amplitude (fast transient test, Figure 7).

It is worth noting that, especially for the assessment of the transient response, several types of tests were studied [40], by varying both amplitude and waveform of the input signal, and the *MBIT.b* demonstrated a good balance between false alarms rejection and fault detection capability.

Once defined, the monitoring frequency (f_{mon}), the error threshold (ε_{th}), and the fault counter threshold ($\text{count}_{\max}$), the algorithm performances were characterized as follows:

i. for a subset of relevant system parameters (par), an acceptable range of deviations from the nominal behavior is defined. In this work, par = (k_3, α_h, β_h), and this selection was made because the hysteresis parameters basically affect the low-frequency behavior, while the tip stiffness also impacts the high-frequency range, by varying the location of structural resonances;

ii. a model of the "degraded" hardware is obtained from the experimentally-validated model, by injecting uniformly-distributed random deviations on the selected model parameters;

iii. *MBIT.a* and *MBIT.b* are simulated by executing the condition-monitoring algorithms;

iv. steps *ii* and *iii* are repeated for *N* times, to cumulate a statistical database.

Figure 11 shows the results of the performance characterization of a condition-monitoring algorithm with the parameters reported in Table 3, by executing 100 simulations on each range of 10% deviations, in terms of number of detected BIT faults (upper plots) and BIT fault probability (lower plots) as functions of the deviation range.

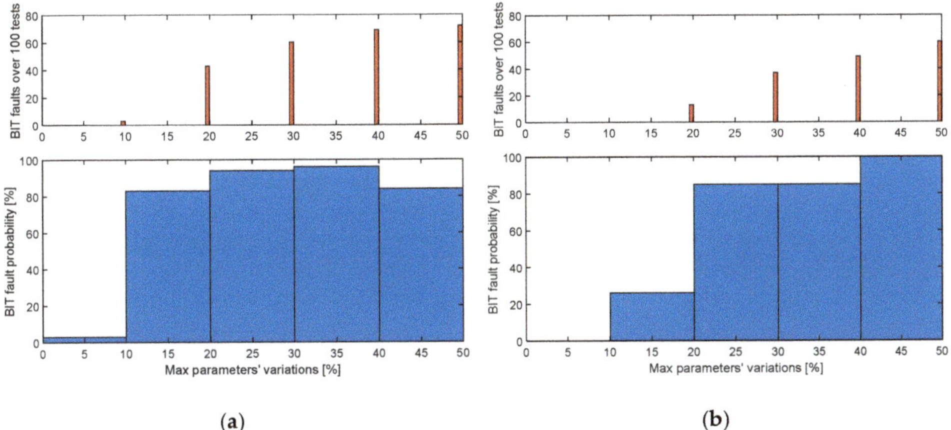

(a) (b)

Figure 11. Performances of the condition-monitoring algorithm: (**a**) *MBIT.a*; (**b**) *MBIT.b*.

Table 3. Condition-monitoring algorithm parameters.

Symbol	Definition	Unit	Value
f_{mon}	Monitoring frequency	Hz	100
ε_{th}	Normalized error threshold	–	0.225
$\text{count}_{\text{max}}$	Fault counter threshold	–	200

Depending on the monitoring objectives, both *MBIT.a* and *MBIT.b* effectively support the maintenance. If the objective is to minimize the performance deviations, *MBIT.a* is more suitable, because its BIT fault probability is 83% for deviations from 10% to 20%, and it increases up to 94% for deviations from 20% to 30% (Figure 11a). On the other hand, if the objective is to accept limited performance deviations, *MBIT.b* is better because no BIT faults occur for deviations up to 10%, while the BIT fault probability is 85% for deviations from 20% to 40%, and 99% for deviations from 40% to 50% (Figure 11b). Clearly, the combined use of the two MBITs would enhance the entire process by minimizing false alarms and undetected anomalies. With particular reference to the false alarms rejection, it is worth noting that the monitoring performances were not characterized in the presence of measurements' noise, because the noise impact on MBITs is expected to be minor (maintenance is performed in a controlled environment). On the other hand, the effects of noise are expected to be more relevant for in-flight monitoring algorithms (not addressed by this work).

5. Conclusions

A prototype of a self-powered SHM system made of a cantilevered composite laminate with six MFC patches was developed, and experimentally characterized in terms of actuation, sensing, and energy harvesting capabilities. A nonlinear dynamic model of the system, including piezoelectric hysteresis and three bending deformation modes, was developed starting from physical first principles, and validated with experiments. High fidelity was obtained in terms of deformation predictions (maximum normalized errors lower than 5% of the actual deformation), while the accuracy on stored voltage, though very good at the first resonance of the laminate (5.8 Hz), was lower at the second resonance (37 Hz). The results demonstrate the feasibility of a self-powered SHM system using the MFC technology, since the acceleration-to-voltage gain of the prototype is 35 V/g at the first resonance, and it achieves 68 V/g at the second resonance. The experimentally-validated model was then used to assess the energy harvesting capability of the prototype system, by demonstrating that the MFC patches are capable of restoring the charge of the 3V-65mAh Maxell ML2032 Li-ion rechargeable battery (commonly used for supplying small electronic boards). In particular, by imposing a 0.10 g sinusoidal base acceleration at 5.8 Hz, and assuming an internal battery resistance of 1 Ohm, the system is capable of harvesting the energy coming from the mechanical vibrations, and to restore the battery charge within 10 min (starting from complete discharge, i.e., 2 V).

Finally, the model is used to test a model-based condition-monitoring algorithm developed by the authors, by simulating two specifically-designed maintenance built-in tests. The "degraded" behavior was obtained from the "nominal" model, by injecting uniformly-distributed random deviations on relevant system parameters (the laminate stiffness and the hysteresis model coefficients), and 500 built-in-test simulations were performed. The results demonstrate that, with a combined use of the two proposed MBITs, the condition-monitoring algorithm effectively supports maintenance by checking the correspondence of performance to the nominal behavior as well as by detecting anomalies (BIT fault probability is 94% for deviations from 20% to 30%, and 99% for deviations from 40% to 50%).

In the paper, the model validation is focused on the actuation behavior, because the model-based condition-monitoring algorithm is designed via MBIT simulations when the patches work as actuators. Validation of the model addressing the sensing/energy harvesting behavior will be a future development of this work.

Author Contributions: The individual contributions of the authors to the present research article is here detailed. Conceptualization, resources, supervision, and funding acquisition: M.R.C. and G.D.R.; methodology, investigation, validation, formal analysis, and visualization: G.D.R.; software and data curation: B.L.; writing—original draft preparation, review and editing: G.D.R. and B.L.; project administration: M.R.C. All authors have read and agreed to the published version of the manuscript.

Acknowledgments: The authors wish to thank Roberto Galatolo from University of Pisa, who contributed to the preliminary conceptualization of the research, and Giuseppe Mattei and Alessandro Matera, who contributed to the initial development of the experimental test system during their MSc Degree Thesis in Aerospace Engineering at the University of Pisa.

Conflicts of Interest: The authors declare no conflicts of interest.

References

1. Bishop, R.; Mossi, K. High displacement, high force piezoelectric actuator and sensor. *J. Acoust. Soc. Am.* **1998**, *104*, 1828. [CrossRef]
2. Jayachandran, V.; Meyer, N.E.; Westervelt, M.A.; Sun, J.Q. Piezoelectrically driven speakers for active interior noise suppression. *Appl. Acoust.* **1999**, *57*, 263–277. [CrossRef]
3. Jayachandran, V.; King, P.; Meyer, N.E.; Li, F.J.; Petrova, M.; Westervelt, M.A.; Hirsch, S.M.; Sun, J.Q. Real-time feed forward control of low-frequency interior noise using shallow spherical shell piezoceramic actuators. *Smart Mater. Struct.* **1999**, *8*, 579–584. [CrossRef]
4. Corsaro, R.D. Lightweight low-frequency woofer for active sound control in payload fairings. In Proceedings of the 8th SPIE Annual International Symposium on Smart Structures and Materials, NDE and Health Monitoring, Newport Beach, CA, USA, 4–8 March 2001; pp. 254–258.
5. Horner, G.C.; Taleghani, B.K. Single axis piezoceramic gimbal. *J. Intell. Mater. Syst. Struct.* **2001**, *12*, 157–160. [CrossRef]
6. Granger, R.; Washington, G.; Kwak, S.K. Modeling and control of a singly curved active aperture antenna using curved piezoceramic actuators. *J. Intell. Mater. Syst. Struct.* **2000**, *11*, 225–233. [CrossRef]
7. Marouzé, J.P.; Cheng, L. A feasibility study of active vibration isolation using THUNDER actuators. *Smart Mater.Struct.* **2002**, *11*, 854–862. [CrossRef]
8. Lee, S.; Cho, B.C.; Park, H.C.; Goo, N.S.; Yoon, K.J. Piezoelectric actuator–sensor analysis using a three-dimensional assumed strain solid element. *J. Intell. Mater. Syst. Struct.* **2004**, *15*, 329–338. [CrossRef]
9. Gao, J.X.; Cheng, L. Modelling of a high performance piezoelectric actuator assembly for active and passive vibration control. *Smart Mater. Struct.* **2004**, *13*, 384–392. [CrossRef]
10. Cox, A.; Monopoli, D.; Cveticanin, D.; Goldfarb, M.; Garcia, E. The development of elastodynamic components for piezoelectrically actuated flapping micro-air vehicles. *J. Intell. Mater. Syst. Struct.* **2002**, *13*, 611–615. [CrossRef]

11. Doherty, K.M.; Nejhad, M.G. Performance of an active composite strut for an intelligent composite modified Stewart platform for thrust vector control. *J. Intell. Mater. Syst. Struct.* **2005**, *16*, 335–354. [CrossRef]
12. Waterfield, G. High performance pre-stressed piezoelectric bender actuator for digital valves. *Ceram. Trans.* **2004**, *105*, 467–482.
13. Munday, D.; Jacob, J. Active control of separation on a wing with conformal chamber. In Proceedings of the 39th AIAA Aerospace Sciences Meeting and Exhibit, Reno, NV, USA, 8–11 January 2001.
14. Mossi, K.M.; Bryant, R.G. Characterization of piezoelectric actuators for flow control over a wing. In Proceedings of the 9th International Conference on New Actuators (Actuator 2004), Breman, Germany, 14–16 June 2004.
15. Mane, P.; Mossi, K.; Bryant, R.G. Synthetic jets with piezoelectric diaphragms. *Proc. SPIE* **2005**, *5761*, 233–243.
16. Mossi, K.M.; Mane, P.; Bryant, R.G. Velocity profiles for synthetic jets using piezoelectric circular actuators. In Proceedings of the 46th AIAA/ASME/ASCE/AHS/ASC Structures, Structural Dynamics and Materials Conference, Austin, TX, USA, 18–21 April 2005.
17. Kymissis, J.; Kendall, C.; Paradiso, J.A.; Gershenfeld, N. Parasitic power harvesting in shoes. In Proceedings of the 2nd IEEE International Conference on Wearable Computing, Pittsburgh, PA, USA, 19–20 October 1998.
18. Mossi, K.M.; Green, C.W.; Ounaies, Z.; Hughes, E. Harvesting energy using a thin unimorph pre-stressed benders: Geometrical effects. *J. Intell. Mater. Syst. Struct.* **2005**, *16*, 249–261. [CrossRef]
19. Ramsay, M.J.; Clark, W.W. Piezoelectric energy harvesting for bio MEMS applications. In Proceedings of the 8th SPIE Annual International Symposium on Smart Structures and Materials, NDE and Health Monitoring, Newport Beach, CA, USA, 4–8 March 2001; pp. 429–438.
20. Dutoit, N.E.; Wardle, B.L.; Kim, S. Design considerations for MEMS scale piezoelectric mechanical vibration energy harvesters. *Integr. Ferroelectr.* **2005**, *1*, 121–160. [CrossRef]
21. Soh, C.K.; Yang, Y.; Bhalla, S. *Smart Materials in Structural Health Monitoring, Control and Biomechanics*; Springer: Berlin/Heidelberg, Germany, 2012.
22. Arnau, A.; Soares, D. *Piezoelectric Transducers and Applications*; Springer: Berlin/Heidelberg, Germany, 2009.
23. Park, S.; Yun, C.B.; Inman, J. Structural health monitoring using electro-mechanical impedance sensors. *Fatigue Fract. Eng. M.* **2008**, *31*, 714–724. [CrossRef]
24. Na, W.S.; Baek, J. A review of the piezoelectric electromechanical impedance-based structural health monitoring technique for engineering structures. *Sensors* **2018**, *18*, 1307. [CrossRef]
25. Mossi, K.M.; Scott, L. Sensor measurements for diagnostic equipment. In Proceedings of the 1st World Congress on Biomimetics and Artificial Muscles, Albuquerque, NM, USA, 9–12 December 2002.
26. Bent, A.A.; Hagood, N.W. Piezoelectric fiber composite with interdigitated eletrodes. *J. Intell. Mater. Syst. Struct.* **1997**, *8*, 903–919. [CrossRef]
27. Kovalovs, A.; Barkanov, E.; Gluhihs, S. Active control of structures using macro-fiber composite (MFC). *J. Phys. Conf. Ser.* **2007**, *93*, 012034. [CrossRef]
28. Prasath, S.S.; Arockiarajan, A. Effective electromechanical response of macro-fiber composite (MFC): Analytical and numerical models. *Int. J. Mech. Sci.* **2013**, *77*, 98–106. [CrossRef]
29. Zhang, S.Q.; Li, Y.X.; Schmidt, R. Modeling and simulation of macro-fiber composite layered smart structures. *Compo. Struct.* **2015**, *126*, 89–100. [CrossRef]
30. Zhang, J.; Tu, J.; Li, Z.; Gao, K.; Xie, H. Modeling on actuation behavior of Macro-Fiber Composite laminated structures based on sinusoidal shear deformation theory. *Appl. Sci.* **2019**, *9*, 2893. [CrossRef]
31. Moses, R.W.; Pototzky, A.S.; Henderson, D.A.; Galea, S.C.; Manokaran, D.S.; Zimcik, D.G.; Wickramasinghe, V.; Pitt, D.M.; Gamble, M.A. Actively controlling buffet-induced excitations. In Proceedings of the RTO/AVT-123 Symposium on Flow Induced Unsteady Loads and the Impact on Military Applications, Budapest, Hungary, 25–29 April 2005.
32. Schiller, N.H.; Perey, D.F.; Cabell, R.H. Development of a practical broadband active vibration control system. In Proceedings of the 2011 ASME International Mechanical Engineering Congress and Exposition, Denver, CO, USA, 11–17 November 2011; pp. 683–690.
33. Giles, A.M.; Machin, J.T.; Geriguis, J.A. Method and Apparatus for Inhibiting Formation of and/or Removing Ice from Aircraft Components. U.S. Patent No. 9,327,839, 3 May 2016.
34. Molinari, G.; Quack, M.; Arrieta, A.F.; Morari, M.; Ermanni, P. Design, realization and structural testing of a compliant adaptable wing. *Smart Mater. Struct.* **2015**, *10*, 105027. [CrossRef]

35. Paradies, R.; Ciresa, P. Active wing design with integrated flight control using piezoelectric macro fiber composites. *Smart Mater. Struct.* **2009**, *3*, 035010. [CrossRef]
36. Chiarelli, M.R.; Binante, V.; Botturi, S.; Massai, A.; Kunzmann, J.; Colbertaldo, A.; Romano, D.G. On the active deformations of hybrid specimens. *Aircr. Eng. Aerosp. Tec.* **2016**, *5*, 676–687. [CrossRef]
37. Chiarelli, M.R.; Cozzolino, A.; Kunzmann, J.; Lanzi, L. Feasibility Analysis on the Use of Piezo-Electric Materials as Actuators of Thin Walled Active Structures. Available online: http://www.futurewings.eu/FILES_linked/Deliverable_D_1_1Pisa_publishable.pdf (accessed on 8 February 2020).
38. Prosser, W.H.; Wu, M.C.; Allison, S.G.; DeHaven, S.L.; Ghoshal, A. Structural health monitoring sensor development at NASA Langley Research Center. In Proceedings of the International Conference on Computational & Experimental Engineering and Sciences (ICCES '03), Corfù, Greece, 25–29 July 2003.
39. Roellig, M.; Schubert, L.; Lieske, U.; Boehme, B.; Frankenstein, B.; Meyendorf, N. FEM assisted development of a SHM-piezo-package for damage evaluation in airplane components. In Proceedings of the 11th International Thermal, Mechanical & Multi-Physics Simulation, and Experiments in Microelectronics and Microsystems (EuroSimE), Bordeaux, France, 26–28 April 2010.
40. Di Rito, G.; Luciano, B.; Chiarelli, M.R.; Galatolo, R. Condition monitoring of a morphing laminate with MFC piezoelectric patches via model-based approach. In Proceedings of the 6th IEEE International Workshop on Metrology for AeroSpace (MetroAeroSpace), Turin, Italy, 19–21 June 2019; pp. 241–246.
41. Green, C.; Mossi, K.M.; Bryant, R.G. Scavenging energy from piezoelectric materials for wireless sensor applications. In Proceedings of the 2005 ASME International Mechanical Engineering Congress and Exposition, Orlando, FL, USA, 5–11 November 2005; pp. 93–99.
42. Yuan, J.; Xie, T.; Shan, X.; Chen, W. Experimental study on a self-powered piezoelectric sensor under vibration environment. In Proceedings of the 18th IEEE International Symposium on the Applications of Ferroelectrics, Xian, China, 23–27 August 2009.
43. Xia, H.; Xia, Y.; Ye, Y.; Qian, L.; Shi, G. Simultaneous wireless strain sensing and energy harvesting from multiple piezo-patches for Structural Health Monitoring applications. In Proceedings of the 11th International Thermal, Mechanical & Multi-Physics Simulation, and Experiments in Microelectronics and Microsystems (EuroSimE), Bordeaux, France, 26–28 April 2010.
44. Chao, P.C.P. Energy harvesting electronics for vibratory devices in self-powered sensors. *IEEE Sens. J.* **2011**, *11*, 3106–3121. [CrossRef]
45. Álvarez-Carulla, A.; Colomer-Farrarons, J.; López-Sánchez, J.; Miribel-Català, P. An adaptative self-powered energy harvester strain sensing device based of mechanical vibrations for structural health monitoring applications. In Proceedings of the 25th International Symposium on Industrial Electronics (ISIE), Santa Clara, CA, USA, 8–10 June 2016; pp. 638–644.
46. Mayergoyz, I.D. The classical Preisach model of hysteresis and reversibility. *J. Appl. Phys.* **1991**, *8*, 4602–4604. [CrossRef]
47. Brokate, M.; Sprekels, J. *Hysteresis and Phase Transitions*; Springer: New York, NY, USA, 1996.
48. Sun, L.; Ru, C.; Rong, W.; Chen, L.; Kong, M. Tracking control of piezoelectric actuator based on a new mathematical model. *J. Micromech. Microeng.* **2004**, *11*, 1439.
49. Xue, X.; Chen, L.; Wu, X.; Sun, Q. Study on electric-mechanical hysteretic model of macro-fiber composite actuator. *J. Intell. Mater. Syst. Struct.* **2014**, *12*, 1469–1483.
50. Kamlah, M.; Jiang, Q. A constitutive model for ferroelectric PZT ceramics under uniaxial loading. *Smart Mater. Struct.* **1999**, *4*, 441. [CrossRef]
51. Goldfarb, M.; Celanovic, N. Modeling piezoelectric stack actuators for control of micromanipulation. *IEEE Control Syst. Mag.* **1997**, *3*, 69–79.
52. Bouc, R. Forced vibration of mechanical systems with hysteresis. In Proceedings of the 4th Conference on Non-Linear Oscillation, Prague, Czechoslovakia, 5–9 September 1967.
53. Wen, Y.K. Method for random vibration of hysteretic systems. *J. Eng. Mech. Div.* **1976**, *2*, 249–263.
54. Ikhouane, F.; Rodellar, J. *Systems with Hysteresis: Analysis, Identification and Control Using the Bouc-Wen Model*; John Wiley & Sons: Hoboken, NJ, USA, 2007.
55. Data SheetModel. Available online: http://www.maxell.com.tw/images/uploads/2015/05/ML2032_DataSheet_table.pdf (accessed on 8 February 2020).
56. Barreras, J.V.; Schaltz, E.; Andreasen, S.J.; Minko, T. Datasheet-based modeling of Li-Ion batteries. In Proceedings of the 2012 IEEE Vehicle Power and Propulsion Conference, Seoul, Korea, 9–12 October 2012.

57. Di Rito, G.; Galatolo, R. Experimental and theoretical study of the electrical failures in a fault-tolerant direct-drive servovalve for primary flight actuators. *Proc. IMechE Part I: J. Syst. Control Eng.* **2008**, *222*, 757–769. [CrossRef]
58. Di Rito, G.; Schettini, F. Health monitoring of electromechanical flight actuators via position-tracking predictive models. *Adv. Mech. Eng.* **2018**, *10*, 1–12. [CrossRef]
59. Di Rito, G.; Schettini, F.; Galatolo, R. Model-based health-monitoring of an electro-mechanical actuator for unmanned aerial system flight controls. In Proceedings of the 4th IEEE International Workshop on Metrology for AeroSpace (MetroAeroSpace), Padua, Italy, 21–23 June 2017; pp. 481–490.
60. Di Rito, G.; Schettini, F.; Galatolo, R. Self-monitoring electro-mechanical actuator for medium altitude long endurance unmanned aerial vehicle flight controls. *Adv. Mech. Eng.* **2016**, *8*, 1–11. [CrossRef]
61. Di Rito, G.; Schettini, F.; Galatolo, R. Model-based prognostic health-management algorithms for the freeplay identification in electromechanical flight control actuators. In Proceedings of the 5th IEEE International Workshop on Metrology for AeroSpace (MetroAeroSpace), Rome, Italy, 20–22 June 2018; pp. 340–345.

© 2020 by the authors. Licensee MDPI, Basel, Switzerland. This article is an open access article distributed under the terms and conditions of the Creative Commons Attribution (CC BY) license (http://creativecommons.org/licenses/by/4.0/).

Article

Characterization of a COTS-Based RF Receiver for Cubesat Applications †

Antonio Lovascio [1,*], Antonella D'Orazio [1] and Vito Centonze [2]

1 Dipartimento di Ingegneria Elettrica e dell'Informazione, Politecnico di Bari, 4, E. Orabona St., 70125 Bari, Italy; antonella.dorazio@poliba.it
2 Space Instrument & Avionic Division, Sitael S.p.A., 21, San Sabino St., 70042 Mola di Bari, Italy; vito.centonze@sitael.com
* Correspondence: antonio.lovascio@poliba.it; Tel.: +39-333-318-7995

† This paper is an extended version of our paper published in: Lovascio, A.; Centonze, V.; D'Orazio, A. Design of COTS-Based Radio-Frequency Receiver for Cubesat Applications. In Proceedings of the 2019 IEEE International Workshop on Metrology for Aerospace (MetroAeroSpace), Torino, Italy, 19–21 June 2019.

Received: 20 November 2019; Accepted: 29 January 2020; Published: 31 January 2020

Abstract: This paper reports the experimental results of a test campaign performed on the radio-frequency (RF) receiver prototype operating at a 2025–2110 MHz frequency range, designed and fabricated for CubeSat applications. The prototype has been tested through a board-level test approach for the verification of the functional requirements and a component-level one for specific characterization measures. The tests have shown the following results: a −115–−70 dBm sensitivity range, 390 MHz intermediate frequency, a 0 dBm output power level with ±1 dB error, a 2.34 dB noise figure, and a 4.86 W power absorption. Such results have been largely achieved implementing an automatic gain control system by cascading two Commercial Off-The-Shelf (COTS) amplifiers. Moreover, an innovative technique based on RF test points has been successfully experimented and validated to measure the S-parameters of a custom low-pass filter integrated on the receiver, showing the possibility of even characterizing the single COTS components exposed to radiation through a unique board-level test setup. The technique may have a great impact on the cost reduction of electronic boards for space applications, since it would avoid using expensive evaluation boards for each COTS component that needs a radiation test.

Keywords: COTS; cubesat; radiation test; receiver; RF; satellite

1. Introduction

Recently, the research on the framework of the aerospace applications has been devoted to CubeSats because they are small and compact, characterized by low launch costs, and suitable to create satellite constellations that can improve the terrestrial network infrastructure, providing better global coverage.

The CubeSats are typically designed using Commercial Off-The-Shelf (COTS) components. The use of COTS components has the great advantage of widely reducing the costs of the satellite subsystems, since additional screening and reliability tests/inspections are not mandatory. Therefore, in recent years, they have generated a great deal of interest from the industry of small satellites, especially for Low-Earth Orbits (LEOs) [1,2].

The LEOs are the best environment for the allocation of hardware based on COTS components. The satellites find a less severe environment in such orbits, within which the components, even if not space-qualified, could "survive" the strong temperature ranges and the ionizing radiations.

The use of COTS components for space missions is well documented in the literature. The COTS components are used to make both the power electronic [3–6] and digital [7–9] boards. The term

"COTS" is also related to the test facilities [10,11] and electrical ground support equipment [1,12]. Finally, the COTS-based design has been even adopted for RF systems [13–20]. They include navigation systems [13,14], transmission data systems for deep-space and remote-sensing applications [15,16], and modules for communication with the earth [2,17–20].

The main problem in the use of COTS components in space applications is the increased risk of failure of the satellite subsystems that implement them. Their suitability for use in space must always be verified by looking for reliability information in the literature, mostly related to radiation tolerance. Some selection criteria of COTS components, even standardized, have been developed in recent years to help designers to make a reasonable choice of component considering the orbit where the satellite will be placed.

The main issue of an approach based on selection criteria is related to the limited available information associated with the radiation tolerance of the COTS components. Since radiation is the main cause of failure in the space, the components are often tested for their tolerance to radiation under appropriate conditions, and the corresponding test reports are filed in databases which are not always freely available. Moreover, most of the components for which a radiation test report is available are components for low-frequency and power applications. Few Radio-Frequency (RF) components are available in such databases. For this reason, the design of COTS-based RF front-ends that have high-performance and reliability at the same time is particularly difficult.

In the literature, there are few papers that report RF front-ends implemented with COTS integrated circuits [21–23], for which reliability information and radiation test reports are more easily available. However, such an approach solves the reliability issue, but does not always allow the creation of applications with very high performance, since the integrated circuits are designed to be general-purpose for terrestrial applications.

The intent of this work is to find an efficient solution to deal with this issue, considering as a case study the design of an RF receiver operating at a 2025–2110 MHz frequency range, conceived for CubeSat applications. The starting idea was to design a completely custom device with a high performance and high reliability using only COTS components, even without having information regarding their behavior under radiation. Such design requirement has been addressed from a purely analytical point of view in our original work, presented at the Metrology for Aerospace 2019 conference [17]. This article is an extension of that work, where the same design requirement has been addressed from an experimental point of view, i.e., through the adoption of an innovative technique based on the use of RF test points. The adoption of this technique has driven the design of the electrical schematic and the layout, up to the fabrication of a functional prototype. The prototype has been tested and the experimental results are reported in this paper. Moreover, a detailed characterization of a component performed by means of the RF test points is presented, proving definitely that the use of RF test points allows the individual testing of the component, insulating it from the rest of the circuit, which is extremely attractive if radiation tests are executed on that component along with the whole board. In other words, by performing a unique radiation test on the board, information regarding the radiation behavior of each single component can be obtained, saving costs and avoiding the use of expensive evaluation boards.

2. The Method Extension

The RF receiver was designed to operate with a 2025–2110 MHz input frequency range to be compliant with the European Cooperation for Space Standardization. A design flow, typically adopted in the aerospace context, was followed. Starting from the mission analysis, a minimum set of system requirements was defined, which constituted the starting point of the design, especially in the choice of the most suitable architecture.

The superheterodyne architecture was adopted for the receiver design. The architecture foresaw a single conversion stage to minimize the number of the components. The RF incoming signal was downconverted to the Intermediate Frequency (IF) by a COTS RF mixer, amplified, filtered, and finally output through a coaxial connector.

With respect to the traditional architecture, some improvements were implemented to meet the noise figure and input dynamic range requirements. The noise performance has been improved by cascading two low-noise and high-gain COTS amplifiers before the preselector filter, which allowed a drastic decrease in the total noise figure. Likewise, two cascaded adjusted gain control COTS amplifiers were placed at the end of the receiver chain, which allowed to obtain a gain variation dynamic of 52 dB that, besides fulfilling the input dynamic range requirement (−115–−70 dBm), also added 7 dB of margin.

In conference paper [17], we focused on the modelling aspects of the system. The receiver model was made considering not only the nominal functional performance but also the performance in the worst-case. We pointed out that the "worst case" is not easily identifiable; therefore, it was implicitly assumed in the system requirements. For example, if the noise figure must be smaller than 3 dB by requirement, such requirements must be valid even in the presence of radiation and thermal changes. Another assumption was that the RF parameters drift from the nominal value because of frequencies, operating temperature, and total ionizing dose. The amount and the signs of such drift were not known a priori, because of the lack of reliability data.

Since the final target of the receiver is to amplify the incoming signal to a fixed output power of 0 dBm ± 1 dB (by requirement) and since the RF parameters suffer a drift, as previously assumed, we implemented a worst-case analysis based on the Monte Carlo analysis that allowed to estimate the total net gain and noise figure of the whole receiver, starting from the variation of the RF parameters of each individual component. Such an estimation was performed assuming a safety margin that considered both lack of reliability information and the lack of the specific tests on the evaluation boards, i.e., using only the data reported in the datasheets, even if they were less accurate. The safety margin was set at 10%, resulting from a reasonable hypothesis on the components' behavior when studying their technology.

The system simulations showed that all requirements were satisfied. The most relevant result was in the output power level, which was equal to 0 dBm even in the presence of net gain variations that exceed the minimum required input dynamic range (45 dB). In other words, the Monte Carlo analysis has shown that the 7 dB of margin obtained by the AGC system can compensate any drift in the output signal power level.

The assumption that RF parameters only suffer a drift may be too conservative. Especially concerning radiation, we cannot exclude a priori the possibility that the ionizing radiation could cause permanent damage to the components. Therefore, the components' behavior can be definitively predicted only by performing a component level radiation test. Such a target has been reached and presented in this manuscript. The adopted method is based on the experimental tests designed to be performed even in the final phases of the project. The innovative element is in using RF test points, simple COTS connectors, not only to guarantee the functional performance of the receiver, but also to enhance the reliability of a device intended to be used in very severe environments. The method extends and improves the analytic approach reported in the conference paper [17] since it allows to complete the system model so that the results' accuracy is improved.

3. RF Receiver Prototype Description

Figures 1 and 2, respectively, show the block diagram of the RF receiver and a photograph of the assembled prototype.

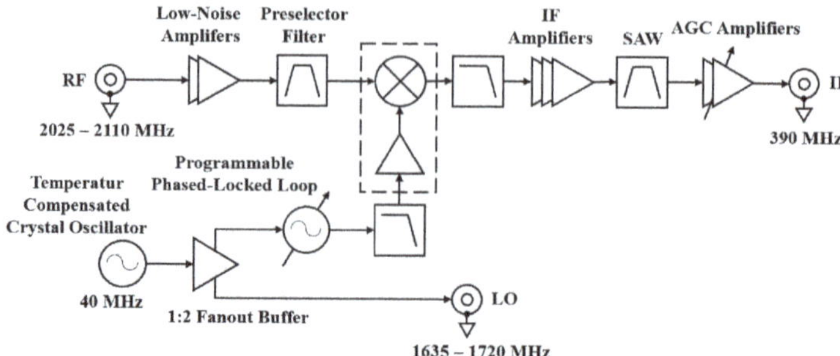

Figure 1. Radio frequency (RF) receiver block diagram.

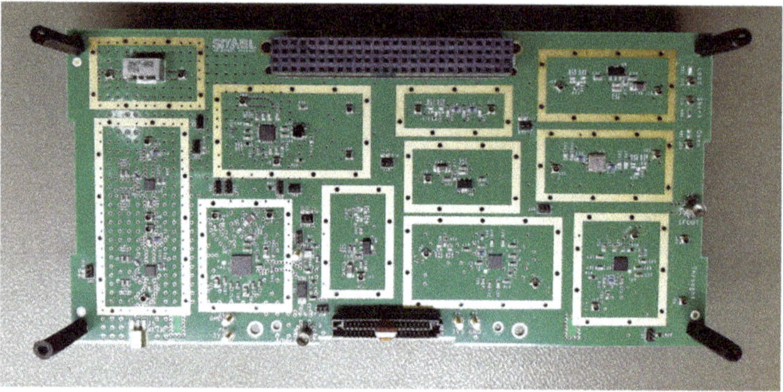

Figure 2. A photograph of the RF receiver prototype.

The Printed Circuit Board (PCB) is made of eight layers, using a 0.254 mm RO4350B substrate as dielectric core for the top and bottom layers and RO4450B substrates for the prepreg inner layers. The stack-up is about 2 mm thick and has 192 mm × 95.9 mm size. One of the two board dimensions has been enlarged (with respect to the 1U CubeSat standard) to allow the placement of the RF test points, which normally are not placed in the flight model of the same system.

All RF paths have been routed on the top layer, while the bottom layer has been reserved for a few secondary components. Moreover, there are three inner ground planes, two planes for the power supply routing (one for the 5 V and the other one for the 3.3 V voltage), and a plane reserved for shielding issues.

In Figure 2, the PC104 and other secondary connectors are evident. Such connectors were the first components to be placed on the board, since they are a mechanical constraint. The schematic and layout of the components were designed by following the guidelines of their reference design, reported in the datasheets, trying to reproduce the placement of the lines, components, and via-holes. The lines that connect the two components have been routed following the logical connection of the high-level system, i.e., of the reception path. The technology of the coplanar microstrips has been chosen to design such lines. Their characteristic impedance has been optimized to 50 Ω because all selected components operate with a 50 Ω reference impedance. In this way, the matching impedance has been guaranteed. With the RO4350B substrate that has 3.48 dielectric constant, a 50.2 Ω characteristic impedance has been estimated, with a 0.52 mm microstrip width and a 0.28 mm gap between the microstrip and coplanar ground plane.

For each active element, a metallic shield has been introduced to reduce the electromagnetic emission and to avoid cross-talk and interference from nearby components. Being a prototype, commercial shields with a prefixed size have been used. They are composed of a frame and a removable cover. The frame is the part of the shield that is mounted on the PCB through via-holes and soldered on copper pads. The cover can be removed, if necessary, for debug operations and to rework the board. The shields are 5.08 mm height and are made of 0.38 mm thick tin-plated steel.

4. Test Method

The experimental characterization of the prototype has been done by adopting two main types of tests: component-level tests and board-level functional tests.

4.1. Component-Level Test Description

The component-level tests have allowed to verify that, for each component, the schematic and layout implementation have been correctly designed and no issues have occurred during the prototype's fabrication and components' mounting. The main purpose of this type of test is to measure the main functional parameters of the RF components, such as the power gain, the input and output return loss, the 1-dB compression point, and the noise figure. The execution of such a test has been made possible thanks to the adoption of the RF test points. They are typically used to monitor the power level along the signal path to control and improve the design of a system.

The RF test points selected for the receiver are surface-mount connectors that have 2.3 mm × 2.3 mm × 1.35 mm size and can operate from Direct Current (DC) to 11 GHz [24]. Figure 3 shows their basic operating principle.

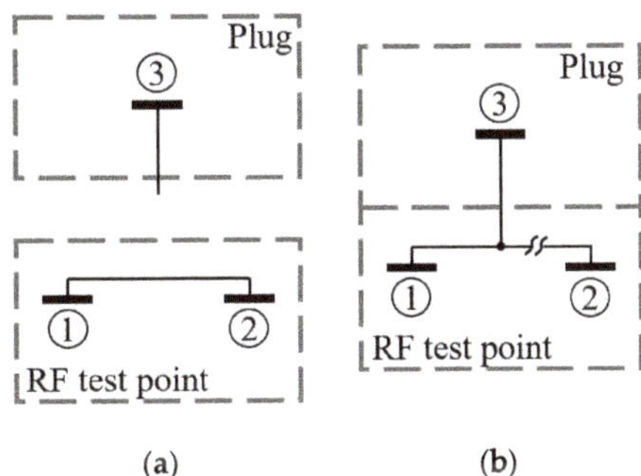

Figure 3. Operating scheme of the RF test point selected for the RF receiver prototype: (**a**) not-mated to plug; (**b**) mated to plug.

When the plug is "not-mated" to the connector, the "1" and "2" pins are short-circuited. From the signal point of view, that short-circuit is a piece of metal (2.3 mm long and 0.35 mm width) that exhibits a small inductance, negligible at the receiver operative frequencies. The voltage standing wave ratio is about 1 at 2 GHz and the attenuation is smaller than 0.1 dB [24].

When the plug is "mated" to the connector, the "1" pin disconnects from the "2" pin and it connects to the "3" pin that is the positive of the plug. If the "1" pin is connected to the input or output of an RF component of the receiver, the "1" pin would allow access to the input or output of that component as if it were an evaluation board. The plug breaks the RF signal path on the board and routes it toward the plug itself, which can be a coaxial connector linked to a measurement instrument.

The basic operation of the RF test point has been exploited to form the input and output ports for each component of the receiver. Each port (regardless if it is an input or output) has been made by connecting the "1" pin of the test point to the input/output pin of the RF component. In this way, the component can be isolated from the rest of the circuit and can be tested individually. For the component-level tests of the active elements, some mechanical switches have been foreseen for debug purposes. The switches allow to power on the active components individually to avoid the influence of the other components under test. Moreover, the switches provide the possibility to shut down the whole board when passive elements are under test.

The RF test points are very effective to measure the S-parameters of the passive components (as the filters) and, as it is known, the S-parameter measurement requires an appropriate calibration kit that must be compatible with the mate-connector used to connect the network analyzer to the RF test points. To perform a more accurate network analyzer calibration, the calibration kit has been custom-designed on the same receiver board, rather than purchasing a commercial one. Such a choice has also been cost-effective because, as the calibration frequencies were smaller than 3 GHz, the custom calibration kit could have been designed using lumped components, especially for the 50 Ω load.

Figure 4 shows the calibration kit schematic and a photograph of its implementation on the prototype board.

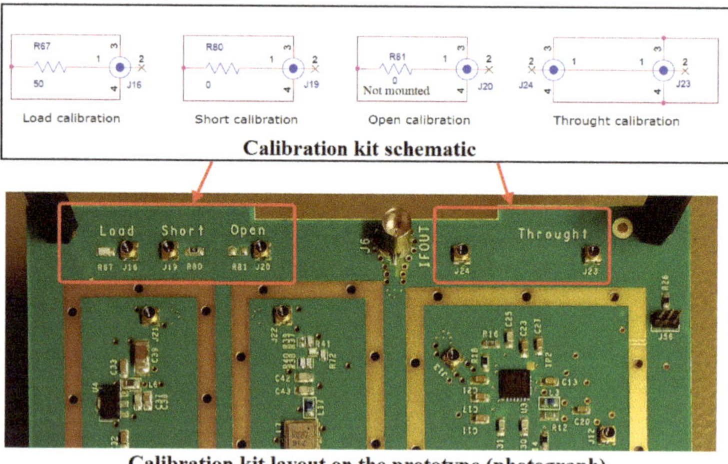

Figure 4. Custom calibration kit for the mate-connector of the RF test points. Top image shows the schematic and bottom image shows a photograph of the its implementation on the prototype board.

The calibration system of the "load" was made using a 50 Ω resistor with very high accuracy, i.e., ±0.1%. The calibration system of the "short" was made using a 0 Ω resistor. The calibration system of the "open" was made foreseeing a simple footprint without any component mounted on the board. Finally, the calibration system of the "through" was made by designing a coplanar microstrip with a 50 Ω characteristic impedance. The length of such a microstrip has been assumed to be equal to the average electrical length of all components placed on the board.

The schematic of the calibration kit, illustrated in Figure 4, also shows that the return signal is not connected to the same receiver ground to prevent all possible eventual errors in calibration caused by the rest of the circuit. The calibration of the network analyzer was always performed with the receiver off.

Besides the aspects related to the calibration, the implementation of the RF test points required attention for the encumbrance and mounting aspects. Regarding the encumbrance aspect, to save space on the board, a 3D analysis has been performed to search for the minimum spacing to place two near to the RF test points (estimated as equal to 0.4 mm) in order to avoid the mate-connector touching the nearby RF test point, preventing the insertion. The 3D analysis has also helped to estimate the right height of the metallic shields (5 mm). Regarding the mounting aspect, since the RF test points have a small size and the mounting has been done by machine, the footprint design of the connectors has been customized in order to provide the correct guidelines for the operator who physically mounted the components, orienting it correctly. An error in the component orientation would have compromised the whole test approach.

4.2. Board-Level Test Description

The board-level tests have allowed to verify the functional performance of the whole receiver from the electrical point of view, i.e., to verify its compliance with the high-level requirements defined in [17]:

- Sensitivity range: −115–−70 dBm;
- Noise figure: <3 dB @ −115 dBm;
- Intermediate Frequency: 390 MHz;
- IF output power: 0 dBm ±1 dB;
- DC power absorption: <3.5 W.

The verification of these requirements needs a test setup that links the instrumentation to the main external interfaces of the receiver, i.e., the coaxial ports RF, IF, Local Oscillator (LO), and the access points for the power supply voltages. The board-level test requires that the whole board is switched on, so that all amplifiers allow the passage of the signal from the input (the RF) up to the output (the IF). Moreover, the power-on of the board allows to measure the DC power absorption.

5. Test Results

5.1. Measure of the S-Parameters of the IF Low-Pass Filter

In this subsection, an example of a component-level test exploiting the RF test points is shown. The S-parameters have been measured on a low-pass filter placed on the receiver IF section. The IF low-pass filter has been custom designed to reject the LO spur that couples on the IF, because of the low insulation of the mixer ports. The measure of the S-parameters has allowed to verify the filter performance and to validate the simulation model.

Figure 5a shows a photograph of the test setup, which is constituted of a network analyzer [25] that has been connected to the filter through two mate-connectors. Figure 5b shows an enlarged photograph of the connection of the mate-connectors to the input and output RF test points foreseen for the filter.

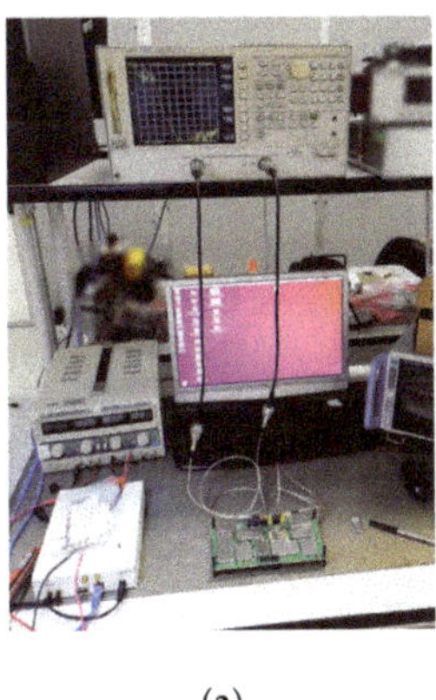

(a)

(b)

Figure 5. Photograph of the test setup implemented to test the custom IF section (390 MHz) low-pass filter: (**a**) The whole test setup; (**b**) A zoom on the mate of the instrument connectors to the filter RF test points.

The network analyzer has been calibrated using the custom calibration kit shown in Figure 4 within the 30 kHz–3 GHz frequency range, setting the test power level equal to −30 dBm.

Figures 6 and 7 report the measured S_{21} and S_{11} parameters, respectively, of the filter compared with the result of the design simulation.

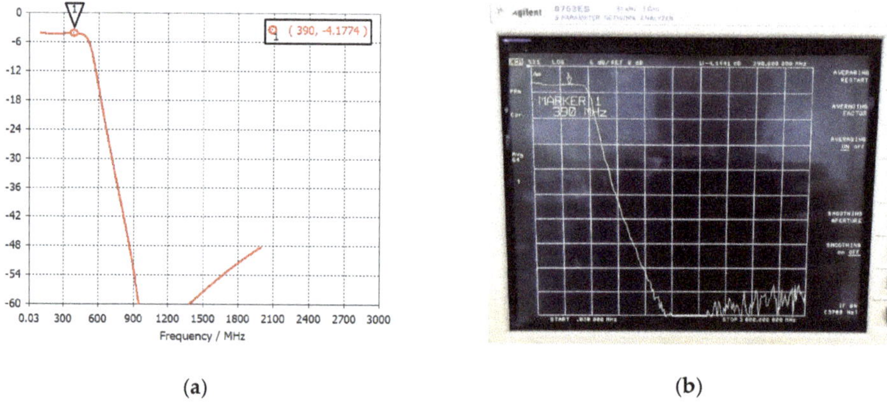

(a) (b)

Figure 6. The low-pass filter S_{21} parameter: (**a**) simulation result; (**b**) result measured with the network analyzer.

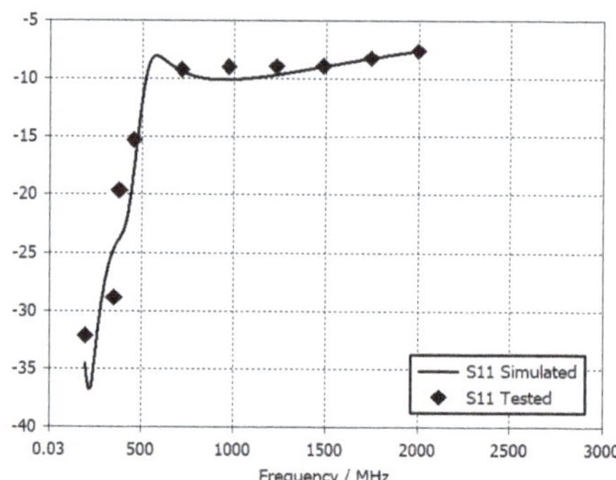

Figure 7. Simulated and tested S_{11} parameters of the low-pass filter.

The measured and simulated S_{21} parameters agree very well. A 4.14 dB in-band attenuation has been measured at 390 MHz frequency, which is exactly equal to the attenuation obtained by simulations at the same frequency. Likewise, the measured and simulated S_{11} parameters agree well. At 390 MHz, a −19.7 dB value has been measured on the S_{11} parameter, which is excellent from the impedance-matching point of view. Therefore, the obtained results show that the layout implementation of the low-pass filter has been done correctly and the RF test points have been effective in performing the experimental characterization of a single component, even if already integrated in a wider system like the receiver board.

5.2. Measure of the Sensitivity Range, Output Power and IF

In this subsection, the results of the board-level functional tests regarding the sensitivity range, the output power and the IF are reported.

Figure 8 shows the general scheme of the test setup adopted for the functional tests. Figure 9 shows a photograph of the implemented test setup.

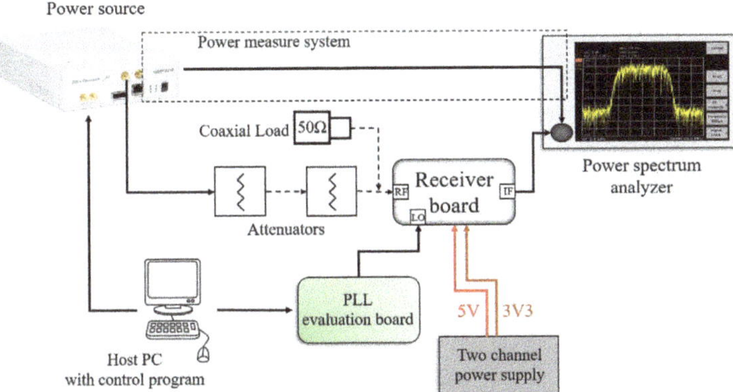

Figure 8. General scheme adopted for the board-level functional tests executed for the RF receiver.

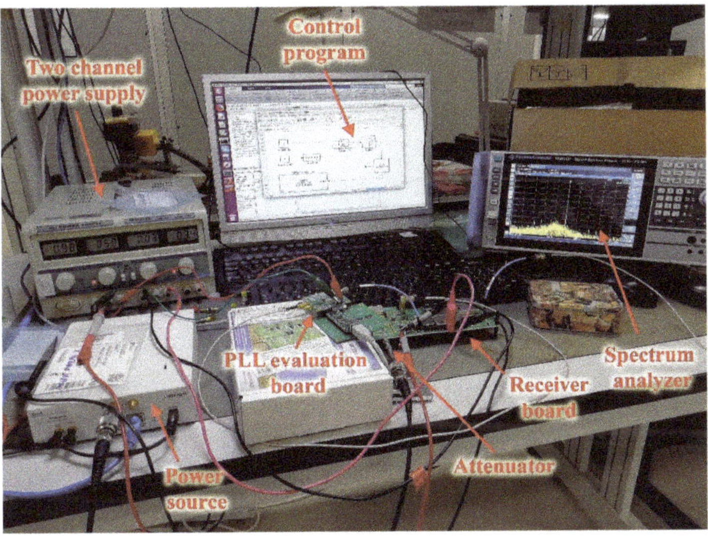

Figure 9. Photograph of the test setup really implemented for the board-level tests executed for the RF receiver.

The test setup is constituted by the device under test (the RF receiver), a power source, a spectrum analyzer, a local oscillator, a bench power supply and a host Personal Computer (PC) with a control program.

The receiver test required the generation of a test signal delivered at the RF input port and the reading the 390 MHz output signal at the IF port. The test signal was generated, exploiting a commercial Software-Defined Radio (SDR) as a power source [26], which was configured to generate a Continuous Wave (CW) signal at 2050 MHz frequency, which is within the receiver input frequency range. The output signal was acquired at the IF port by means of a spectrum analyzer [27], which has provided information on frequency, power level, and noise floor.

Regarding the local oscillator, although a COTS Phased-Locked Loop (PLL) was implemented directly on the receiver board, in the first test phase the 1660 MHz LO signal was delivered externally to the mixer LO input using the evaluation board of the same PLL. The signal was brought to the receiver exploiting a debug LO input, which was made through an RF test point. The PLL evaluation board was power-supplied with a unique 5 V voltage and was configured using a dedicated software provided by the manufacturer [28] along with the board. An integer-N configuration was set configuring a 0 dBm power level signal, carefully measured on the spectrum analyzer. The PLL configuration was executed with the same host PC, connected to the evaluation board through a USB A-B cable and an interface board.

The receiver board was power supplied with a 30 V-5 A two-channel bench power supply [29], adjusted to deliver the 5 and 3.3 V voltages. The 5 V voltage was used to supply both the receiver and the PLL evaluation board while the 3.3 V voltage was delivered to the receiver board to power supply some auxiliary functions of the AGC amplifiers. The 5 V and 3.3 V voltages were brought to the receiver by means of dedicated test points directly implemented on the board.

The test was executed in two steps: in the former, the input power level was finely adjusted to generate CW signals from −115 dBm to −70 dBm with a 2.5 dB step; in the latter, the CW signal was applied at the receiver RF input and measured the IF output with the spectrum analyzer. Tuning of the input power level was obtained by combining the SDR gain control (performed by the control program) with the use of one or two 30 or 60 dB external coaxial attenuators. The correct tuning was carefully checked, connecting the SDR directly on the spectrum analyzer input through a coaxial cable and measuring the power level. For each input power level, the IF output signal was measured.

Figure 10 shows a graph that reports the output power level (vertical axis) as a function of the input power level (horizontal axis).

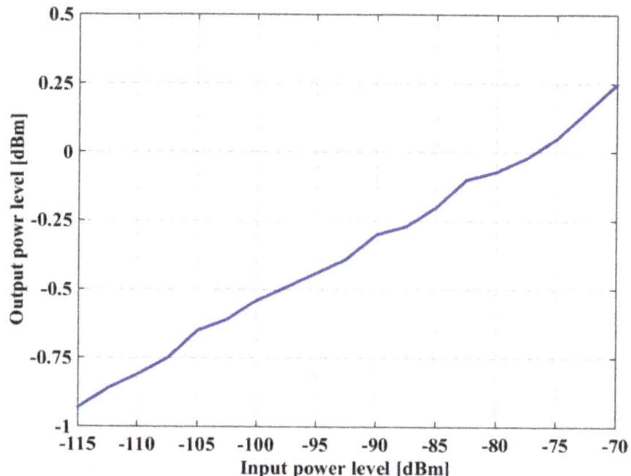

Figure 10. Measured input dynamic range of the receiver board for from −115 dBm to −70 dBm with a 2.5 dB step.

The graph shows that, for input power levels from −115 to −70 dBm, the output power level is within the −1–+1 dBm range. Moreover, for each input power level delivered at 2050 MHz, the IF output signal appeared at 390 MHz frequency on the spectrum analyzer.

These results prove that the system-level requirements regarding the sensitivity range, output power, and the value of the IF have been fulfilled.

5.3. Measure of the DC Power Absorption

In this subsection, the measure of the receiver DC power absorption is reported. The test setup is the same as shown in Figure 9. As shown in the figure, the absorbed current is equal to 0.98 A on the 5 V output channel, and 30 mA on the 3.3 V one. Therefore, the total DC power absorption is about 5 W. However, the absorbed power also considers the presence of the PLL evaluation board, which is implemented inside a linear voltage regulator to generate the 3.3 V voltage, starting from the 5 V input voltage. This 3.3 V voltage is not used to power supply the receiver, and is only used for the PLL normal operation and other minor components. Therefore, the current absorption, due to this 3.3 V voltage (which has been estimated equal to 84 mA reading the datasheet), must not be considered in the total DC power absorption of the receiver.

The DC power dissipated (P_D) by the voltage regulator was estimated as equal to 142.8 mW using Equation (1)

$$PD \cong (VIN - VOUT) * IOUT, \tag{1}$$

where V_{IN} is the regulator input voltage, and V_{OUT} and I_{OUT} are the regulator output voltage and current, respectively. Therefore, the real DC power absorbed by the receiver has been calculated by subtracting to 5 W the power dissipated by the PLL evaluation board (142.8 mW), obtaining a total DC power absorption of about 4.86 W. This value is higher than the system requirement (3.5 W).

5.4. Measure of the Noise Figure

The Noise Figure (NF) measurement was performed by adopting the same scheme depicted in Figure 8. In this case, a 50 Ω coaxial load was applied at the 2050 MHz RF input connector. The receiver board was normally power supplied and the external LO signal was delivered through the PLL evaluation board.

Figure 11a shows a photograph of the test setup. Figure 11b shows a screenshot of the noise power measured with the spectrum analyzer [30]. A marker was placed at the 390 MHz central frequency.

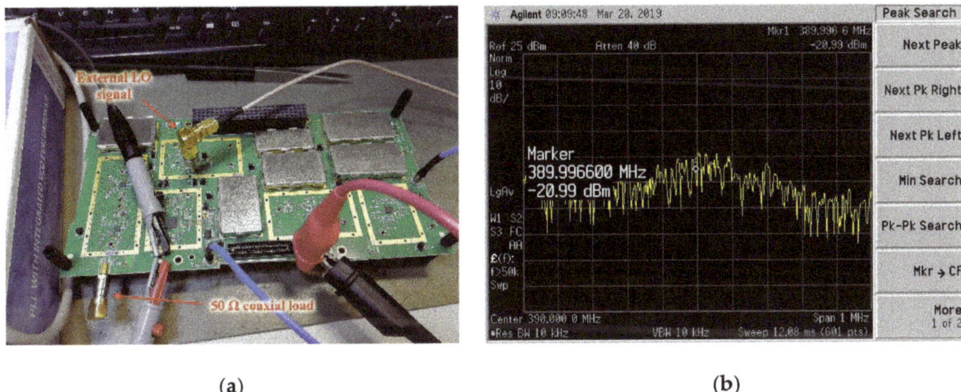

(**a**) (**b**)

Figure 11. Photograph of the noise figure test setup (**a**) and output noise power screenshot at 390 MHz of the receiver board (**b**) with the RF input terminated with a 50 Ω coaxial load.

Starting from the noise power evaluation, the noise figure has been calculated using Equation (2),

$$NF = NOUT - 10\log_{10}(RBW) + 174 - GS, \tag{2}$$

where N_{OUT} is the noise power, RBW is the instrument resolution bandwidth, and G_S is the system gain. The system gain has been calculated, starting from the theoretical gain obtained by system simulations (i.e., 112.4 dB when the input power level is −115 dBm). The total loss of the 50 Ω impedance lines analyzed in [17] and of the RF test points has been subtracted from the gain value. Assuming a 0.05 dB loss for each RF test point [24], and as there are 20 RF test points on the signal path, the system gain has been assumed to be equal to 112.4 − 0.73 − (20 × 0.05), i.e., 110.67 dB. Using Figure 11b and Equation (2), the receiver noise figure has been estimated to be equal to 2.34 dB, which is smaller than the system requirement (3 dB).

6. Discussion

The results obtained by the experimental tests have highlighted both the effectiveness of the modeling methodology reported in [17] and the ability to make the board suitable for the characterization of single components. The compliance of the experimental results with the requirements is summarized in Table 1.

Table 1. Compliance of the receiver test results with the high-level requirements.

Requirement	Test Result
Sensitivity range: −115–−70 dBm	$P_{IN} = -115$ dBm → $P_{OUT} = -0.93$ dBm $P_{IN} = -70$ dBm → $P_{OUT} = +0.25$ dBm
Noise figure: <3 dB @ −115 dBm	NF = 2.34 dB
IF: 390 MHz	IF = 390 MHz
IF output power: 0 dBm ± 1 dB	−0.93 dBm ≤ P_{OUT} ≤ +0.25 dBm
DC power absorption: <3.5 W	$P_A = 4.86$ W

Except for the DC power absorption, all high-level requirements have been satisfied. The DC power absorption requirement has not been fulfilled because the receiver was designed without considering the reception system of the signal at the intermediate frequency, which digitalizes it by an ADC. In other words, the output power requirement is set a priori without considering the ADC noise performance, although even the ADC can be characterized in terms of noise figure, like the other RF components [2]. If its noise model is included into the receiver system model, the amplification net gain of the receiver could decrease. A lower amplification gain would reduce the number of the active elements and, therefore, satisfy the DC power absorption requirement.

Regarding the component-level tests, the adopted methodology is the most innovative aspect, especially if it is adopted for the radiation tests. When the COTS components are used, the radiation tests are foreseen in the final phases of the project, when all boards of the whole system are assembled. In this phase of the project, a fault in the system caused by the ionizing radiations would prevent the single component that has generated the failure from being found. A typical approach to discover the radiation behavior of the components is to test them using test boards (or the same evaluation boards) and to characterize the components with functional tests. The parameters obtained by functional tests after radiation help to improve the accuracy of the system model. Such an approach has a high cost because the radiation tests should be repeated for each element used to design the system.

Instead, proving that the RF test points can be used to test a single component (as shown in the example of the filter characterization presented in this paper), even if it is already assembled on the board, shows that the test approach can be used to characterize individual components after the exposition to radiation. The great advantage of this is the fact that the board can be irradiated only once, whilst still maintaining the possibility of discovering the behavior of all components individually, which is very attractive for the reduction in costs. Therefore, by means of the RF test points, the RF components can be checked in order to perform debug operations and to find the component that has caused the fault after a radiation test. Moreover, RF components could be characterized without purchasing evaluation boards, which can be 10 times more expensive than the component itself and are not always available on the market.

Finally, to complete the test methodology, a system of mechanical switches has been also foreseen on the board to allow to switch on and off only components under test. In addition, the tests are possible even in the presence of the shields, since they have been selected with a removable cover.

7. Conclusions

In this paper, the experimental results of the receiver prototype designed for Cubesat applications have been reported. The tests have been executed both at a component-level and at board-level. Regarding the component-level tests, an example of the characterization of a custom low-pass filter has been reported. The characterization has been performed measuring the S_{21} parameter with a network analyzer connected to the input and output port of the component by means of the RF test points. The measured S_{21} parameter agrees very well with the simulated one, exhibiting a 4.14 dB in-band attenuation at 390 MHz frequency, which is exactly equal to the attenuation obtained by simulations. Such a result has shown that the design approach allows to test the components individually after they are mounted on the board, which could be used to perform radiation tests and to discover the behavior of the COTS RF components for debugging purposes. Performing the component-level radiation tests through a unique board-level radiation test is extremely attractive to reduce design costs and avoid the purchase of expensive evaluation boards.

Regarding the board-level tests, compliance with high-level requirements has been successfully verified. A CW signal with a power level within −0.93–+0.25 dBm has been measured at the IF output port at a 390 MHz frequency for a sensitivity range of −115––70 dBm. Moreover, a 2.34 dB noise figure and a 4.86 W DC power absorption have been measured.

Author Contributions: A.L. performed the experimental tests and drafted the manuscript. A.D.; V.C. supervised and reviewed the manuscript. All authors have read and agreed to the published version of the manuscript.

Funding: The Ph.D. A. Lovascio has benefited from a PhD MIUR fellowships for the 2018/2019 academic year, course XXXII, awarded within the framework of the "Programma Operativo Nazionale Ricerca e Innovazione" (PON RI 2014/2020) Axis I "Investments in Human Capital" - Action I.1– "Innovative PhDs with industrial characterization." Funding FSE-FESR.

Conflicts of Interest: The authors declare no conflict of interest.

References

1. Lovascio, A.; D'Orazio, A.; Centonze, V. An Innovative EGSE Approach for Satellite RF system characterization based on Real-Time Monitoring and Space Environment Oriented Analysis. In Proceedings of the 2017 IEEE International Workshop on Metrology for Aerospace (MetroAeroSpace), Padua, Italy, 21–23 June 2017.
2. Lovascio, A.; D'Orazio, A.; Centonze, V. Design of a Telemetry, Tracking, and Command Radio-Frequency Receiver for Small Satellites Based on Commercial Off-The-Shelf Components. *Int. J. Aerosp. Mech. Eng.* **2019**, *13*, 564–573.
3. Gopakumar, S.; Eega, S.; Suresh, S.V.; Reddy, M.S.S.; Antony, A.; Ramachandran, H.; Koilpillai, D. Design of Electrical Power Subsystem for IITMSAT. In Proceedings of the 2015 International Conference on Space Science and Communication (IconSpace), Langkawi, Malaysia, 10–12 August 2015.
4. Khan, M.U.; Ali, A.; Ali, H.; Khattak, M.S.; Ahmad, I. Designing efficient electric power supply system for micro-satellite. In Proceedings of the 2016 International Conference on Computing, Electronic and Electrical Engineering (ICE Cube), Quetta, Pakistan, 11–12 April 2016.
5. Adnan, S.M.; Bhutto, M.S. Modular & COTS based power system for small LEO satellite. In Proceedings of the 2013 International Conference on Aerospace Science & Engineering (ICASE), Islamabad, Pakistan, 21–23 August 2013.
6. Schirone, L.; Macellari, M.; Schiaratura, A. On the reliability of modular power conversion systems for small spacecraft. In Proceedings of the 2009 IEEE Aerospace conference, Big Sky, MT, USA, 7–14 March 2009.
7. Sabripour, S.; Haque, J.; Ciszmar, A.; Magesacher, T. A COTS-based software-defined communication system platform and applications in LEO. In Proceedings of the 36th International Communications Satellite Systems Conference (ICSSC 2018), Niagara Falls, ON, Canada, 15–18 October 2018.
8. Nagarajan, C.; D'souza, R.G.; Karumuri, S.; Kinger, K. Design of a cubesat computer architecture using COTS hardware for terrestrial thermal imaging. In Proceedings of the 2014 IEEE International Conference on Aerospace Electronics and Remote Sensing Technology, Yogyakarta, Indonesia, 13–14 November 2014.
9. Dunham, M.E.; Baker, Z.; Stettler, M.; Pigue, M.; Graham, P.; Schmierer, E.N.; Power, J. High Efficiency Space-Based Software Radio Architectures: A Minimum Size, Weight, and Power TeraOps Processor. In Proceedings of the 2009 International Conference on Reconfigurable Computing and FPGAs, Quintana Roo, Mexico, 9–11 December 2009.
10. Cunningham, A.; Kass, M. Development, integration, and test architecture for a software-based hardware-agnostic Fault Tolerant Flight Computer. *IEEE Instrum. Meas. Mag.* **2016**, *19*, 38–43. [CrossRef]

11. Ross, R.W. Integrated component-based data acquisition systems for aerospace test facilities. In Proceedings of the ICIASF 2001 Record, 19th International Congress on Instrumentation in Aerospace Simulation Facilities (Cat. No. 01CH37215), Cleveland, OH, USA, 27–30 August 2001.
12. Cordero, F.; Mendes, J.; Kuppusamy, B.; Dathe, T.; Irvine, M.; Williams, A. A cost-effective Software Development and Validation environment and approach for LEON based satellite & payload subsystems. In Proceedings of the 5th International Conference on Recent Advances in Space Technologies-RAST2011, Istanbal, Turkey, 9–11 June 2011.
13. Bedmutha, N.D.; Biraris, P.N.; Shah, J.P. A low cost GNSS software receiver design with SEE mitigation approach for microsatellites. In Proceedings of the 2013 IEEE International Conference on Space Science and Communication (IconSpace), Melaka, Malaysia, 1–3 July 2013.
14. Grelier, T.; Ries, L.; Bataille, P.; Perrot, C.; Richard, G. A new operational low cost GNSS Software receiver for Microsatellites. In Proceedings of the 2012 6th ESA Workshop on Satellite Navigation Technologies (Navitec 2012) & European Workshop on GNSS Signals and Signal Processing, Noordwijk, The Netherlands, 5–7 December 2012.
15. Aguirre, F.H. X-Band electronics for the INSPIRE Cubesat deep space radio. In Proceedings of the 2015 IEEE Aerospace Conference, Big Sky, MT, USA, 7–14 March 2015.
16. Kobayashi, M. Iris Deep-Space Transponder for SLS EM-1 CubeSat Missions. In Proceedings of the Small Satellite Conference, Logan, UT, USA, 23 June 2017.
17. Lovascio, A.; Centonze, V.; D'Orazio, A. Design of COTS-Based Radio-Frequency Receiver for Cubesat Applications. In Proceedings of the 2019 IEEE International Workshop on Metrology for Aerospace (MetroAeroSpace), Torino, Italy, 19–21 June 2019.
18. Addaim, A.; Kherras, A.; Zantou, E.B. Design of a Telecommand and Telemetry System for use on Board a Nanosatellite. In Proceedings of the 2007 14th IEEE International Conference on Electronics, Circuits and Systems, Marrakech, Morocco, 11–14 December 2007.
19. Cinarelli, D.; Tortora, P. TT&C system for the ALMASat-EO microsatellite platform. In Proceedings of the 2012 IEEE First AESS European Conference on Satellite Telecommunications (ESTEL), Rome, Italy, 2–5 October 2012.
20. Davalle, D.; Cassettari, R.; Saponara, S.; Fanucci, L.; Cucchi, L.; Bigongiari, F.; Errico, W. Design, implementation and testing of a flexible fully-digital transponder for low-earth orbit satellite communications. *J. Circuits Syst. Comput.* **2014**, *23*, 1450148. [CrossRef]
21. Budroweit, J. Design of a highly integrated and reliable SDR platform for multiple RF applications on spacecrafts. In Proceedings of the GLOBECOM 2017—2017 IEEE Global Communications Conference, Singapore, 4–8 December 2017.
22. Ali, H.; Ali, A.; Mughal, M.R.; Reyneri, L.; Sansoe, C.; Praks, J. Modular Design of RF Front End for a Nanosatellite Communication Subsystem Tile Using Low-Cost Commercial Components. *Int. J. Aerosp. Eng.* **2019**, *2019*. [CrossRef]
23. Renaudie, C.; Markgraf, M.; Montenbruck, O.; Garcia, M. Radiation testing of commercial-off-the-shelf GPS technology for use on low earth orbit satellites. In Proceedings of the 2007 9th European Conference on Radiation and Its Effects on Components and Systems, Deauville, France, 10–14 September 2007.
24. MS-156C3 datasheet. Subminiature Coaxial Switch 1.35 mm High, DC to 11 GHz. Available online: https://www.hirose.com/product/p/CL0358-0340-0-00?lang=en#/ (accessed on 30 January 2020).
25. Keysight. 8753ES S-Parameter Network Analyzer, 30 kHz–3 GHz. Available online: https://www.keysight.com/en/pd-1000002292%3Aepsg%3Apro-pn-8753ES/s-parameter-network-analyzer?cc=IT&lc=ita/ (accessed on 30 January 2020).
26. National Instruments. USRP N210 Software Defined Radio by Ettus Research. Available online: https://www.ettus.com/all-products/un210-kit/ (accessed on 30 January 2020).
27. Rohde & Schwarz. FSVA3007 Signal & Spectrum Analyzer, 10 Hz–7.5 GHz. Available online: https://www.rohde-schwarz.com/it/prodotto/fsva3000-pagina-iniziale-del-prodotto_63493-601504.html?rusprivacypolicy=0/ (accessed on 30 January 2020).
28. Analog Devices. HMC830 datasheet (Rev. 03.0512). HMC830LP6GE, Fractional-N PLL with Integrated VCO, 25–3000 MHz. Available online: https://www.analog.com/en/products/hmc830.html#product-overview/ (accessed on 30 January 2020).

29. O. Fi. Electronics Ltd. DF1731SB Dual-Output DC Power Supply, 30V-5A. Available online: https://www.elektrolinna.fi/kauppa/index.php?ryhma=000000000&infosivukoodi=DF1730SB-5A (accessed on 30 January 2020).
30. Keysight. E4443A PSA Spectrum Analyzer, 3 Hz–6.7 GHz. Available online: https://www.keysight.com/en/pdx-x201709-pn-E4443A/psa-spectrum-analyzer-3-hz-to-67-ghz?nid=-32514.1150235&cc=IT&lc=ita&pm=ov/ (accessed on 30 January 2020).

© 2020 by the authors. Licensee MDPI, Basel, Switzerland. This article is an open access article distributed under the terms and conditions of the Creative Commons Attribution (CC BY) license (http://creativecommons.org/licenses/by/4.0/).

Article

Improving Depth Resolution of Ultrasonic Phased Array Imaging to Inspect Aerospace Composite Structures †

Reza Mohammadkhani *, Luca Zanotti Fragonara, Janardhan Padiyar M., Ivan Petrunin, João Raposo, Antonios Tsourdos * and Iain Gray

School of Aerospace, Transport and Manufacturing (SATM), Cranfield University, Cranfield MK43 0AL, UK; l.zanottifragonara@cranfield.ac.uk (L.Z.F.); m.padiyar@cranfield.ac.uk (J.P.M.); i.petrunin@cranfield.ac.uk (I.P.); j.raposo@cranfield.ac.uk (J.R.); i.gray@cranfield.ac.uk (I.G.)

* Correspondence: r.mohammadkhani@cranfield.ac.uk (R.M.); a.tsourdos@cranfield.ac.uk (A.T.)

† This paper is an extended version of our conference paper Mohammadkhani, R.; Zanotti Fragonara, L.; Janardhan, P.M.; Petrunin, I.; Tsourdos, A.; Gray, I. Ultrasonic Phased Array Imaging Technology for the Inspection of Aerospace Composite Structures. In Proceedings of the 2019 IEEE 5th International Workshop on Metrology for AeroSpace (MetroAeroSpace), Torino, Italy, 19–21 June 2019.

Received: 18 December 2019; Accepted: 16 January 2020; Published: 20 January 2020

Abstract: In this paper, we present challenges and achievements in development and use of a compact ultrasonic Phased Array (PA) module with signal processing and imaging technology for autonomous non-destructive evaluation of composite aerospace structures. We analyse two different sets of ultrasonic scan data, acquired from 5 MHz and 10 MHz PA transducers. Although higher frequency transducers promise higher axial (depth) resolution in PA imaging, we face several signal processing challenges to detect defects in composite specimens at 10 MHz. One of the challenges is the presence of multiple echoes at the boundary of the composite layers called structural noise. Here, we propose a wavelet transform-based algorithm that is able to detect and characterize defects (depth, size, and shape in 3D plots). This algorithm uses a smart thresholding technique based on the extracted statistical mean and standard deviation of the structural noise. Finally, we use the proposed algorithm to detect and characterize defects in a standard calibration specimen and validate the results by comparing to the designed depth information.

Keywords: ultrasonic NDE; autonomous inspection; ultrasonic phased array; NDT; composite materials; depth resolution; defect sizing

---

1. Introduction

This work is part of the EU-H2020 FET-OPEN CompInnova project [1] that aims to develop an innovative solution for the automatic Non-Destructive Testing (NDT) inspection, sizing, localization and repair of damages on aircraft composite structures [2,3]. For the NDT inspection phase, the CompInnova solution employs two different and complementary technologies: (i) Infrared Thermography (IRT) to detect near-surface defects, and (ii) ultrasonic Phased Array (PA) for sub-surface defects. The combination of these two methods in the overlapping areas is also considered as the future work in CompInnova, in order to improve the accuracy of detection. These modules are mounted on a vortex robot for autonomous inspection of composite structures [4].

There is an increasing use of ultrasonic phased array [5–12] in recent years in comparison to conventional single transducer ultrasonic inspections for NDT, due to their flexibility, speed of operation and good imaging performance. Ultrasonic phased arrays are able to capture multiple A-scans (full waveform) and provide a B-scan (a cross-sectional view of the specimen to show depth and size information of defects) at each measurement [5,6]. The authors in [6] showed that ultrasonic wheel arrays can produce C-scans (a top-view of the material to illustrate location and size of defects) with a comparable quality to an immersion system, with a much shorter scanning time. A recent study has compared the ultrasonic inspection of composite materials using single element and phased array ultrasonic testing methods [13]. This study highlighted that, although both methods can be used for inspecting materials with a thickness of up to 25 mm, the phased array provides more stable signal parameters and a higher chance of detection for lower signal strength.

The use of autonomous systems for non-destructive testing utilising a wide range of sensing techniques has been explored by several researchers in the literature. For instance, robotic platforms have been designed and developed for the inspections of long welded lines [14]. A relatively recent literature review about climbing robots was carried out by Schmidt et al. [15]. The use of unmanned aerial systems for the visual inspection of aircraft wing panels was described by Malandrakis et al. [16]. The application of ultrasonic phased array for automated inspections has been recently explored in [17,18], leveraging different technologies such as robotic manipulators and robots for in-pipe inspections [19,20]. The solution proposed in the CompInnova project aims to step change NDT inspections during scheduled maintenance C-check and D-check (heavy maintenance checks), allowing faster repeatable inspections and accurate localization, sizing and classification of defects. Having these maintenance checks automated, we can achieve a significant reduction in both costs and inspection time [3]. A vortex-based robotic platform [21] is the solution chosen for the PA and IRT automation of scanning process.

However, when the inspected material is not homogeneous, conventional ultrasonic imaging techniques are not effective, and we need to employ signal processing techniques to improve the temporal resolution of the ultrasonic signal [22]. Some studies suggest to use a combination of Wiener filtering and autoregressive (AR) spectral extrapolation to improve signal-to-noise ratio (SNR) and temporal (axial) resolution of the ultrasonic inspection [23–25]. They use Wiener filter for deconvolution, then, using a part of the deconvolved spectrum with high SNR, an AR model of the process is built. Next, the rest of the spectrum is extrapolated by the obtained AR model. However, performance of this method strongly depends on the width of the frequency window and order of the AR filter [23,25].

Some other works assume that back-scattered ultrasonic echoes can be modeled as superimposition of multiple Gaussian echoes and try to estimate multiple unknown parameters of such model from the received signals [26,27]. This solution needs a multi-dimensional signal optimization over multiple parameters, which requires extra processing power and may not converge to the optimal solution [28].

Wavelet analysis is another method that has been widely used in many signal processing applications for signal estimation, classification, compression and de-noising. [29]. Due to its multi-resolution characteristics for signal decomposition, it has attracted significant attention from many researchers in the area of ultrasonic NDT [30–36].

1.1. Challenges of Automated Post-Processing of PA Data

In general, phased arrays are able to electronically control their beam to scan, steer or sweep their beams. One of the main advantages of the PA inspection with full waveform data capture and storage is the ability to extract the depth information using sliding gate (window) analysis. Depth information is crucial in the assessment of the defect, the state of the component and to decide whether repair, replacement or no-action is needed. In addition, for the CompInnova concept, the depth information is very valuable because it allows an accurate calculation of the volume and area of material that the laser will remove during repair process.

Dead zone and limited axial (depth) resolution are the main drawbacks related to the use of lower frequency PA ultrasonic transducers for defect detection and characterization of thin composite structures [37]. The physical background of this limitation is related to the length of the wave packet transmitted into the material by the PA transducer. When the distance between the echoes in the analysed A-scan is comparable or less than the width of the wave packet, the separation of these echoes becomes a challenging task for the operator or automation of defect detection. The problem can be illustrated by the following examples. Let us consider a pristine case, characterized by the A-scan shown in Figure 1a, where the front surface and back wall echoes are clearly separated. In the presence of a defect, with a large distance between the front surface and the back wall echoes (comparable or greater than the width of the wave packet), we can make a clear distinction between all echoes, i.e., localize each echo correctly. However, when echoes are partial overlapping, for example as it is shown in Figure 1b, there is overlap between the defect echo and the back wall echo and proper defect localization is either not possible or it will have a significant error when performed by conventional gating methods. Because of this phenomenon, the depth of defects that are close to the front or back walls is difficult to estimate. In addition, even when there is no echo overlap, due to the limited time resolution of A-Scan signals, the defects in C-Scan obtained from unprocessed A-Scans may appear to be much larger than the actual defects. In order to solve this problem, time resolution of A-Scan signals needs to be improved.

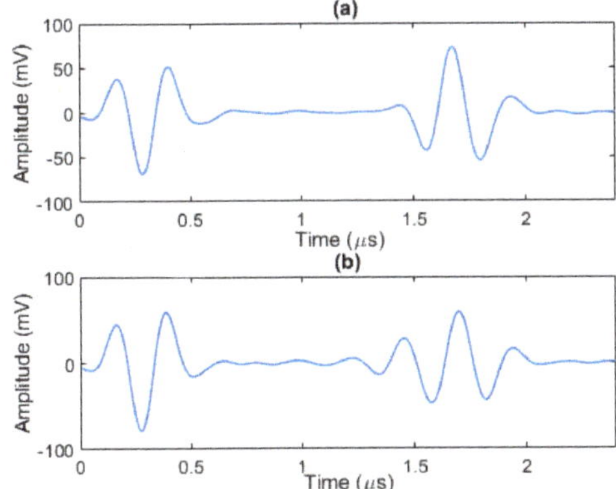

Figure 1. Two different A-scans using 5 MHz PA transducer: (**a**) at a non-defective location, and (**b**) at a location with defect close to back wall.

1.2. Necessity of High Frequency 10 MHz PA Transducer

There are fundamentally two resolutions in two different scanning axes which determine the imaging performance of a PA C-scan image. One is the lateral resolution, and another is the axial resolution. Furthermore, for thin composite structures, near surface resolution is also considered as an important factor in PA transducer selection for estimating the defect depth accurately. Near surface resolution depends on the extent of the dead zone of the transducer which in turn depends on the pulse-width. The lateral resolution of the transmitted ultrasound beam depends upon the centre frequency, bandwidth, and active aperture, near field distance from the transducer, material thickness and material attenuation properties. Having a larger aperture and higher frequencies results in a smaller beam width, which in turn causes a higher lateral resolution and sharper image [38,39]. The axial resolution is directly affected by the excitation frequency of PA transducers and the pulse width. For example, given the velocity of longitudinal ultrasonic wave of 3000 m/s in a specimen made of carbon fiber reinforced polymer (CFRP) and using 5, 10, and 15 MHz excitation frequencies, the wavelength of longitudinal waves are 0.6, 0.3 and 0.2 mm, respectively. For a layer thickness of 0.18 mm, we have an axial (depth) resolution for the above frequencies in C-scan approximately equal to the thickness of 3, 2, and 1 layers, respectively. We have observed that, with 5 MHz frequency, we were able to detect and accurately quantify the lateral extent of embedded inserts and impact damages in thin composites [37]. However, we were not able to resolve individual layers or plies at near surface or far surface for such a frequency, without signal processing (because the excitation pulse contains 2 to 3 cycles). Hence, a PA transducer frequency of at least 10 MHz is needed for defect depth estimation in thin composites.

However, increasing the frequency of PA transducer results in the increase of the ultrasonic attenuation and scattering noise from the material. The scattering noise is due to wavelength being much closer to ply thickness, which causes internal reflections at each resin-ply interface and results in a train of continuous noise-like signatures between the front wall and back wall of the specimen [40]. This noise is known as *structural noise* and has the same spectral characteristic as the defect echoes [41,42], and adds coherently as the ultrasound propagates in the material. Therefore, it is necessary to improve the performance of the echo localization algorithm [37] in the presence of noise by dedicated signal processing.

Another concern is about the interpretation of the PA data. Although manual interpretation is the de facto standard for composite inspections, it is time-consuming, expensive, and represents the major bottleneck of a PA inspection process. Advances in PA, full waveform data acquisition and ease of application of signal processing during post-processing, have led to a new area of research in semi-automated defect detection. Using this approach vast amount of data collected can be processed efficiently using dedicated signal processing algorithms with only the most important information presented to the NDT inspector. Additionally, 3D visualisation of defect in composite structures can lead to better defect characterisation. One further challenge of post-processing PA ultrasonic data are that the acquired phased array data of a larger area is presented by multiple strips of data. In the case of manual phased array inspection, an experienced technician spends significant amount of time for merging, aligning, stitching and adjustment of the individual strips before displaying a C-scan image of the large structure. In manual inspection, re-scanning is sometimes required, due to lack of couplant, variation of load applied on the probe or not capturing data along a straight line. However, scanning the structure by a manipulator or robot is less prone to these problems since required pressure on the wedge and uniform distribution of couplant is ensured. Thus, post processing of PA data can help to automate or eliminate some of the manual post processing needs.

In this paper, we propose an algorithm based on wavelet transform for detection and localization of defect echoes during the inspection of composite components using the high frequency ultrasonic phased array. The proposed solution offers improved axial resolution while maintaining sensible performance. We analyse two sets of scan data using 5 MHz and 10 MHz PA transducers, and address the benefit of using higher frequency ultrasonic PA transducers. Subsequently, the signal processing challenges of this choice is presented. Then, we propose solutions for the challenges and smooth the way toward semi-automated PA inspection of composite structures at high frequency 10 MHz. Finally, the paper is concluded in Section 4. The remainder of the paper is organized as follows. Section 2 begins with a short review of our echo localization algorithm with 5 MHz PA transducer presented in [37], followed by the necessity and signal processing challenges for the use of high frequency 10 MHz PA transducers. Then, a modified version of our algorithm is proposed for 10 MHz. Next, Section 3 illustrates how the algorithm is capable of visualizing echo location and post-processing results in three-dimensional plots, which can be very helpful and easy to understand by the inspector.

2. Echo Detection and Localization

2.1. Composite Material and Phased Array Configuration

The material used for the experiment consists of a carbon reinforced epoxy resin laminate of size 200×200 mm which is manufactured through a hand layup process using cross-ply $(0/90)_{12}$ as stacking sequence with uni-directional pre-preg IMS-977-2-34-24KIMS-196 material (with the following material properties: $E_1 = 125$ GPa, $E_2 = 8.68$ GPa, $G_{12} = 4.7$ GPa) from CYCOM and cured in an autoclave. The phased array module for data acquisition consisted of Sonatest (16:64) VEO+ series. A Sonatest PA wheel probe 5 MHz with 64 elements and 0.8 mm pitch along with a longitudinal wave 10 MHz 64 element linear phased array transducer with a pitch of 0.6 mm from Sonatest X3 series with a custom developed flat elastomer wedge, are used to acquire data at two different frequencies. Electronic scanning is performed with a group of eight active elements as active aperture, and full waveform data are acquired at 125 MHz sampling frequency during NDT inspection. The total number of A-scans that can be acquired with one line-scan of a PA transducer is equal to M = (number of elements – active sub-aperture +1). Therefore, for a 64-element transducer and choosing 8 elements as an active sub-aperture, the total number of A-scans will be $M = 64 - 8 + 1 = 57$.

2.2. Baseline Echo Localization Algorithm

The baseline echo localization algorithm extracts the peak information for all echoes, following the stages according to the flowchart shown in Figure 2 was initially presented in [37]. It builds a *reference echo model* based on back wall echoes from a scanned area with no defects (later in the text, we may call it *reference model*). Then, with the help of wavelet transform, it searches for all echoes having an absolute amplitude above a certain threshold. Next, it finds the actual phase information of the echo and calculates exact location of the defect by analysing the phase information from the defect echo and the reference echo model. For the threshold, we use the following definition:

$$\text{Threshold} = \frac{\alpha}{N} \sum_{n=1}^{N} |x(n)|, \tag{1}$$

where α is a scaling parameter to control the threshold and N is the total number of samples of the original signal $x(n)$.

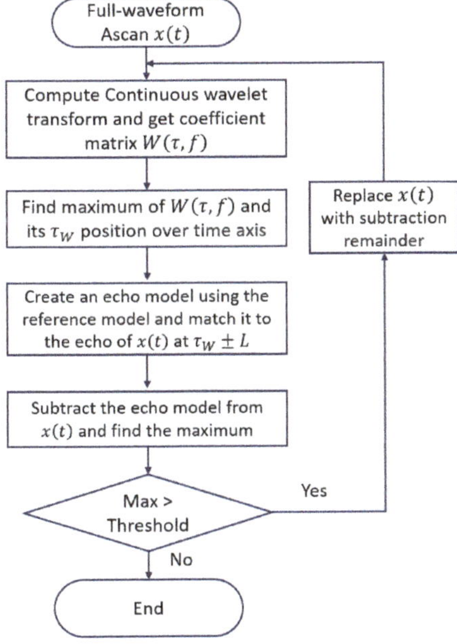

Figure 2. Flowchart of our baseline echo localization algorithm.

Having the echo reference model, the algorithm operates as follows:

- Compute the complex continuous wavelet transform of the signal $x(t)$ to obtain the wavelet coefficients - scalogram $W(\tau, f)$;
- The maximum value in the scalogram $W(\tau, f)$ and its τ_W position is located over time axis;

- An echo estimate is created by scaling the reference echo model and shifting to $\tau_W \pm L$ (for a small L) to find the best possible approximate of the echo of $x(t)$. In other words, we have a window with a width of $2L+1$.
- The echo estimate is removed from the original signal and the maximum amplitude (absolute) value within the residue (remainder signal) is found and compared to the threshold. If this is larger than the threshold, the resulting signal will become the input of the next iteration until the maximum value of the residue is smaller than the threshold.

The above algorithm offers the following benefits:

- *All echoes with an amplitude above the defined threshold are detected and localized considering their phase information.* In comparison, the conventional C-scan image generation method used only one maximum peak using an absolute value of the rectified signal in the defined gate. In the best case, this gate covers the whole distance between front wall and back wall echoes. When multiple echoes are present in this gate, manual processing extracts only the strongest echo, other echoes are missed. Furthermore, phase information of the strongest echo is missed.
- *Overlapped echoes are also localized by the algorithm.* Figure 3 shows an example of using our algorithm for data obtained with 5 MHz PA transducer and its capability of localizing echoes, even partially overlapped ones. Two cases of (i) having well separated echoes and (ii) echoes with overlap are shown in Figure 3a,c. Comparing the results in Figure 3b,d, we can find the relative depths of inserts by measuring the time of flight for the front wall (τ_f), defect (τ_d) and back wall (τ_b) echoes, and knowing the velocity of sound in the material. For example, assuming the ultrasonic wave velocity is 3000 m/s, the thickness of the test specimen in the areas with no inserts from Figure 3b, can be calculated as

$$x = v(\tau_b - \tau_f)/2 = 2.1 \text{ mm}. \tag{2}$$

The defect depth can be obtained by using τ_d instead of τ_b and it equals to 1.8 mm for the sample with insert in Figure 3d. Having the insert thickness less than 0.1 mm, and each layer thickness of 0.183 mm, and knowing that each insert is placed between layer 10 and 11 (specimen consists of 12 layers), the above measures are reasonable.

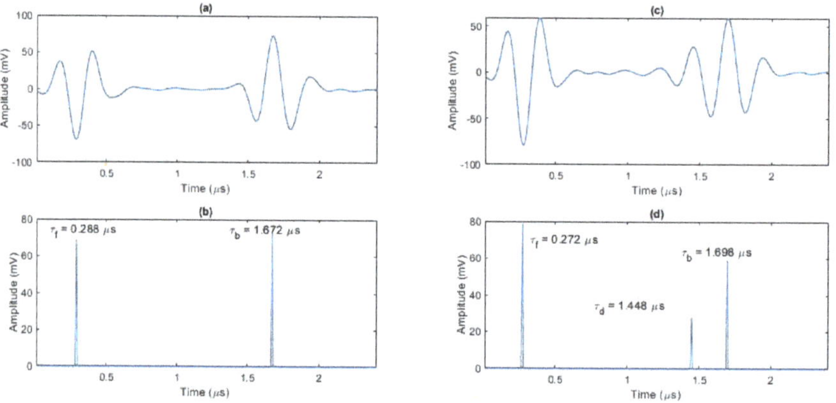

Figure 3. Two different A-scans using 5 MHz PA transducer: (**a**) one at non-defective location, and (**c**) an A-scan at a location with defect close to back wall; localisation results of the baseline localisation algorithm for (**a**,**c**) are shown in (**b**,**d**) respectively, providing time-of-flight for the front wall (τ_f), defect (τ_d) and back wall (τ_b) echoes.

In order to provide a clearer view for the inspector on the defect profile over the material depth, visualization of the localization results has been implemented. Section 3 describes this part in more detail.

2.3. Processing Data from the 10 MHz PA Transducer

The use of 10 MHz ultrasonic transducers promises higher resolution of defect (depth) detection, which will be useful in thin skin-stiffer panel and step-repaired panel to separate the echoes and characterize damages.

It was found, however, that post-processing of inspection data using 10 MHz transducer is more challenging in comparison with inspection data at 5 MHz. This is mostly due to structural noise, affected also by higher sensitivity to distribution of couplant (sprayed water) and surface roughness. A previously proposed algorithm was not performing well for this data from transducer with 10 MHz and required some modifications to address the above issues.

Figure 4 illustrates some of the challenges for automated defect characterization with ultrasonic scanning at 10 MHz. Figure 4a shows 3D plot of a frame of scan data, and Figure 4b presents three different A-scans from the above set. We note that one frame corresponds to a single linear scan of the PA probe over the specimen. Each frame consists of 57 A-scans in our PA device set up, and B-scan can be provided as a two-dimensional image of each frame information. We use longitudinal full waveform (A-scans) data captured by a portable linear PA transducer having 64 elements with 0.8 mm pitch. In comparison to 5 MHz scan data (an example shown in Figure 3), we observe the following:

- The first challenge is the existence of multiple echoes between front wall and back wall, i.e., structural noise, corresponding to the boundaries of layers. Due to this noise, detection of defect echoes is much harder compared to the 5 MHz case, and we can only detect echoes stronger than the structural noise (i.e., having higher amplitudes).
- Another issue is the higher attenuation of the propagating wave at a higher frequency. As a result, defect echoes closer to the back wall have lower amplitudes. One practical solution found in the literature is to apply a time gain compensation that increases signal amplification with depth.
- With the 10 MHz PA transducer, the amount of couplant (water) should be as small as possible and it should be evenly distributed. Furthermore, a constant pressure should be applied to the wedge. Excess of couplant causes reverberation of the signals and increases the front wall echo length and hence the dead zone.

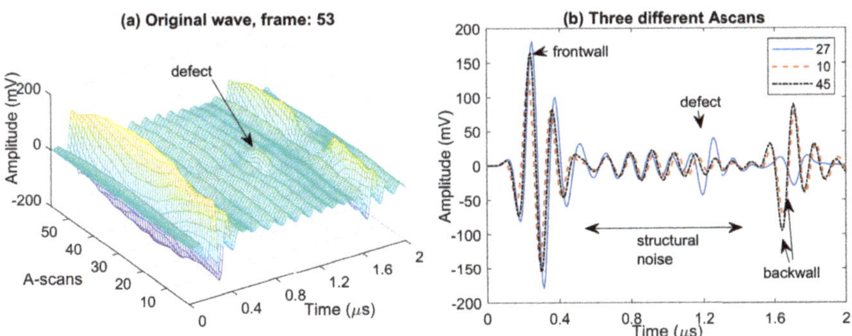

Figure 4. (**a**) 3D plot of full waveform scan data using 10 MHz PA transducers for a frame (one line-scan) having defect echoes, and (**b**) three selected A-scans from this frame.

Due to increased sensitivity to a surface roughness at 10 MHz, we can see from Figure 5a that misalignment of A-scans is amplified in comparison to 5 MHz results. Therefore, the error in depth estimation is amplified too.

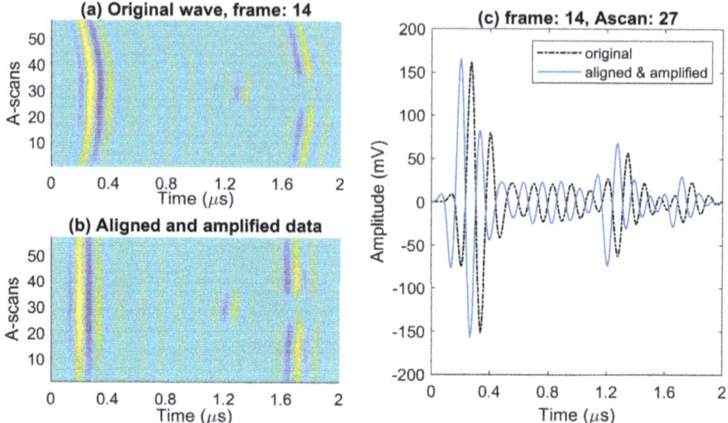

Figure 5. Post processing of a frame (with defect echoes) at 10 MHz: (**a**) before, and (**b**) after applying initial processing techniques (alignment, and linear slope gain), and (**c**) A-scan from defective region before and after alignment and amplification.

To address this issue in our signal processing, prior to the echo localization algorithm, all A-scans are aligned in time domain and amplified using time gain compensation (0.85 to 0.95 dB per mm). This is to compensate the higher attenuation we observe in 10 MHz ultrasonic wave. Figure 5b,c illustrate the alignment and amplification, respectively, for all A-scans of a frame that include defect echoes. As a result, the extracted depth information for defect echoes is more consistent.

2.4. Modified Echo Localization Algorithm

In this section, we present a modified echo localization algorithm that is better suited to work with 10 MHz ultrasonic data. It is worth highlighting that, prior to our echo localization algorithm, we use data pre-processing steps (namely the alignment and time-gain-compensation of the original data). Successively, the output data are fed to the localization algorithm for further processing. The following changes have been implemented in the modified algorithm:

- A threshold to ignore structural noise,
- Search Window width,
- Better echo-fit search methodology.

The algorithm searches for best echo fit is changed as follows. First, the position of the strongest echo is found from the wavelet, similar to the baseline algorithm. Then, we define a window over the max position with a width that corresponds to envelope width where its values are above the threshold. Next, we use cross-correlation between the reference echo model and the windowed signal followed by search for a combination of strongest echoes. These echoes are removed from the signal resulting in the lowest residual signal in this window. After that, the algorithm analyses the rest of A-scan signal (called residue) that has a maximum of its envelope above the threshold.

Flowchart of the modified algorithm is presented in Figure 6. We describe these changes and modification with more detail in what follows:

1. Calculate wavelet transform of the signal,
2. Find the location of strongest peak from the wavelet scalogram,
3. Define a window at the location of the peak showing the area where envelope is above the threshold,
4. Calculate cross-correlation of the windowed signal $x_W(t)$ and the reference signal $s(t)$, find the two strongest peaks with correlation coefficients a_1, a_2 and their lag information τ_1 and τ_2 for the next step. It is worth noting that in general we may need superimposition of more than

two echoes to model the detected defect echo. However, for our measured data, we found that two echoes corresponding to the first two peaks of the cross-correlation are sufficient for echo representation.

5. Define two echo models for the signal from the cross-correlation peaks information by applying coefficients and time-lags to the reference model, i.e., $\hat{s}_i(t) = a_i s(t - \tau_i)$ for $i = 1, 2$,
6. Subtract $\hat{s}_1(t)$ and $\hat{s}_2(t)$ from the windowed signal and select the echo model that gives us lower residue value,
7. The procedure in steps 1–6 is iteratively applied for all remaining peaks above the threshold.

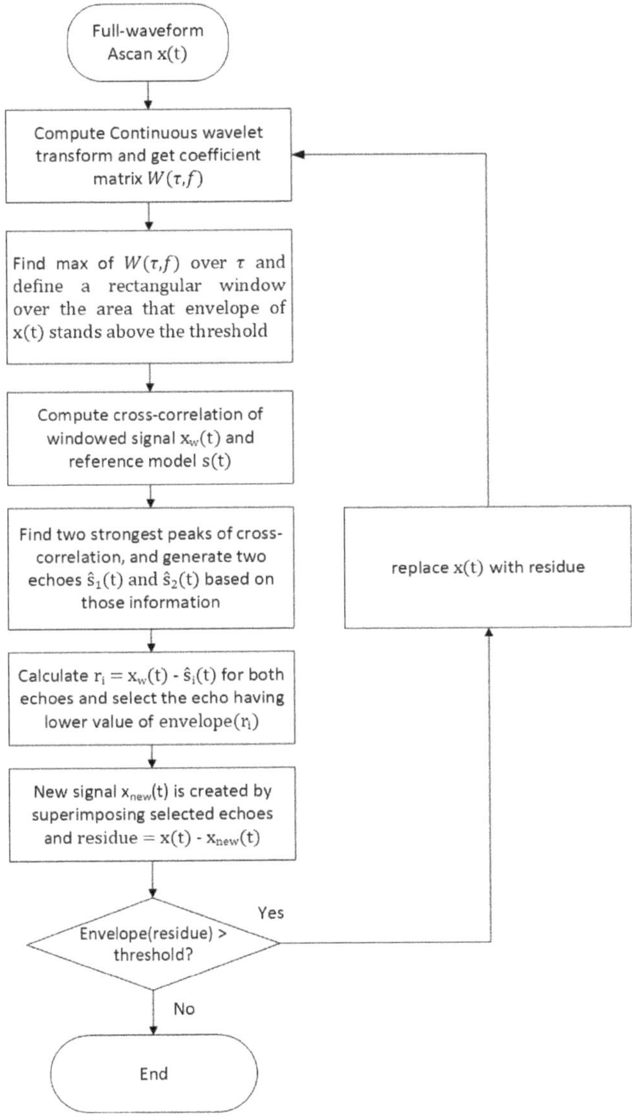

Figure 6. Flowchart of the modified echo localization algorithm. We note that the reference echo model and the threshold are provided as input for this algorithm.

We note that the reference echo model is a normalized version of the back wall echo that is obtained by averaging over multiple back wall echoes taken from a scan area with no defect. Figures 7–9 illustrate the above steps of the algorithm.

This procedure continues iteratively until the peak echo amplitude values after subtraction in the analysis window fall below the threshold. This is presented in Figure 8 and we see that, between the two models, echo model 1 will be selected because it produces the lowest residual signal. As depicted in Figure 9, after completing the echo localization in the first window, we have a new signal showing detected echoes of the signal (solid black line), and one signal consists of extracted echoes' peaks information (peak value as a positive or negative real number, and its location is time) plotted in solid red line. As seen in Figure 9b, the residual signal is updated and if it has a peak above the threshold, the algorithm starts again to localize remaining echoes.

An important factor here is the right value of the threshold. If it is too low, we will see wrong peaks detected due to having structural noise above the threshold. On the other side, if we select a too high threshold, we may miss some defect echoes.

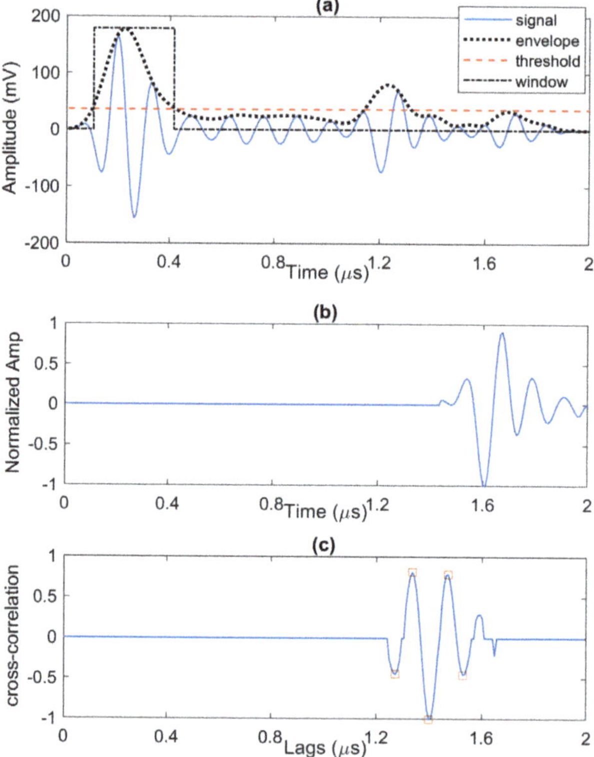

Figure 7. (**a**) Full wave A-scan with its envelope, threshold, and a window showing the area where envelope is above the threshold, (**b**) reference echo model, and (**c**) cross-correlation between the windowed signal and the reference model. A few largest peaks (based on absolute value of cross-correlation) are shown with red markers.

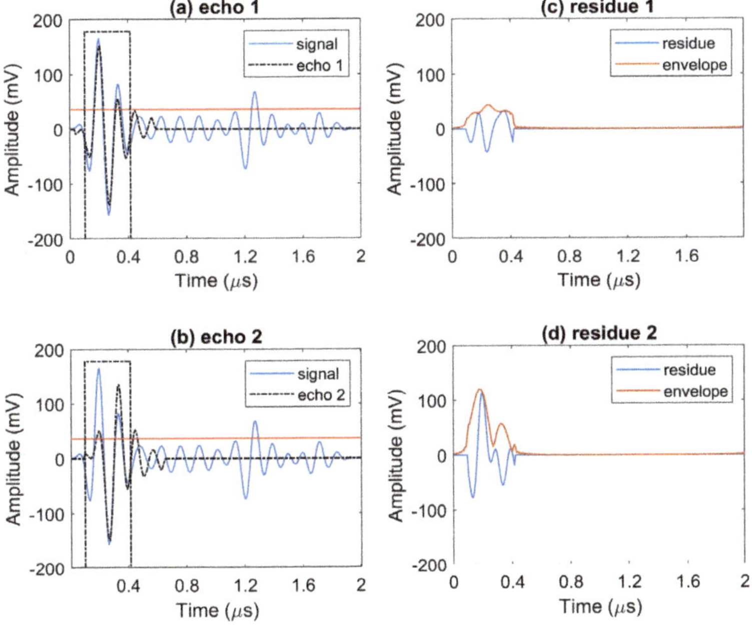

Figure 8. Full wave A-scan with two possible choices of echo models that differ by location, amplitude and phase in (**a**,**b**), and corresponding residue signals after subtracting echo models from the signal in (**c**,**d**).

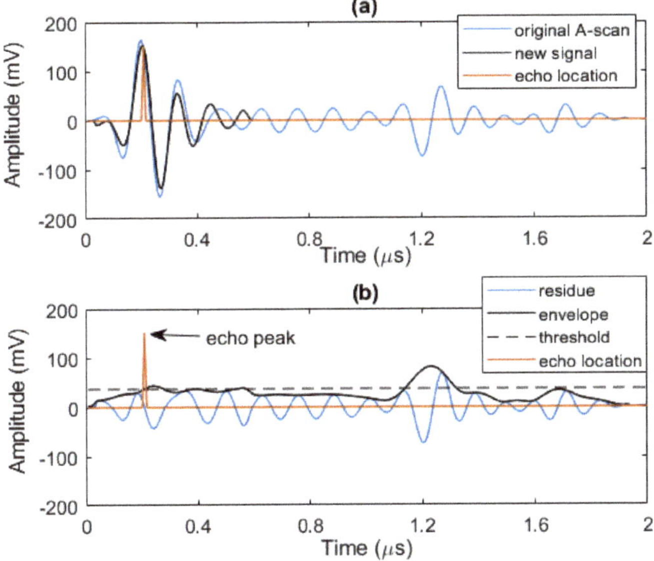

Figure 9. (**a**) Original A-scan and its approximate (new signal) after the 1st step of performing echo localization, (**b**) residue of the 1st step and its envelope that feeds back to the algorithm for next step, and the peak echo information.

2.5. Smart Thresholding

In order to optimise the detection process, we are looking for a solution that gives us a threshold that is adaptive with respect to the structural noise level to avoid being wrongly detected as defect echoes. We note that the structural noise should have the same shape as front wall and back wall echoes except having lower amplitudes.

In this study, we select threshold following a data driven strategy. To realise this approach, we analyze the distribution of the structural noise of the A-scans in the part of material with minimal influence from the defect-related echoes. This allows for effective threshold selection due to various conditions imposed by different amplitudes of the wave packet propagating in the media and variations in the structure of the inspected material.

In order to analyze the distribution of peak absolute value amplitudes of the structural noise, each A-scan is divided into three zones (see Figure 10): (i) Zone 1 or the dead zone that includes front wall echo, (ii) Zone 2 that only include back wall echo, and (iii) Zone 3 or the area between Zone 1 and Zone 2, and it contains structural noise and any possible defects. It is worth noting that these zones can be defined by having the prior knowledge of the dead zone and the thickness of the inspected composite material. Whenever there is any defect in Zone 3, the amplitude of front wall and/or back wall in Zones 1 and 2 will be affected. Analysing the amplitudes in Zones 1 and 2, we have been able to remove A-scans that include any defect in Zone 3 and obtain the distribution of structural noise with minimal influence from the defect echoes. Figure 11 shows an example of applying such procedure for a calibration sample (we analyse the results in more details in the next section). Comparing those two histograms, we realize that the structural noise has a distribution on the left-hand side of the histogram (with a peak approximately at 20 mV), with longer right tail as some of the few defect echoes are still remaining in the search domain.

Observing the shape of the distribution in Figure 11a,b we can assume that the underlying distribution is close to Gaussian. Therefore, to remove the anomalous right-tail of the structural noise distribution, we have decomposed the empirical distribution of the noise into the mixture of two Gaussian distributions. The decomposed distribution with lower mean is believed to be a close representation of the true structural noise distribution in an area with no defects. The mean value (μ) and standard deviation (σ) of this distribution are used to define the smart threshold. For example, we use ($\mu + 4\sigma$) as the smart threshold in the modified echo localization algorithm.

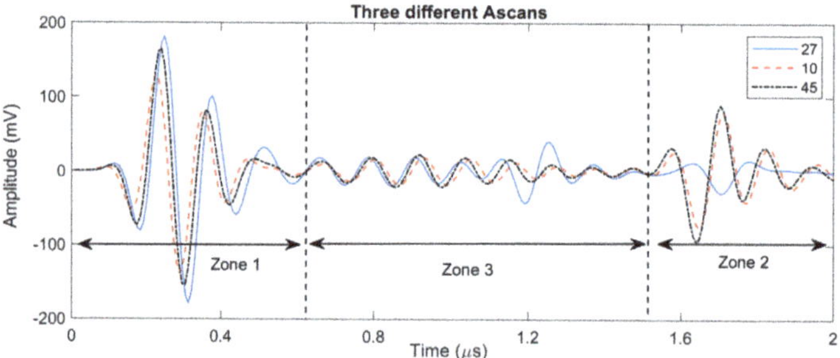

Figure 10. Zones 1, 2 and 3 for three different A-scans with and without defects.

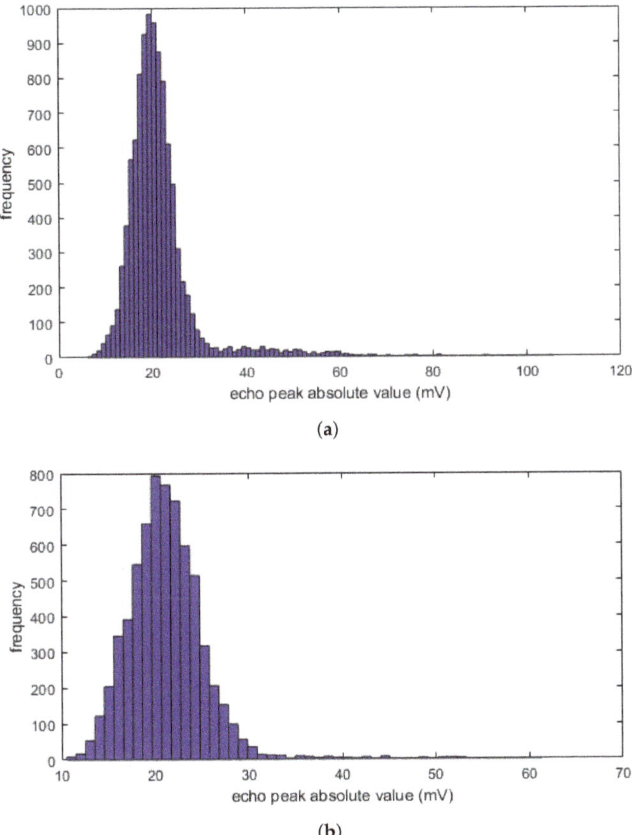

Figure 11. Histogram of the absolute peak values in Zone 3 for (**a**) all A-scans, and (**b**) A-scans with no defects.

3. Three-Dimensional (3D) Visualization

Our algorithm extracts depth information of defects (based on time of flight information) for all A-scans and visualize the results in 3D plot. It is also capable of slicing the depth information into several zones and customizing each slice size which can be set, conveniently, equal to thickness of the composite layer. Detection is performed in each slice separately, for example, based on amplitude information. The depth of the zone is calculated according to time of flight information and a 3D plot containing detected defects from each zone (slice) is created. Using this technique, it is possible to quantify the defects layer-by-layer.

Figure 12 shows the result of using our signal processing, on the 10 MHz scan data in the calibration sample with 12 layers and different types of inserts of size 6 × 6 mm located between 8th and 9th layers of the specimen, i.e., at a depth of eight layers [37]. Figure 12a shows a schematic diagram of the inserts between 8th and 9th layers of the calibration sample with the following material types from left to right: Teflon, Paper, Release tape, Bag tape, and Peel ply. Figure 12b shows a volumetric plot of all echoes (with an amplitude above the threshold) versus thickness. It includes echoes from front wall, back wall and defects. The colour map shows the amplitude of the echo at each point considering the actual amplitude information from the full waveform A-scan signal (i.e., it can be either positive or negative). Front wall peak position is taken as the reference for depth measurement.

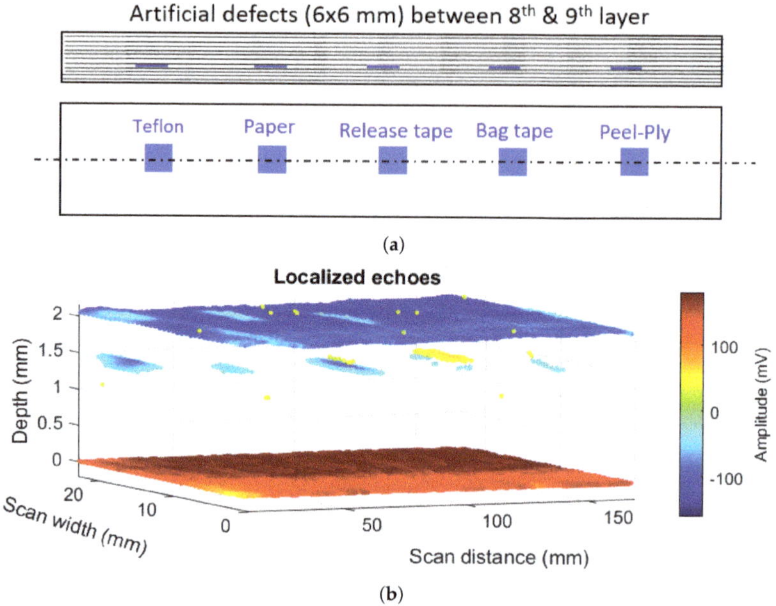

Figure 12. (**a**) Schematic of inserts (artificial defects) in the calibration sample and (**b**) 3D visualization of localized echoes using our modified algorithm and the smart thresholding.

Figure 13a,b plots the size and depth information of defects. Size and depth estimates of these artificial defects (6 × 6 mm inserts) are shown in Table 1 in more details. Sizing is performed by a part of our code and it calculates the surface of detected defects in Figure 13b in mm^2. We see that estimated depth is quite accurate. Regarding the sizing at the first look, we see that some inserts are oversized. One can notice that different materials have different reflectivity. For example, Teflon and Release tape inserts have stronger reflections compared to other types. This effect is reflected in Table 1, where size estimation for these two types are larger compared with others. Furthermore, Figure 13a shows that outer borders of detected defects have lower amplitude compared to the inside areas.

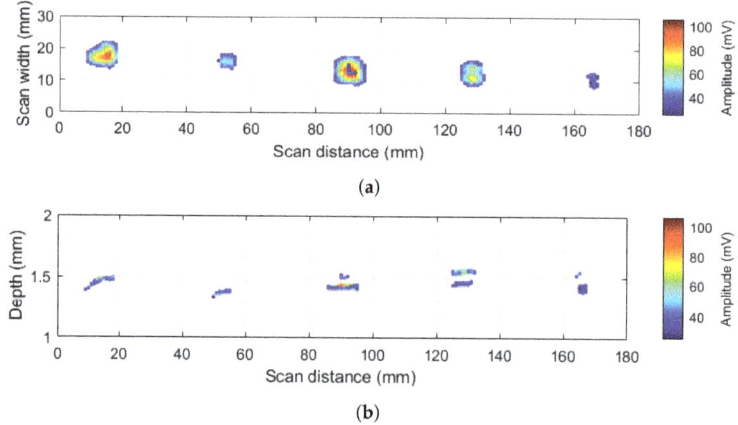

Figure 13. (**a**) sizing, and (**b**) depth information of 6x6 mm inserts in the calibration sample.

We believe that the observed error in detected defect sizes is due to different reflection properties of different insert materials. We address this issue by removing a lower 40th percentile of the cumulative distribution function (CDF) of absolute peak values of detected defect echoes shown in Figure 14a. Results of such procedure are shown in Figure 14b. For this case, we achieve following defect sizes (in mm^2): Teflon (reduced to 36.6), Paper (no change, 21), Release tape (reduced to 44.4), Bag tape (reduced to 30.6), and Peel-ply (no change, 12).

Table 1. Size and depth estimation of different types of defects with the size (6 × 6 mm) at the design depth 1.47 mm in the reference standard specimen.

	Teflon	Paper	Release Tape	Bag Tape	Peel Ply
Automated defect depth (deepest) estimate (mm)	1.50	1.39	1.52	1.57	1.54
Automated defect depth estimate error (%)	2%	5.4%	3.4%	6.8%	4.8%
C-scan Manual defect size estimate (mm^2)	30	9	38.4	faintly detected	8.4
Automated defect size estimate (mm^2)	53.4	21	59.4	43.2	12
Automated defect size estimate error (%)	43.8%	−41.7%	65%	20%	−66.7%
Manual defect size estimate error (%)	−17%	−75%	7%	not applicable	−77%

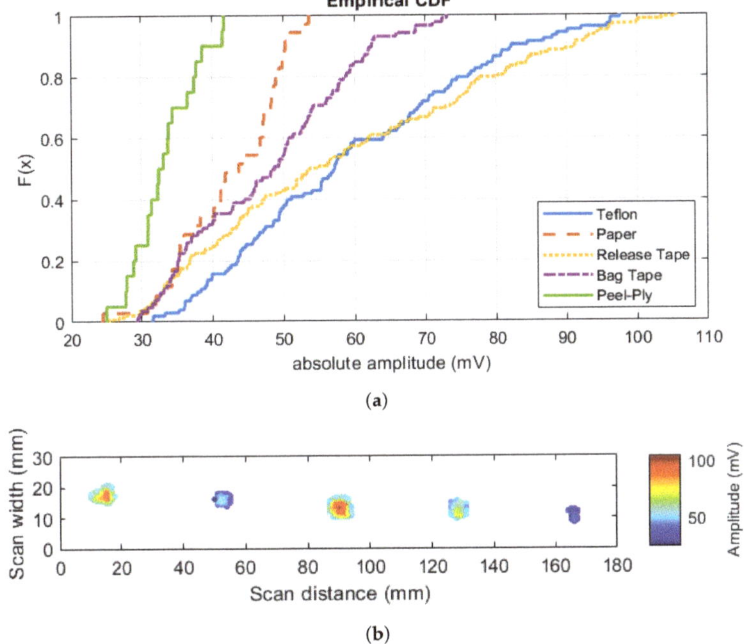

Figure 14. (**a**) Empirical CDF of absolute peak of defect echoes, and (**b**) refined size estimation after resizing defects with a wide range of amplitude spread.

4. Conclusions

In this paper, we analyzed signal processing challenges of NDT of aerospace grade composite using an ultrasonic PA module in an autonomous inspection platform designed for the EU-H2020 FET-OPEN CompInnova project. The post processing of the PA data with the proposed algorithm leads to resolving overlapped echoes and improving defect detection and axial (depth) resolution. We demonstrated an approach that allows achieving good results at low hardware and computational costs. We addressed the signal processing challenges for the use of high frequency 10 MHz PA transducers, and proposed solutions to overcome these problems. The proposed hardware solution is relatively inexpensive due to a conservative selection of the frequency (10 MHz), and it is going to be deployed using the robotic platform developed within the CompInnova project. This solution allows for eliminating some of the adverse effects of manual scanning related to non-uniform pressure distribution and couplant application, and increased repeatability of the scanning process.

The proposed modification of the resolution enhancement algorithm allows for robust detection of the defects close to either front or back walls of the inspected structure and resolution improvement up to the ply thickness of composite material while remaining simple in implementation. The algorithm also benefits from the smart thresholding technique that allows for keeping good performance in terms of peaks due to structural noise overlapping with defect echoes, while keeping defect detection sensitivity at or above the level of conventional immersion-based ultrasonic scanning. Another advantage of the proposed approach is intuitive 3D visualization of the inspected material volume that is highly suitable for semi-automated data analysis and interpretation, where routine post-processing tasks are automated and an inspector is only evaluating results presented in easy to perceive 3D form.

Author Contributions: Conceptualization, R.M., L.Z.F., I.P., and J.P.M.; methodology, R.M., I.P., L.Z.F., and J.P.M.; software, R.M.; validation, L.Z.F., I.P., and J.P.M.; formal analysis, R.M.; investigation, R.M.; resources, A.T., I.G.; data curation, R.M. and J.P.M.; writing—original draft preparation, R.M.; writing—review and editing, R.M., L.Z.F., I.P., and J.P.M.; visualization, R.M.; supervision, L.Z.F. and I.P.; project administration, J.R., I.G., and A.T.; funding acquisition, I.G. and A.T. All authors have read and agreed to the published version of the manuscript.

Funding: The authors wish to acknowledge the funding contribution from the European Union's H2020 Framework Program, call FET-OPEN under REA Grant No. 665238 CompInnova.

Acknowledgments: We would like to thank Spyridon Psarras and George Sotiriadis from the research team of Vassilis Kostopoulos, Laboratory of Applied Mechanics and Vibrations, University of Patras, Greece for manufacturing composite laminate for the NDT experiments.

Conflicts of Interest: The authors declare no conflict of interest.

Abbreviations

The following abbreviations are used in this manuscript:

AR	Auto-Regressive
CDF	Cumulative Distribution Function
CFRP	Carbon Fiber Reinforced Polymer
IRT	Infrared Thermography
NDT	Non-Destructive Testing
PA	Phased Array
SNR	Signal-to-Noise-Ratio
ToF	Time of Flight

References

1. CompInnova: An Advanced Methodology for the Inspection and Quantification of Damage on Aerospace Composites and Metals using an Innovative Approach. Available online: http://compinnova.net/ (accessed on 17 December 2019).
2. Gray, I.; Padiyar, M.; Petrunin, I.; Raposo, J.; Zanotti Fragonara, L.; Kostopoulos, V.; Loutas, T.; Psarras, S.; Sotiriadis, G.; Tzitzilonis, V.; et al. A Novel Approach for the Autonomous Inspection and Repair of Aircraft Composite Structures. In Proceedings of the 18th European Conference on Composite Materials, Athens, Greece, 24–28 June 2018.
3. Kostopoulos, V.; Psarras, S.; Loutas, T.; Sotiriadis, G.; Gray, I.; Padiyar, M.; Petrunin, I.; Raposo, J.; Zanotti Fragonara, L.; Tzitzilonis, V.; et al. Autonomous Inspection and Repair of Aircraft Composite Structures. *IFAC-PapersOnLine* **2018**, *51*, 554–557. [CrossRef]
4. Brusell, A.; Andrikopoulos, G.; Nikolakopoulos, G. Vortex Robot Platform for Autonomous Inspection: Modeling and Simulation. In Proceedings of the IECON 2019-45th Annual Conference of the IEEE Industrial Electronics Society, Lisbon, Portugal, 14–17 October 2019; pp. 756–762.
5. Smith, R.A.; Bending, J.M.; Jones, L.D.; Jarman, T.R.; Lines, D.I. Rapid ultrasonic inspection of ageing aircraft. *Insight-Non Test. Cond. Monit.* **2003**, *45*, 174–177. [CrossRef]

6. Brotherhood, C.; Drinkwater, B.; Freemantle, R. An ultrasonic wheel-array sensor and its application to aerospace structures. *Insight-Non Test. Cond. Monit.* **2003**, *45*, 729–734. [CrossRef]
7. Drinkwater, B.W.; Wilcox, P.D. Ultrasonic arrays for non-destructive evaluation: A review. *NDT E Int.* **2006**, *39*, 525–541. [CrossRef]
8. Zhang, J.; Drinkwater, B.W.; Wilcox, P.D. The use of ultrasonic arrays to characterize crack-like defects. *J. Nondestruct. Eval.* **2010**, *29*, 222–232. [CrossRef]
9. Tant, K.M.; Mulholland, A.J.; Gachagan, A. A model-based approach to crack sizing with ultrasonic arrays. *IEEE Trans. Ultrason. Ferroelectr. Freq. Control* **2015**, *62*, 915–926. [CrossRef]
10. Bai, L.; Velichko, A.; Drinkwater, B.W. Ultrasonic characterization of crack-like defects using scattering matrix similarity metrics. *IEEE Trans. Ultrason. Ferroelectr. Freq. Control* **2015**, *62*, 545–559. [CrossRef]
11. Van Pamel, A.; Huthwaite, P.; Brett, C.R.; Lowe, M.J. Numerical simulations of ultrasonic array imaging of highly scattering materials. *NDT E Int.* **2016**, *81*, 9–19. [CrossRef]
12. Safari, A.; Zhang, J.; Velichko, A.; Drinkwater, B.W. Assessment methodology for defect characterisation using ultrasonic arrays. *NDT E Int.* **2018**, *94*, 126–136. [CrossRef]
13. Taheri, H.; Hassen, A.A. Nondestructive Ultrasonic Inspection of Composite Materials: A Comparative Advantage of Phased Array Ultrasonic. *Appl. Sci.* **2019**, *9*, 1628. [CrossRef]
14. Shang, J.; Bridge, B.; Sattar, T.; Mondal, S.; Brenner, A. Development of a climbing robot for inspection of long weld lines. *Ind. Robot. Int. J.* **2008**, *35*, 217–223. [CrossRef]
15. Schmidt, D.; Berns, K. Climbing robots for maintenance and inspections of vertical structures—A survey of design aspects and technologies. *Robot. Auton. Syst.* **2013**, *61*, 1288–1305. [CrossRef]
16. Malandrakis, K.; Savvaris, A.; Domingo, J.A.G.; Avdelidis, N.; Tsilivis, P.; Plumacker, F.; Fragonara, L.Z.; Tsourdos, A. Inspection of aircraft wing panels using unmanned aerial vehicles. In Proceedings of the 2018 5th IEEE International Workshop on Metrology for AeroSpace (MetroAeroSpace), Rome, Italy, 20–22 June 2018; pp. 56–61.
17. Mineo, C.; MacLeod, C.; Morozov, M.; Pierce, S.G.; Lardner, T.; Summan, R.; Powell, J.; McCubbin, P.; McCubbin, C.; Munro, G.; et al. Fast ultrasonic phased array inspection of complex geometries delivered through robotic manipulators and high speed data acquisition instrumentation. In Proceedings of the 2016 IEEE International Ultrasonics Symposium (IUS), Tours, France, 18–21 September 2016; pp. 1–4.
18. Mineo, C.; Summan, R.; Riise, J.; MacLeod, C.N.; Pierce, S.G. Introducing a new method for efficient visualization of complex shape 3D ultrasonic phased-array C-scans. In Proceedings of the 2017 IEEE International Ultrasonics Symposium (IUS), Washington, DC, USA, 6–9 September 2017; pp. 1–4.
19. Zhong, H.J.; Ling, Z.W.; Miao, C.J.; Guo, W.C.; Tang, P. A New Robot-Based System for In-Pipe Ultrasonic Inspection of Pressure Pipelines. In Proceedings of the 2017 Far East NDT New Technology & Application Forum (FENDT), Xi'an, China, 22–24 June 2017; pp. 246–250.
20. Svejda, M. New Robotic Architecture for NDT Applications. *IFAC Proc. Vol.* **2014**, *47*, 11761–11766. [CrossRef]
21. Andrikopoulos, G.; Nikolakopoulos, G. Vortex Actuation via Electric Ducted Fans: An Experimental Study. *J. Intell. Robot. Syst.* **2019**, *95*, 955–973. [CrossRef]
22. Abdessalem, B.; Ahmed, K.; Redouane, D. Signal Quality Improvement Using a New TMSSE Algorithm: Application in Delamination Detection in Composite Materials. *J. Nondestruct. Eval.* **2017**, *36*, 16. [CrossRef]
23. Honarvar, F.; Sheikhzadeh, H.; Moles, M.; Sinclair, A.N. Improving the time-resolution and signal-to-noise ratio of ultrasonic NDE signals. *Ultrasonics* **2004**, *41*, 755–763. [CrossRef]
24. Karslı, H. Further improvement of temporal resolution of seismic data by autoregressive (AR) spectral extrapolation. *J. Appl. Geophys.* **2006**, *59*, 324–336. [CrossRef]
25. Jiao, J.; Ma, T.; Hou, S.; Wu, B.; He, C. A Pulse Compression Technique for Improving the Temporal Resolution of Ultrasonic Testing. *J. Test. Eval.* **2018**, *46*, 1238–1249. [CrossRef]
26. Demirli, R.; Saniie, J. Model-based estimation of ultrasonic echoes. Part I: Analysis and algorithms. *IEEE Trans. Ultrason. Ferroelectr. Freq. Control* **2001**, *48*, 787–802. [CrossRef]
27. Abbate, A.; Nguyen, N.; LaBreck, S.; Nelligan, T.; Carruthers, J.B. Ultrasonic signal processing algorithms for the characterization of thin multilayers. *J. Nondestruct. Test.* **2002**, *7*, 8.
28. Benammar, A.; Drai, R.; Guessoum, A. Detection of delamination defects in CFRP materials using ultrasonic signal processing. *Ultrasonics* **2008**, *48*, 731–738. [CrossRef] [PubMed]

29. Najmi, A.H.; Sadowsky, J. The continuous wavelet transform and variable resolution time-frequency analysis. *Johns Hopkins APL Tech. Dig.* **1997**, *18*, 134–140.
30. Patterson, D.; DeFacio, B.; Neal, S.P.; Thompson, C.R. Wavelets and their application to digital signal processing in ultrasonic NDE. In *Review of Progress in Quantitative Nondestructive Evaluation*; Springer: Boston, MA, USA, 1993; pp. 719–726.
31. Abbate, A.; Frankel, J.; Das, P. Wavelet transform signal processing applied to ultrasonics. In *Review of Progress in Quantitative Nondestructive Evaluation*; Springer: Boston, MA, USA, 1996; pp. 741–748.
32. Chen, C.; Hsu, W.L.; Sin, S.K. A comparison of wavelet deconvolution techniques for ultrasonic NDT. In Proceedings of the ICASSP-88, International Conference on Acoustics, Speech, and Signal Processing, New York, NY, USA, 11–14 April 1988; pp. 867–870.
33. Izzetoglu, M.; Onaral, B.; Bilgutay, N. Wavelet domain least squares deconvolution for ultrasonic backscattered signals. In Proceedings of the 22nd Annual International Conference of the IEEE Engineering in Medicine and Biology Society (Cat. No. 00CH37143), Chicago, IL, USA, 23–28 July 2000; Volume 1, pp. 321–324.
34. Cardoso, G.; Saniie, J. Data compression and noise suppression of ultrasonic NDE signals using wavelets. In Proceedings of the IEEE Symposium on Ultrasonics, Honolulu, HI, USA, 5–8 October 2003; Volume 1, pp. 250–253.
35. Praveen, A.; Vijayarekha, K.; Abraham, S.T.; Venkatraman, B. Signal quality enhancement using higher order wavelets for ultrasonic TOFD signals from austenitic stainless steel welds. *Ultrasonics* **2013**, *53*, 1288–1292. [CrossRef] [PubMed]
36. Liao, X.; Wang, Q.; Yan, T. Data-processing for ultrasonic phased array of austenitic stainless steel based on wavelet transform. In *Mechatronics and Automatic Control Systems*; Springer: Cham, Switzerland, 2014; pp. 1041–1046.
37. Mohammadkhani, R.; Zanotti Fragonara, L.; Janardhan, P.M.; Petrunin, I.; Tsourdos, A.; Gray, I. Ultrasonic Phased Array Imaging Technology for the Inspection of Aerospace Composite Structures. In Proceedings of the 2019 IEEE 5th International Workshop on Metrology for AeroSpace (MetroAeroSpace), Torino, Italy, 19–21 June 2019; pp. 203–208. [CrossRef]
38. O'Brien, W.D., Jr. Single-element transducers. *Radiographics* **1993**, *13*, 947–957. [CrossRef] [PubMed]
39. Braconnier, D.; Yoon, B.; Lee, H. *Understanding of Key Number in Phased Array;* In Proceedings of the 8th International Conference on NDE in Relation to Structural Integrity for Nuclear and Pressurised Components, Berlin, Germany, 29 October–1 November 2010..
40. Smith, R.; Bruce, D.; Jones, L.; Marriott, A.; Scudder, L.; Willsher, S. Ultrasonic C-scan standardization for polymer-matrix composites-Acoustic considerations. In *Review of Progress in Quantitative Nondestructive Evaluation*; Springer: Boston, MA, USA, 1998; pp. 2037–2044.
41. Kachanov, V.K.; Kartashev, V.G.; Popko, V.P. Application of signal processing methods to ultrasonic non-destructive testing of articles with high structural noise. *Nondestruct. Test. Eval.* **2001**, *17*, 15–40. [CrossRef]
42. Pagodinas, D. Ultrasonic signal processing methods for detection of defects in composite materials. *Ultragarsas Ultrasound* **2002**, *45*, 47–54.

© 2020 by the authors. Licensee MDPI, Basel, Switzerland. This article is an open access article distributed under the terms and conditions of the Creative Commons Attribution (CC BY) license (http://creativecommons.org/licenses/by/4.0/).

Article

Monitoring the Risk of the Electric Component Imposed on a Pilot During Light Aircraft Operations in a High-Frequency Electromagnetic Field †

Joanna Michałowska [1,*], Arkadiusz Tofil [2], Jerzy Józwik [2], Jarosław Pytka [2], Stanisław Legutko [3], Zbigniew Siemiątkowski [4] and Andrzej Łukaszewicz [5]

1 The Institute of Technical Sciences and Aviation, The State School of Higher Education in Chelm, Pocztowa 54, 22-100 Chełm, Poland
2 Faculty of Mechanical Engineering, Lublin University of Technology, 20-618 Lublin, Poland; atofil@pwsz.chelm.pl (A.T.); j.jozwik@pollub.pl (J.J.); j.pytka@pollub.pl (J.P.)
3 Institute of Mechanical Technology, Poznan University of Technology, 60-965 Poznań, Poland; stanislaw.legutko@put.poznan.pl
4 Faculty of Mechanical Engineering, Kazimierz Pulaski University of Technology and Humanities, 26-600 Radom, Poland; z.siemiatkowski@uthrad.pl
5 Faculty of Mechanical Engineering, Bialystok University of Technology, 15-351 Bialystok, Poland; a.lukaszewicz@pb.edu.pl
* Correspondence: jmichalowska@pwsz.chelm.pl
† This paper is an extended version of the conference paper: Michalowska, J.; Józwik, J.; Tofil, A. Monitoring of the intensity of the electromagnetic field during the aircraft operation in the field of high frequencies. In Proceedings of the 2019 IEEE 5th International Workshop on Metrology for AeroSpace (MetroAeroSpace), Torino, Italy, 19–21 June 2019; pp. 366–370.

Received: 17 November 2019; Accepted: 9 December 2019; Published: 14 December 2019

Abstract: High-frequency electromagnetic fields can have a negative effect on both the human body and electronic devices. The devices and systems utilized in radio communications constitute the most numerous sources of electromagnetic fields. The following research investigates values of the electric component of electromagnetic field intensification determined with the ESM 140 dosimeter during the flights of four aircrafts—Cessna C152, Cessna C172, Aero AT3 R100, and Robinson R44 Raven helicopter—from the airport in Depultycze Krolewskie near Chelm, Poland. The point of reference for the obtained results were the normative limits of the electromagnetic field that can affect a pilot in the course of a flight. The maximum value registered by the dosimeter was E = 3.307 V/m for GSM 1800 frequencies.

Keywords: electromagnetic fields (EMFs); dosimeter; high frequencies; aircraft operation; environment

1. Introduction

The effect of electromagnetic fields on human health and various areas of human activity is studied in numerous scientific and research centers worldwide. Despite researchers' widespread interest, the methodology for the assessment of the effect of electromagnetic fields has not been standardized. Considerable activities aimed at decreasing the negative influence of electromagnetic fields are being undertaken [1–3]. It is worth mentioning that there are plentiful positive applications of electromagnetic fields, particularly in medicine, such as its therapeutic effect in the treatment of such disorders as neoplasms, burns, circulatory system diseases, or arthritis, as well as modern diagnostic imaging methods that utilize electromagnetic fields, including computed tomography, optical tomography, and microwave tomography. Moreover, research on the effect of electromagnetic fields on human health, which is concerned with the operation of electric and electronic devices, are being conducted [3–5]. Except for the direct protection of health, there are numerous technical aspects regarding the hazards resulting from electromagnetic fields exceeding their normal values. High-frequency electromagnetic fields can disrupt or even cause permanent damage to electronic equipment used in aviation, among other areas. Therefore, some special attention is paid to the issue of electromagnetic fields that can disturb the operation of communication, radar, and location devices or systems. The research investigates avionic measuring equipment that is subjected to high risk during its operation [6–10].

Research on electronic devices used in aviation is performed in specially screened chambers where avionic equipment is subjected to high-frequency electromagnetic fields from 10 kHz to 40 GHz. The research is conducted in order to determine the sensitivity of the devices to the effects of external fields and whether the instruments affect the environment. The requirements, which must be met by all onboard electronic devices and systems used in aviation, are depicted in the EU Directive 2013/35/ [11–16]. Onboard devices, radio-navigation devices, aviation communication, and navigation systems constitute basic avionic equipment.

High-frequency electromagnetic fields are most frequently produced by radio communication devices. Radio stations, telecommunication devices, mobile telephony base stations, WiFi devices, radars, and communication systems are indispensable nowadays. Tests on the effect of electromagnetic fields on electronic devices are commonly conducted when designing the device [17–19].

The increasing load of the electromagnetic field in light aircrafts can lead to negative effects on the pilots' health and mental condition, especially when it comes to instructors flying many-hour flights daily, which can lead to safety risks. Studies on the impact of electromagnetic fields on humans result from the EU Directive 2013/35/ directive. The continuous increase in radio infrastructure, including mobile telephony, is associated with an increasing electromagnetic field strength. In particular, in the area of Global System for Mobile Communications (GSM) and Universal Mobile Telecommunications System (UMTS) relays, the electromagnetic field values can reach high values. The second source of electromagnetic fields that affects the pilot during the flight is avionics. Integrated avionics, which use one large display instead of many single indicators, is a great convenience for the pilot due to the presentation of a great deal of flight information on one display. Glasscocpit is currently the standard in aircraft communication. Thus, flight training organizations (FTO) are increasingly willing to use training aircrafts with the Glasscocpit system [20–24].

The purpose of the present study was to conduct electromagnetic field measurements on selected aircrafts from the Aviation Training Center in Royal Depultycze, near Chelm, Poland. The center is an integral part of the higher vocational school in Chelm and conducts training for airline pilots on airplanes and helicopters as a part of engineering studies.

The results were depicted in the form of graphs as a function of time. Due to the stochastic character of the sample, the analysis of the data obtained, with the emerging characteristic trends and parameters, was performed using the Statistica 13. 3 software (JPZ803D036327AR-2, StatSoft Poland, Cracow, Poland).

2. Materials and Methods

2.1. Measuring Method

The method used during the research involved broadband measurements using broadband meters. The meters utilized for environmental measurements of an electromagnetic field (EMF) in the vicinity of the objects researched and the far field are commonly used in protective measurements. Such measurements were used in order to obtain a single result that corresponds to the field intensity values of all sources within the measuring range of the probe.

2.2. The Device Used for the Measurements

The Maschek ESM 140 (Serial No: 30171, Maschek Electronic, Bad Wörishofen, Germany) dosimeter was used for the flight test measurements. The device measures the electromagnetic field for broadband high frequencies in real time and the measured data was stored in the device's memory. We transferred the data to a personal computer for analysis after each flight. The measuring range was 0.01 to 70 V/m, with a sensitivity of 10 mV/m and an accuracy of ±2 dB in a free field and ±4 dB when the device was installed on the pilot's arm for in–flight measurements (see Figure 1, right) [1,3,4].

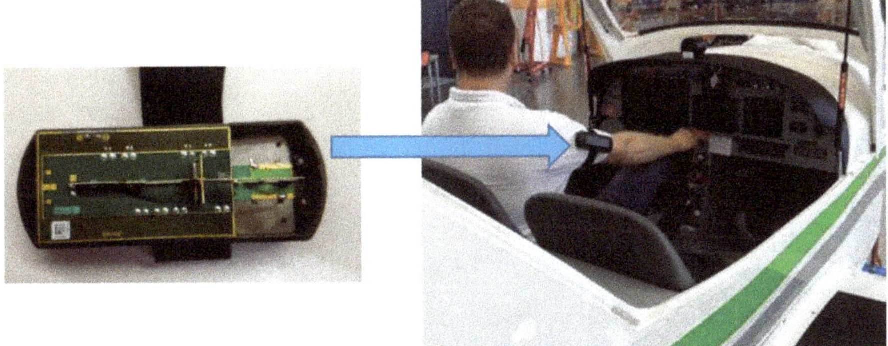

Figure 1. Measuring system for obtaining the intensity of the electromagnetic field from electrical components using an EMS 140 dosimeter during the flight: (**left**) the ESM 140 dosimeter used for the measurements, and (**right**) the location of the device during test flights.

The ESM 140 dosimeter with a battery weighs 87 g, which is relatively light for a professional measuring device. As a result, the device is almost imperceptible for the pilot during many-hour flights. The ESM 140 m were applied in tests on several different flight routes performed by four types of aircrafts.

The measuring process was initiated the moment the pilot left the service room. The entire work space was subjected to the measurement, including all areas where the electromagnetic fields could be present. Such a method ensured full reproduction of the exposure to which the pilot was subjected to.

The aircrafts used for the tests were fixed on a rotary wing, as shown in Figure 2. Three fixed-wing aircrafts were: the Cessna 152, the Cessna 172, and the AT3. The rotary-wing aircraft used in this study was the Robinson R44 Raven. All the aircrafts were used in every day flight training in the Center of Aviation of The State School of Higher Education in Chełm, East Poland.

All the aircrafts were of typical monocoque construction, made of duralumin. They had a piston engine drive and a set of typical on-board instruments for Visual Flight Rules (VFR) or Instrument Flight Rules (IFR) flights [25,26]. Selected technical data of the aircraft used in the test flights are collected in Table 1.

The Cessna 152 and Cessna 172 aircrafts had Glasscocpit avionics, with the Cessna 152 aircraft having been installed as part of the retrofitting of an existing panel. The Aero AT3 aircraft was the lightest among those tested and also had Glasscocpit avionics. The Robinson R44 Raven helicopter featured the classic Helicopter In Flight Refuelling (H-IFR) flight instrument set together with the Aspen Avionics EFIS (see Figure 2).

Table 1. Basic technical data of the aircrafts used for the study.

Type	No. of Seats	Wingspan/Rotor Diameter (m)	Engine (kW)	Avionics	Other
Fixed-Wing Aircrafts					
Cessna 152	2	10.11	86	Garmin G5	
Cessna 172	4	11.00	125	GarminG1000	
AERO AT3	2	7.55	75	Garmin G500TXI	VLA—Very Light Aircraft Class
Rotary-Wing Aircraft					
R44 Raven	4	9.00	183	Mixed: Analog/Aspen electronic flight instrument system (EFIS)	

Figure 2. Avionics of the Robinson R44 Raven helicopter.

3. Results and Discussion

The measurements were performed during the flights. In order to measure the intensity of the electric field, the first measurement was conducted on the ground in the Air Traffic Control Tower of Depułtycze Królewskie airfield (Figure 3).

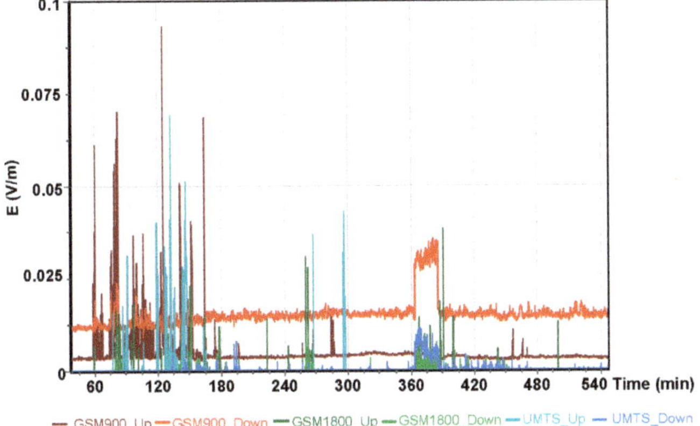

Figure 3. The intensity of the electric field for high frequencies on the Air Traffic Control Tower.

Maximum values of the electric component of the electromagnetic field obtained during measurements in the ground service station were measured using the GSM 900 frequency band. What is of great significance is that the values were several-fold higher than the ones recorded in other frequency bands. The intensity of the electromagnetic field in this location was relatively low.

The next step involved the measurement during the flights. The flights presented in the study were performed in the vicinity of such towns as: Krasnystaw, Bilgoraj, and Tomaszow Lubelski where antennas of mobile telephony are located. The selected flight route is shown in Figure 4.

Figure 4. The flight route of the Cessna C152.

The results for the GSM 900 system are shown in Figure 5. The values depicted in Figures 5–7 refer to the measurement results obtained using the dosimeter in the Cessna 152.

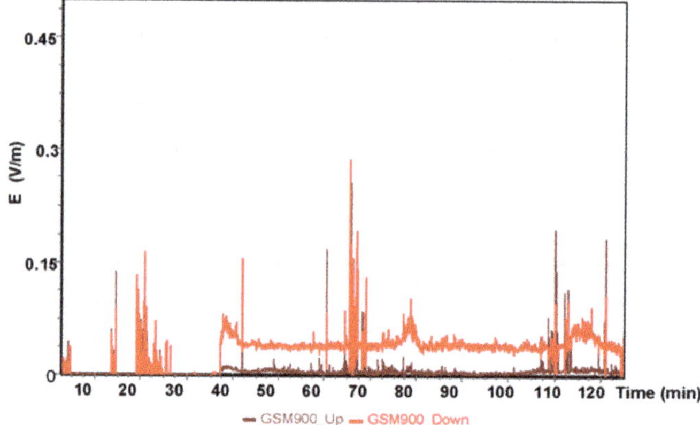

Figure 5. The intensity of the electrical component of the electromagnetic field in the course of a flight for the GSM 900 system.

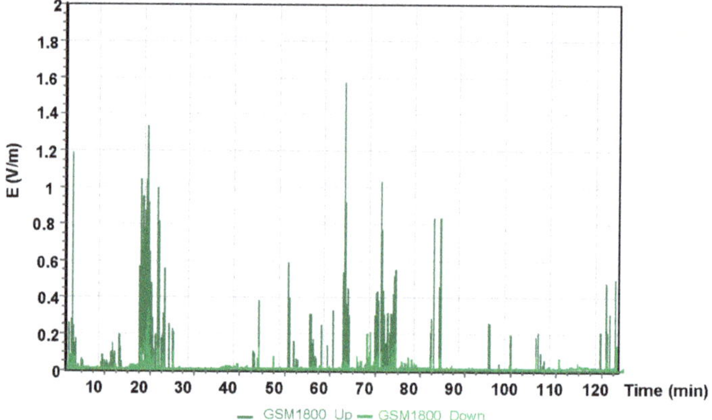

Figure 6. The intensity of the electrical component of the electromagnetic field in the course of a flight for the GSM 1800 system.

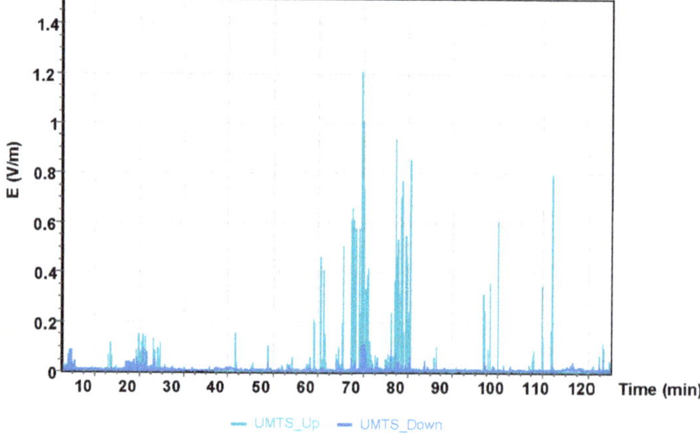

Figure 7. The intensity of the electrical component of the electromagnetic field in the course of a flight for the UMTS frequency.

The highest measured value using the GSM 900 frequency band obtained was E = 0.29 V/m for the oldest mobile telephony system. GSM 900 frequency was the first system utilized in mobile telephony communication. The next GSM 1800 frequency band obtained in the measurement with the ESM 140 dosimeter is shown in Figure 6.

The maximum result of the intensity of the electric component of the electromagnetic field using the GSM 1800 obtained during the measurement with the dosimeter was E = 1.58 V/m.

Communication systems, such as UMTS, used for frequencies of 1920–2170 MHz, and GSM differ in terms of the implementation of various modern multimedia services. Moreover, UMTS uses services that are available both on the ground and through the satellite system. The system enables the simultaneous transmission of audio, video, and data in real time.

The maximum value obtained during the Cessna C152 flight for the UMTS communication system was E = 1.21 V/m (Figure 7). It can be observed that the maximum values of the electric field for all frequency bands analyzed refer to the point of time from the 68th to the 70th minute of the flight, which corresponds to the location of the aircraft over Bilgoraj.

The measurement error was recorded for the frequencies researched during a flight, which are presented in Figure 8.

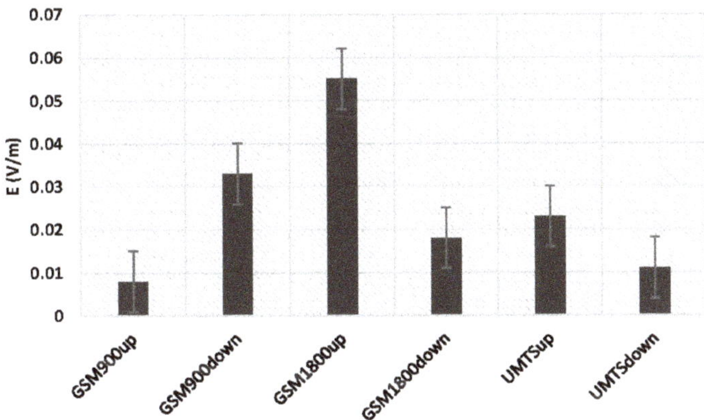

Figure 8. Measurement error of the electrical component of the electromagnetic field intensity during a flight for the frequencies researched.

The limits of measurement uncertainty for the data obtained are given in Figure 8. Systematic error was chosen as the basis for uncertainty, whose sources were mainly: instrument class, characteristics of the band filters in the tested frequencies, flight altitude and heading relative to electromagnetic field sources, and meteorological factors. Moreover, the measurement error was also affected by the installation of the measuring unit, which, mainly for practical reasons, was placed on the pilot's arm (see Figure 1, right), which meant that the measurement was not carried out in the free field. In future test, we are planning to use an on-board electromagnetic field-monitoring sensor in order to minimize the installation effects. This sensor is described in Section 4 of the present paper.

Although the majority of high-frequency electromagnetic field measurements are performed in scientific centers on land, this study was concerned with the comparison of the results obtained both before take-off and during the flight. The results confirmed that the pilot was more exposed to the effects of the high-frequency electromagnetic field during the flight.

The analysis presented next was for a two-hour training flight by the Cessna C172 on the following route: Depułtycze Królewske–Lublin Airport–Krasnystaw–Depułtycze Królewski. During the flight, the approach to landing on runway 25 at Lublin Airport was performed. Measurements of electromagnetic field for individual frequency ranges are shown in Figures 9–11.

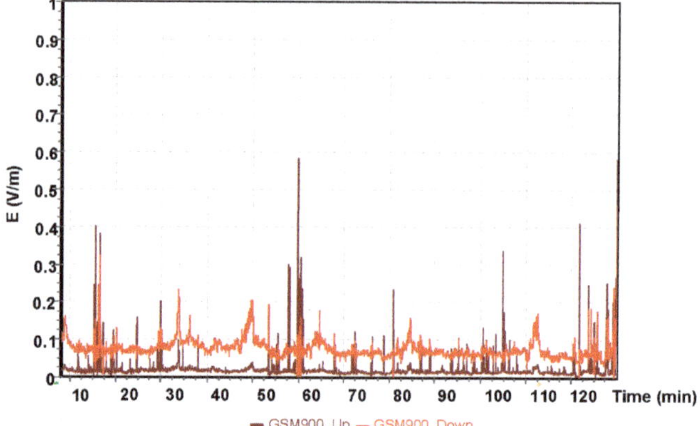

Figure 9. The intensity of the electrical component of the electromagnetic field during a flight of the Cessna C172 for the GSM 900 system.

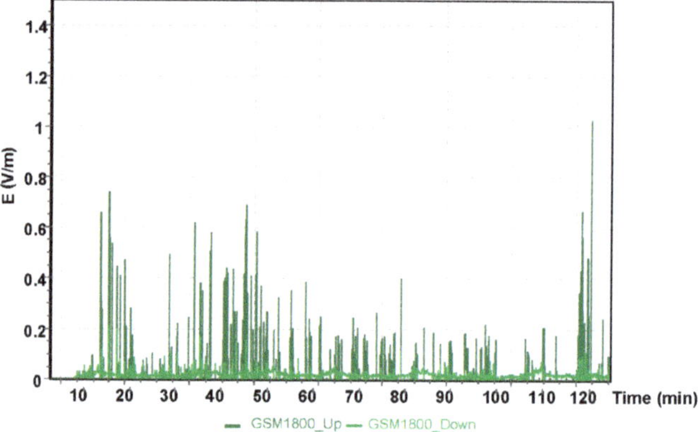

Figure 10. The intensity of the electrical component of the electromagnetic field during a flight of the Cessna C172 for the GSM 1800 system.

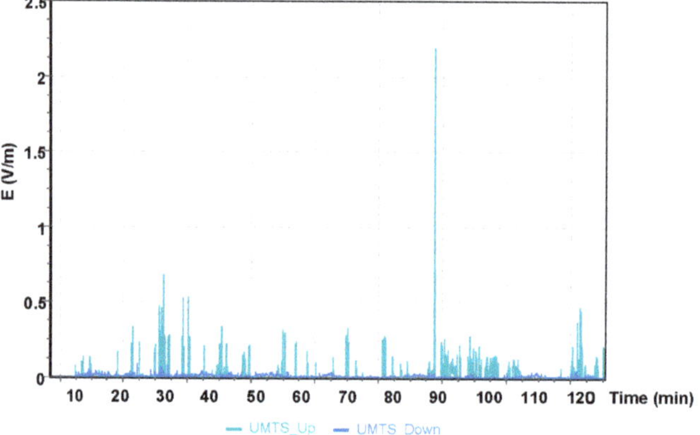

Figure 11. The intensity of the electrical component of the electromagnetic field during a flight of the Cessna C172 for the UMTS frequency.

The highest value of the electric component of the electromagnetic field was recorded using the UMTS frequency band with E = 2.30 V/m. At this point, the aircraft had entered the instrument landing system (ILS) approach path. It is a radio navigation system that supports the landing of an aircraft in conditions of limited visibility. For the other frequencies tested, the electric field values obtained that affect the pilot was exposed are much lower. For the GSM 900 frequency band, the electric component of the electromagnetic field was E = 0.6 V/m, and for the GSM 1800, it was E = 1.05 V/m.

The next aircraft analyzed was the AT3. The flight took place on the route Depułtycze Królewskie–Lublin–Mielec–Depułtycze Królewskie (Figures 12–14). The total flight duration was 1.40 h.

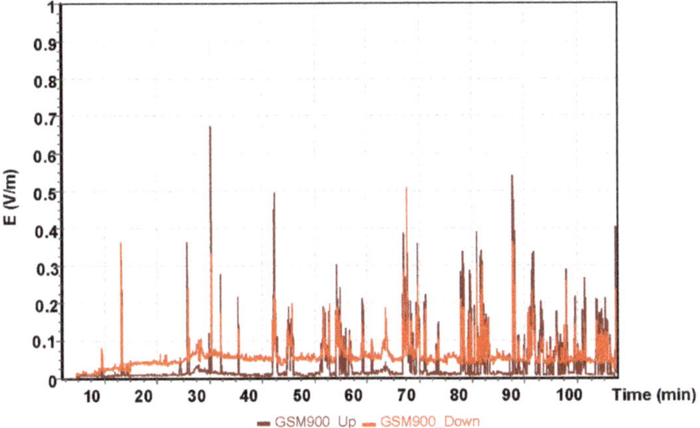

Figure 12. The intensity of the electrical component of the electromagnetic field during a flight of the AT3 for the GSM 900 system.

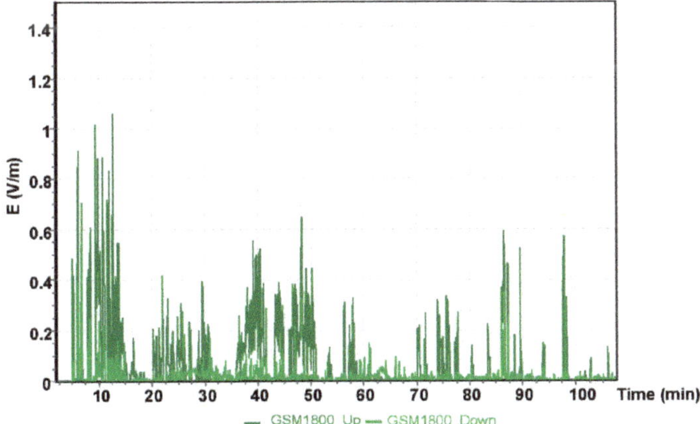

Figure 13. The intensity of the electrical component of the electromagnetic field during a flight of the AT3 for the GSM 1800 system.

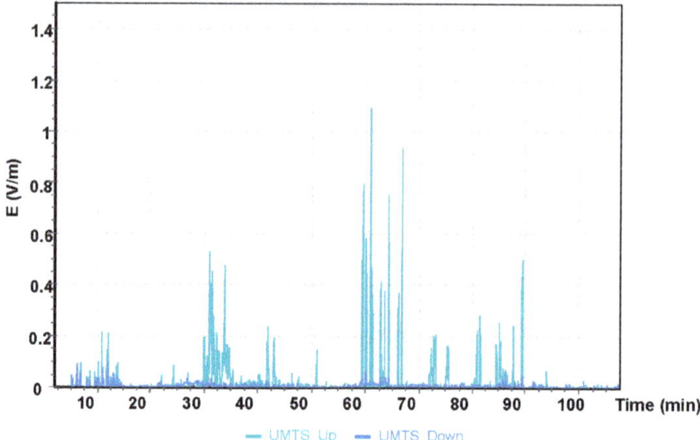

Figure 14. The intensity of the electrical component of the electromagnetic field in the course of a flight of the AT3 for the UMTS frequency.

For the training flight with the AERO AT3 aircraft, the highest value was also recorded using the UMTS frequency band with E = 1.15 V/m (Figure 14). A similar value was recorded for the GSM 900 frequency with E = 1.14 V/m (Figure 13) and the lowest was for the GSM 1800 with E = 0.68 V/m.

Another analysis was carried out for the Robinson R44 helicopter flight (Figures 15–17).

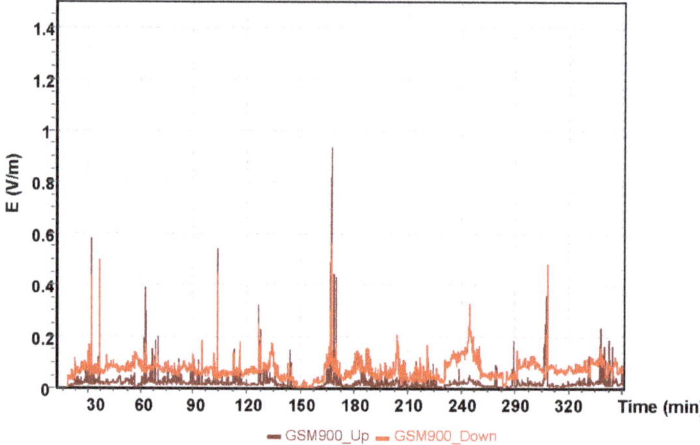

Figure 15. The intensity of the electrical component of the electromagnetic field during a flight of the Robinson R44 for the GSM 900 frequency.

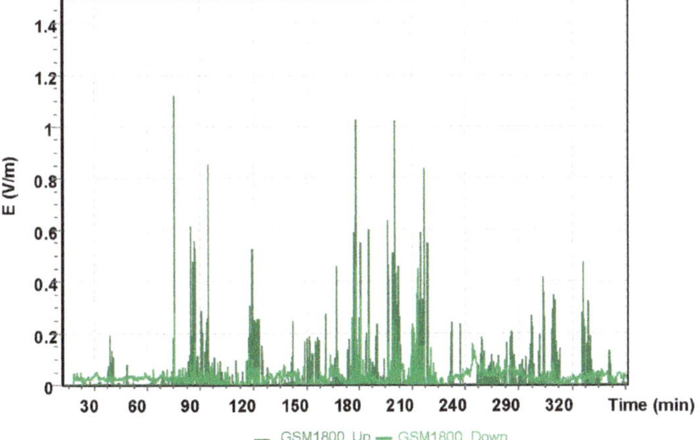

Figure 16. The intensity of the electrical component of the electromagnetic field during a flight of the Robinson R44 for the GSM 1800 frequency

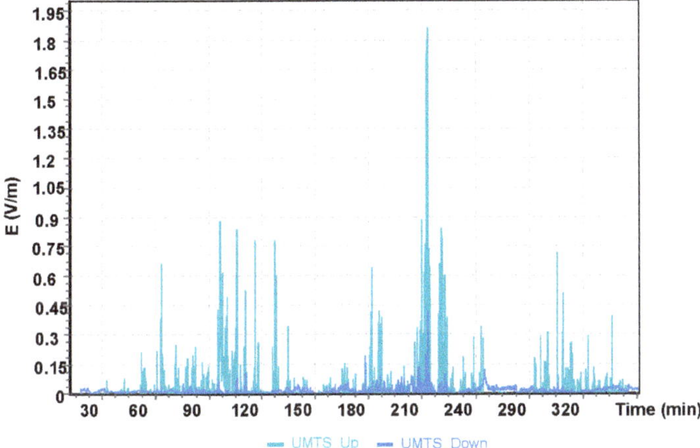

Figure 17. The intensity of the electrical component of the electromagnetic field during a flight of the Robinson R44 for the UMTS frequency.

For the Robinson R44 helicopter flight, the highest value of the electric component of the electromagnetic field E = 1.89 V/m was read using the UMTS frequency. For the GSM 1800 frequency band, the maximum value of the electrical component was E = 1.16 V/m. For the oldest GSM 900 communication system, the maximum values (electric field) were two–fold lower than values obtained from measurements for other frequencies and equaled E = 0.89 V/m.

The statistical analysis involved the introduction of the values of the electric field intensity recorded by the ESM 140 dosimeter into the Statistica 13.0 software. The values of the analyzed parameters measured in the normal scale were characterized by the number and percentage, while those measured on a ratio scale by means of average, medians, standard deviation (SD), and range of variation. A 5% error of inference and associated significance level of $p < 0.05$ were adopted.

The data used in the statistical analysis were obtained during flights with all four aircrafts. First, results from a flight with the Cessna 152 aircraft were analyzed. The flight took 1.42 h on the route 1 from Depultycze Krolewskie to Bilgoraj and Tomaszow Lubelski using all frequency bands, from UMTS installations on the ground, as well as from onboard instruments (see Table 2). The following table presents the characteristics of the electric field intensity E for individual frequency bands of the selected route. A sample group included a total of 15641 measurements registered for each frequency band.

Table 2. Characteristic of the intensity of the electrical component of the electromagnetic field E (V/m) for route 1 with the Cessna C152.

Variable	Mean	Minimum	Maximum	SD
GSM900up	0.0072	0.000	0.4539	0.0194
GSM900down	0.0315	0.000	0.5178	0.0260
GSM1800up	0.0480	0.000	3.3078	0.2080
GSM1800down	0.0174	0.000	0.9989	0.0595
UMTSup	0.0158	0.000	1.9821	0.0736
UMTSdown	0.0103	0.000	2.5000	0.0340

The mean value of the electric field intensity using the individual GSM and UMTS frequency bands ranged from 0.007 V/m to 0.048 V/m. The range of the variable analyzed was from 0.000 V/m to 3.30 V/m. Differences for the indicated frequency bands obtained with the ESM 140 dosimeter proved to be statistically significant and are depicted in Figure 18.

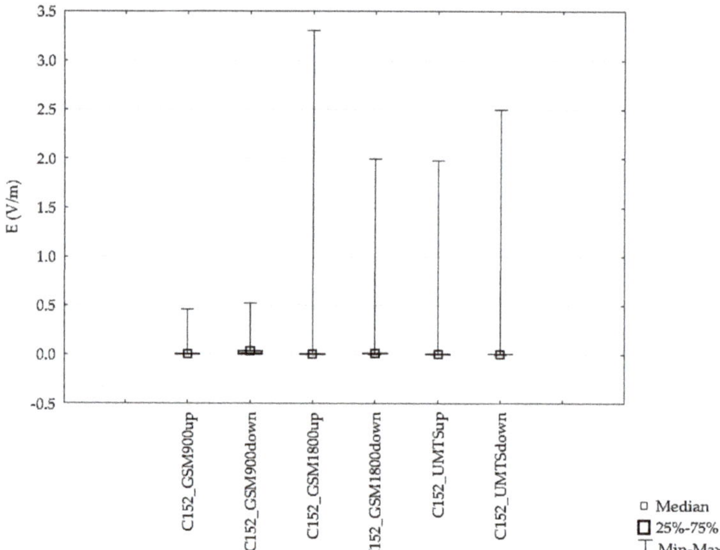

Figure 18. Range of the variability and median the intensity of electric field for high-frequency on the pilot during an airplane flight for one route.

Selected results for one GSM1800 frequency range in which the highest values of the intensity of the electric component of the electromagnetic field were recorded during flights with one type of Cessna 152 aircraft was made on four routes. The variation of the electric field intensity for various aircraft routes is shown in Figure 19.

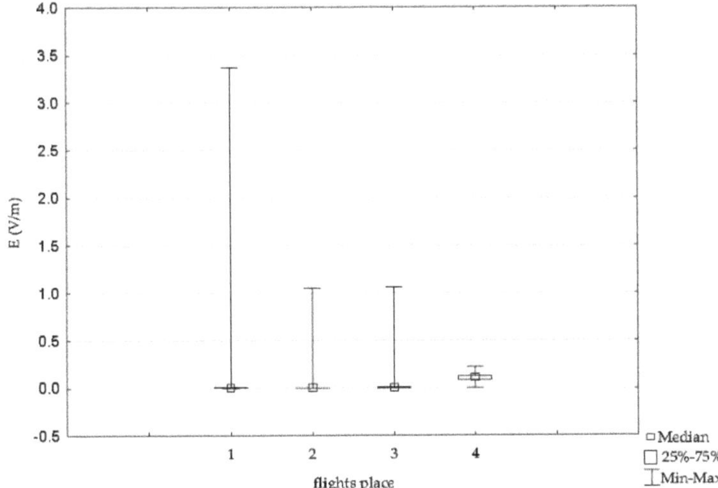

Figure 19. Range of variability and median the intensity of electric field for high-frequency on the pilot during an airplane flight of the Cessna C152 for four routes.

The data presented in Figure 19 consider the highest values of the electric field recorded for four aircraft routes for the GSM 1800up frequency band, which justifies the choice of the frequency band.

The numbers from one to four represent the flight routes researched, namely route 1: Depultycze Krolewskie-Bilgoraj-Tomaszow Lubelski, route 2: Depultycze Krolewskie–Rejowiec–Siennica–Pokrowka, route 3: Depultycze Krolewskie–Cycow–Rejowiec, and route 4: Depultycze Krolewski–Krasnystaw–Frampol.

Variation in the intensity of the electric field observed proved to be statistically significant and ranged from 0.000 V/m to 3.30 V/m. The mean value in time ranged from a minimum of 0.054 V/m to maximum of 0.101 V/m. The differences described are reflected in the values of the test function (Table 3).

Table 3. The characteristics of the electric field intensity E (V/m) for the routes 1, 2, 3, and 4.

Variable	Mean	Minimum	Maximum	SD
1	0.0540	0.0002	3.3078	0.2675
2	0.0142	0.0000	1.0467	0.0670
2	0.0192	0.0000	1.0593	0.0568
4	0.1019	0.0002	0.2174	0.0374

Student's *t*-test was performed in order to determine statistically significant differences between the values of electric component of the electromagnetic field for the aircraft routes (1,2,3,4) compared to the values obtained for the aircraft located on the ground (background) and is presented in Table 4.

Table 4. Test results for individual comparisons (GSM1800).

Test Results for Individual Comparisons		Statistical Significance
Background vs. 1	$p = 0.000$	Relevant
Background vs. 2	$p = 0.000$	Relevant
Background vs. 3	$p = 0.000$	Relevant
Background vs. 4	$p = 0.000$	Relevant

Having compared the values of the electric component of the electromagnetic field for the GSM1800 band, which presented the highest values recorded on different aircraft routes compared to the background (the airport), statistically significant differences were observed.

The next analysis presented included the results obtained during flights with the Cessna C172 aircraft on the route Depułtycze Królewske–Lublin–Krasnystaw–Depułtycze Królewskie for individual frequency ranges. Table 5 presents the characteristics of the electric field intensity E for individual frequency bands of the Cessna C172. A sample group included a total of 7372 measurements registered for each frequency band.

Table 5. Characteristics of the intensity of the electrical component of the electromagnetic field E (V/m) with the Cessna C172.

Variable	Mean	Minimum	Maximum	SD
GSM900up	0.0229	0.0031	0.5821	0.0311
GSM900down	0.0749	0.0000	0.4582	0.0246
GSM1800up	0.0602	0.0602	1.1602	0.0000
GSM1800down	0.0197	0.0000	0.3463	0.0133
UMTSup	0.0184	0.0000	1.8870	0.0704
UMTSdown	0.0114	0.0000	0.1360	0.0074

The mean value of the electric field intensity for individual GSM and UMTS frequency bands ranged from 0.011 V/m to 0.074 V/m. The range of the variable analyzed was from 0.000 V/m to 2.18 V/m.

Next, the statistical analysis of the results obtained during flights on AT3 aircraft on the route Depułtycze Królewskie-Lublin-Mielec-Depułtycze Królewskie was performed. The characteristics of the electric field intensity E for individual frequency bands of the AT3 is presented in Table 6. A sample group included a total of 7287 measurements registered for each frequency band.

Table 6. Characteristics of the intensity of the electrical component of the electromagnetic field E (V/m) for a flight with the AERO AT3 aircraft.

Variable	Mean	Minimum	Maximum	SD
GSM900up	0.0335	0.0000	0.6838	0.0311
GSM900down	0.0612	0.0000	0.5155	0.0246
GSM1800up	0.0629	0.0000	1.1507	0.0000
GSM1800down	0.0098	0.0000	0.5669	0.0133
UMTSup	0.0195	0.0000	1.1697	0.0704
UMTSdown	0.0046	0.0000	0.2169	0.0074

The mean value of the electric field intensity for individual GSM and UMTS frequency bands ranged from 0.0046 V/m to 0.0629 V/m. The range of the variable analyzed was from 0.000 V/m to 1.17 V/m.

The results obtained during the flight of a Robinsson R44 aircraft were also statistically analyzed. The flight was made on the route Depułtycze Królewskie–Lublin–Kielce–Pyszkowice–Czestochowa–Krasnystaw–Depułtycze Królewskie. A sample group included a total of 14653 measurements registered for each frequency band (Table 7).

Table 7. Characteristics of the intensity of the electrical component of the electromagnetic field E (V/m) with the Robinsson R44.

Variable	Mean	Minimum	Maximum	SD
GSM900up	0.0210	0.0000	0.9839	0.0357
GSM900down	0.0687	0.0000	0.5182	0.0401
GSM1800up	0.0205	0.0000	1.1545	0.0861
GSM1800down	0.0235	0.0000	0.5748	0.0313
UMTSup	0.0230	0.0000	1.8976	0.1063
UMTSdown	0.0127	0.0000	0.4411	0.0186

The mean value of the electric field intensity for individual GSM and UMTS frequency bands ranged from 0.012 V/m to 0.0687 V/m. The range of the variable analyzed was from 0.000 V/m to 1.89 V/m.

Differences for the indicated frequency bands obtained with the ESM 140 dosimeter proved to be statistically significant and are depicted in Figure 17. Results from flights with different types of aircraft were analyzed for each of the tested frequency ranges. The variations of the electric field intensity for various types of aircraft are shown in Figures 20–22.

The data presented in the figures relate to the highest electric field values recorded for four routes made with different types of aircraft in the GSM 900, GSM 1800, and UMTS frequency bands.

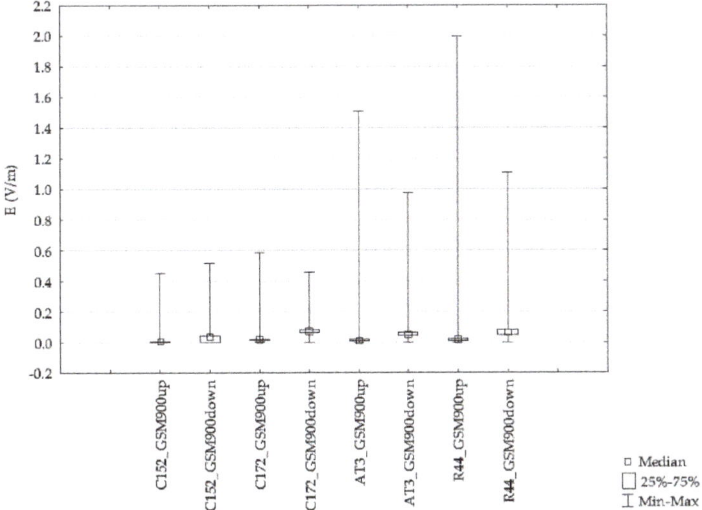

Figure 20. Range of the variability and median during a GSM 900 frequency airplane flight for different routes.

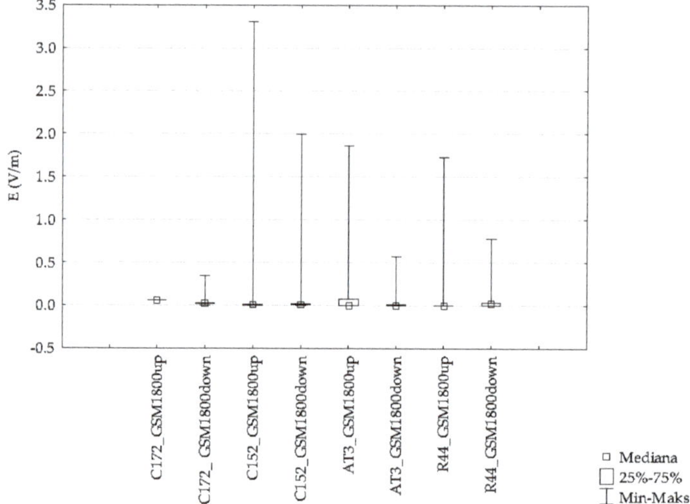

Figure 21. Range of the variability and median during a GSM 1800 frequency airplane flight for different routes.

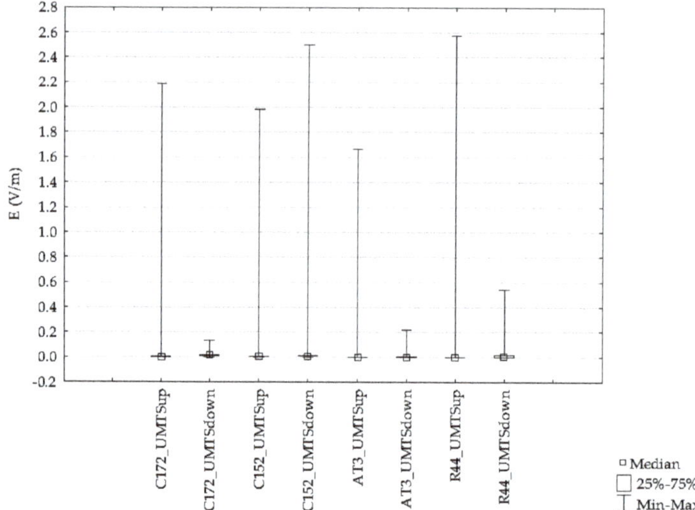

Figure 22. Range of the variability and median during a UMTS frequency airplane flight for different routes.

4. Design of an Onboard EMF Monitoring Sensor System

Although the results obtained confirm that the measured values of the electromagnetic fields in the tested aircraft did not exceed the permissible values, there is a risk related to the high load on the pilot's body, especially the instructor, who performed a significant number of flights per day. As such, the radiation doses add up and this may cause premature fatigue and other negative effects on the instructor's mental condition. Bearing in mind the need to increase the level of flight operations safety, it is reasonable to undertake research and development activities to develop a system for monitoring electromagnetic radiation and its impact on pilots during a flight. The authors' idea is an onboard electromagnetic field monitoring system, shown schematically in Figure 23.

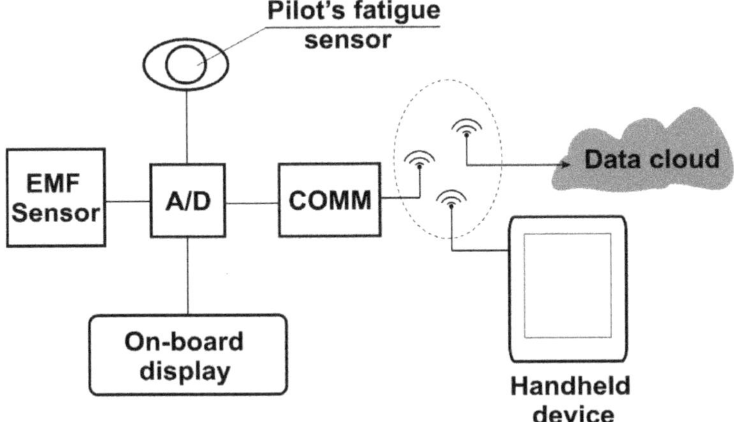

Figure 23. Onboard electromagnetic field-monitoring system.

The system uses an electromagnetic radiation sensor and a measurement data recorder built based on a microprocessor system. Communication with the environment is possible via a Bluetooth- and GSM-based radio system. In addition, the system includes an optical pilot fatigue sensor using an iris image of the eye. The whole system functions as part of an Internet of Things, thanks to which, it is possible to remotely read, configure, set alerts, etc. Adjusting the system to the needs of a particular case is also an important feature, taking into account various types of aircraft, pilots' workload, and other factors.

The system can be installed anywhere in the aircraft cabin, though the radiation sensor should be placed close to the pilot, for example, on the back of the seat.

Developmental works on the presented idea of the electromagnetic field-monitoring system will be the subject of the authors' continued and future research.

5. Conclusions

The results of the measurements were compared to the 2013/35/UE directive described in Michalowska et al. [1]. The maximum values recorded in the bands, which ranged from 2 GHz to 6 GHz (GSM 1800, UMTS) and from 400 MHz to 2 GHz (GSM 900), were within safe limits accepted by the directive, and some of the values obtained were several-fold lower. The ESM 140 dosimeter indicated the maximum value for the GSM 1800 band of E = 3.30 V/m during the flight of the Cessna C152, which was registered during flights with all types of aircraft. The readings were taken when the Cessna C152 aircraft was located above Bilgoraj. The maximum value of the electric component of the electromagnetic field E = 2.0 V/m was recorded using the GSM 900 frequency during the Robinson R44 flight. Values registered with the electric component dosimeter for the UMTS frequency band were for all types of aircraft of similar order. The maximum value of E = 2.59 V/m was also registered during the longest flight of the Robinson R44 helicopter. The dosimeter showed the lowest values for the GSM 900 frequency with E = 1.98 V/m, also for the R44 helicopter.

It is worth highlighting that the values of the electric component intensity of the electromagnetic field recorded close to the ground were considerably lower compared to the readings taken during the flight. Similar findings were confirmed in the course of the statistical analysis.

Field intensity levels in the vicinity of antenna masts were caused by radio communication of a divergent character. The levels registered depended to a great extent on the power of the antennas within the base stations. Due to the character of the transmission and construction of the antennas, the distribution of the field intensity was directional. Although it is difficult to determine the effect of these levels on the environment and its living organisms, they should be constantly monitored. Most of the results presented refer to the measurement performed on board of Cessna C152 aircrafts. It is worth emphasizing that similar results were found for four similar aircrafts. Quantitative measurements of the electromagnetic field intensity were most frequently conducted on the ground or in its close vicinity at a maximum of several dozen meters. A different methodology was utilized in the study where the electric component of the electromagnetic field was analyzed at height ranging from 0 to 860 m.

Measurement with the dosimeter did not influence board instruments nor avionic units. Measurement antennas were placed outside this type of aircraft. The ESM 140 dosimeter did not generate any high-frequency radiation and thus is suitable for electro-sensitive persons and epidemiological studies considering a minimum field intensity.

Author Contributions: Conceptualization, J.M., J.P.; Methodology, J.M. and J.P.; Software, J.M., A.T., J.J.; Formal Analysis, J.M., J.P.; Investigation, J.M.; Resources, J.M., and J.J.; Data Curation, J.J.; Writing-Original Draft Preparation, J.M., J.P.; Writing-Review & Editing, J.M., J.P.; Visualization, J.M. Supervision, S.L., Z.S., A.Ł.; Project Administration, J.J.; Funding Acquisition, J.J. All authors provided critical feedback and collaborated in the research.

Funding: The project/research was financed in the framework of the project Lublin University of Technology–Regional Excellence Initiative, funded by the Polish Ministry of Science and Higher Education (contract no. 030/RID/2018/19).

Conflicts of Interest: The authors declare no conflict of interest.

References

1. Michalowska, J.; Tofil, A.; Józwik, J. Analysis of the radiated interferences emission of small household appliances. In Proceedings of the 2019 Applications of Electromagnetics in Modern Engineering and Medicine (PTZE), Janow Podlaski, Poland, 9–12 June 2019; pp. 117–120.
2. Michałowska, J.; Tofil, A.; Józwik, J.; Pytka, J.; Budzyński, P.; Korzeniewska, E. Measurement of high–frequency electromagnetic fields in CNC machine tools area. In Proceedings of the 2018 IEEE 4th International Symposium on Wireless Systems within the International Conferences on Intelligent Data Acquisition and Advanced Computing Systems (IDAACS-SWS), Lviv, Ukraine, 20–21 September 2018; pp. 162–165.
3. Michalowska, J.; Józwik, J.; Tofil, A. Monitoring of the intensity of the electromagnetic field during the aircraft operation in the field of high frequencies. In Proceedings of the 2019 IEEE 5th International Workshop on Metrology for AeroSpace (MetroAeroSpace), Torino, Italy, 19–21 June 2019; pp. 366–370.
4. Michałowska, J.; Mazurek, P.; Gad, R.; Chudy, A.; Kozieł, J. Identification of the Electromagnetic Field Strength in Public Spaces and During Travel. In Proceedings of the 2019 Applications of Electromagnetics in Modern Engineering and Medicine (PTZE), Janow Podlaski, Poland, 9–12 June 2019; pp. 121–124.
5. Gas, P.; Wyszkowska, J. Influence of multi–tine electrode configuration in realistic hepatic RF ablative heating. *Arch. Electr. Eng.* **2019**, *68*, 521–533.
6. Kurgan, E.; Gas, P. Treatment of tumors located in the human thigh using RF hyperthermia. *Prz. Elektrotechniczn.* **2011**, *87*, 103–106.
7. Balal, N.; Gad, A.; Yosef Pinhasi, Y. Atmospheric and fog effects on ultra–wide band radar operating at extremely high frequencies. *Sensors* **2016**, *16*, 751. [CrossRef] [PubMed]
8. Bieńkowski, P.; Zubrzak, B.; Cała, P. Mitigation measures of electromagnetic field exposure in the vicinity of high frequency welders. *Med. Pr.* **2017**, *68*, 693–703.
9. Grodzki, W.; Łukaszewicz, A. Design and manufacture of unmanned aerial vehicles (UAV) wing structure using composite materials. *Materialwiss. Werkst.* **2015**, *46*, 269–278. [CrossRef]
10. Kruk, A.; Wusatowska-Sarnek, A.M.; Zietara, M.; Jemielniak, K.; Siemiatkowski, Z.; Czyrska-Filemonowicz, A. Characterization on white etching layer formed during ceramic milling of inconel 718. *Met. Mater. Int.* **2018**, *24*, 1033–1045. [CrossRef]
11. Leccese, F.; Cagnetti, M.; Ferrone, A.; Pecora, A.; Maiolo, L. An infrared sensor Tx/Rx electronic card for aerospace applications. In Proceedings of the 2014 IEEE Metrology for Aerospace (MetroAeroSpace), Benevento, Italy, 29–30 May 2014; pp. 353–357.
12. Leccese, F.; Cagnetti, M.; Sciuto, S.; Scorza, A.; Torokhtii, K.; Silva, E. Analysis, design, realization and test of a sensor network for aerospace applications. In Proceedings of the 2017 IEEE International Instrumentation and Measurement Technology Conference (I2MTC), Turin, Italy, 22–25 May 2017; pp. 1–6.

13. Józwik, J.; Pytka, J.; Tofil, A. Dynamic analysis of aircraft landing gear wheel. In Proceedings of the 2018 5th IEEE International Workshop on Metrology for AeroSpace (MetroAeroSpace), Rome, Italy, 20–22 June 2018; pp. 462–467.
14. Siemiatkowski, Z.; Rucki, M.; Lavrynenko, S. Investigations of the shrink–fitted joints in assembled crankshafts. In Proceedings of the 7th International Conference on Mechanics and Materials in Design (M2D), Albufeira, Portugal, 11–15 June 2017; pp. 1155–1158.
15. Sidun, P.; Łukaszewicz, A. Verification of ram–press pipe bending process using elasto–plastic FEM model. *Acta. Mech. Autom.* **2017**, *11*, 47–52. [CrossRef]
16. Pytka, J.; Tarkowski, P.; Budzyński, P.; Józwik, J. Method for testing and evaluating grassy runway surface. *J. Aircr.* **2017**, *54*, 229–234. [CrossRef]
17. Łukaszewicz, A. Nonlinear numerical model of friction heating during rotary friction welding. *J. Frict. Wear* **2018**, *39*, 476–482. [CrossRef]
18. Mazurek, P.; Michalowska, J.; Koziel, J.; Gad, R. The intensity of the electromagnetic fields in the coverage of GSM 900, GSM 1800 DECT, UMTS, WLAN in built–up areas. In Proceedings of the 2018 Applications of Electromagnetics in Modern Techniques and Medicine (PTZE), Racławice, Poland, 9–12 September 2018; pp. 159–162.
19. Wang, J.; Kochan, O.; Przystupa, K.; Su, J. Information–measuring system to study the thermocouple with controlled temperature field. *Meas. Sci. Rev.* **2019**, *19*, 161–169. [CrossRef]
20. Kozieł, J.; Przystupa, K. Using the FTA method to analyze the quality of an uninterruptible power supply unitreparation UPS. *Prz. Elektrotechniczn.* **2019**, *95*, 65–68. [CrossRef]
21. Przystupa, K. The methods analysis of hazards and product defects in food processing. *Czech. J. Food Sci.* **2019**, *37*, 44–50. [CrossRef]
22. Przystupa, K. Reliability assessment method of device under incomplete observation of failure. In Proceedings of the 2018 18th International Conference on Mechatronics–Mechatronika (ME), Brno, Czech Republic, 5–7 December 2018; pp. 1–6.
23. Pongsakornsathien, N.; Lim, Y.; Gardi, A.; Hilton, S.; Planke, L.; Sabatini, R.; Kistan, T.; Ezer, N. Sensor networks for aerospace human–machine systems. *Sensors* **2019**, *19*, 3465. [CrossRef] [PubMed]
24. Siemiatkowski, Z.; Gzik–Szumiata, M.; Szumiata, T.; Rucki, M.; Martynowski, R. Metallurgical quality evaluation of the wind turbine main shaft 42CrMo4 steel: Microscopic and mssbauer studies. *Nukleonika* **2017**, *62*, 171–176. [CrossRef]
25. Michalowska, J.; Tofil, A.; Józwik, J. Prognosis of the electromagnetic field parameters of numerically controlled machine tools in industrial conditions. In Proceedings of the 2018 Applications of Electromagnetics in Modern Techniques and Medicine (PTZE), Racławice, Poland, 9–12 September 2018; pp. 167–170.
26. Michalowska, J.; Józwik, J. Prediction of the parameters of magnetic field of CNC machines tools. *Przegląad Elektrotechniczny* **2019**, *1*, 136–138. [CrossRef]

© 2019 by the authors. Licensee MDPI, Basel, Switzerland. This article is an open access article distributed under the terms and conditions of the Creative Commons Attribution (CC BY) license (http://creativecommons.org/licenses/by/4.0/).

Article

A Complete EGSE Solution for the SpaceWire and SpaceFibre Protocol Based on the PXI Industry Standard †

Luca Dello Sterpaio [1,*], Antonino Marino [1,*], Pietro Nannipieri [1,*], Gianmarco Dinelli [1,*], Daniele Davalle [2] and Luca Fanucci [1,2]

1 Department Information Engineering, University of Pisa, 56122 Pisa (PI), Italy; luca.fanucci@unipi.it
2 Space Division, IngeniArs S.r.l., 56121 Pisa (PI), Italy; daniele.davalle@ingeniars.com
* Correspondence: luca.dellosterpaio@ing.unipi.it (L.D.S.); antonino.marino@ing.unipi.it (A.M.); pietro.nannipieri@ing.unipi.it (P.N.); gianmarco.dinelli@ing.unipi.it (G.D.); Tel.: +39-346-843-6731 (L.D.S.); +39-349-321-4423 (P.N.)

† This paper is an extended version of our paper published in Luca, D.S.; Pietro, N.; Antonino, M.; Luca, F. Design of a SpaceWire/SpaceFibre EGSE system based on PXI industry standard. In Proceedings of the 2019 IEEE International Workshop on Metrology for AeroSpace (MetroAeroSpace), Torino, Italy, 19–21 June 2019.

Received: 25 October 2019; Accepted: 12 November 2019; Published: 16 November 2019

Abstract: This article presents a complete test equipment for the promising on-board serial high-speed SpaceFibre protocol, published by the European Committee for Space Standardization. SpaceFibre and SpaceWire are standard communication protocols for the latest technology sensor devices intended for on-board satellites and spacecrafts in general, especially for sensors based on image acquisition, such as scanning radiometers or star-tracking devices. The new design aims to provide the enabling tools to the scientific community and the space industry in order to promote the adoption of open standards in space on-board communications for current- and future-generation spacecraft missions. It is the first instrument expressly designed for LabVIEW users, and it offers tools and advanced features for the test and development of new SpaceFibre devices. In addition, it supports the previous SpaceWire standard and cross-communications. Thanks to novel cutting-edge design methods, the system complex architecture can be implemented on natively supported LabVIEW programmable devices. The presented system is highly customizable in terms of interface support and is provided with a companion LabVIEW application and LabVIEW Application Programming Interface (API) for user custom automated test-chains. It offers real-time capabilities and supports data rates up to 6.25 Gbps.The proposed solutions is then fairly compared with other currently available SpaceFibre test equipment. Its comprehensiveness and modularity make it suitable for either on-board device developments or spacecraft system integrations.

Keywords: SpaceFibre; SpaceWire; instrumentation; test equipment; EGSE; PXI; LabVIEW; FPGA; field programmable gate arrays; spacecraft; communications; serial communications

1. Introduction

This work is an in-depth discussion of the design and implementation of a novel test equipment for the SpaceFibre and SpaceWire protocols, the first natively supported and expressly designed protocols for the LabVIEW framework. Preliminary results of this work were presented at the 2019 IEEE International Workshop on Metrology for AeroSpace (MetroAeroSpace) [1]. If compared to those preliminary results, in this article, the system architecture is presented comprehensively and in depth, along with the innovative and cutting-edge methodologies for its implementation; also, the proposed system is compared with other currently available SpaceFibre instruments to provide a fair analysis.

1.1. Background Knowledge

According to [2], an on-board communication system denotes any on-board architecture that permits performing information transmission on-board a spacecraft. It can be implemented with different technologies, physical architectures, and redundancy schema. Such a communication system is currently demanded to withstand substantial data rates, mainly due to the usage of image based sensors: Earth observation missions are often equipped with imager sensors, Synthetic Aperture Radars (SARs), and other sensing payloads which, can require a large bandwidth in the order of one or even several Gbps to perform the routing of payloads related to data traffic throughout the rest of the spacecraft (data storage, telemetry transmitter, etc.).

SpaceWire is a communication protocol developed by the University of Dundee [3] for the European Space Agency (ESA) and standardized by ECSS in 2003 with a revision of the standard released in 2019 [4]. It has been designed to improve the on-board communication line from different points of view: it is a low error rate line, with low resource utilization (5–8 Kgates). The communication line is full duplex. SpaceWire is able to reach a data rate up to 400 Mbps (with dedicated hardware) in on-board application, making it appropriate for some payload-to-platform communication lines. The SpaceWire network solution has been already adopted for payload data handling in several ESA, NASA, and the Japan Aerospace eXploration Agency (JAXA) missions, such as Rosetta, Mars-Express, Galileo, and many others, making it a highly reliable solution. It also implements fault tolerance mechanisms, which switch automatically from one link to another in the case of failure with very small loss of information: no error recovery feature is included. SpaceWire is not a deterministic protocol itself, also due to the presence of routers inside a SpaceWire network, which creates even greater time uncertainty. This is not always acceptable, especially when determinism is a key application requirement. This led to the release of an additional standard named SpaceWire-D, which implements deterministic features on top of the SpaceWire protocol. A wide set of solutions of codecs and routers is available on the market, both Application Specific Integrated Circuit (ASIC) and Field Programmable Gate Array (FPGA) [5]. At the time of writing, SpaceWire is still a hot topic in the space community as almost all the recent space missions are equipped with such protocols. Therefore, the protocol is still relevant from the test equipment point of view as well. In current missions, most of the data traffic is made of packets including readings of peripheral sensor payload devices. Platform devices generate traffic as well for control or house-keeping, often aggregating readings of various monitoring sensors (i.e., voltage, current, and temperature). However, a set of future missions is currently under design or under proposal. The use of new sophisticated imaging sensors, for example camera instruments capturing hyper-spectrum frames at very high resolution, will definitively require payload related data traffic bandwidth in the order of several Gbps. SpaceWire will not therefore be able to fulfill such requirements. For this reason, the European Space Agency started working with the University of Dundee and all the SpaceWire community on the evolution of the protocol: SpaceFibre.

SpaceFibre standardization by the European Committee for Space Standardisation (ECSS) [6] was completed in May 2019. It is a high speed communication protocol developed by the University of Dundee under ESA supervision [3].Compared to other proprietary solutions derived from information technology to address the data rate issue, SpaceFibre remains an open standard. It is a multi-lane protocol (up to 16 lanes can operate in parallel) able to provide data rates up to 6.25 Gbps per single lane, overcoming all previous limitations in terms of maximum achievable bandwidth. It can run both on copper cables and on optical medium. The innovative Quality of Service (QoS) mechanism implemented in the SpaceFibre standard provides concurrent bandwidth reservation, packet priority, and scheduling. Advanced Fault Detection, Isolation, and Recovery (FDIR) techniques are also included in the SpaceFibre protocol, improving system reliability. SpaceFibre supports multiple accesses with a virtual bus mechanism, named virtual channels. Error containment has been introduced in both virtual channels and frames, along with transparent recovery from transient errors, galvanic isolation, and "babbling idiot" protection. This relieves the upper layers to check the data integrity; thus, it reduces both development and system validation in terms of time and cost. In the literature, there are also reduced feature SpaceFibre interfaces, which benefit from the high performances of SpaceFibre in terms of bandwidth, with very low resource utilization [7] in exchange for non-critical (for the particular application) advanced features. Furthermore, SpaceFibre is backward compatible at the packet level with SpaceWire: this easily enables interconnection of SpaceWire devices into a SpaceFibre network, gaining the QoS and FDIR advantages of SpaceFibre. The SpaceFibre technology is getting more and more mature: not only single interfaces, but also routing switches [8] have been developed, together with higher level instruments such as a mixed SPW/SPFI network simulator [9].

SpaceFibre is an open standard. Being openly accessible is not a characteristic to be overlooked considering that aerospace is nevertheless a strategic sector. Indeed, other solutions exist on the market to address the data rate and time-determinism requirements of modern space missions, yet they are proprietary. TTEthernet [10] is a technology developed by TTTech. It is an extension of the Institute of Electrical and Electronics Engineering (IEEE) 802.3 standard, making TTEthernet fully compatible with the Ethernet standard. TTEthernet (TTE) implements along the best-effort Ethernet the time triggered communication paradigm of the SAE AS6802 standard and the reserved-bandwidth protocol of ARINC664p7. This makes it perfectly suitable for safety critical applications like space missions. TTE is a noteworthy example of Ethernet based solutions, derived from Gigabit-Ethernet (thus limited to a 1 Gbps data rate) and implementing bounded-delay strategies [11]. This technology, due to its compatibility with the regular Ethernet, is oriented toward launchers, manned missions, and space exploration missions, but its applicability for telecommunication and Earth observation missions is now under investigation.

1.2. Satellite On-Board Data-Handling Electrical Ground Segment Equipment

Such advancements in terms of data rates require also appropriate control and testing equipment. Electrical Ground Segment Equipment (EGSE) is meant to be used in both the system development phase and system operation. Focusing on system development, proper EGSE shall enable device testing in the least obtrusive way possible. It shall intensively perform a series of tests by injecting a large number of inputs into the target Device Under Test (DUT) (i.e., the payload) while continuously recording, analyzing, saving, and processing the outputs.

According to the schema of a generic EGSE shown in Figure 1, this kind of equipment is composed of a monitoring and control block, which is responsible for the generation of both control signals and the input stimuli for the DUT. The DUT outputs are analyzed and, eventually, stored to be used by the control block. There are generally three interfaces: one with the host PC, which controls EGSE operations, and two with the payload (one input, one output). Of course, this is a very abstract conceptual visualization of the system, which may vary significantly, depending on the particular mission requirements.

Ideally, an SPW/SPFI EGSE should be able to have full control, be unobtrusive, and produce and consume data in Real Time (RT). Indeed, RT data generation, consuming, and processing is the most challenging of the capabilities to achieve for this kind of system given the high Gbps throughput requirements.

Current state-of-the-art test and EGSE SPW/SPFI equipment requires system developers and system integrators to rely on many different instruments, in order to address all the following tasks:

- data processing, recording, and playback: to emulate a SpaceWire or SpaceFibre device, which will be interacting with the DUT;
- bandwidth saturation: producing and/or consuming data at a controllable rate;
- unobtrusive monitoring to watch over two nodes without meddling with the link;
- error-injection: to verify corner scenarios either of very circumstantial occurrences (i.e., radiation- induced Bit Error Rate (BER)) or hazardous scenarios for people or property.

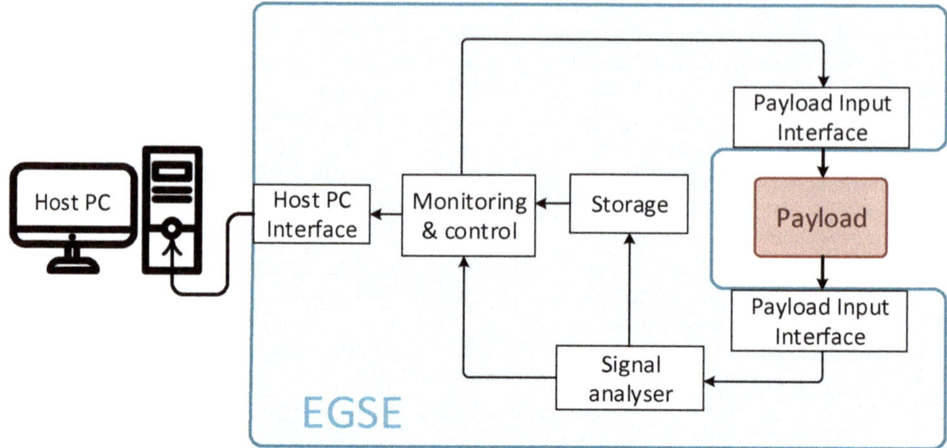

Figure 1. Electrical ground segment equipment generic block scheme.

EGSE systems can be completely custom designs, or instead, they can be realized on commercial off-the-shelf boards or can even be extensions of general purpose testing systems. Among standard platforms, the Peripheral Component Interconnect (PCI) eXtension for Instrumentation (PXI) is an extension of the PCI specification created to address specifically industry needs for instrumentation and automation in terms of performance and reliability [12]. Modularity is a key feature of PXI based systems, which aims at maximizing integrability and ease of use. One of the most successful applications of this standard is the LabVIEW integration. Indeed, National Instruments is one of the main actors in the PXI standard committee with all its related workflows, such as the National Instruments Hardware In the Loop (NI HIL) platform. Moreover, test equipment is usually tailored for each specific application. Sometimes, their support for the LabVIEW environment or derived workflows is patched up, wrapping driver libraries with all the incompatibility issues and the overhead of the case.

Currently, no solutions based on commercial PXI FPGA peripheral modules for SpaceWire/SpaceFibre are available [13], and only a few are available for similar protocols [14]. For the SpaceFibre protocol in particular, there are only two pioneering early standard adopter companies offering test equipment: STAR Dundee, with its StarFire Mk3 [15], and the SpaceWire and SpaceFibre Analyser Real Time (SpaceART) from IngeniArs [16]. Both are custom proprietary designs. Furthermore, an early attempt of multi-lane SpaceFibre test equipment was documented in [17]. In this work, a brief comparison between these solution and the proposed one is presented. However, the reader shall consider how the proposed work is the first and only PXI based implementation for LabVIEW. System designers will benefit from the PXI based EGSE: modularity, re-usability, and native LabVIEW support; all features that will lower system development time and costs. The new system based on National Instruments PXI FPGA peripheral modules presented in this paper addresses all the above-mentioned points, for both SpaceWire and SpaceFibre standards. Given the employment of third-party SpaceFibre and SpaceWire IPs and the complexity of a soft-core architecture, a novel cutting-edge LabVIEW FPGA implementation workflow has been followed [18].

This concludes Section 1 of this document, where background knowledge is provided in order to understand the topic and this work's goals properly.

Section 2 on page 5 discusses in depth the whole system architecture, its hardware/software partitioning, and major design choices.

Section 3 on page 8 provides the reader an accurate report of the challenges to overcome when implementing a complex design into PXI FPGA devices.

Section 4 on page 10 presents the final implementation results in terms of target device resource utilization, along with the verification tests carried out and a noteworthy real use case.

Section 5 on page 14 draws conclusions together with possible speculation on further work's improvement and impact on the industry. A brief comparison between the proposed system and a major alternative for SPFI equipment on the market is provided.

2. Proposed Architecture and Design

The aim of this work is to offer to the scientific community and the space industry an innovative, proper, and complete instrument for both the SpaceFibre and SpaceWire protocols. This is the first of its kind to be natively supported within the LabVIEW environment for test and measurement at a very high level, yet retaining the finest and deepest controllability and observability over the links. Figures 2 and 3 illustrate the overall system hardware and software architecture, how it is partitioned at a higher level, and of what functional blocks it consists.

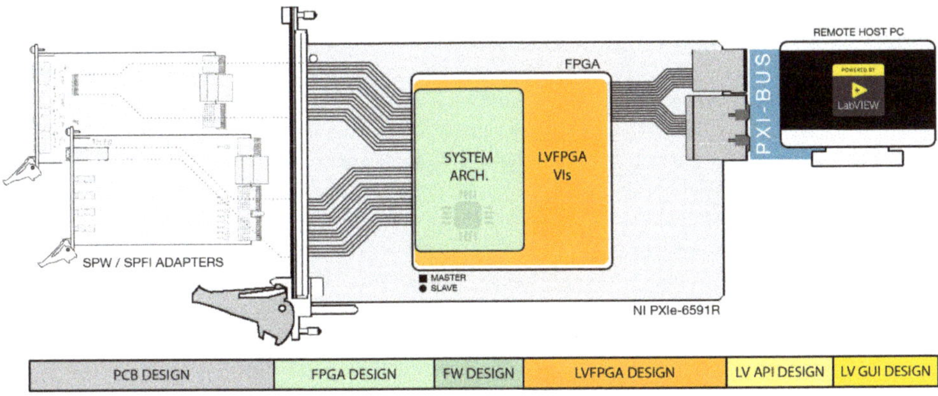

Figure 2. System overall organization and highlight of all encompassed design domains.

The test equipment presented in this paper is new and unique for its kind, because it is endowed with all the following features in one single instrument:

- up to four standard compliant SPW ports;
- up to eight standard compliant SPFI ports;
- eight SPFI Virtual-Channels (VC) per port;
- point-to-point codec interfaces or router IP configurations;
- real-time link status monitoring and control;
- real-time link protocol error report and counter by type;
- SPW/SPFI bridged communication;
- in-hardware packet generation/consumption;
- in-link error injection capabilities;
- word-replacement capabilities over traffic;
- Transmission/Reception (TX/RX) trace memory with triggerable events;
- unobtrusive low level link-analysis monitoring;
- real-time communication with remote-host PC via PXI connection;
- LabVIEW API library package;
- a ready-to-use general purpose LabVIEW Graphical User Interface (GUI) application.

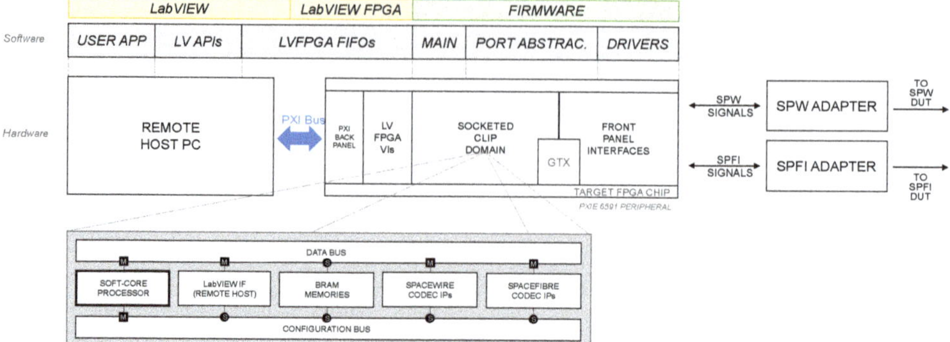

Figure 3. Block diagram of the proposed architecture.

The proposed EGSE solution is organized as a soft-core processor system with dedicated interface hardware accelerators, as shown in the block-diagram at the bottom of Figure 3. The system features multiple buses implementing the Advanced eXtensible Interface (AXI) of the Advanced Microcontroller Bus Architecture (AMBA) 4 specification [19]. Hardware accelerators have a dedicated master interface for DMA concurrent capabilities over the interconnect. The system is completely independent and requires a remote-host PC system running the LabVIEW environment for commands and controls. To add support for the current-generation SpaceWire standard as well seems the most obvious and clear design choice, given the backward compatibility of SPW and SPFI at the packet level.

The system design is rather complex and spans across multiple domains (as highlighted in Figure 2): FPGA design, firmware and drivers' programming, LabVIEW programming, LabVIEW FPGA programming, and PCB design. hardware/software design partitioning has been decided to maximize performance, avoid any design effort overhead, promote design reuse, and thus, reduce overall development time and costs. All EGSE functionalities and the front-end interfaces are implemented in hardware in order to offer Real-Time RT analysis capabilities and in order to take full advantage of either FPGA on-board resources and previously fully developed Hardware Description Language (HDL) Intellectual Property (IP) cores. Physical PCB adapters are also needed to provide SPW and SPFI standard connectivity. LVFPGA abstracts conveniently the PXI interconnect between the FPGA board and the remote-host PC system into a simpler three-wire handshake protocol. The firmware and drivers are designed to act as coordinator and controller among all system interface ports for concurrent operations. LabVIEW APIs are provided to operate the EGSE system from a remote-host correctly. LabVIEW APIs can be used to design user custom test chains or custom GUI applications. Furthermore, a general-purpose GUI application for Windows hosts is provided together to let the user immediately start using the instrument. This application can control all analysis tools and features of the instrument. It can address most of the test and measurement scenarios on SPW and SPFI links. It is to be highlighted that this GUI application was developed upon the very same LabVIEW APIs mentioned earlier.

2.1. Hardware Design

The proposed system implementation is based on the National Instruments PXIe-6591R Peripheral Module (National Instrument, Roscoe, IL, USA). This board has been chosen because it is a commercially available general-purpose FPGA board by National Instruments with LabVIEW native support. It features a Xilinx Kintex-7 410T FPGA unit [20], with on-chip GTX transceivers to implement the SPFI physical layer. This target requires the LabVIEW FPGA module to be programmed.

2.1.1. PCB Design

The target board has general I/O connectors. To offer instead standard compliant connectivity and connect SPW or SPFI DUTs to the EGSE system, additional adapter modules were designed specifically: the SPFI_eSATA to miniSAS-HD board is a passive physical adapter; the SPW_microD9 to VHDCI-68 board has active buffer components to operate single ended to LVDS signal conversion. The adaptations are bi-directional, and all PCB adapters are made on a FR4 support, sized to fit in the very same host PXI chassis that will host the target NI PXIe-6591 board. The final result is shown in Figure 4. Adapter modules allow expanding the PXIe-6591R VHDCI-68 port to four SPW ports and to convert each miniSAS-HD port into four SPFI ports. Of the 20 single ended I/O signals available on the PXIe-6591R port, 16 were employed for four SPW ports of four single ended data/strobe coded signals each, reserving the four left I/O signals for hardware debugging. The main challenges to overcome for the adapters' design were careful routed to reduce SPFI signal loss, following best practices like path equalization of differential signals. Given the fragility of such a signal path, using fast-prototyping techniques was simply not viable, and thus, the production was outsourced to a trusted and experienced PCB manufacturing company.

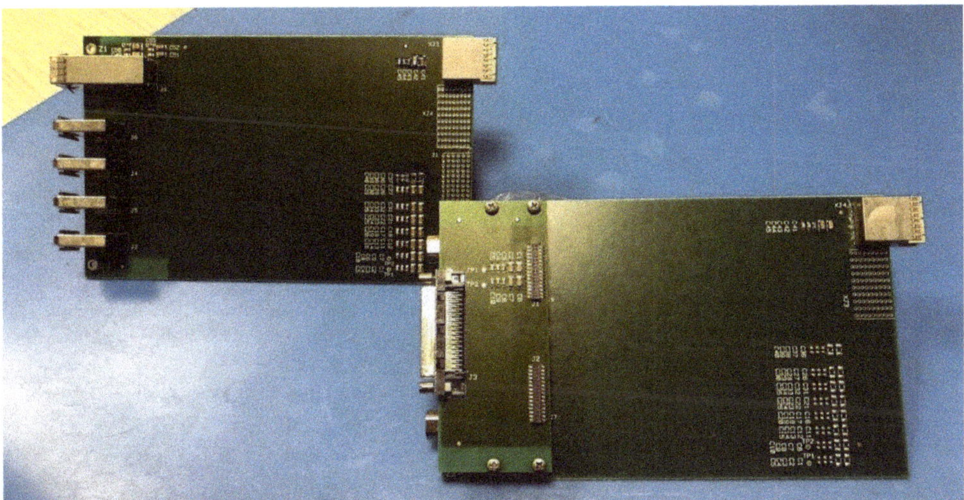

Figure 4. Picture of the adapters for the SpaceFibre and SpaceWire compliant connectors.

2.1.2. FPGA Design

The hardware architecture implemented in the FPGA target is shown at the bottom of Figure 3 as a block-diagram. It is based on a MicroBlaze™ soft-core processor aided by dedicated interface IPs. Interface IPs are capable of operating DMA transactions on an AXI-4 interconnect. The system is designed as a multiple bus architecture in order to keep conveniently separated interconnect fabrics for TX/RX traffic and configuration data, preventing traffic congestion. This also improves robustness because the MicroBlaze™ processor unit is the only bus master of the configuration bus and, thus, the only one responsible to configure the port interfaces. On-chip block-RAM resources were used as memory buffers in data communications.

The remote host interface through PXI interconnect is supervised by LabVIEW FPGA environment and abstracted as simpler FIFO-like macros with a data-ready-valid handshake protocol. A Direct Memory Access (DMA) capable bridge interface IP was designed specifically in order to translate with minimal latency and maximum throughput such transfers into AMBA 4 AXI burst transactions and initiate them on the system data bus [21].

The front-panel interfaces' configuration can vary with specific mission needs: SPW point-to-point codec IP, SPW router IP, or SPFI point-to-point codec IP can be instantiated and used as the external port. The only limitation is the number of physically available pins, which limits consequently the number of external ports to a maximum of four SPW and eight SPFI. All SPW and SPFI interface IP cores are derived from currently-flying or flight-ready designs that have been modified in order to obtain as much as possible the observability and controllability over links:

- expose link related statuses and internal signals;
- SPFI/SPW traffic generation/consumption, forwarding, and recording;
- control/data character sniffing and trace back recording;
- low level and high level error injection capabilities (bit flips, control/data characters replacement, error packet termination);
- unobtrusive monitoring of a link between two ports.

One additional feature of this SPFI/SPW test equipment is the interesting possibility to operate a cross-standard communication, since the two protocols are compatible at the packet level [6].

2.2. Software Design

Embedded software (also referred to as firmware in the following) is a bare-metal executable created for this system specifically. The design tool of reference is the Software Development Kit (SDK), Version 2018.2, Xilinx, USA in particular. The bare-metal firmware application is executed on the MicroBlaze^{TM} processor instance in order to coordinate the whole system operation. The reasons to opt for a bare-metal solution are essentially three: (1) maximize the performance of the MicroBlaze processor, given its limited resources, (2) minimize the memory footprint, since no dynamic RAM is available within the FPGA, and (3) the operations to carry out are all things considered short and of limited complexity. The firmware architecture is hierarchically organized by consecutive layers of abstraction, as per Figure 2. This encapsulation allows breaking down tasks into simpler steps: the main application executes a general and simple schema for HW initialization and transfer data and control messages following a round-robin scheduling; port and BRAM buffer abstraction offers a general interface to let the main program invoke functional routines; the hardware IP C/C++ drivers underneath carry out HW specific task to achieve the generalized functionality. DMA transfers are handled by exceptions. The abstracted port interface as a more general port class allows updating the system with ease and adding support for more interface protocols in future developments. The abstracted generic ports follow a header-Ack-payload-Ack paradigm with a TX-credit mechanism to operate flow control and avoid bottlenecks. The schema is an original work, and it was conceived to be the most general as possible. Generality is needed in order to cope with a wider range of interface hardware IPs. This paradigm provides the needed flexibility to wrap port interfaces of any hardware kind into a common abstracted interface wrapper. Some hardware may offer some feature intrinsically, and others may not. The features not provided, flow control credit mechanics for example, can be mimicked by means of this paradigm. Unnecessary steps instead can be skipped with dummy functions. Send and receive data transactions are split into two steps: order the transaction, and then, carry out the transaction.

The software architecture also spans the LabVIEW domain in addition to firmware design. End-users can issue commands and configure the system with remote-host LabVIEW APIs or a LabVIEW GUI application. The APIs follow the action engine schema, while the GUI application follows the queued message handler architecture [22].

3. PXI Workflow

A key aspect of the proposed system is the design choice to implement the architecture described in Section 2 on a market available FPGA that is natively supported within the LabVIEW environment.

Speaking of the LVFPGA design methodology, it is important to highlight first of all how the LabVIEW FPGA is a tool directed to final users without any Hardware Description Language (HDL) knowledge or skill, who want to benefit anyway from hardware acceleration for their own LabVIEW data processing or test setups [18]. The LVFPGA approach preserves the LV peculiar graphical approach of data flow block designs. HW/SW partitioning is carried out by the tool itself without user control over all the synthesis and implementation processes.

Given that LVFPGA is not addressed to digital HW designers, but final users not fluent in HDL instead, it is a very closed environment in order to reduce human design errors. In particular, the target programmable chip cannot be configured outside the LVFPGA framework. Since reproducing system functionalities as LVFPGA VIs is simply nonviable, a new implementation methodology needs to be defined in order to overcome these limitations.

A special macro function in LVFPGA is intended to import small VHDL sub-designs in an LVFPGA VI, called "Socketed-CLIP". Despite being carried out within the LVFPGA environment, LVFPGA project VI synthesis and implementation operations fall back to Xilinx tools running in the background (i.e., Vivado for all 7-Series devices). The exploit takes advantage of this dependency: it was found that it is not possible to import in a SckCLIP all kinds of source files that are usually supported in Vivado, yet more than the sole VHDL sources [18] for which the SckCLIP macro is meant. Specifically, the whole hardware HDL design can be imported in a SckCLIP as a single, large, technology dependent EDIF netlist of primitive FPGA resource elements.

Figure 5 summarizes the workflow to target National Instruments PXI FPGA peripheral modules taking advantage of the SckCLIP exploit introduced above. FPGA design can be conducted as usual in Vivado with no limitation on source types, but the tool's support (1), on the workflow diagram shown in Figure 5). Functional simulation can be accomplished as usual as well. A first in-context (in-CTX) synthesis run is carried out (2), necessary to generate block-RAM descriptor and HW hand-off files for SW development. The SW development tool (Xilinx SDK) will compile an executable file for the processor to run as firmware (3). Since there is no means to deploy the generated executable file directly, it is converted into a constraints file to initialize every block-RAM location value according to the generated memory descriptor file (4). An additional Out-Of-Context (OOC) synthesis run shall be carried out (5) to obtain the actual EDIF netlists to be included (6) with its VHDL instantiation wrapper (7) in the upper hierarchy of the whole LVFPGA design. SckCLIP top level ports will be available in LVFPGA as controls and indicators (of the appropriate type). LVFPGA compilation (10) will complete the design implementation, obtaining the bitstream configuration for deployment and remote-host execution (11).

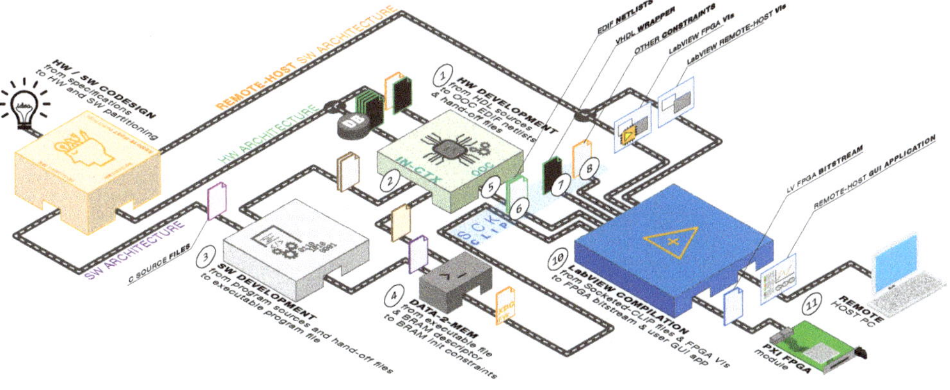

Figure 5. New design workflow exploiting the LVFPGA SckCLIP macro [18].

4. Results and Use Cases

The proposed EGSE solution implements four SPW point-to-point codec instances and four SPFI point-to-point codec instances. This has been appointed as the reference configuration because it is the most likely configuration to be requested on the market.

The following tests were carried out with an NI PXIe-8880 Controller unit as remote-host PC in an NI PXI-1085 chassis. The instrument was controlled through the provided LabVIEW GUI user application, in order to send/receive data to/from the SpaceFibre and SpaceWire ports, check the test results, and monitor the links in real time, as well as provide higher flexibility in the number of SPFI ports offered.

4.1. Results

Implementation results are shown in Table 1. The resource utilization was not negligible; however, no more than 30–40% of the FPGA was occupied, which means that there were resources still available for possible tailored customization. The rightmost columns of Table 1 report indeed the usage per single SPW and SPFI interface respectively in the configuration. The utilization took into account all codec IP resources and all extra resources to provide EGSE functionalities to that IF (for example, the tracing memory core). These data provided a quick estimation of the usage for different configurations in terms of interface number or kind. The EGSE design was indeed modular from that point of view. Router configurations should take into account that codec cores underneath were required anyway.

Table 1. Implementation results on the target NI PXIe-6591R Kintex-7 FPGA, system reference configuration.

Resource	Usage	Availability	Ratio	Each SPW Codec	Each SPFI Codec
Block RAM	420	795	52.83%	32 (4.03%)	40 (5.03%)
LUT	85,380	254,200	33.59%	3752 (1.48%)	11,621 (4.57%)
GTX	4	16	25.00%	0 (0%)	1 (6.25%)
FF Registers	83,644	508,400	16.45%	2416 (0.48%)	8958 (1.76%)

Tables 2 and 3 summarize the proposed EGSE system characteristics in comparison with the main design solutions available on the market at the time this document was redacted.

Table 2. Comparison table of the available SpaceFibre EGSE solutions by features: maximum bandwidth per link, number of SPW IFs, number of SPFI IFs, type of remote host IF, type of control user experience.

EGSE System	SPFI/SPW Bandwidth	SPW Ports	SPFI Ports	Host IF	User Experience
Proposed SPFI/SPW EGSE	6.25 Gbps/200 Mbps	Up to 4	Up to 8	PXI	LV API or GUI
IngeniArs SpaceART	2.50 Gbps/400 Mbps	4	0 or 2	Ethernet or PCIe	Windows/Linux API or GUI
StarDubdee StarFire MK3	3.20 Gbps/400 Mbps	2	2	USB 3.0	Windows/Linux GUI

The proposed EGSE system is not only the very first test equipment for SpaceFibre technology natively integrated into LabVIEW. It possesses all analysis capabilities of the most advanced EGSE system available on the market, including unobtrusive monitoring and error injection functionalities. Furthermore, as per Tables 2 and 3, it is the only instrument capable of offering support up to eight SPFI interfaces and capable of reaching the full bandwidth of SPFI links, up to 6.25 Gbps, at the time this document was redacted. The instrument's capability of reaching a higher SPW bandwidth, as other market available solutions state they can do, is currently under investigation. The triggering feature is not directly supported in hardware by the PXI FPGA peripheral module, yet can be achieved under the

LabVIEW environment, if so programmed. The presented architecture's high flexibility and modularity allow the better configuration of a proper EGSE setup for each specific space segment to be verified. As a down-side, the PXI setup requires higher hardware costs and initial investments for chassis and controller procurement.

Table 3. Comparison table of the available SpaceFibre EGSE solutions by capabilities: in-hardware packet generation/consumption, data record, and playback from remote host, SPFI/SPW cross bridging, link error injection, tracing memory, external trigger signals for synchronization.

EGSE System	HW Data Generator/ Consumer	Host Reception and Play	SPW/SPFI Bridging	Error Injection	Trace Memory	External Triggers
Proposed SPFI/SPW EGSE	Yes	Yes	Yes	Yes	Yes	No *
IngeniArs SpaceART	Yes	Yes	Yes	Yes	Yes	Yes
StarDundee StarFire MK3	Yes	Yes	Yes	Yes	No	Yes

* Triggering feature can be achieved with the mean of additional signal acquisition PXI Peripheral modules.

4.2. Testing

Various in-field tests were carried out to prove the reliability, strength, and potential of the proposed test equipment.Tests were carried out with the aid of a SpaceART unit, an instrument by IngeniArs.

4.2.1. SPFI Validation Tests

As pointed out, no other test equipment solutions could keep up with the 6.25 Gbps data rate capabilities of the presented EGSE system. Thus, SPFI testing at such high transfer rate was carried out in loop-back configuration. The IngeniArs SpaceART unit was used for 2.5 Gbps tests instead. This also verified that the instrument could correctly communicate with another SPFI device.

A meaningful example of all the extensive tests carried out to validate SPFI capabilities is the tracing memory and error injection the "NACK Test", where the tracing memory trigger condition was set to catch the first occurring Non-Acknowledgment (NACK) control word. The test aimed to verify the correctness of TX/RX data streaming, either generated/consumed by hardware or by the remote-host, and the effectiveness of the error injection feature as well. The data stream flowed without errors in both directions, and the trigger condition did not occur until the error injection was enabled. Once error injection was activated, the triggering condition was fired, and the dump of tracing memory content clearly showed on screen the traffic data window centered on the triggering NACK word.

4.2.2. SPW Real-Time CRC Calculation and Fly Back

This test aimed to estimate the latency for which the analyzer system could process and reply. In this test, the system was closed in a loop-back configuration as shown in Figure 6a, transmitting from SPW-Port 1 to SPW-Port 2. The EGSE system was used simultaneously as producer and as consumer with the first SPW port on the SPW adapter connected to the second one. An SPW packet of known size (1000 words of 32 bits)was prepared and sent over the link to the transmission port along with its Cyclic Redundancy Check (CRC) value; once received, the CRC of the packet content was calculated again, compared, and appended. Eventually, the whole packet was sent back. The overall measured latency was 0.4 ms, from the packet dispatch to the packet reception.

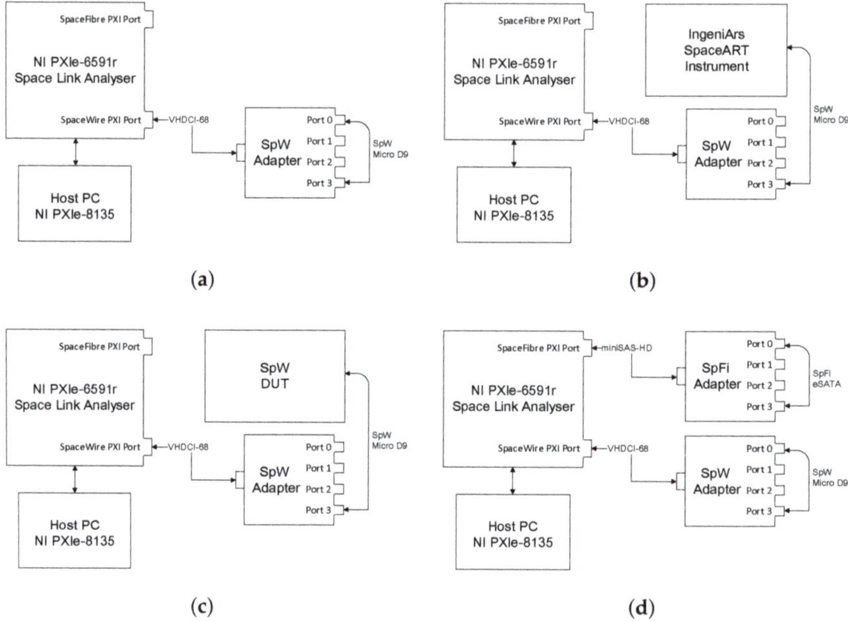

Figure 6. Various setups of the tests carried out. From top to bottom and left to right: (**a**) loop-back configuration, (**b**) SpaceART (IngeniArs) test configuration, (**c**) SpaceWire DUT test configuration, and (**d**) SpaceFibre test configuration.

4.2.3. SPW Real-Time Full Bandwidth Saturation and Data Logging

The aim of this test was to demonstrate system capability to sustain multiple real-time SpaceWire links without affecting the actual bandwidth. All four SPW links were tested while receiving packets, one at a time. The test setup is shown in Figure 6b. Data were generated with an IngeniArs SpaceART unit, in order to saturate the bandwidth and keep the link under control. Running the test over one hour, the system recorded 67 GB of data on an SSD support. Considering that just 80% of all the SPW transferred bits actually carried information, due to the fact that an SPW data character was ten bits long with a control and a parity bit appended, which means that just eight out of these ten bits contained data [4], we can acknowledge the capability of sustaining a 200 Mbps throughput. Link speed readings from the attached SpaceART unit were congruent.

4.2.4. Device Emulation

In this functional test, an FPGA development board was used in place of flight hardware. A common scenario while developing the spacecraft subsystems is indeed the unavailability of other devices with which to communicate. EGSE systems like the proposed system can act in place of missing devices, replicating the same behavior. Figure 6c shows the direct connection between the instrument and the DUT. Once connected to the unit to be tested, the instrument was capable of letting the user emulate a hypothetical control unit (in this case), preparing telecommand packets, and waiting to receive the telemetry data stream back from flight hardware.

4.2.5. SPFI/SPW Performance Test

This series of functional test was performed with the hardware configuration of Figure 6d. Two SPW ports and two SPFI ports were close in a loop-back configuration. Verifying the SPFI capabilities of the instrument was more complicated because neither the IngeniArs SpaceART nor the StarDundee StarFire MK3 units could cope with the full 6.25 Gbps bandwidth. For this reason, only loop-back tests could be carried out at a 6.25 Gbps link speed.

The SPFI links were tested at different speeds, both 2.5 Gbps and 6.25 Gbps. The FDIR feature of the SpaceFibre protocol was tested and verified. Tests were performed by injecting over the link a wide range of possible Bit Error Rate (BER) values, from 10^{-9} to 10^{-5} values. The SPW links were tested similarly, from 10 to 200 Mbps communication speeds. The error injection feature was verified as well, introducing low-level SPW character errors, bit flips, and replacing the end of packet terminations with the error end of packet.

4.2.6. SPFI/SPW Bridging Test

With the very same test setup of Figure 6d described above, SPFI/SPW cross-protocol compatibility was verified bridging SPW and SPFI data stream. SpaceWire traffic was generated from the first SPW port and received on the second SPW port, then successfully forwarded internally to the first SPFI port on a virtual channel and received on the second SPFI port. No errors, data corruption, nor protocol errors were detected during the performance tests.

4.3. Real In-Field Use Case

The presented EGSE system was tested in-the-field in a real case study. The chance to prove its effectiveness and ease of use came from an important company in the aerospace field. Researchers of this company were in need of addressing an unexpected behavior of a device under development. The case scenario was the following: The SPW device after receiving a TeleCommand (TC) at 50 Mbps shall reply with a corresponding TeleMetry (TM) data message at 100 Mbps. TM shall be transmitted without generating bottlenecks on the SPW link. This implies that the receiver SPW node shall consume incoming data assuring that the data rate is sustained in real time. The developers were experiencing sudden and abrupt transmission interruption of the TM stream, but just randomly, once in a while. The developers were using a full custom EGSE solution. Thanks to the new EGSE system presented in this work, it was possible to point out that the device was actually working correctly and that the unexpected disconnections were due to the custom EGSE system: under certain circumstances, the custom EGSE solution in use was not able to consume the stream fast enough, and thus, the device was experiencing overflow.

Figure 7 shows the custom Test-Chain VI prepared with the provided instrument LV API library, including both the block-diagram and user friendly GUI. Compared to reference test solutions available on the market, the EGSE equipment presented in this paper is natively supported within LabVIEW, thus integrated into the National Instruments environment. This unique characteristic allowed significantly reducing the test setup time, from hardware setup to custom LV test-chain preparation and actual test execution.

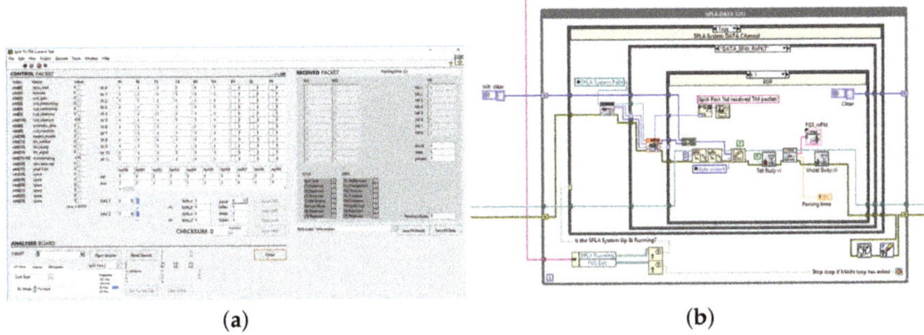

(a) (b)

Figure 7. Real Use-Case VI. (**a**) Front-panel view and (**b**) block-diagram view.

5. Conclusions

A novel SpaceFibre and SpaceWire test equipment for the LabVIEW environment was presented in this article. The proposed EGSE system was the first SpaceFibre solution to be natively integrated into the LabVIEW environment, thanks to the research in the field of synthesis and implementation methodologies for LVFPGA [18]. Early adopters of the SpaceFibre protocol, as well as experienced SpaceWire developers can benefit from the fine degrees of controllability and observability over the link of interest. The proposed EGSE system is suitable both for on-board device development and spacecraft integration, fulfilling its purposes of offering to the scientific community and to the space industry an innovative and complete equipment for both the SpaceFibre and SpaceWire protocols. All features of the proposed instrument system can be operated with either the provided ready-to-use GUI application or the LabVIEW API library. The system can be configured with up to four SpaceWire ports and up to eight SpaceFibre ports, for point-to-point connections or, for SpaceWire, routers. This modularity allows the EGSE test equipment to replicate as close as possible the real use scenario and to interface with the widest range of spacecraft on-board peripheral, sensors, and other mission specific payload instruments.

This article also provides the reader with a brief survey about the EGSE solutions available on the market for the newly published SpaceFibre protocol, providing a fair comparison in terms of features and performance with the newly developed instrument. From this analysis, SpaceFibre emerged as a very promising technology for the space missions of today and tomorrow, because it addressed data rate requirements in the Gbps order of magnitude.

The EGSE system discussed in these pages is the very first test equipment solution on the market for the SpaceFibre standard implemented on National Instruments COTS PXI FPGA peripheral modules. It is the only current solution capable of operating at SpaceFibre full bandwidth. Its impact on the industry is hard to predict, but hopefully, it will provide the enabling technology in order to promote the diffusion and adoption of the SPFI standard.

Author Contributions: The research and the article preparation were carried out by L.D.S., A.M., G.D., and P.N. under the supervision of L.F. and D.D.

Funding: IngeniArs SpaceFibre technologies were developed in the framework of the SIMPLE project (SpaceFibre IMPLementation design & test Equipment). This project received funding from the European Unions Horizon 2020 research and innovation program under Grant Agreement No. 757038.

Acknowledgments: The authors want to express their gratitude to National Instruments Italy for lending one unit of the PXIe-6591 FPGA peripheral module for the purpose of this work.

Conflicts of Interest: The authors declare no conflict of interest.

Abbreviations

The following abbreviations are used in this manuscript:

API	Application Programming Interface
BD	Block Diagram
CLIP	Component-Level IP
COTS	Commercial-Off-The-Shelf
CRC	Cyclic Redundancy Check
CTX	Context
DMA	Direct Memory Access
ECSS	European Cooperation for Space Standardization
EDIF	Electronic Design Interchange Format
EGSE	Electrical Ground Segment Equipments
ESA	European Space Agency
FCT	Flow Control Token
FP	Front-Panel
FDIR	Fault Detection Isolation and Recovery

FPGA	Field Programmable Gate Array
GUI	Graphical User Interface
HDL	Hardware Description Language
HW	Hardware
IF	Interface
IP	Intellectual Property
JAXA	Japan Aerospace eXploration Agency
LUT	Look-Up-Tables
LV	LabVIEW
LVDS	Low-Voltage Differential Signaling
LVFPGA	LabVIEW FPGA
MAC	Medium Access Controller
NASA	National Aeronautics and Space Administration
PC	Personal Computer
PCB	Printed Circuit Board
OOC	Out Of Context
QoS	Quality of Service
RAM	Random Access Memory
Reg	Register
RX	Reception, received
SAR	Synthetic Aperture Radars
SckCLIP	Socketed CLIP
SPFI	SpaceFibre
SPW	SpaceWire
SW	Software
TX	Transmission, transmitter
VC	Virtual Channel
VCB	Virtual Channel Buffers
VI	Virtual Instrument

References

1. Dello Sterpaio, L.; Nannipieri, P.; Marino, A.; Fanucci, L. Design of a SpaceWire/SpaceFibre EGSE System Based on PXI Industry Standard. In Proceedings of the IEEE International Workshop on Metrology for AeroSpace (MetroAeroSpace), Turin, Italy, 19–21 June 2019.
2. Space AVionics Open Interface aRchitecture (SAVOIR). *SAVOIR On-Board Communication System Requirement Document*; SAVOIR-GS-008; SAVOIR, ESA. Available online: http://savoir.estec.esa.int/SAVOIRDocuments.htm (accessed on 15 November 2019).
3. Parkes, S.; Armbruster, P. SpaceWire A Spacecraft Onboard Network for Real-Time Communications. In Proceedings of the 14th IEEE-NPSS Real Time Conference, Stockholm, Sweden, 4–10 June 2005.
4. SpaceWire Standard ECSS-E-ST-50-12C. Available online: https://ecss.nl/standard/ecss-e-st-50-12c-spacewire-links-nodes-routers-and-networks/ (accessed on 4 October 2019).
5. Saponara, S.; Fanucci, L.; Tonarefli, M.; Petri, E. Radiation Tolerant SpaceWire Router for Satellite On-Board Networking. *IEEE Aerosp. Electron. Syst. Mag.* **2017**, *22*, 3–12. [CrossRef]
6. SpaceFibre Standard, ECSS-E-ST-50-11C. Available online: http://ecss.nl/standard/ecss-e-st-50-11c-dir1/ (accessed on 4 October 2019).
7. Dinelli, G.; Nannipieri, P.; Davalle, D.; Fanucci, L. Design of a reduced SpaceFibre interface: An enabling technology for low-cost spacecraft high-speed data-handling. *Aerospace* **2019**, *6*, 101. [CrossRef]
8. Leoni, A.; Nannipieri, P.; Fanucci, L. VHDL Design of a SpaceFibre Routing Switch. *IEICE Trans. Fundam. Electron. Commun. Comput. Sci.* **2019**, *E102-A*, 729–731. [CrossRef]
9. Leoni, A.; Nannipieri, P.; Davalle, D.; Fanucci, L.; Jameux, D. SHINe: Simulator for Satellite on-Board High-Speed Networks Featuring SpaceFibre and SpaceWire Protocols. *Aerospace* **2019**, *6*, 43. [CrossRef]

10. Eramo, V.; Lavacca, F.G.; Listanti, M.; Caporossi, S. Definition and performance evaluation of an Advanced Avionic TTEthernet Architecture for the support of Launcher Networks. *IEEE Aerosp. Electron. Syst. Mag.* **2018**, *33*, 30–43. [CrossRef]
11. Eramo, V.; Lavacca, F.G.; Valente, F.; Pisculli, A.; Caporossi, S. Simulation and experimental evaluation of a flexible time triggered ethernet architecture applied in satellite Nano/Micro Launchers. *Aerospace* **2018**, *5*, 84. [CrossRef]
12. PXISA. *PXI-1 Hardware Specifications, Rav. 2.2*; PXISA: Niwot, CO, USA, 2004.
13. Marino, A.; Dello Sterpaio, L.; Fanucci, L. Design and implementation of a complete test equipment solution for SpaceWire links. In Proceedings of the 2018 14th Conference on Ph.D. Research in Microelectronics and Electronics (PRIME), Prague, Czech Republic, 2–5 July 2018.
14. Nannipieri, P.; Dello Sterpaio, L.; Marino, A.; Fanucci, L. A PXI based implementation of a TLK2711 equivalent interface. In *Application in Electronics Pervading Industry, Environment and Society (ApplEPIES)*; Elsevier: Pisa, Italy, 2018.
15. StarFire MK3. Available online: https://www.star-dundee.com/products/star-fire-mk3 (accessed on 7 October 2019).
16. SpaceART. SpaceWire/SpaceFibre Analyser Real-Time. Available online: https://www.ingeniars.com/english/products/space/spaceart-en.html (accessed on 7 October 2019).
17. Nannipieri, P.; Dinelli, G.; Davalle, D.; Fanucci, L. A SpaceFibre multi lane codec System on a Chip: Enabling technology for low cost satellite EGSE. In Proceedings of the PRIME 2018 14th Conference on Ph.D. Research in Microelectronics and Electronics, Prague, Czech Republic, 2–5 July 2018; pp. 173–176.
18. Dello Sterpaio, L.; Marino, A.; Nannipieri, P.; Fanucci, L. Exploiting LabVIEW FPGA Socketed CLIP to Design and Implement Soft-Core Based Complex Digital Architectures on PXI FPGA Target Boards. In Proceedings of the 2019 International Conference on Synthesis, Modeling, Analysis and Simulation Methods and Applications to Circuit Design (SMACD), Lausanne, Switzerland, 15–18 July 2019.
19. *AMBA AXI and ACE Protocol Specifications, Issue, D*; ARM Holdings: Cambridge, UK, 2011.
20. PXIe-6591. PXI High-Speed Serial Instrument. Available online: https://www.ni.com/en-us/support/model.pxie-6591.html (accessed on 7 October 2019).
21. Dello Sterpaio, L.; Marino, A.; Nannipieri, P.; Dinelli, G.; Fanucci, L. AXI4LV: Design and Implementation of a Full-Speed AMBA AXI4-Burst DMA Interface for LabVIEW FPGA. In Proceedings of the 2019 Applications in Electronics Pervading Industry, Environment and Society (ApplePies), Pisa, Italy, 11–13 September 2019.
22. LabVIEW Core 3. Available online: http://sine.ni.com/tacs/app/overview/p/ap/of/lang/en/pg/1/sn/n24:12754/id/1584/ (accessed on 7 October 2019).

© 2019 by the authors. Licensee MDPI, Basel, Switzerland. This article is an open access article distributed under the terms and conditions of the Creative Commons Attribution (CC BY) license (http://creativecommons.org/licenses/by/4.0/).

MDPI
St. Alban-Anlage 66
4052 Basel
Switzerland
Tel. +41 61 683 77 34
Fax +41 61 302 89 18
www.mdpi.com

Sensors Editorial Office
E-mail: sensors@mdpi.com
www.mdpi.com/journal/sensors

www.ingramcontent.com/pod-product-compliance
Lightning Source LLC
LaVergne TN
LVHW071620170726
843515LV00009B/2439
9783036500584